STRUCTURE–FUNCTION ANALYSIS OF G PROTEIN-COUPLED RECEPTORS

RECEPTOR BIOCHEMISTRY AND METHODOLOGY

SERIES EDITORS

David R. Sibley
Molecular Neuropharmacology Section
Experimental Therapeutics Branch
NINDS
National Institutes of Health
Bethesda, Maryland

Catherine D. Strader
Department of CNS and
  Cardiovascular Research
Schering-Plough Research
  Institute
Kenilworth, New Jersey

New Volumes in Series

Receptor Localization: Laboratory Methods and Procedures
Marjorie A. Ariano, *Volume Editor*

Identification and Expression of G Protein–Coupled Receptors
Kevin R. Lynch, *Volume Editor*

Structure–Function Analysis of G Protein–Coupled Receptors
Jürgen Wess, *Volume Editor*

Founding Series Editors
J. Craig Venter   Len C. Harrison

STRUCTURE–FUNCTION ANALYSIS OF G PROTEIN-COUPLED RECEPTORS

Edited by

JÜRGEN WESS
Laboratory of Bioorganic Chemistry
National Institutes of Health, NIDDK
Bethesda, Maryland

A JOHN WILEY & SONS, INC., PUBLICATION

New York • Chichester • Weinheim • Brisbane • Singapore • Toronto

Copyright © 1999 by Wiley-Liss, Inc. All rights reserved.

Published simultaneously in Canada.

For ordering and customer service, call 1-800-CALL-WILEY.

Library of Congress Cataloging-in-Publication Data:
Structure-function analysis of G protein–coupled receptors / edited by
    Jürgen Wess.
        p.   cm.   — (Receptor biochemistry and methodology)
    Includes index.
        ISBN 0-471-25228-X (cloth : alk. paper). — ISBN 0-471-25227-1
        1. G proteins—Receptors—Research—Methodology.   2. Cell
    receptors—Research—Methodology.   3. Cellular signal transduction—
    Research—Methodology.   I. Wess, Jürgen, 1958– .   II. Series.
    QP552.G16S77   1999
    571.6—dc21   98-48496

Printed in the United States of America.

10 9 8 7 6 5 4 3 2 1

CONTENTS

SERIES PREFACE   vii

PREFACE   ix

CONTRIBUTORS   xi

1 **Overview of Mutagenesis Techniques**   1
 Tung Ming Fong

2 **The Substituted-Cysteine Accessibility Method**   21
 Jonathan A. Javitch

3 **Using Mutagenesis to Map the Binding Domains of Small Molecule Receptors**   43
 Jean E. Lachowicz and Catherine D. Strader

4 **Mapping Binding Sites for Peptide G Protein–Coupled Receptors: The Receptor for Thyrotropin-Releasing Hormone**   59
 Roman Osman, Anny-Odile Colson, Jeffrey H. Perlman, Liisa J. Laakkonen, and Marvin C. Gershengorn

5 **Lessons From Rhodopsin**   85
 Thomas P. Sakmar and K. Christopher Min

6 **Split Receptors as Tools for Analyzing G Protein–Coupled Receptor Structure**   109
 Masahiro Kono and Daniel D. Oprian

7 **Metal-Ions as Atomic Scale Probes of G Protein–Coupled Receptor Structure**   121
 John A. Schetz and David R. Sibley

8 **Genetic Approaches for Studying the Structure and Function of G Protein–Coupled Receptors in Yeast**   141
 Christine M. Sommers and Mark E. Dumont

9 **Constitutively Active Receptor Mutants as Probes for Studying the Mechanisms Underlying G Protein–Coupled Receptor Activation**   167
 Susanna Cotecchia, Francesca Fanelli, Alexander Scheer, and Pier G. De Benedetti

10 β$_2$-Adrenoceptor–Gsα Fusion Protein as a Model for the Analysis of Receptor–G Protein Coupling   185

Roland Seifert and Brian K. Kobilka

11 Peptides as Tools for the Study of Receptor–G Protein Interactions   205

Richard R. Neubig

12 Electron-Crystallographic Analysis of Two-Dimensional Rhodopsin Crystals   233

Gebhard F. X. Schertler

13 Site-Directed Spin-Labeling (SDSL) Studies of the G Protein–Coupled Receptor Rhodopsin   289

David L. Farrens

14 Fluorescence Spectroscopy Analysis of Conformational Changes in the β$_2$-Adrenergic Receptor   315

Ulrik Gether

15 Biosynthetic Incorporation of Unnatural Amino Acids into G Protein–Coupled Receptors   335

André Chollet and Gerardo Turcatti

16 Use of Nuclear Magnetic Resonance Techniques to Study G Protein–Coupled Receptor Structure   355

Philip L. Yeagle

17 Lead Discovery and Development for G Protein–Coupled Receptors   375

Dennis J. Underwood and Margaret A. Cascieri

INDEX   403

SERIES PREFACE

The activation of cell surface receptors serves as the initial step in many important physiological pathways, providing a mechanism for circulating hormones or neurotransmitters to stimulate intracellular signaling pathways. Over the past 10–15 years, we have witnessed a new era in receptor research, arising from the application of molecular biology to the field of receptor pharmacology. Receptors can be classified into families on the basis of similar structural and functional characteristics, with significant sequence homology shared among members of a given receptor family. By recognizing parallels within a receptor family, our understanding of receptor-mediated signaling pathways is moving forward with increasing speed. The application of molecular biological tools to receptor pharmacology now allows us to consider the receptor–ligand interaction from the perspective of the receptor as a complement to the classic approach of probing the binding pocket from the perspective of the ligand.

Against this background, the newly launched Receptor Biochemistry and Methodology series will focus on advances in molecular pharmacology and biochemistry in the receptor field and their application to the elucidation of the mechanism of receptor-mediated cellular processes. The previous version of this series, published in the mid-1980s, focused on the methods used to study membrane-bound receptors at that time. Given the rapid advances in the field over the past decade, the new series will focus broadly on molecular and structural approaches to receptor biology. In this series, we interpret the term *receptor* broadly, covering a large array of signaling molecules including membrane-bound receptors, transporters and ion channels, as well as intracellular steroid receptors. Each volume will focus on one aspect of receptor biochemistry and will contain chapters covering the basic biochemical and pharmacological properties of the various receptors, as well as short reviews covering the theoretical background and strategies underlying the methodology. We hope that the series will provide a valuable overview of the status of the receptor field in the late 1990s, while also providing information that is of practical utility for scientists working at the laboratory bench. Ultimately, it is our hope that this series, by pulling together molecular and biochemical information from a diverse array of receptor fields, will facilitate the integration of structural and functional insights across receptor families and lead to a broader understanding of these physiologically and clinically important proteins.

DAVID R. SIBLEY
CATHERINE D. STRADER

PREFACE

The activity of virtually every body cell is regulated by extracellular signals (e.g., neuro-transmitters, hormones, sensory stimuli) that are transmitted into the cell interior via different classes of plasma membrane receptors. The vast majority of such receptors belongs to the superfamily of G protein–coupled receptors (GPCRs), which is thought to consist of approximately 1,000 different members. Despite the great structural diversity of their activating ligands, all GPCRs are predicted to share a common molecular architecture and mechanism of activation.

GPCRs play key roles in a remarkably large number of physiological and pathophysiological conditions. For this reason, it is not surprising that GPCRs or GPCR-dependent signaling pathways are the targets of numerous clinically useful drugs. Detailed knowledge about the molecular structure of GPCRs is therefore likely to open new therapeutic perspectives, including the design of novel drugs with increased specificity for distinct receptor subtypes and/or signal transduction pathways.

This volume focuses on methodologies useful for studying structure–function relationships in GPCRs. It consists of 17 chapters, contributed by different authors (or groups of authors) who are all leaders in their respective fields. Each chapter starts with a short introduction, followed by a more detailed description of basic procedures and a discussion of critical experimental parameters. However, each chapter also contains sufficient background information to provide the reader with an up-to-date overview about important findings. In most cases, actual protocols are embedded in the descriptions of specific experiments designed to address a particular question of interest.

The approaches described in the different chapters are aimed to address the following questions: What does the three-dimensional structure of a GPCR look like? How do GPCRs bind ligands? Which conformational changes occur upon GPCR activation? What are the structural characteristics of the intracellular receptor surface predicted to be involved in G protein recognition?

Currently, a high-resolution x-ray structure is not available for any GPCR. However, a very comprehensive description of the usefulness of electron crystallography of two-dimensional GPCR (rhodopsin) crystals to study GPCR structure is given in Chapter 12. Several other chapters (Chapters 1–6, 8, and 9) describe mutagenesis approaches to elucidate different aspects of GPCR structure, including the arrangement of the transmembrane receptor core, the structural features of the ligand binding pocket, and the conformational differences between the resting and activated GPCR states. More specific topics covered in these chapters include the usefulness of cysteine scanning mutagenesis (Chapter 2), "split" receptors (Chapter 6), and yeast expression technology (Chapter 8). Two of these chapters (Chapters 4 and 9) also provide a detailed description of molecular modeling techniques. In addition, Chapter 7 describes how metal ion binding sites can provide information about GPCR structure and activity.

Three chapters outline methods that have been applied successfully to study the structural basis of receptor–G protein interactions. Specifically, these chapters describe the usefulness of receptor–G protein fusion proteins (Chapter 10) and of short receptor peptides (Chapters 11 and 16) to study the receptor–G protein interface by biochemical and biophysical (nuclear magnetic resonance) techniques.

Chapters 14–16 outline various spectroscopic strategies useful for studying the molecular architecture of GPCRs, including the dynamic conformational changes that accompany receptor activation. These chapters cover site-directed spin-labeling studies (Chapter 13) as well as fluorescence spectroscopic techniques (Chapter 14), including an approach that involves the synthesis of GPCRs carrying "unnatural" amino acids (Chapter 15). Chapter 17 provides several examples for the potential impact of structural studies of GPCRs on rational drug design.

Finally, I thank all chapter authors for their insightful contributions; the series editors, Drs. Catherine Strader and David Sibley, for many helpful discussions; and Ms. Colette Bean at Wiley-Liss for efficient editorial support. I am also grateful to Ms. June Yun, one of my graduate students, for excellent assistance throughout all proofreading and editing stages. I hope that this volume will provide a useful reference guide for everybody interested in learning more about the GPCR protein family.

JÜRGEN WESS
Bethesda, MD

CONTRIBUTORS

PIER G. DE BENEDETTI, Institut de Pharmacologie et Toxicologie, Universite de Lausanne, Lausanne, Switzerland

MARGARET A. CASCIERI, Merck Research Laboratories, Rahway, NJ, USA

ANDRÉ CHOLLET, Serono Pharmaceuticals Research Institute, Geneve, Switzerland

ANNY-ODILE COLSON, Department of Physiology and Biophysics, Mount Sinai School of Medicine of the City University of New York, New York, NY, USA

SUSANNA COTECCHIA, Institut de Pharmacologie et Toxicologie, Universite de Lausanne, Lausanne, Switzerland

MARK E. DUMONT, Department of Biochemistry and Biophysics, University of Rochester, School of Medicine and Dentistry, Rochester, NY, USA

FRANCESCA FANELLI, Institut de Pharmacologie et Toxicologie, Universite de Lausanne, Lausanne, Switzerland

DAVID L. FARRENS, Departments of Biochemistry and Molecular Biology, Oregon Health Sciences University, Portland, OR, USA

TUNG MING FONG, Merck Research Laboratories, Rahway, NJ, USA

MARVIN C. GERSHENGORN, Division of Molecular Medicine, Department of Medicine, Cornell University Medical College and The New York Hospital, New York, NY, USA

ULRIK GETHER, Division of Molecular and Cellular Physiology, Department of Medical Physiology, The Panum Institute, Copenhagen, Denmark

JONATHAN A. JAVITCH, Columbia University College of Physicians and Surgeons, Departments of Psychiatry and Pharmacology, Center for Molecular Recognition, New York, NY, USA

BRIAN K. KOBILKA, Howard Hughes Medical Institute, Division of Cardiovascular Medicine, Stanford University Medical School, Stanford, CA, USA

MASAHIRO KONO, Department of Biochemistry and Volen Center for Complex Systems, Brandeis University, Waltham, MA, USA

LIISA J. LAAKKONEN, Department of Physiology and Biophysics, Mount Sinai School of Medicine of the City University of New York, New York, NY, USA

JEAN E. LACHOWICZ, Department of Central Nervous System and Cardiovascular Research, Schering-Plough Research Institute, Kenilworth, NY, USA

STRUCTURE–FUNCTION ANALYSIS OF G PROTEIN-COUPLED RECEPTORS

CHAPTER I

OVERVIEW OF MUTAGENESIS TECHNIQUES

TUNG MING FONG

I. INTRODUCTION   I
2. STRATEGIC ISSUES IN DESIGNING MUTATIONS   2
    A. Static and Dynamic Structures of GPCRs   2
    B. Ligand Binding Site in GPCRs   3
    C. Receptor Activation Mechanism   4
    D. Receptor–G Protein Interactions   5
3. DATA INTERPRETATION   7
4. PROTOCOLS   7
    A. Prediction of Helical Orientation   7
    B. Uracil-Replacement Method of ssDNA-Based Site-Directed
       Mutagenesis   9
    C. PCR-Based Mutagenesis   I I
    D. Deletion, Insertion, and Multiple-Residue Substitution
       Mutagenesis   13
    E. Chimeric Receptors   14
5. FUTURE PERSPECTIVES   14

I. INTRODUCTION

G protein–coupled receptors (GPCRs) are a class of membrane receptors that transduce extracellular chemical information into the cells via G proteins. Upon binding agonists, GPCRs catalyze the dissociation of GDP from the α subunit of G proteins, thus allowing the binding of GTP and the subsequent dissociation of the G protein α subunit from the $\beta\gamma$ subunit. The dissociated α subunit and/or the $\beta\gamma$ subunit will, in turn, modulate the activities of downstream enzymes or ion channels.

To establish structure–function correlations for any protein, a frequently used approach is to perturb the protein structure and observe the consequential functional changes. Structure can be perturbed by either chemical modification

Structure–Function Analysis of G Protein-Coupled Receptors, Edited by Jürgen Wess.
ISBN 0-471-25228-X   Copyright © 1999 Wiley-Liss, Inc.

or site-directed mutagenesis, and, if possible, one should confirm the intended structural perturbation by direct structural analysis. In the case of GPCRs, however, high-resolution structural determination is not yet feasible. Therefore, it is important to stress that mutational analysis of GPCRs lacks the component of direct three-dimensional structural determination. Structure–function analysis of GPCRs is unique in that, without direct structural information, one can only attempt to infer the structural basis of a functional change due to a mutation. Certain assumptions have to be made during this exercise, and it is the experimenter's responsibility to ensure that inferences are reasonable and structure–function relationships are testable. While mutational analysis will never replace direct structural determination, clever experimental design and continuous testing of proposed models should allow the generation of meaningful results.

Another consideration is the magnitude of structural perturbations brought about by mutations. If the change is too small, one frequently does not observe any functional change. If the perturbation is too large, receptor conformation may be so severely distorted that receptor function is abolished. In the latter case, even when one can document the presence of receptor mutants on the cell surface, no meaningful conclusions can be drawn. Therefore, one should always attempt to create and compare a series of mutants with a variety of measurable functional changes.

2. STRATEGIC ISSUES IN DESIGNING MUTATIONS

Mutational analysis can be employed to study several aspects of the molecular properties of GPCRs. In general, two design principles should be considered: Which residue should be mutated and what should be the new residue at the chosen position? The following sections discuss information that can be inferred from mutational analysis, experimental designs for mutational analysis of GPCRs, and the advantages and disadvantages of various approaches.

2.A. Static and Dynamic Structures of GPCRs

Several features of the static structure of a GPCR can be inferred from mutational analysis, including the determination of the membrane boundary of transmembrane helices (Altenbach et al., 1990), the lateral proximity of helices (Jung et al., 1993), and the lateral orientation of each helix (Altenbach et al., 1990). These studies usually involve mutating a series of residues into Cys or Lys for site-directed labeling, followed by determination of the location or environment of the labeled probe by various spectroscopic methods. In addition to direct measurements of probe location, functional properties (such as binding affinity) can be monitored after systematically introducing Cys residues followed by sulfhydryl modification (Javitch et al., 1995) or after complementary double mutation (Liu et al., 1995). Both of these approaches can supplement spectroscopic approaches to determine helical orientation, helical proximity, and location of the ligand binding site.

All proteins have a dynamic structure that forms the basis of biochemical catalysis, and GPCRs are no exception (Neubig and Thomson, 1989; Wessling-

Resnich et al., 1987). Site-directed spin labeling has been successfully applied to rhodopsin to detect movements of loops (Farahbakhsh et al., 1993) or helices (Farrens et al., 1996; Hubbell and Altenbach, 1994) upon rhodopsin activation. The experimentally detected helical movements provide direct support for a helical rotation hypothesis based solely on computer simulation (Luo et al., 1994), demonstrating the power of combining mutagenesis, spectroscopy, and computer modeling to elucidate the structure–function relationships of GPCRs.

In addition, receptor folding and assembly can be studied by splitting GPCRs into two parts and co-expressing the two separate units. In the case of the muscarinic M_3 receptor, some of the combinations of split receptors can bind ligand and even activate downstream effectors (Schöneberg et al. 1995). Combining split receptor studies with site-directed mutagenesis should allow one to determine the relative association affinity among the seven helices.

2.B. Ligand Binding Site in GPCRs

Another aspect amenable to mutational analysis is the localization of the ligand binding site. Such studies require intelligent design of mutations to maximize the amount of information that can be obtained from a minimal number of mutants. A strategic design of binding site localization through mutational analysis always begins with a model or working hypothesis. For most GPCRs, the central cavity defined by the upper half of the transmembrane region is predicted to contain the ligand binding site or part of the ligand binding site (Fong and Strader, 1994; Strader et al., 1994). To determine which residues are in contact with the bound ligand, one needs to build a low-resolution model and target those residues thought to project into the interior of the helical bundle. Specifically, one needs to analyze the sequence of each transmembrane helix and identify its most hydrophilic face, which is expected to face the interior of the protein. While sophisticated programs exist to predict helix polarity (Eisenberg et al., 1984; Rees et al., 1989), any spreadsheet program can accomplish the same task (Protocol 4.A).

For those GPCRs (such as the metabotropic glutamate receptors) in which the large amino-terminal extracellular domain is the most likely ligand binding domain, identification of residues to be targeted for site-directed mutagenesis usually requires prior deletion or chimeric receptor analysis. It should be pointed out that such mutations frequently cause significant conformational perturbations, and caution should therefore be exercised when interpreting results from such studies. Conclusions drawn from deletion or chimeric receptor analyses should always be tested by site-directed mutagenesis (Huang et al., 1994b). Nonetheless, chimeric receptor analysis can be extremely valuable in guiding subsequent site-directed mutagenesis studies. For example, the molecular basis of species-dependent pharmacological properties of the human and rat NK_1 receptors was attributed to two phylogenetically divergent residues based on initial chimeric receptor analysis followed by the construction of point mutations and double mutants (Fong et al., 1992b). This analysis also suggested that these two residues are not part of the ligand binding site, but rather that they affect the conformation of the binding pocket.

Another strategy for selecting residues for site-directed mutagenesis is based on sequence alignment of homologous receptors. A conserved residue may contribute to the interaction with a critical group on a ligand acting on related receptor subtypes. For example, the conserved Asp residue in helix 3 of adrenergic receptors has been found to contribute to the specific binding of biogenic amines (Strader et al., 1994)

As soon as candidate residues for mutational analysis have been identified, one needs to decide which replacement residues to choose. It is crucial that the structural changes introduced by the mutation do not lead to a gross conformational change (Ackers and Smith, 1985). Changing a residue capable of noncovalent interactions to Ala usually gives an initial indication of whether a particular side chain is required for high-affinity ligand binding. However, the cavity created by Ala substitution sometimes leads to unstable proteins (Eriksson et al., 1992). The substitution of a chemically related residue may also provide information on the mode of interaction. Thus, creating multiple mutations at the same position is beneficial in controlling for conformational effects and for providing hints on what type of molecular interactions occur at that position (Fong et al., 1993). Detailed knowledge of the physicochemical character of individual amino acids is required for proper data interpretation. For example, changing a Glu residue to Gln not only eliminates a potential negative charge and H-bond acceptor, but also introduces a new H-bond donor. Similarly, in the case of a Cys residue that does not participate in a disulfide bond, a Cys to Ser substitution (or vice versa) can still cause significant conformational changes because of the dihedral angle difference of sulfhydryl and hydroxyl in H-bond formation (He and Quiocho, 1991).

One important issue is whether the observed effect of a mutation is due to elimination of direct binding interactions, alteration of direct binding as a result of a small local conformational change within the binding site, or long-range conformational effects. If a mutation affects the binding of all ligands tested, one can assume a global conformational effect. If a mutation only affects the binding of a subclass of ligands, a generalized structural change is less likely. The most obvious way to test for direct ligand–receptor interactions is through complementary modifications in which a series of receptor mutants is analyzed with a series of ligand analogs (Fong et al., 1993; Strader et al., 1989; Swain et al., 1995). This strategy is based on the nonadditivity principle (Table 1.1), and it generally has a minimal effect on the gross receptor conformation because it involves only elimination of functional groups. A more drastic form of this strategy involves charge reversal or charge-to-H bond conversion (Strader et al., 1991). This approach is more problematic because it involves introduction of new charges or new functional groups that are more likely to affect the gross conformation of both the ligand and the receptor.

2.C. Receptor Activation Mechanism

Some mutations can lead to an impairment or enhancement (constitutive activity) of receptor activation, without affecting the ligand binding site directly. A rational design of mutations to probe receptor activation involves targeting of those residues that are highly conserved in GPCRs; such residues are located in the bottom half of the transmembrane domain, such as D/E in helix 2, DRY in helix 3, and P/W in helices 4–7 (Fong, 1996). In many cases, substitution of these highly conserved residues leads to a substantial loss of G protein–cou-

TABLE 1.1. Use of the Nonadditivity Principle To Confirm a Direct Ligand–Receptor Interaction

Ligand	Receptor	K_d	
Ligand-R	Xxx Receptor	K1	
Ligand-H	Xxx Receptor	K2	K2 > K1
Ligand-H	Ala Receptor	K3	K3 = K2

In this scheme, prior mutagenesis should have shown that mutation of the residue Xxx leads to a reduction in binding affinity, and the functional group R of the ligand is also required for high-affinity binding. Removal of R from the ligand should disrupt a noncovalent interaction, leading to a reduction in binding affinity for the wild-type receptor (K2 > K1). If this interaction is directly mediated by R and the residue Xxx of the receptor, it follows that substituting Xxx with Ala should not lead to a further reduction in binding affinity (K3 = K2). If R does not interact with Xxx, the mutation of Xxx and the removal of R should have an additive effect.

pling efficiency (Huang et al., 1994a; Oliveira et al., 1994; Perlman et al., 1992; Rosenthal et al., 1993; Strader et al., 1988; Wess et al., 1993). Other less conserved residues have also been found to be required for receptor activation, but such findings are often chance encounters (Huang et al., 1995). Similarly, mutations that confer constitutive activity were discovered by chance (Kjelsberg et al., 1992), and a consensus has not yet emerged as to which mutations generally generate constitutive activity.

Despite the lack of direct structural information, simple modeling efforts can facilitate the design of mutations that modify receptor activation. Because GPCR activation is likely to be a conformational relaxation event, residues involved in interhelical packing may play a major role in activation. To identify such residues, the first step is to orient each helix by maximizing the number of hydrophobic residues that face the bilayer and maximizing the number of hydrophilic residues that face the protein interior (Protocol 4.A). The seven transmembrane helices are expected to form a helical bundle, arranged sequentially in a counterclockwise fashion when viewed from the extracellular space (Unger et al., 1997). Residues located at helix–helix interfaces can then be identified using a helical projection map (Fig. 1.1). A similar approach has been used to identify two interhelical residues the mutation of which can modulate receptor activation presumably by affecting helical packing (Liu et al., 1995).

2.D. Receptor–G Protein Interactions

The molecular basis of receptor–G protein coupling selectivity can be studied best by hybrid receptor analysis. For example, the third intracellular loop of one receptor activating Gi can be substituted into another receptor activating Gq, thereby conferring to the second receptor the ability to couple to Gi (Lechleiter et al., 1990). Chimeric receptor analysis of receptor–G protein interactions usually leads to the identification of a considerable number of residues controlling G protein–coupling specificity (Hedin et al., 1993; Wess, 1997) because the

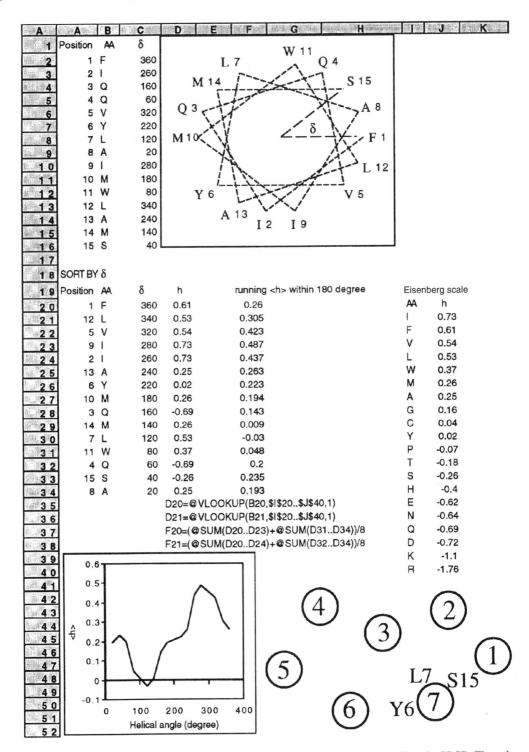

Figure 1.1. Prediction of the orientation of transmembrane helices in GPCR. The calculation can be done in any spreadsheet program. The example shown was created in Lotus 1-2-3 (see text for details).

conformational complementarity of the protein–protein interface is a likely de-terminant of G protein–coupling specificity.

On the other hand, site-directed mutations may be designed to probe the G protein activation mechanism using information derived from chimeric receptor analysis. "Random" site-directed mutagenesis within the third intracellular loop has led to the identification of a single residue where any substitution can lead to constitutive G protein activation. These data were interpreted in terms of a re-ceptor–G protein activation mechanism through conformational relaxation (Kjelsberg et al., 1992). Alternatively, designing a perfect α helix within this re-gion, which presumably increases order and stability, can lead to a receptor mu-tant with reduced G protein–coupling efficiency (Cheung et al., 1992).

3. DATA INTERPRETATION

Mutational analysis has been proven to be a powerful tool in our efforts to un-derstand how receptors function, how ligands interact with receptors, and how receptors interact with G proteins. Mutagenesis techniques will continue to be at the center stage of structure–function analyses even after the three-dimen-sional structure of some receptors will become available. However, it should be emphasized that the structural information deduced from mutational analysis is low resolution at best, and the deduced structural model cannot be proven based on mutational analysis alone. Thus, mutagenesis data should be inter-preted cautiously. It is generally not meaningful to interpret data obtained with a single mutant measuring only one functional parameter. In the case of chimeric receptors or insertion/deletion mutants, conclusions should also be tested by site-directed mutagenesis.

In the case of site-directed mutagenesis studies, multiple mutations at any given position and multiple functional indicators can help to reduce the risk of overinterpretation. When studying ligand–receptor interactions, the complemen-tary modification approach (Table 1.1) represents the most powerful and gener-ally applicable method to assign ligand–receptor contacts. The major limitation of mutagenesis studies is that one can never be sure about the precise conforma-tional consequences of amino acid sequence changes in the absence of a high-res-olution structure. Relationships between primary structure and function can easily be established, while relationships between three-dimensional structure and function can only be extrapolated. With the understanding of these caveats, mutational analysis of GPCR allows one to get an early view of structure–func-tion relationships before the availability of three-dimensional structures.

4. PROTOCOLS

4.A. Prediction of Helical Orientation

1. Create a spreadsheet as shown in Figure 1.1. Enter the helical sequence in column B (e.g., the top half of helix 7 from the human NK_1 receptor is shown in cells B2 to B16). If there is a Pro residue within the region,

obviously there will be a kink in a nonidealized helix. For practical purposes, however, such a kink would not affect the low-resolution modeling described here. Even a helical tilt does not affect this modeling because the present protocol only considers part of a helix, which is equivalent to a cross section of the electron density map in a high-resolution structure (Unger et al., 1997).

2. A helical wheel projection is drawn (upper insert in Fig. 1.1, as viewed from the extracellular side), each residue having a set of (x,y) coordinates in which x = diameter · COS(δ · 2 · 3.14/360), y = diameter · SlN (δ · 2 · 3.14/360), and δ = helical angle. The diameter value is arbitrarily set. Cells C2 to C16 show the helical angle for residues of TM1, TM3, TM5, or TM7 in which the helical wheel is clockwise for right-handed helices when viewed from the extracellular space. For TM2, TM4, or TM6, the helical wheel is counterclockwise, and the helical angle should be 0°, 100°, 200°, 300°, 40°, 140°, and so forth in cells C2 to 616.

3. The concept behind the subsequent calculation is to calculate an average hydrophobicity within a window of 180° and assign the value to the residue at the center of the window, then rotate to the next residue on the wheel and continue the calculation. For example, for F at position 1 (δ = 0°), a 180° window would encompass residues with δ in the range of −90° to 90°. This calculation is somewhat analogous to the Kyte-Doolittle plot except that the present calculation is a hydrophobicity plot along the perimeter of a two-dimensional projection for a helix. The following description of a spreadsheet will automate the calculation.

4. To convert a linear sequence into a string along the helical wheel projection, sort the data in the spreadsheet range of A2 to C16 based on the helical angle. The result is shown in A20 to C34.

5. Look up the hydrophobicity value h for each amino acid from the table in I21 to J40. The VLOOKUP command is available in both Lotus 1-2-3 and Microsoft Excel. Two examples are shown for entries in D20 and D21 (Fig. 1.1) in which the first argument is the cell address containing the amino acid symbol, the second argument is the cell address range of the hydrophobicity table, and the third argument is the offset number indicating column location of the value to be looked up. The consensus hydrophobicity scale is used here (Eisenberg et al., 1984).

6. Calculate the running average of hydrophobicity (h) for a 180° window, using each residue as the center of such a window consecutively. The equation for calculating h is illustrated for entries in F20 and F21 (Fig. 1.1).

7. Plot h as a function of helical angle (lower left inset in Fig. 1.1). A minimum is seen at position 7, indicating that the helical half encompassing M10, Q3, M14, L7, W11, Q4, and S15 is the most hydrophilic half and is most likely facing the receptor interior. In addition, Y6, M10, S15, A8, and F1 are most likely to be at the interhelical interface, adjacent to helix 1 and helix 6.

8. By repeating the same procedure for each helix, a projection model can be developed (lower right insert in Fig. 1.1).

4.B. Uracil-Replacement Method of ssDNA-Based Site-Directed Mutagenesis

This method is particularly convenient when making many point mutations on the same receptor because the same single-stranded (ss) DNA can be used repeatedly and the amount of sequencing is relatively small. Many commercial vendors provide ssDNA mutagenesis kits, such as the Muta-Gene *in vitro* mutagenesis kit from Bio-Rad (Geisselsoder et al., 1987). If one intends to make only a few mutants, the PCR method (Protocol 4.C) would be the preferred choice.

Preparation of ssDNA

1. Insert the gene of interest into a phagemid such as pCDM9 (Fong et al., 1992b) or any cytomegalovirus expression phagemid containing the M13 replication origin.

2. Transform 1 ng of the phagemid (prepared by standard plasmid mini-prep methods) into the special *Escherichia coli* strain CJ236 (Bio-Rad, Richmond, CA) and incubate the plate (containing ampicillin) overnight.

3. Pick a single colony, inoculate 3 ml Luria-Bertani (LB) media (containing chloramphenicol), and incubate overnight with shaking.

4. Inoculate 50 ml of $2\times$ YT media containing chloramphenicol with 1 ml of the overnight culture and continue shaking until $OD_{600} = 0.3$ (approximately 2–4 hours), corresponding to 1×10^7 cells/ml. Add helper phage R408 (Stratagene, San Diego, CA) to achieve a multiplicity of infection (MOI) of 20 (i.e., 20 phages/cell). Incubate overnight with shaking.

5. Centrifuge the overnight culture ($17,000g$) and transfer the phage-containing supernatant to a fresh tube. Repeat the centrifugation once. Precipitate the phage by adding 1/4 volume of 20% PEG-6000/2.5 M NaCl.

6. Prepare the ssDNA from the phage pellet by standard phenol-chloroform extraction and ethanol precipitation.

7. Determine the ssDNA concentration and quality by running a sample on an agarose gel with a standard ssDNA marker. This ssDNA can be used for all mutagenesis experiments because only the oligonucleotide needs to be changed for different mutants.

Mutagenesis

8. Synthesize an oligonucleotide complementary to the ssDNA prepared above, but encoding the desired mutation. A 30-mer oligo with one to three base mismatches in the middle is usually appropriate. The GC content of the oligonucleotide should preferentially be within the range of 40%–60%. A string of four or more identical nucleotides should be avoided.

9. Purify the oligonucleotide using a NENSOPB cartridge (NEN, Boston, MA), and phosphorylate the 5'-OH with T4 polynucleotide kinase.

10. Anneal the oligonucleotide (6 pmol) with ssDNA (0.3 pmol) in T7 polymerase buffer by placing the reaction tube in 200 ml of 70°C water in a beaker, and let the beaker cool to room temperature (this should take

approximately 30 minutes). DNA can then be extended from the oligonucleotide with T7 DNA polymerase (0.05 units/μl), followed by ligation with T4 DNA ligase (0.3 units/μl).

11. Transform the heteroduplex DNA into an *E. coli* strain such as SURE cells (Stratagene, San Diego, CA), pick four to five colonies, and prepare miniprep DNA from each of them.

12. Sequence the region from which the oligonucleotide is derived to confirm the success of mutation. Double-stranded DNA sequencing is fast and convenient. If all clones are wild type (which is not common), sequence another 20 clones or screen a larger number of clones by hybridization using the ^{32}P-labeled mutagenic oligonucleotide.

Expression

13. Prepare a larger quantity of the mutated DNA (usually a 100–200 ml culture is sufficient for multiple transient expression experiments in COS cells).

14. Transfect 10 μg of DNA into 1×10^7 COS cells (or other appropriate mammalian cells) by electroporation (Sambrook et al., 1989), followed by appropriate assays for the receptor of interest. Alternatively, lipofection can be used for transient expression experiments, using a kit from a commercial vendor. In most cases, analysis of receptor mutants does not require the generation of stable cell lines. The choice of cells for transfection depends on the receptors being studied and whether the non-transfected cells express any endogenous receptor that may interfere with the planned assay. COS and CHO cells are two frequently used cell lines because of their ease of culturing and nonhuman origin, requiring less biohazard containment.

There are several other considerations. The ssDNA templates must be of high quality. Contaminating DNA, RNA, polyethylene glycol, or salts may inhibit the hybridization of the mutant oligonucleotide to the template. The best way to ensure a high quality ssDNA preparation is to achieve complete separation at each step of the protocol. With a final preparation of ssDNA, only proteins and RNA can be easily removed, not other contaminating DNA. The quality of ssDNA can be checked by using it as a template for dideoxy sequencing. An unsuccessful sequencing reaction indicates that the ssDNA is not pure and probably will not generate mutants.

Difficulties in achieving mutagenesis can arise due to intermolecular self-hybridization of the oligonucleotide. One solution is to introduce another silent mutation simultaneously, thereby changing the sequence of the oligonucleotide without affecting the sequence of the resulting mutant protein. Another common problem is that some positions in the DNA sequence are difficult to mutate due to either a high GC content at those positions or to the presence of a secondary structure in the ssDNA. A balance between achieving hybridization and reducing the secondary structure can be accomplished by experimenting with several annealing temperatures near the calculated melting temperature

(T_m). For oligonucleotides shorter than 20-mer (Wallace et al., 1979), T_m (°C) = 4 · (number of G+C) + 2 · (number of A+T). For oligonucleotides longer than 50-mer (Meinkoth and Wahl, 1984), T_m (°C) = 81.5 + 16.6 · log(ionic strength, M) + 0.41 · (% GC) − 500/length − 0.61 · (% formamide).

4.C. PCR-Based Mutagenesis

Polymerase chain reaction (PCR) techniques can be used for almost any kind of mutagenesis experiment. The following protocol uses a single residue mutation as an example. However, simple variations of the general scheme can generate deletions, insertions, multiple-residue substitutions, or chimeric receptors. The success of the PCR technique appears to be limited only by the length of the oligonucleotide. A design that calls for oligonucleotides longer than approximately 80-mer is not recommended.

Design of Oligonucleotides

1. Four oligonucleotides are needed to generate one mutation. Two are the mismatch oligonucleotides encoding the mutation, one being sense (A) and one being antisense (B). The other two are perfectly matching oligonucleotides, one being sense (C) and upstream of A/B and the other being antisense (D) and downstream of A/B (Fig. 1.2). The sequence between C and A/B and the sequence between A/B and D should contain two unique restriction sites (site x and site y, respectively) so that the mutated PCR fragment can be exchanged with the corresponding wild-type fragment. The oligonucleotide design considerations are the same as those described in step 8 of Protocol 4.2. In addition, one should avoid consecutive G/C bases at the 3′ end of any oligonucleotide to prevent primer–dimer formation.

Mutagenesis

2. The first PCR uses C and B as primers (1 μM final concentration) and a plasmid containing wild-type cDNA as template (1 μg) in 100 μl of 10 mM Tris, pH 8.3, 50 mM KCl, 1.5 mM $MgCl_2$, 800 μM dNTP, and 2.5 U of Taq DNA polymerase. Standard PCR conditions recommended by Perkin-Elmer should be followed (e.g., denature at 94°C for 1 minute, anneal at 37°C for 1 minute, extend at 72°C for 1 minute, for a total of 20–30 cycles).

3. A second PCR uses A and D as primers and a plasmid containing the wild-type cDNA as template.

4. Gel purify the PCR products from steps 2 and 3, and recover the PCR fragments using GeneClean (Bio101, La Jolla, CA).

5. The third PCR (i.e., overlap PCR) uses the first and second PCR products from steps 2 and 3 as template and C and D as primers. Extract the final PCR product with phenol/$CHCl_3$, followed by ethanol precipitation and resuspension in water. Cleave the PCR product with enzymes recognizing

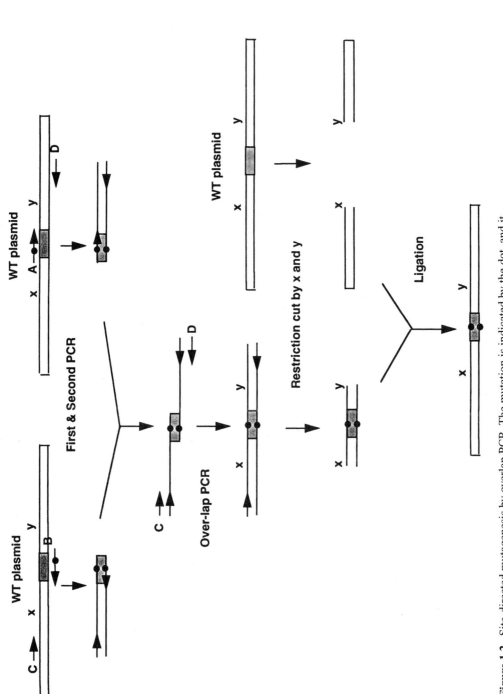

Figure 1.2. Site-directed mutagenesis by overlap PCR. The mutation is indicated by the dot, and it can be a substitution, deletion, or insertion of one or more nucleotides (see text for details).

sites x and y. Gel purify the correct fragment, which should contain the mutated sequence.

6. Cleave the plasmid containing the wild-type cDNA with enzymes recognizing sites x and y. Gel purify the fragment that does not contain the sequence corresponding to the PCR fragment in step 5.

7. Ligate the two purified fragments from steps 5 and 6. Transform the ligation product into *E. coli*. Pick several colonies and prepare mini-prep plasmid DNA.

8. Sequence the entire region between sites x and y to confirm the presence of the designed mutation and the absence of any unwanted mutation caused by the PCR.

Expression

9. See Protocol 4.B.

Other considerations include the following. Although PCR-based mutagenesis usually will eliminate the problem of secondary structure, denaturing agents such as formamide and dimethylsulfoxide can also be used if one is trying to amplify DNA with a high GC content. Formamide should be titrated between 1% and 5% and dimethylsulfoxide between 2.5% and 15%, and the purity of the reagents is critical. To reduce the secondary structure of the template, "hot start" PCR can be used instead of denaturing agents. This technique involves melting the template DNA in buffer in the presence of primers, followed by a quick cool before adding the nucleotides and polymerase. Finally, when PCR fails to produce the desired fragment, optimization of the PCR buffer pH or Mg^{2+} concentration may be necessary (e.g., PCR optimization kit from Invitrogen).

4.D. Deletion, Insertion, and Multiple-Residue Substitution Mutagenesis

Three experimental designs are possible using this approach: deletion of a stretch of amino acids, insertion of a new sequence, and combined deletion and insertion (which is equivalent to a substitution) of a stretch of residues. The last approach is sometimes referred to as *chimeric receptor analysis* when the substitution is performed between two homologous receptors. The deletion or insertion analysis should be performed in GPCR loop regions only, and even in those cases caution should be exercised to avoid gross conformational effects as a result of the mutation. The substitution analysis can be performed either in helical regions (Gether et al., 1993; Kikkawa et al., 1998; Pittel and Wess, 1994) or loop regions (Fong et al., 1992a; Gerszten et al., 1994), and the choice of substitution boundaries is extremely critical in avoiding gross conformational distortions.

Protocol 4.C can be used with minor modifications to insert, delete, or substitute multiple residues. The mutagenic oligonucleotides should be designed according to the general principles in Figure 1.2. If one intends to insert or substitute more than 15 residues (i.e., more than 45 nucleotides), it will be necessary to include an extra PCR step (Fig. 1.3).

4.E. Chimeric Receptors

A chimeric receptor is usually created between two structurally related receptors by selecting a restriction enzyme site that is present in both receptors at homologous positions and exchanging one part between the two receptors. A more complex chimeric receptor can be constructed by selecting two common restriction enzyme sites, each present at homologous positions of the two receptors, and exchanging the center part.

1. Scan the sequence to identify unique restriction enzyme sites in both receptors. If a common restriction enzyme site cannot be found, it may be possible to mutate one or two residues in one of the two receptors to create a common restriction enzyme site (Fig. 1.4).
2. Cut both receptor plasmids with the common restriction enzymes. Gel purify the large plasmid fragment from one receptor and the small receptor fragment from the other using the GeneClean kit.
3. Ligate the two fragments with T4 DNA ligase at 12°–15°C overnight.
4. Transform the ligated plasmid into *E. coli.*
5. Select several colonies to verify the ligation junction by sequencing.
6. Perform expression analysis as described in Protocol 4.B.

In cases where no common restriction enzyme site can be created, a PCR-based approach can be used, although the disadvantage is that the entire mutant receptor has to be sequenced to confirm the absence of unintended mutations. Alternatively, a restriction site–independent method of chimera generation can be used, but in this case one cannot control the exact site of exchange, and extensive screening is required to isolate functional chimeras (Moore and Blakely, 1994).

5. FUTURE PERSPECTIVES

Mutational analysis, low-resolution spectroscopic analysis (see following chapters), and computer modeling (Ballesteros and Weinstein, 1995; Underwood and Prendergast, 1997) are frequently used techniques to study the structure–function relationship of GPCR. Incorporation of unnatural amino acids should become more routine in the future. This new technique can provide more options for subtly changing the chemical properties of amino acid side chains or the protein backbone (Ellman et al., 1992; Nowak et al., 1995) or achieving site-directed incorporation of probes without chemical labeling (Turcatti et al., 1996). It can be anticipated that the first high-resolution structure will be that of rhodopsin. However, a true correlation between structure and function will require structural determination of not only the resting receptor but also the liganded or activated receptor.

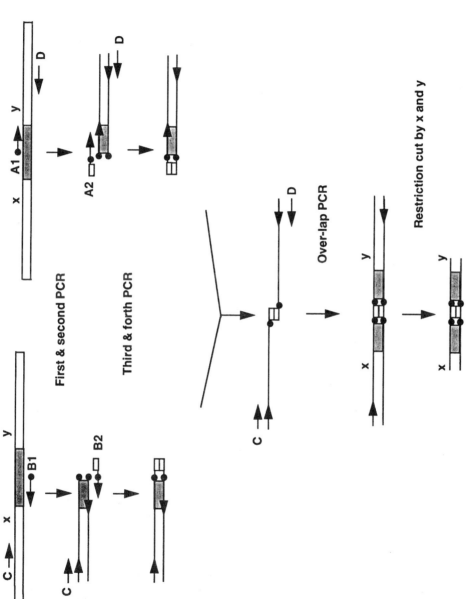

Figure 1.3. Extended overlap PCR to insert, delete, or substitute a long stretch of residues. Only the earlier part of Protocol 4.D is shown; the ligation part is identical to that in Figure 1.2 (see text for details).

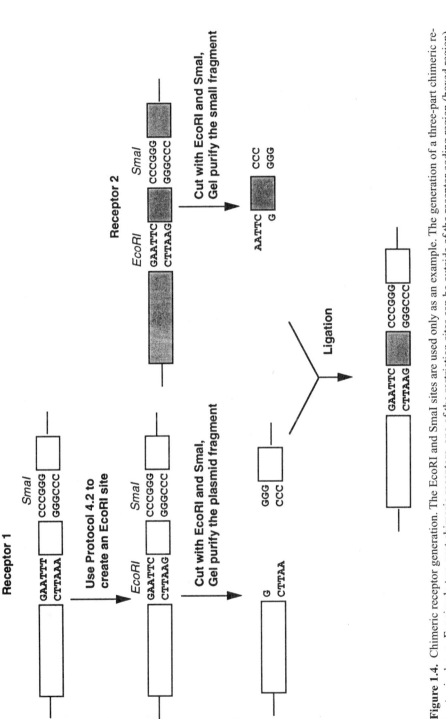

Figure 1.4. Chimeric receptor generation. The EcoRI and SmaI sites are used only as an example. The generation of a three-part chimeric receptor is shown. For simple two-part chimeric receptors, one of the restriction sites can be outside of the receptor coding region (boxed region). The rest of the plasmid is represented by a short line.

At this time, one needs to validate and refine current structural models through a continuous process of hypothesis testing. For example, the antagonist binding site model of the NK_1 receptor was based on single residue mutational analysis and was then further supported by studies using conformationally constrained ligands and triple mutant analysis (Fong et al., 1996; Swain et al., 1995). Mutagenesis studies can also complement pharmacological studies, for example, by confirming the competitive mechanism of antagonists when an antagonist exhibits a pseudoirreversible effect (Cascieri et al., 1997). The full power of mutational analysis is realized when one can determine both three-dimensional structure and biochemical function, as has been demonstrated for the D96N mutant of bacteriorhodopsin, which slows proton pumping without perturbing the ground-state structure (Mitra et al., 1993), the role of Asp^{398} and ATP binding in triggering folding within the chaperonin complex (Lorimer, 1997), and the design of double-mutant cycles to detect structural change (Carter et al., 1984).

REFERENCES

Ackers GK, Smith FR (1985): Effects of site-specific amino acid modification on protein interactions and biological function. Annu Rev Biochem 54:597–629.

Altenbach C, Marti T, Khorana HG, Hubbell WL (1990): Transmembrane protein structure: Spin labeling of bacteriorhodopsin mutants. Science 248:1088–1092.

Ballesteros JA, Weinstein H (1995): Integrated methods for the construction of three-dimensional models and computational probing of structure–function relations in G protein coupled receptors. In Sealfon SC (ed): Receptor Molecular Biology. San Diego: Academic Press, pp 366–428.

Carter PJ, Winter G, Wilkinson AJ, Fersht AR (1984): The use of double mutants to detect structural changes in the active site of the tyrosyl-tRNA synthase. Cell 38: 835–840.

Cascieri MA, Ber E, Fong TM, Hale J, Tang F, Shiao LL, Mills SG, MacCoss M, Sadowski S, Tota MR, Strader CD (1997): Characterization of the binding and activity of a high affinity pseudoirreversible morpholino tachykinin NK_1 receptor antagonist. Eur J Pharmacol 325:253–261.

Cheung AH, Huang RRC, Strader CD (1992): Involvement of specific hydrophobic, but not hydrophilic, amino acids in the third intracellular loop of the beta-adrenergic receptor in the activation of Gs. Mol Pharmacol 41:1061–1065.

Eisenberg D, Weiss RM, Terwilliger TC (1984): The hydrophobic moment detects periodicity in protein hydrophobicity. Proc Natl Acad Sci USA 81:140–144.

Ellman JA, Mendel D, Schultz PG (1992): Site-specific incorporation of novel backbone structures into proteins. Science 255:197–200.

Eriksson AE, Baase WA, Zhang XJ, Heinz DW, Blaber M, Balwin EP, Matthews BW (1992): Response of a protein structure to cavity-creating mutations and its relation to the hydrophobic effect. Science 255:178–183.

Farahbakhsh ZT, Hideg K, Hubbell WL (1993): Photoactivated conformational changes in rhodopsin: A time-resolved spin label study. Science 262:1416–1419.

Farrens DL, Altenbach C, Yang K, Hubbell WL, Khorana HG (1996): Requirement of rigid-body motion of transmembrane helices for light activation of rhodopsin. Science 274:768–770.

Fong TM (1996): Mechanistic hypotheses for the activation of G protein–coupled receptors. Cell Signal 8:217–224.

Fong TM, Cascieri MA, Yu H, Bansal A, Swarn C, Strader CD (1993): Amino-aromatic interaction between histidine-197 of the human neurokinin-1 receptor and CP-96,345. Nature 362:350–353.

Fong TM, Huang R-RC, Strader CD (1992a): Localization of agonist and antagonist binding domains of the human neurokinin-1 receptor. J Biol Chem 267:25664–25667.

Fong TM, Strader CD (1994): Functional mapping of the ligand binding sites of G-protein coupled receptors. Med Res Rev 14:387–399.

Fong TM, Yu H, Huang RRC, Cascieri MA, Swain CJ (1996): Relative contribution of polar interactions and conformational compatibility to the binding of neurokinin-1 receptor antagonists. Mol Pharmacol 50:1605–1611.

Fong TM, Yu H, Strader CD (1992b): Molecular basis for the species selectivity of the neurokinin-1 receptor antagonist CP-96,345 and RP67580. J Biol Chem 267:25668–25671.

Geisselsoder J, Witney F, Yuckenberg P (1987): Efficient site-directed in vitro mutagenesis. BioTechniques 5:786–791.

Gerszten RE, Chen J, Ishli M, Ishli K, Wang L, Nanevicz T, Turck CW, Vu TH, Coughlin SR (1994): Specificity of the thrombin receptor for agonist peptide is defined by its extracellular surface. Nature 368:648–651.

Gether U, Johansen TE, Snider RM, Lowe JA III, Nakanishi S, Schwartz TW (1993): Different binding epitopes on the NK_1 receptor for substance P and a non-peptide antagonist. Nature 362:345–348.

He JJ, Quiocho FA (1991): A nonconservative serine to cysteine mutation in the sulfate-binding protein, a transport receptor. Science 251:1479–1481.

Hedin KE, Duerson K, Clapham DE (1993): Specificity of receptor–G protein interactions: Searching for the structure behind the signal. Cell Signal 5:505–518.

Huang RRC, Huang D, Strader CD, Fong TM (1995): Conformational compatibility as a basis of differential affinities of tachykinins for the neurokinin-1 receptor. Biochemistry 34:16467–16472.

Huang RRC, Yu H, Strader CD, Fong TM (1994a): Interaction of substance P with the second and seventh transmembrane domains of the neurokinin-1 receptor. Biochemistry 33:3007–3013.

Huang RRC, Yu H, Strader CD, Fong TM (1994b): Localization of the ligand binding site of the neurokinin-1 receptor: Interpretation of chimeric mutants and single residue substitutions. Mol Pharmacol 45:690–695.

Hubbell WL, Altenbach C (1994): Investigation of structure and dynamics in membrane proteins using site-directed spin labeling. Curr Opin Struct Biol 4:566–573.

Javitch JA, Fu D, Chen J, Karlin A (1995): Mapping the binding-site crevice of the dopamine D2 receptor by the substituted-cysteine accessibility method. Neuron 14:825–831.

Jung K, Jung H, Wu J, Prive GG, Kaback HR (1993): Use of site-directed fluorescence labeling to study proximity relationships in the lactose permease of E. coli. Biochemistry 32:12273–12278.

Kikkawa H, Isogaya M, Nagao T, Kurose H (1998): The role of the seventh transmembrane region in high affinity binding of a beta2-selective agonist TA-2005. Mol Pharmacol 53:128–134.

Kjelsberg MA, Cotecchia S, Ostrowski J, Caron MG, Lefkowitz RJ (1992): Constitutive activation of the alpha-1b adrenergic receptor by all amino acid substitutions at a single site. J Biol Chem 267:1430–1433.

Lechleiter J, Hellmiss R, Duerson K, Ennulat D, David N, Clapham D, Peralta E (1990): Distinct sequence elements control the specificity of G protein activation by muscarinic acetylcholine receptor subtypes. EMBO J 9:4381–4390.

Liu J, Schöneberg T, van Rhee M, Wess J (1995): Mutational analysis of the relative orientation of transmembrane helices I and VII in G protein coupled receptors. J Biol Chem 270:19532–19539.

Lorimer G (1997): Folding with a two-stroke motor. Nature 388:720–723.

Luo X, Zhang D, Weinstein H (1994): Ligand-induced domain motion in the activation mechanism of a G protein coupled receptor. Prot Eng 7:1441–1448.

Meinkoth J, Wahl G (1984): Hybridization of nucleic acids immobilized on solid supports. Anal Biochem 138:267–284.

Mitra AK, Miercke LJW, Turner GJ, Shand RF, Betlach MC, Stroud RM (1993): Two-dimensional crystallization of E. coli–expressed bacteriorhodopsin and its D96N variant: High resolution structural studies in projection. Biophys J 65:1295–1306.

Moore KR, Blakely RD (1994): Restriction site–independent formation of chimeras from homologous neurotransmitter–transporter cDNAs. BioTechniques 17:130–135.

Neubig RR, Thomson WJ (1989): How does a key fit a flexible lock? Structure and dynamics in receptor function. BioEssays 11:136–141.

Nowak MW, Kearney PC, Sampson JR, Saks ME, Labarca CG, Silverman SK, Zhong W, Thorson J, Abelson JN, Davidson N, Schultz PG, Dougherty DA, Lester HA (1995): Nicotinic receptor binding site probed with unnatural amino acid incorporation in intact cells. Science 268:439–442.

Oliveira L, Paiva ACM, Sander C, Vriend G (1994): A common step for signal transduction in G protein–coupled receptors. Trends Pharmacol Sci 15:170–172.

Perlman JH, Nussenzveig DR, Osman R, Gershengorn MC (1992): Thyrotropin-releasing hormone binding to the mouse pituitary receptor does not involve ionic interactions. J Biol Chem 267:24413–24417.

Pittel Z, Wess J (1994): Intramolecular interactions in muscarinic acetylcholine receptors studied with chimeric m2/m5 receptors. Mol Pharmacol 45:61–64.

Rees DC, DeAntonio L, Eisenberg D (1989): Hydrophobic organization of membrane proteins. Science 245:510–513.

Rosenthal W, Antaramian A, Gilbert S, Birnbaumer M (1993): Nephrogenic diabetes insipidus. A V2 vasopressin receptor unable to stimulate adenylyl cyclase. J Biol Chem 268:13030–13033.

Sambrook J, Fritsch EF, Maniatis T (1989): Molecular Cloning, 2nd ed. Cold Spring Harbor, NY: Cold Spring Harbor Laboratory.

Schöneberg T, Liu J, Wess J (1995): Plasma membrane localization and functional rescue of truncated forms of a G protein coupled receptor. J Biol Chem 270:18000–18006.

Strader CD, Candelore MR, Hill WS, Sigal IS, Dixon RAF (1989): Identification of two serine residues involved in agonist activation of the beta-adrenergic receptor. J Biol Chem 264:13572–13578.

Strader CD, Fong TM, Tota MR, Underwood D, Dixon RAF (1994): Structure and function of G protein coupled receptors. Annu Rev Biochem 63:101–132.

Strader CD, Gaffney T, Sugg EE, Canderlore MR, Keys R, Patcher AA, Dixon RAF (1991): Allele-specific activation of genetically engineered receptors. J Biol Chem 266:5–8.

Strader CD, Sigal IS, Canderlore MR, Rands E, Hill WS, Dixon RAF (1988): Conserved aspartic acid residues 79 and 113 of the beta-adrenergic receptor have different roles in receptor function. J Biol Chem 263:10267–10271.

Swain CJ, Fong TM, Haworth K, Owen SN, Seward EM, Strader CD (1995): Quinucli-dine-based NK1 antagonists, the role of the benzhydryl. Bioorg Med Chem Lett 5:1261–1264.

Turcatti G, Nemeth K, Edgerton MD, Meseth U, Talabot F, Peitsch M, Knowles J, Vogel H, Chollet A (1996): Probing the structure and function of the tachykinin neu-rokinin-2 receptor through biosynthetic incorporation of fluorescent amino acids at specific sites. J Biol Chem 271:19991–19998.

Underwood D, Prendergast K (1997): Getting it together: Signal transduction in G-pro-tein coupled receptors by association of receptor domains. Curr Biol 4:239–248.

Unger VM, Hargrave PA, Baldwin JM, Schertler GFX (1997): Arrangement of rhodopsin transmembrane alpha-helices. Nature 389:203–206.

Wallace RB, Shaffer J, Murphy RF, Bonner J, Hirose T, Itakura K (1979): Hybridization of synthetic oligonucleotides to $\phi\chi$174 DNA: The effect of single base pair mis-match. Nucleic Acids Res 6:3543–3656.

Wess J (1997): G protein–coupled receptors: Molecular mechanisms involved in recep-tor activation and selectivity of G protein recognition. FASEB J 11:346–354.

Wess J, Nanavati S, Vogel Z, Maggio R (1993): Functional role of proline and trypto-phan residues highly conserved among G protein coupled receptors studied by mu-tational analysis of the m3 muscarinic receptor. EMBO J 12:331–338.

Wessling-Resnick M, Kelleher DJ, Weiss ER, Johnson GL (1987): Enzymatic model for receptor activation of GTP-binding regulatory proteins. Trends Biochem Sci 12:473–477.

CHAPTER 2

THE SUBSTITUTED-CYSTEINE ACCESSIBILITY METHOD

JONATHAN A. JAVITCH

1. INTRODUCTION   22
   A. The Substituted-Cysteine Accessibility Method   22
   B. Detection of Reaction   23
   C. Endogenous Cysteines   23
2. METHODS   24
   A. Application of the Substituted-Cysteine Accessibility Method   24
   B. Site-Directed Mutagenesis   24
   C. Construction of Epitope-Tagged D2 Receptor Construct   25
   D. Transient and Stable Transfection   25
   E. Harvesting Cells   25
   F. ^{3}H-N-Methylspiperone Binding   25
   G. Use of the MTS Reagents   26
   H. Reactions With MTS Reagents   26
   I. Protection   26
3. INTERPRETATION OF RESULTS   27
   A. Assumptions of SCAM   27
   B. Mechanisms of Altered Binding   28
   C. Cysteine Substitution   30
   D. Secondary Structure   30
   E. Distinctions Between SCAM and Classic Mutagenesis   32
   F. Protection   32
   G. Differentiating Dopamine D2 Ligands by Their Sensitivity
     to Modification of Engineered Cysteines   33
   H. Electrostatic Potential   33
   I. Comparison of Reactions With MTSEA and MTSET   34
   J. The Binding Site and the Aromatic Cluster   34
   K. Conformational Changes Associated With Receptor Activation   35
4. FUTURE PERSPECTIVES   37
   A. Transduction of Agonist Binding Into Receptor Activation   37
   B. Structural Basis of Pharmacological Specificity   37

Structure–Function Analysis of G Protein-Coupled Receptors, Edited by Jürgen Wess.
ISBN 0-471-25228-X   Copyright © 1999 Wiley-Liss, Inc.

I. INTRODUCTION

Cysteine substitution and covalent modification have been used to study structure–function relationships and the dynamics of protein function in a variety of membrane proteins (Altenbach et al., 1990; Careaga and Falke, 1992; Jakes et al., 1990; Jung et al., 1993; Pakula and Simon, 1992; Todd et al., 1989). Moreover, charged hydrophilic, lipophobic sulfhydryl reagents have been used to probe systematically the accessibility of substituted cysteines in putative membrane-spanning segments of a number of proteins. This approach, the substituted-cysteine accessibility method (SCAM) (Karlin and Akabas, 1998), has been used to map channel-lining residues in the nicotinic acetylcholine receptor (Akabas and Karlin, 1995; Akabas et al., 1992, 1994a), the $GABA_A$ receptor (Xu and Akabas, 1993; Xu et al., 1995), the cystic fibrosis transmembrane conductance regulator (Akabas et al., 1994b), the UhpT transporter (Yan and Maloney, 1995), and potassium channels (Pascual et al., 1995), among others. We have adapted this approach to map the surface of the binding-site crevice in the dopamine D2 receptor and the β_2-adrenergic receptor, members of the G protein–coupled receptor superfamily (Fu et al., 1996; Javitch et al., 1994, 1995a,b, 1996, 1997, 1998).

I.A. The Substituted-Cysteine Accessibility Method

SCAM provides an approach to map systematically the residues on the water-accessible surface of a protein. These residues are identified by substituting them with cysteine and assessing for the reaction of charged, hydrophilic sulfhydryl reagents with the engineered cysteines. Consecutive residues in putative membrane-spanning segments are mutated to cysteine, one at a time, and the mutant proteins are heterologously expressed in cells. If ligand binding to a cysteine-substitution mutant is near normal, we assume that the structure of the mutant receptor is similar to that of wild type and that the substituted cysteine lies in an orientation similar to that of the wild-type residue. In the membrane-spanning segments, the sulfhydryl of a cysteine can face the binding-site crevice, the interior of the protein, or the lipid bilayer; sulfhydryls facing the water-accessible binding-site crevice should react much faster with charged, hydrophilic, lipophobic sulfhydryl-specific reagents.

For such polar sulfhydryl-specific reagents, we use derivatives of methanethiosulfonate (MTS): positively charged MTS ethylammonium (MTSEA) and MTS ethyltrimethylammonium (MTSET) and negatively charged MTS ethylsulfonate (MTSES) (Fig. 2.1) (Stauffer and Karlin, 1994). These reagents differ somewhat in size with MTSET > MTSES > MTSEA. The largest, MTSET, fits into a cylinder 6 Å in diameter and 10 Å long; thus the reagents are approximately the same size as dopamine. The MTS reagents form mixed disulfides with the cysteine sulfhydryl, covalently linking $-SCH_2CH_2X$, where X is NH_3^+, $N(CH_3)_3^+$, or SO_3^-. The MTS reagents are specific for cysteine sulfhydryls and do not react with disulfide-bonded cysteines or with other residues. The reagents are charged and quite hydrophilic. Moreover, they react with the ionized thiolate (RS^-) more than 1 billion times faster than with the un-ionized thiol (RSH) (Roberts et al., 1986), and only cysteines accessible to water are

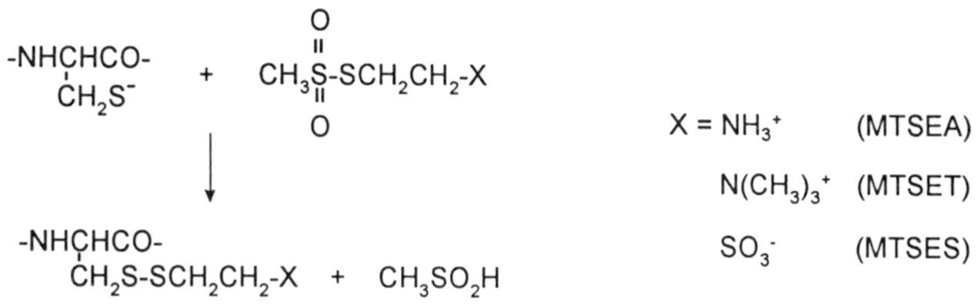

Figure 2.1. The structures of the methanethiosulfonate derivatives and their reaction with cysteine.

likely to ionize to a significant extent. The hydrophilic, negatively charged, organomercurial *p*-chloromercuribenzenesulfonate (pCMBS) also has been used to probe the accessibility of substituted cysteines in membrane-spanning segments of a number of membrane proteins (Olami et al., 1997; Yan and Maloney, 1993, 1995).

I.B. Detection of Reaction

Reaction with sulfhydryl reagents can be detected either directly or indirectly by measuring the effect of the reaction on a functional property of the protein. Because of the very small quantities of protein produced in most heterologous expression systems, we cannot rely on the direct detection of reaction. Instead, we use the irreversible modification of function to assay the reaction. In a receptor, the reaction of an MTS reagent with an engineered cysteine in the binding-site crevice should alter binding irreversibly (Fig. 2.2). Additionally, reaction with a cysteine near the binding site should be retarded by the presence of antagonist or agonist.

I.C. Endogenous Cysteines

The function of the protein used as the background for SCAM must not be affected by the sulfhydryl reagents. In some cases endogenous cysteines are not accessible to reaction with the MTS reagents, while in other cases endogenous cysteines are accessible and must first be identified and mutated to other residues. The ideal starting point would be to create a cysteine-less protein with normal expression and function. Such a construct has been created with lactose permease (Jung et al., 1993), the NhaA-Na^+/H^+ antiporter (Olami et al., 1997), and a glutamate transporter (Seal and Amara, 1996), but a dopamine D2 receptor with all five cysteines in the membrane-spanning segments simultaneously substituted by other residues expressed too poorly for further study. Nonetheless, in the dopamine D2 receptor, replacement of a single endogenous cysteine (Cys[118]) with serine resulted in a 100-fold decrease in the reactivity of the receptor with MTSEA and MTSET (Javitch et al., 1994). C118S expresses normally and has unaltered binding properties; this mutant was used as the background for further cysteine substitutions (Fu et al., 1996; Javitch et al., 1995a,b, 1998).

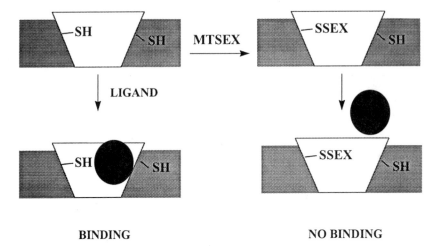

BINDING   NO BINDING

Figure 2.2. Schematic representation of the reaction of the MTS reagents with a cysteine exposed in the binding-site crevice. The membrane is represented by a stippled rectangle; the binding-site crevice is indicated within the plane of the membrane, and the solid oval represents ligand. SSEX represents $-SSCH_2CH_2X$ (where X is NH_3^+, $N[CH_3]_3^+$, or SO_3^-), which is covalently linked to the water-accessible cysteine sulfhydryl. In the bound state, represented in the lower left, ligand is reversibly bound within the binding-site crevice. In the unbound state, represented in the upper left, the binding site is unoccupied. After irreversible reaction with MTSEX, represented in the upper right, ligand binding is altered. The cysteine sulfhydryl facing lipid or the interior of the protein does not react with MTSEX. Ligand retards the rate of reaction of receptor with MTSEX, thereby protecting subsequent ligand binding.

2. METHODS

2.A. Application of the Substituted-Cysteine Accessibility Method

In a background of C118S, the mutant D2 receptor relatively insensitive to the MTS reagents, we have used SCAM to determine residues that form the surface of the binding-site crevice. Cysteines are substituted, one at a time, for residues in a putative membrane-spanning segment (see Site-Directed Mutagenesis). The mutant receptor is expressed in heterologous cells (see Transient and Stable Transfection), and ligand binding is assayed (see ^{3}H-N-Methylspiperone Binding). If the receptor binds ligand, then the effect of reaction with the MTS reagents is determined (see Use of the MTS Reagents; and Reactions With MTS Reagents), always comparing the effects with those on the receptor used as the background for the cysteine substitutions. We then test for the ability of ligand to retard the rate of reaction of an MTS reagent at the reactive positions (see Protection).

2.B. Site-Directed Mutagenesis

Cysteine mutations were generated using the Altered Sites mutagenesis kit (Promega) with oligonucleotides incorporating a change in a restriction site as well as the desired mutation. Mutants were screened by restriction mapping, and the mutations were confirmed by DNA sequencing.

2.C. Construction of Epitope-Tagged D2 Receptor Construct

The plasmid encoding the β_2-adrenergic receptor that was epitope tagged at the amino terminus with the cleavable influenza–hemagglutinin signal sequence followed by the "FLAG" epitope (IBI, New Haven, CT), was a gift from Dr. B. Kobilka (Kobilka, 1995). The sequence encoding the epitope tag was excised and ligated in-frame to the D2 receptor cDNA, thereby creating a fusion protein in which the epitope tag led directly into the sequence of the D2 receptor (Javitch et al., 1998). The affinity of the epitope-tagged receptor for N-methylspiperone, YM-09151-2, and sulpiride was unchanged from that of the wild-type receptor. This epitope-tagged D2 receptor fragment was then subcloned into the bicistronic expression vector pcin4 (a gift from Dr. S. Rees, Glaxo) (Rees et al., 1996).

2.D. Transient and Stable Transfection

HEK 293 cells in DMEM/F12 (1:1) with 10% bovine calf serum (Hyclone) and 293-TSA cells (a clonal line of HEK 293 cells stably expressing the SV40 large-T antigen) in DMEM with 10% fetal calf serum were maintained at 37°C and 5% CO_2. For transient transfection, 35-mm dishes of 293-TSA cells at 70%–80% confluence were transfected with 2 µg of wild type or mutant D2 receptor cDNA in pcDNA/Amp (Invitrogen) or pcin4 (see above) using 9 µl of lipofectamine (GIBCO) and 1 ml of OPTIMEM (GIBCO). Five hours after transfection, the solution was removed and fresh medium added. Twenty-four hours after transfection the medium was changed. Forty-eight hours after transfection, cells were harvested as described below.

For stable transfection, HEK 293 cells were transfected with D2 receptor cDNA in pcin4 as described above. Twenty-four hours after transfection the cells were split to a 100-mm dish and 700 µg/ml geneticin was added to select for a stably transfected pool of cells. No differences in binding affinity or accessibility to the MTS reagents were detected between receptor from transiently and stably transfected HEK 293 cells.

2.E. Harvesting Cells

Cells were washed with phosphate-buffered saline (PBS; 8.1 mM NaH_2PO_4, 1.5 mM KH_2PO_4, 138 mM NaCl, 2.7 mM KCl, pH 7.2), briefly treated with PBS containing 1 mM EDTA, and then dissociated in PBS. Cells were pelleted at 1,000g for 5 minutes at 4°C and resuspended for binding or treatment with MTS reagents.

2.F. ^{3}H-N-Methylspiperone Binding

Whole cells from a 35-mm plate were suspended in 450 µl of buffer A (25 mM HEPES, 140 mM NaCl, 5.4 mM KCl, 1 mM EDTA, 0.006% bovine serum albumin, pH 7.4). Cells were then diluted 20-fold with buffer A. ^{3}H-N-methylspiperone (Dupont/NEN) binding was performed as described previously (Javitch et al., 1995b). For saturation binding, duplicate polypropylene minitubes contained six different concentrations of ^{3}H-N-methylspiperone between 5 and 800 pM in buffer A with 300 µl of cell suspension in a final volume of 0.5 ml. The mixture was incubated at room temperature for 60 minutes and then filtered using a Brandel cell harvester through Whatman 934AH glass fiber filters (Brandel). The filter was washed three times with 1 ml of 10 mM Tris HCl and 120 mM NaCl, pH 7.4, at 4°C. Specific ^{3}H-N-methylspiperone binding was defined as total binding less nonspecific binding in the presence of 1 μM (+)butaclamol (Research Biochemicals). Competition assays were conducted in a final volume of 1 ml with 10 different concentrations of the tested drug and 150 pM ^{3}H-N-methylspiperone. Depending on the level of expression of the various mutants, adjustments in the number of cells per assay tube were made as necessary to prevent depletion of ligand in the case of very high expression or to increase the signal in the case of low expression.

2.G. Use of the MTS Reagents

At pH 7 and 22°C, MTSEA, MTSET, and MTSES (Toronto Research Chemicals) rapidly hydrolyze with a half-time of 5–20 minutes (Karlin and Akabas, 1998). At lower pH and lower temperature, hydrolysis is appreciably slower. Stock reagents should be stored desiccated at 4°C. A frequently used stock can be kept desiccated at room temperature, and it can be replenished from the 4°C stock after appropriate warming to room temperature. The reagents should be weighed but kept dry until immediately before use. The reagents are relatively stable unbuffered in water at 4°C; if it is necessary to dissolve the reagents or perform an intermediate dilution in buffer at a physiological pH, this should be done immediately before starting the reaction.

2.H. Reactions With MTS Reagents

Whole cells from a 35-mm plate were suspended in 400 μl buffer A. Aliquots (50 μl) of cell suspension were incubated with freshly prepared MTS reagents at the stated concentrations at room temperature for 2 minutes. Cell suspensions were then diluted 16-fold, and 200-μl aliquots were used to assay for ^{3}H-N-methylspiperone (200 pM) binding as described above. The fractional inhibition was calculated as 1 − [(specific binding after MTS reagent)/(specific binding without reagent)]. We used the SPSS for Windows (SPSS, Inc.) statistical software to analyze the effects of the MTS reagents by one-way ANOVA according to Dunnett's post hoc test ($p < 0.05$).

The second-order rate constant (k) for the reaction of an MTS reagent with each susceptible mutant was estimated by determining the extent of reaction after a fixed time, 2 min, with six concentrations of reagent (typically 0.025 to 10 mM, 0.25 mM in the case of MTSEA) (all in excess over the reactive sulfhydryls). The fraction of initial binding, Y, was fit to (1-plateau)$*$e^{-kct} + plateau, where plateau is the fraction of residual binding at saturating concentrations of MTSEA, k is the second-order rate constant (in $M^{-1}s^{-1}$), c is the concentration of MTSEA (M), and t is the time (120 s).

2.I. Protection

Dissociated cells were incubated in buffer A for 20 minutes at room temperature in the presence or absence of (±)sulpiride. Subsequently, MTSEA was added, in the continued presence or absence of sulpiride, for 2 minutes at a concentration chosen to produce approximately 75% of the maximal inhibition of specific ^{3}H-N-methylspiperone binding in the absence of sulpiride. For most mutants, sulpiride was used at a concentration of 10 μM. To compensate for changes in the K_i of particular mutants, sulpiride concentrations were adjusted as necessary. Cells were washed four times for 5 minutes each and then filtered though 96-well multiscreen plates containing GF/B filters (Millipore). In the wash buffer, sodium was replaced by choline to facilitate removal of residual sulpiride. ^{3}H-N-methylspiperone binding to the washed cells was performed in buffer A in the multiscreen plates in a final volume of 0.25 ml. Protection was calculated as $1 - $ [(inhibition in the presence of sulpiride)/(inhibition in the absence of sulpiride)]. The statistical significance of protection by sulpiride was assessed by a paired t test.

3. INTERPRETATION OF RESULTS

3.A. Assumptions of SCAM

To interpret the results of SCAM we make a number of assumptions. First, we assume that the highly polar MTS reagents react much faster at the water-accessible surface of the protein than in lipid or in the protein interior. As discussed above, the rate of reaction of the MTS reagents with ionized thiolate anion (RS$^-$) is more than 1 billion times faster than with un-ionized thiol (RSH) (Roberts et al., 1986), and only water-accessible thiols are likely to ionize. Furthermore, the MTS reagents are very hydrophilic with a relative solubility in water:octanol greater than 2,500:1 (Akabas et al., 1992; Stauffer and Karlin, 1994). Experimental support for this assumption comes from a recent study in the aspartate receptor of the accessibility of engineered cysteines to reaction with another aqueous, sulfhydryl-specific alkylating agent (Danielson et al., 1997): In the α2 helix of the periplasmic domain, a striking correlation was observed between the measured chemical reactivity of each engineered cysteine and the calculated solvent accessibility of the β carbon at the corresponding position in the crystal structure.

We further assume that in the membrane-spanning segments, access of highly polar reagents to side chains is only through the binding-site crevice, that the addition of –SCH$_2$CH$_2$X to a cysteine at the surface of the binding-site crevice is likely to alter binding irreversibly, and, reciprocally, that for substituted cysteines that line the binding site, agonists and antagonists should retard the reaction with the MTS reagents.

Another assumption is that the engineered cysteine is an accurate reporter for the water accessibility of the corresponding wild-type residue. If a cysteine-substitution mutant is functional, its overall three-dimensional structure is likely to be similar to the structure of wild-type receptor. Nonetheless, local changes at the site of the engineered cysteine could, in principle, alter the accessibility of the residue relative to the accessibility of the wild-type residue. A strength of SCAM is that entire membrane-spanning segments can be studied and regular patterns of accessibility can be identified. Given the general consistency of the results obtained in the various receptors and channels studied to date, it is likely that, in most cases, the position of the cysteine residue is similar to that of the replaced wild-type residue. In several cases, irregular patterns have been observed. We cannot be certain whether the secondary structure in such regions is irregular in the native structure, the protein structure fluctuates in such a region alternately exposing multiple residues, or the cysteine substitution has disrupted the local secondary structure, making the cysteine accessible when the wild-type residue is not. Thus, we must be cautious when the accessibility of an individual residue breaks a regular pattern of accessibility. Nonetheless, such cases can also provide important information about the structure that is not available through other experimental methods (see below for examples).

If an MTS reagent has no effect on a mutant, the interpretation of the results must be made with caution. The temptation is to infer that the engineered cysteine is inaccessible to the MTS reagents and is therefore not on the water-accessible surface of the protein. While this is the most likely explanation, there are other possibilities. First, electrostatic or steric factors may alter the reactivity of the MTS reagents with a water-accessible residue. Second, while it seems unlikely that a residue forming the surface of the binding-site crevice could be covalently modified by the addition of the charged $-SCH_2CH_2X$ without interfering with binding, such a result is nonetheless possible. In the D2 receptor, we have observed that reaction of MTSEA at certain positions has a much greater effect on the binding of particular ligands: For example, reaction of MTSEA with the highly reactive endogenous cysteine Cys[118], causes a negligible decrease in the affinity of the receptor for particular ligands but a large decrease in its affinity for other ligands (Javitch et al., 1996). To reduce the likelihood of such a false-negative determination, we typically screen for effects with antagonists from two different structural classes. Nonetheless, these potential complications further demonstrate the importance of systematically mutating to cysteine consecutive residues along an entire membrane-spanning segment; while mutation of any individual residue might be subject to potential pitfalls due to steric or electrostatic factors or "silent" reaction, this is unlikely to be a systematic problem affecting the overall pattern of accessibility of multiple residues in a membrane-spanning segment.

3.B. Mechanisms of Altered Binding

The effects of the addition of $-SCH_2CH_2X$ to the engineered cysteine could be a result of steric block, electrostatic interaction, or indirect structural changes. Regardless, although we do not know the detailed mechanism of the alterations in binding, an irreversible effect is evidence of reaction, and, therefore, of the accessibility of the engineered cysteine. While reaction usually inhibits binding, it can also potentiate binding. This can be illustrated in the dopamine D2 receptor by the mutation of Asp[108] (Javitch et al., 1995b). Mutation to cysteine of this residue at the extracellular end of the third membrane-spanning segment (M3) reduced the receptor's affinity for antagonist binding about threefold. Reaction of the positively charged MTSEA or MTSET at this position significantly inhibited binding. In contrast, reaction of the negatively charged MTSES restored the negative charge at this position and shifted the affinity toward that of wild-type receptor, thereby increasing occupancy and potentiating binding.

The fact that reaction can potentiate function necessitates care in experimental design; a potentiation of binding could be missed by measuring binding at too high a ligand concentration relative to the K_d (Fig. 2.3). For example, if assayed at a ligand concentration equal to the original K_d, a fivefold increase in affinity would result in an approximately 67% increase in binding. In contrast, if assayed at a ligand concentration 10 times the original K_d, the potentiation would be only about 8%, a change that might not be statistically significant, especially if the background receptor were not completely unreactive. In contrast, a fivefold decrease in affinity can be readily detected even at a ligand concentration 10 times the original K_d.

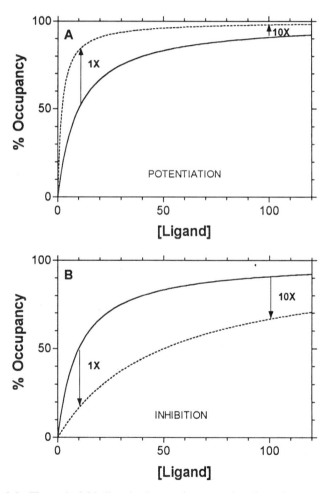

Figure 2.3. Theoretical binding isotherms demonstrating the effects of a change in binding affinity on binding measured at different ligand concentrations. **A:** The dotted line represents a fivefold increase in affinity compared with the solid line. At $L = K_{d(unreacted)}$, binding is increased by 67%, while at $L = 10 \times K_{d(unreacted)}$, binding is increased by only 8%. **B:** The dotted line represents a fivefold decrease in affinity compared with the solid line. At $L = K_{d(unreacted)}$, binding is decreased by 67%, while at $L = 10 \times K_{d(unreacted)}$, binding is decreased by 27%.

3.C. Cysteine Substitution

The ability to substitute cysteine residues for other residues and still obtain functional receptors is central to this approach. In the dopamine D2 receptor, 91 of 96 cysteine-substitution mutants tested to date bound antagonists with near-normal affinity (Fu et al., 1996; Javitch et al., 1995a,b, 1998). These tolerated substitutions were for hydrophobic residues (alanine, leucine, isoleucine, methionine, valine), polar residues (asparagine, serine, threonine), neutral residues (proline), acidic residues (aspartate), aromatic residues (phenylalanine, tryptophan, tyrosine), and glycine.

There are several reasons why cysteine substitution may be so well tolerated. Cysteine is a relatively small amino acid with a volume of 108 Å^3; only glycine, alanine, and serine are smaller (Creighton, 1993). In globular proteins, roughly half of nondisulfide-linked cysteines are buried in the protein interior and half are on the water-accessible surface of the protein (Chothia, 1976). Cysteine substitution, therefore, is likely to be tolerated in both of these environments. Furthermore, cysteine has little preference for a particular secondary structure (Chou and Fasman, 1977; Levitt, 1978), and cysteine substitution, therefore, is unlikely to systematically alter the secondary structure of the region into which it is substituted.

A cysteine-substitution mutant that does not function cannot be studied by SCAM (or by traditional site-directed mutagenesis). The residues that cannot be mutated to cysteine either are accessible in the binding-site crevice and make a crucial contribution to binding or are critical for maintaining the structure of the site and/or to the folding and processing of the receptor. The determination by SCAM of the accessibility of the neighbors of a crucial residue may allow us to infer the secondary structure of the segment containing this residue and, thus, whether it is likely to be accessible as well. If the residue is not accessible, then the functional effect of its mutation is likely due to an indirect effect on structure.

The use of an epitope-tagged receptor allows us to determine whether the rare mutant receptor that does not bind antagonist is expressed at the cell surface. We have constructed a D2 receptor with a cleavable signal peptide and a FLAG epitope at its amino terminus (see Methods); the signal sequence slightly increases expression of some mutants, and the epitope tag has no effect on the function of the receptor. Even the presence in the membrane of a nonbinding receptor mutant, however, does not prove that the mutated residue contacts ligand, as the residue could still interfere with binding indirectly.

3.D. Secondary Structure

To infer a secondary structure, we must assume that if binding to a mutant is not affected by the MTS reagents, then no reaction has occurred, and that the side chain at this position is not accessible in the binding-site crevice (but see above). In an α-helical structure one would expect the accessible residues to form a continuous stripe when the residues are represented on a helical net. For example, in M3 of the dopamine D2 receptor, the pattern of accessibility is consistent with this membrane-spanning segment forming an α helix with a stripe of about 140° facing the binding-site crevice (Javitch et al., 1995b)

(Fig. 2.4). In contrast, in an antiparallel β strand, one would expect every other residue to be accessible to the reagents.

More complex or irregular patterns of accessibility are more difficult to interpret, but these findings can also be rather informative. For example, we have found that cysteines substituted, one at a time, for 10 consecutive residues in the fifth membrane-spanning segment (M5) of the dopamine D2 receptor are accessible to MTSEA and MTSET (Javitch et al., 1995a). This pattern of accessibility is not consistent with M5 being a fixed α helix with one side facing the binding-site crevice. The exposed region of M5, which contains the serines likely to bind agonist, might loop out into the lumen of the binding-site crevice and be completely accessible to water and thus to MTSEA. Alternatively, the exposed region of M5 might be embedded in the membrane and also be in contact with other membrane-spanning segments but move rapidly to expose different sets of residues. In such a scenario, at any instant only a limited set of residues in M5 might be exposed in the binding-site crevice.

A recent 4.5 Å projection structure of rhodopsin revealed a band of density leading from the presumed M5 toward the binding site (Schertler and Hargrave, 1995). This band of density could represent a deviation in M5 of rhodopsin from a fixed α-helical structure. Furthermore, recent experiments suggest that substituted cysteines at the extracellular end of M5 in rhodopsin crosslinked

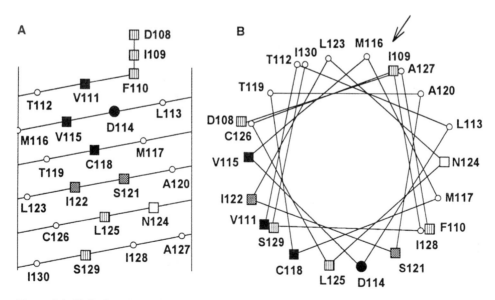

Figure 2.4. Helical net (**A**) and helical wheel (**B**) representations of the residues in and flanking the third membrane-spanning segment (M3) of the dopamine D2 receptor, summarizing the effects of MTSEA on ^{3}H-YM-09151-2 binding. Reactive residues are represented by squares, where the fill indicates the range of the second-order rate constants in $M^{-1} s^{-1}$ for reaction with MTSEA: solid squares, $k \geq 20$; hatched squares, $20 > k \geq 10$; striped squares, $10 > k \geq 3$; open squares, $3 > k > 1$. Small open circles indicate that MTSEA had no effect on binding. Solid circles indicate no binding after cysteine substitution. D108 and I109 are represented outside of the α helix in the preceding loop. The arrow indicates where the helix is "split" to create the helical net.

with multiple substituted cysteines in neighboring membrane-spanning segments (D. Oprian, personal communication), a finding that is also difficult to reconcile with a fixed α-helical structure.

We also observed an unusual pattern of accessibility in the seventh membrane-spanning segment (M7) of the dopamine D2 receptor. Again, the overall pattern of exposure was not consistent with a simple secondary structure of either α helix or β strand. M7, however, contains the highly conserved residues Asn and Pro in the middle of the putative membrane-spanning segment. In soluble proteins, these residues have been observed to introduce kinks and twists in α helices. In molecular modeling work, we found that the pattern of exposure of the cysteine-substitution mutants to MTSEA can be explained if M7 is a kinked and twisted α helix (Fu et al., 1996). Thus, "irregular" patterns of accessibility can lead to new insights or directions for further experimental pursuit.

3.E. Distinctions Between SCAM and Classic Mutagenesis

A distinction between our approach and that of classic mutagenesis experiments is that SCAM does not rely on the functional effects of a given mutation. The interpretation of the results of typical mutagenesis experiments requires one to assume that perturbations caused by a mutation, such as changes in binding affinity, are due to local effects at the site of the mutation rather than to indirect effects on protein structure. The validity of this assumption is rarely assessed for individual mutations. The structure of the λ phage receptor maltoporin, however, showed that 50% of the mutated residues that had been implicated in λ phage recognition are actually located in the protein interior (Schirmer et al., 1995). Thus, mutation of these buried residues alters λ phage binding indirectly. Likewise, the crystal structures of several dihydrofolate reductase mutants have demonstrated that a mutation approximately 15 Å from the substrate binding pocket exerts an effect on catalytic activity through an extended structural perturbation (Brown et al., 1993).

In contrast to mutagenesis approaches that only detect perturbations in protein function, SCAM allows one to determine whether a residue is on the water-accessible surface of the binding-site crevice when the mutant has near-normal function. Other advantages of the approach include the ability to probe binding sites by assessing the ability of different ligands to retard the reaction of the MTS reagents with particular substituted cysteines and the ability to probe the steric constraints and electrostatic potential of sites by comparing the rates of reaction of reagents of varying sizes and charges.

3.F. Protection

If reaction with the sulfhydryl reagents is slowed by antagonist or agonist, we infer that the residue is accessible in the binding-site crevice. Each and every residue that is protected, however, need not contact ligand; ligand could protect residues deeper in the crevice by binding above them and blocking the passage of the MTS reagent from the extracellular medium to the cytoplasmic end of the crevice. In addition, we cannot rule out indirect protection through ligand-mediated propagated structural rearrangement. Because propagated structural changes are less likely induced by antagonists than by agonists, we typically use the antagonist sulpiride to screen for protection. Two additional advantages of this compound are that it is hydrophilic and that its binding is sodium dependent, two factors that facilitate its removal and thereby decrease interference of residual drug with the final determination of binding.

Ligands from different structural classes might be expected to protect a different but overlapping set of engineered cysteines in the binding-site crevice. In practice, however, it has been very difficult to study protection with other antagonists due to the difficulty in completely removing these relatively lipophilic compounds prior to the determination of binding.

3.G. Differentiating Dopamine D2 Ligands by Their Sensitivity to Modification of Engineered Cysteines

As discussed above, Cys^{118} in M3 of the D2 receptor reacts with MTSEA and MTSET, and this reaction is retarded by the presence of antagonists and agonists (Javitch et al., 1994). The reaction of MTSEA covalently attaches $-SCH_2CH_2NH_3^+$ to the cysteine sulfhydryl, producing a lysine-like side chain. The reaction of MTSEA with Cys^{118} decreased the affinity of substituted benzamide antagonists by 3,000-fold, whereas the affinities of other antagonists were minimally decreased (Javitch et al., 1996). Thus, although all of these compounds are "competitive" antagonists, their set of contact residues and the space they occupy within the binding-site crevice differ in that only substituted benzamides are markedly affected by the presence of the modified side chain of Cys^{118}. Furthermore, agonists, which must activate receptor and not simply block activation, also have different sets of contact residues, as the affinity of dopamine decreased 12,000-fold, whereas that of bromocriptine decreased only threefold. Similar studies with cysteine-substitution mutants in other positions have provided additional clues to the detailed interaction of ligands within the binding site (Fu et al., 1996; Javitch et al., 1995a, 1998).

3.H. Electrostatic Potential

Because positively charged MTSET and negatively charged MTSES are similar in size, differences in their reactivities with engineered cysteines are likely to be due to differences in the electrostatic potential of the binding-site crevice. For example, in M3 of the D2 receptor, MTSES did not react with any engineered cysteines more cytoplasmic than Val^{111}, while MTSET reacted with several residues more cytoplasmic than this position (Javitch et al., 1995b). This reflects the negative electrostatic potential deeper in the binding-site crevice, likely due in part to Asp^{114}. By subsequent application of positively and negatively charged reagents, we can rule out the possibility that reaction has occurred without alteration of function in the case of addition of one but not the other charged moiety. For example, we determined that MTSES did not react with Cys^{118} in D2 receptor because subsequent application of MTSET still inhibited binding. If MTSES had reacted silently with Cys^{118}, it would have prevented the cysteine from further reaction with MTSET.

In contrast, the effects of reaction of MTSET and MTSES with F110C and V111C, residues located near the extracellular end of M3, are similar. This indicates that the electrostatic potential near these residues is not as negative as it is "below" Val[111]. This negative electrostatic potential has also been less apparent in the other membrane-spanning segments studied (Fu et al., 1996; Javitch et al., 1995a, 1998). This suggests that Asp[114] is a significant contributor to the negative potential in M3 and that the potential is not uniform throughout the crevice at similar depths. This distribution of the field likely helps to orient ligand within the binding-site crevice with the protonated amine toward Asp[114].

3.I. Comparison of Reactions With MTSEA and MTSET

When adjusted for the rate constants for their reactions with simple thiols in solution (Stauffer and Karlin, 1994), the reaction of MTSEA with cysteines in the binding-site crevice of the dopamine D2 receptor is typically accelerated approximately 10-fold relative to that of MTSET. MTSEA is smaller than MTSET, and its access to substituted cysteines may be less sterically hindered. Moreover, MTSEA, like dopamine, contains an ethylammonium group, and it could be the affinity of this group for the dopamine binding site that accelerates the reaction of MTSEA relative to that of MTSET. This pattern is broken, however, by T412C in M7 (Fu et al., 1996) and by F389C and F390C in M6 (Javitch et al., 1998), which are nearly equireactive with MTSEA and MTSET. Thus, at these positions, the reaction of MTSET is accelerated approximately 10-fold relative to that of MTSEA. For T412C in M7 (Fu et al., 1996), we noted that the specific increase may be the result of an interaction of the aromatic side chains of Trp[413] and Tyr[416] with the hydrophobic quaternary ammonium group of MTSET (Dougherty, 1996), which would favor the reaction with MTSET at T412C (Fig. 2.5). The cluster of accessible aromatic residues in M6, Phe[382], Trp[386], Phe[389], and Phe[390], could have a similar effect. In particular, favorable interactions of the MTSET cation with Trp[386] and other nearby aromatic side chains are likely to be responsible for the increase in reactivity seen at F389C and F390C (Javitch et al., 1998).

3.J. The Binding Site and the Aromatic Cluster

Phe[382], Trp[386], Phe[389], and Phe[390], the four accessible aromatic residues in the aromatic cluster, are completely conserved within related neurotransmitter G protein–coupled receptors, and some of these residues have been shown to be important for ligand binding and/or receptor activation (Cho et al., 1995; Roth et al., 1997). Molecular modeling simulations have suggested that the conformation of the side chains of these aromatic residues are interdependent and that ligand binding may induce coordinated movements of these residues, resulting in a rotational/translational movement about the proline kink in M6 (Javitch et al., 1998).

Ligand–receptor interactions that can be supported by such an "aromatic cluster" have been suggested from structure–activity data for dopamine agonists acting on the D2 receptor (Seeman, 1980). Suggested requirements for

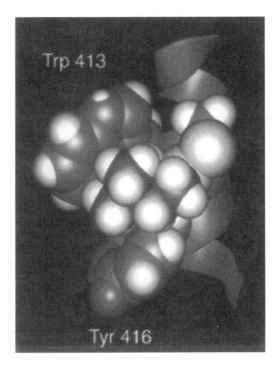

Figure 2.5. Molecular model showing the quaternary ammonium of MTSET after reaction with T412C in Van der Waals contact with the aromatic side chains of Trp^{413} and Tyr^{416}. (See color plates.)

potency included an electrostatic interaction between the protonated amine and a negative subsite that has been established to be Asp^{114} in M3 (Javitch et al., 1995b; Mansour et al., 1992; Strader et al., 1988), a hydrogen bonding group or groups that interact with one or more serines in M5 (Cox et al., 1992; Javitch et al., 1995a; Mansour et al., 1992; Strader et al., 1989b), and an aromatic ring that interacts with a hydrophobic site. This hydrophobic site is likely to be the aromatic cluster in M6 that extends to the adjacent M5, where it includes Phe^{198}, and to the adjacent M7, where it includes Tyr^{416}, both of which (Phe^{198} and Tyr^{416}) are also accessible in the binding-site crevice (Fu et al., 1996; Javitch et al., 1995a).

3.K. Conformational Changes Associated With Receptor Activation

To further explore the unexpected pattern of accessibility of the substituted-cysteine mutants in M5 of the D2 receptor, we determined the accessibilities of the aligned residues in the homologous β_2-adrenergic receptor (Javitch et al., manuscript in preparation). Surprisingly, the pattern of accessibility of M5 of the β_2 receptor, unlike M5 of the D2 receptor, is compatible with a fixed α-helical structure. This may reflect a difference in the structures and/or packing of membrane-spanning segments in the two receptors, although a major difference in the structures of the two receptors seems unlikely given the similarities of their primary sequence. Alternatively, at rest, the D2 receptor may be more dynamic than the β_2 receptor, which may be more constrained. Thus, in the absence of ligand, the D2 receptor may undergo sufficient conformational change to alternately expose multiple residues that are not exposed simultaneously. Indeed, a number of D2 antagonists have recently been found to act as inverse agonists at the D2 receptor, suggesting that there may be significant native constitutive activity of this receptor (Hall and Strange, 1997; Kozell and Neve, 1997). Movement of the extracellular portion of the M5, which contains the serines likely to bind agonist (Strader et al., 1989a), might be part of the mechanism of receptor activation, and the accessibility of these residues may change with the functional state of the receptor.

Conformational changes in a protein may result in changes in the accessibility of substituted cysteines as assessed by their rates of reaction with polar sulfhydryl-specific reagents. For example, residues lining the channel of the nicotinic acetylcholine receptor change in accessibility upon activation of the receptor and opening of the channel (Akabas et al., 1994a, 1992). Similarly, it should be possible to determine changes in the accessibility of residues in G protein–coupled receptors in different functional states.

To identify activation-induced structural changes in the residues forming the surface of the binding-site crevice, we sought to determine the relative accessibilities of a series of engineered cysteines in the resting and activated receptor. Agonist cannot be used to activate receptor, however, because the presence of a ligand within the binding site would interfere with access of the MTSEA to the engineered cysteines. Alternatively, the activated state of the receptor can be achieved by using a constitutively active mutant (CAM) receptor as a background for further cysteine substitution. A CAM receptor is intrinsically active and has a higher affinity for agonist than does the wild-type receptor. The high affinity state for agonist is typically associated with the activated receptor–G protein complex. The higher agonist affinity in the CAM even in the absence of G protein suggests that the structure of the binding site of the CAM is likely to be similar to that of the agonist-activated wild-type receptor binding site (or more easily isomerizes to the active state). Thus, we can compare the resting and active forms of the receptor by determining the accessibility of substituted cysteines in the binding-site crevice in these two states using wild-type receptor and a CAM as background constructs.

We have chosen to pursue initial studies in the β_2-adrenergic receptor because of the availability of a well-characterized CAM (Samama et al., 1993) (kindly provided by R. Lefkowitz). MTSEA had no effect on the binding of agonist or antagonist to wild-type β_2 receptor expressed in HEK 293 cells. This suggested that no endogenous cysteines are accessible in the binding-site crevice. In contrast, in the CAM β_2 receptor MTSEA significantly inhibited antagonist binding, and isoproterenol slowed the rate of reaction of MTSEA (Javitch et al., 1997). This implies that at least one endogenous cysteine becomes accessible in the binding-site crevice of the CAM β_2 receptor.

We found that Cys^{285}, in M6, is responsible for the inhibitory effect of MTSEA on ligand binding to the CAM (Javitch et al., 1997). The acquired accessibility of Cys^{285} in the CAM may result from a rotation and/or tilting of M6 associated with activation of the receptor. This rearrangement could bring Cys^{285}

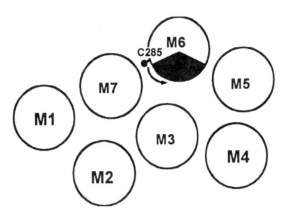

Figure 2.6. Rotation and/or tilting of the sixth membrane-spanning segment (M6) associated with the activation of the β_2 receptor. The indicated rearrangement brings Cys^{285} to the margin of the binding-site crevice and allows it to react with MTSEA to inhibit ligand binding. The arrangement of the membrane-spanning segments is based on the projection structure of rhodopsin (Schertler et al., 1993). The accessible surface of the M6 determined in the homologous dopamine D2 receptor (Javitch et al., 1998) is shaded.

to the margin of the binding-site crevice where it becomes accessible to MTSEA (Fig. 2.6). Such a movement of M6 on receptor activation is consistent with the results of fluorescence spectroscopy studies in the β_2 receptor (Gether et al., 1995, 1997) and spin-labeling studies in rhodopsin (Farrens et al., 1996), suggesting that the substituted-cysteine accessibility method in a CAM background is a powerful approach for probing conformational changes in these receptors.

4. FUTURE PERSPECTIVES

4.A. Transduction of Agonist Binding Into Receptor Activation

The membrane-spanning segments not only form the binding-site crevice but also constitute the transduction pathway from the binding site to the intracellular loops that interact with G proteins. Agonist binding in the crevice must therefore alter the conformations or orientations of at least some of the membrane-spanning segments. As described above, we have identified a cysteine in M6 that becomes accessible in a CAM background. By extending this method, it should now be possible to map the activation-related changes in accessibility of all the residues that form the surface of the binding-site crevice. Indeed, in preliminary work in the M5 of the β_2 receptor, we have observed significant changes in the rates and extents of reaction of a number of engineered cysteines in the CAM background compared with the wild-type background.

4.B. Structural Basis of Pharmacological Specificity

From mutagenesis and SCAM studies in the D2 receptor, it appears that a number of conserved residues critical for ligand recognition in the β_2-adrenergic receptor are also critical for binding in the dopamine receptors as well. These include Asp^{114} in the M3, the serines in the M5, and aromatic residues in the M6. Completely conserved residues, however, cannot account for the profound differences in binding specificities among the catecholamine receptors, or even among the D2-like receptors. Additional residues must contribute to binding specificity, either directly or indirectly. These residues are likely to be among the nonconserved residues identified by SCAM as accessible in the binding-site crevice.

Identification of these accessible, nonconserved residues therefore may be useful to determine the structural basis of pharmacological specificity among related catecholamine receptors. For example, although 37 residues in M3, M5, M6, and M7 differ between the dopamine D2 and D4 receptors, only 10 of these face the binding-site crevice in the D2 receptor. If M3, M5, M6, and/or M7 contribute to the pharmacological differences between these receptors, then one or more of these 10 accessible nonconserved residues, or their counterparts in the D4 receptor, might be critical determinants of the pharmacological differences between the D2 and D4 receptors. Thus, the information obtained from SCAM decreases dramatically the number of residues that must be examined for such contributions to pharmacological specificity.

ACKNOWLEDGMENTS

I thank Myles Akabas and Arthur Karlin, my colleagues in the Center for Molecular Recognition, for much helpful discussion and advice. I thank my collaborators Juan Ballesteros and Harel Weinstein for their critical insights into molecular modeling and protein structure. I thank the current and previous members of my laboratory for their hard work and important contributions to the work described in this manuscript: Jiayun Chen, Victor Chiappa, Dingyi Fu, Joshua Kaback, Xiochuan Li, George Liapakis, and Merrill M. Simpson. This work was supported in part by NIH grants MH01030 and MH54137, by the G. Harold & Leila Y. Mathers Charitable Trust, and by the Lebovitz Trust.

REFERENCES

Akabas MH, Karlin A (1995): Identification of acetylcholine receptor channel–lining residues in the M1 segment of the α-subunit. Biochemistry 34:12496–12500.

Akabas MH, Kaufmann C, Archdeacon P, Karlin A (1994a): Identification of acetylcholine receptor channel–lining residues in the entire M2 segment of the α subunit. Neuron 13:919–927.

Akabas MH, Kaufmann C, Cook TA, Archdeacon P (1994b): Amino acid residues lining the chloride channel of the cystic fibrosis transmembrane conductance regulator. J Biol Chem 269:14865–14868.

Akabas MH, Stauffer DA, Xu M, Karlin A (1992): Acetylcholine receptor channel structure probed in cysteine-substitution mutants. Science 258:307–310.

Altenbach C, Marti T, Khorana HG, Hubbell WL (1990): Transmembrane protein structure: Spin labeling of bacteriorhodopsin mutants. Science 248:1088–1092.

Brown KA, Howell EE, Kraut J (1993): Long-range structural effects in a second-site revertant of a mutant dihydrofolate reductase. Proc Natl Acad Sci USA 90:11753–11756.

Careaga CL, Falke JJ (1992): Structure and dynamics of *Escherichia coli* chemosensory receptors. Engineered sulfhydryl studies. Biophys J 62:209–216.

Cho W, Taylor LP, Mansour A, Akil H (1995): Hydrophobic residues of the D2 dopamine receptor are important for binding and signal transduction. J Neurochem 65:2105–2115.

Chothia C (1976): The nature of the accessible and buried surfaces in proteins. J Mol Biol 105:1–12.

Chou PY, Fasman GD (1977): β-Turns in proteins. J Mol Biol 115:135–175.

Cox BA, Henningsen RA, Spanoyannis A, Neve RL, Neve KA (1992): Contributions of conserved serine residues to the interactions of ligands with dopamine D2 receptors. J Neurochem 59:627–635.

Creighton TE (1993): Proteins: Structures and Molecular Properties, 2nd ed. New York: WH Freeman & Co.

Danielson MA, Bass RB, Falke JJ (1997): Cysteine and disulfide scanning reveals a regulatory alpha-helix in the cytoplasmic domain of the aspartate receptor. J Biol Chem 272:32878–32888.

Dougherty DA (1996): Cation–pi interactions in chemistry and biology: A new view of benzene, Phe, Tyr, and Trp. Science 271:163–168.

Farrens DL, Altenbach C, Yang K, Hubbell WL, Khorana HG (1996): Requirement of rigid-body motion of transmembrane helices for light activation of rhodopsin. Science 274:768–770.

Fu D, Ballesteros JA, Weinstein H, Chen J, Javitch JA (1996): Residues in the seventh membrane–spanning segment of the dopamine D2 receptor accessible in the binding-site crevice. Biochemistry 35:11278–11285.

Gether U, Lin S, Ghanouni P, Ballesteros JA, Weinstein H, Kobilka BK (1997): Agonists induce conformational changes in transmembrane domains III and VI of the beta2 adrenoceptor. EMBO J 16:6737–6747.

Gether U, Lin S, Kobilka BK (1995): Fluorescent labeling of purified beta 2 adrenergic receptor. Evidence for ligand-specific conformational changes. J Biol Chem 270:28268–28275.

Hall DA, Strange PG (1997): Evidence that antipsychotic drugs are inverse agonists at D2 dopamine receptors. Br J Pharmacol 121:731–736.

Jakes KS, Abrams CK, Finkelstein A, Slatin SL (1990): Alteration of the pH-dependent ion selectivity of the colicin E1 channel by site-directed mutagenesis. J Biol Chem 265:6984–6991.

Javitch JA, Ballesteros JA, Weinstein H, Chen J (1998): A cluster of aromatic residues in the sixth membrane–spanning segment of the dopamine D2 receptor is accessible in the binding-site crevice. Biochemistry 37:998–1006.

Javitch JA, Fu D, Chen J (1995a): Residues in the fifth membrane–spanning segment of the dopamine D2 receptor exposed in the binding-site crevice. Biochemistry 34:16433–16439.

Javitch JA, Fu D, Chen J (1996): Differentiating dopamine D2 ligands by their sensitivities to modification of the cysteine exposed in the binding-site crevice. Mol Pharmacol 49:692–698.

Javitch JA, Fu D, Chen J, Karlin A (1995b): Mapping the binding-site crevice of the dopamine D2 receptor by the substituted-cysteine accessibility method. Neuron 14:825–831.

Javitch JA, Fu D, Liapakis G, Chen J (1997): Constitutive activation of the beta2 adrenergic receptor alters the orientation of its sixth membrane–spanning segment. J Biol Chem 272:18546–18549.

Javitch JA, Li X, Kaback J, Karlin A (1994): A cysteine residue in the third membrane–spanning segment of the human D2 dopamine receptor is exposed in the binding-site crevice. Proc Natl Acad Sci USA 91:10355–10359.

Jung K, Jung H, Wu J, Prive GG, Kaback HR (1993): Use of site-directed fluorescence labeling to study proximity relationships in the lactose permease of *Escherichia coli.* Biochemistry 32:12273–12278.

Karlin A, Akabas MH (1998): Substituted-cysteine accessibility method. Methods Enzymol 293:123–145.

Kobilka BK (1995): Amino and carboxyl terminal modifications to facilitate the production and purification of a G protein–coupled receptor. Anal Biochem 231:269–271.

Kozell LB, Neve KA (1997): Constitutive activity of a chimeric D2/D1 dopamine receptor. Mol Pharmacol 52:1137–1149.

Levitt M (1978): Conformational preferences of amino acids in globular proteins. Biochemistry 17:4277–4285.

Mansour A, Meng F, Meador WJH, Taylor LP, Civelli O, Akil H (1992): Site-directed mutagenesis of the human dopamine D2 receptor. Eur J Pharmacol 227:205–214.

Olami Y, Rimon A, Gerchman Y, Rothman A, Padan E (1997): Histidine 225, a residue of the NhaA-Na$^+$/H$^+$ antiporter of *Escherichia coli* is exposed and faces the cell exterior. J Biol Chem 272:1761–1768.

Pakula AA, Simon MI (1992): Determination of transmembrane protein structure by disulfide cross-linking: The *Escherichia coli* Tar receptor. Proc Natl Acad Sci USA 89:4144–4148.

Pascual JM, Shieh CC, Kirsch GE, Brown AM (1995): Multiple residues specify external tetraethylammonium blockade in voltage-gated potassium channels. Biophys J 69:428–434.

Rees S, Coote J, Stables J, Goodson S, Harris S, Lee MG (1996): Bicistronic vector for the creation of stable mammalian cell lines that predisposes all antibiotic-resistant cells to express recombinant protein. BioTechniques 20:102–110.

Roberts DD, Lewis SD, Ballou DP, Olson ST, Shafer JA (1986): Reactivity of small thiolate anions and cysteine-25 in papain toward methyl methanethiosulfonate. Biochemistry 25:5595–5601.

Roth BL, Shomam M, Choudhary MS, Khan N (1997): Identification of conserved aromatic residues essential for agonist binding and second messenger production at 5-hydroxytryptmaine$_{2A}$ receptors. Mol Pharmacol 52:259–266.

Samama P, Cotecchia S, Costa T, Lefkowitz RJ (1993): A mutation-induced activated state of the beta 2-adrenergic receptor. Extending the ternary complex model. J Biol Chem 268:4625–4636.

Schertler GFX, Hargrave PA (1995): Projection structure of frog rhodopsin in two crystal forms. Proc Natl Acad Sci USA 92:11578–11582.

Schertler GF, Villa C, Henderson R (1993): Projection structure of rhodopsin. Nature 362:770–772.

Schirmer T, Keller TA, Wang YF, Rosenbusch JP (1995): Structural basis for sugar translocation through maltoporin channels at 3.1 Å resolution. Science 267:512–514.

Seal RP, Amara SG (1996): Residues involved in substrate interactions with a sodium-dependent glutamate transporter identified using cysteine scanning mutagenesis. Soc Neurosci Abstr 22:1575.

Seeman P (1980): Brain dopamine receptors. Pharmacol Rev 32:229–313.

Stauffer DA, Karlin A (1994): Electrostatic potential of the acetylcholine binding sites in the nicotinic receptor probed by reactions of binding-site cysteines with charged methanethiosulfonates. Biochemistry 33:6840–6849.

Strader CD, Candelore MR, Hill WS, Sigal IS, Dixon RA (1989a): Identification of two serine residues involved in agonist activation of the beta-adrenergic receptor. J Biol Chem 264:13572–13578.

Strader CD, Sigal IS, Candelore MR, Rands E, Hill WS, Dixon RA (1988): Conserved aspartic acid residues 79 and 113 of the beta-adrenergic receptor have different roles in receptor function. J Biol Chem 263:10267–10271.

Strader CD, Sigal IS, Dixon RA (1989b): Genetic approaches to the determination of structure–function relationships of G protein–coupled receptors. Trends Pharmacol Sci Suppl 26–30.

Todd AP, Cong J, Levinthal F, Levinthal C, Hubbell WL (1989): Site-directed mutagenesis of colicin E1 provides specific attachment sites for spin labels whose spectra are sensitive to local conformation. Proteins 6:294–305.

Xu M, Akabas MH (1993): Amino acids lining the channel of the gamma-aminobutyric acid type A receptor identified by cysteine substitution. J Biol Chem 268:21505–21508.

Xu M, Covey DF, Akabas MH (1995): Interaction of picrotoxin with GABA$_A$ receptor channel–lining residues probed in cysteine mutants. Biophys J 69:1858–1867.

Yan RT, Maloney PC (1993): Identification of a residue in the translocation pathway of a membrane carrier. Cell 75:37–44.

Yan RT, Maloney PC (1995): Residues in the pathway through a membrane transporter. Proc Natl Acad Sci USA 92:5973–5976.

CHAPTER 3

USING MUTAGENESIS TO MAP THE BINDING DOMAINS OF SMALL MOLECULE RECEPTORS

JEAN E. LACHOWICZ and CATHERINE D. STRADER

1. INTRODUCTION   43
2. PROCEDURAL CONSIDERATIONS REGARDING RECEPTOR MUTAGENESIS AND PHARMACOLOGY   44
   A. Determination of Membrane Incorporation   44
   B. Radioligand Binding and Data Analysis for Large Experiments   45
   C. Analysis of Receptor Function   47
3. MAPPING OF THE BIOGENIC AMINE BINDING SITE   48
   A. Scanning Deletion Mutagenesis   48
   B. Construction of a Pharmacophore Map   49
   C. Specific Point Mutations   50
   D. Two-Dimensional Mutagenesis   51
      a. Plan A: Design of a Novel Binding Site   51
      b. Plan B: Additivity Analysis   51
   E. Conservation Pattern Analysis   52
4. IDENTIFICATION OF SITES FOR ALLOSTERIC MODULATION   54
   A. Assessment of Degree of Cooperativity   54
   B. Mutations That Affect Allosteric Binding of Muscarinic Receptors   55

1. INTRODUCTION

G protein–coupled receptors (GPCRs) are activated by a diverse family of ligands ranging in size from small molecules to large proteins. Many of the most thoroughly studied of these receptor systems are those whose endogenous ligands are biogenic amines. Biogenic amine receptors serve as targets for important classes of drugs, most notably the β-blockers for the treatment of cardiovascular disease and the antihistamines for allergy. The importance of this area for medicinal chemistry is shown by the award of the 1988 Nobel Prize in Physiology or Medicine to Sir James Black, who was honored for his

Structure–Function Analysis of G Protein-Coupled Receptors, Edited by Jürgen Wess.
ISBN 0-471-25228-X   Copyright © 1999 Wiley-Liss, Inc.

apparently unrelated successes in discovering drugs for not only one but both of these therapeutically important targets. What is now apparent is that these breakthrough discoveries *are,* in fact, related: The receptor targets for antihistamines and β-blockers both belong to the family of GPCRs. This chapter focuses on techniques used to probe the binding sites in these receptors for small molecule agonists and antagonists.

The GPCRs whose endogenous agonists are small molecules can be thought of as a specialized subclass of receptors representing unique challenges and opportunities for understanding receptor–ligand interactions. As is the case for all GPCRs, the potential for direct biophysical measurement of these interactions using techniques such as x-ray crystallography or nuclear magnetic resonance imaging remains limited at present. Difficulties in the isolation of sufficient quantities of pure protein, coupled with the requirement of lipid or detergent to maintain activity of the transmembrane-spanning GPCRs, have thus far prevented high-resolution techniques from providing a direct view of the receptor's ligand binding pocket. However, the small size of these ligands, in contrast to the larger peptides or proteins that activate other GPCRs, makes exhaustive structure–activity profiles of the binding pockets for small molecule ligands an achievable goal. The optimal strategy, as exemplified below, combines the traditional medicinal chemical approach of developing structure–activity relationships for the ligand with the molecular biological approach of developing structure–function relationships for the receptor. We have termed this approach *two-dimensional mutagenesis* because it provides a dual view of the binding pocket from the perspectives of both the ligand and the receptor. This technique, described in more detail in Chapter 1, is especially powerful for the small molecule subset of GPCRs because the small size of the ligand improves the resolution of the medicinal chemistry arm of the two-dimensional mutagenesis experiment. In addition, the rich medicinal chemical history from drug discovery efforts prior to molecular characterization of the receptor targets has provided a wealth of structure–activity information from which the probe of the binding site can be launched.

2. PROCEDURAL CONSIDERATIONS REGARDING RECEPTOR MUTAGENESIS AND PHARMACOLOGY

2.A. Determination of Membrane Incorporation

Most studies designed to identify amino acids critical to ligand binding begin with observation of loss of function when specific residues are mutated. To accurately interpret such results, it is critical to determine whether the mutant receptor is properly expressed. Several examples illustrate the ability of point mutations and chimeras to affect GPCR expression, folding, glycosylation, and membrane insertion (Zhang et al., 1991; Weiss et al., 1994, Liu et al., 1995). To ascertain whether the receptor is expressed, an antibody that recognizes a portion of the receptor distant from the mutated residues can be used to immunoprecipitate the protein. Immunoprecipitation confirms expression and allows for comparison of relative mobilities of mutant and wild-type receptors.

In an early study of point mutations of the β2-adrenergic receptor, substitution of a seventh transmembrane domain (TM7) proline with a serine failed to produce a protein of the expected molecular mass (Strader et al., 1987). Immunoprecipitation produced two smaller polypeptides that corresponded to nonglycosylated and partially glycosylated β_2 receptors. This receptor did not bind iodocyanopindolol, probably due to incomplete processing. In the case of an incorrectly processed mutant receptor, no assertions can be made about the contribution of the mutated amino acid(s) to ligand binding.

Various lines of experimentation have been used to confirm that mutant receptors are expressed on the cell surface. Detection of the receptor by immunoprecipitation of the membrane fraction may not indicate that the receptor resides in the plasma membrane; the receptor may be present in membrane-associated vesicles within the cytoplasm. If antibodies to the receptor are available, indirect immunofluorescence microscopy can be used to visualize the receptor on the cell surface (Maneckjee et al., 1988). Receptors can also be epitope tagged for visualization. A Green Fluorescent Protein β_2-adrenergic receptor (β_2AR) was shown to exhibit pharmacological and functional properties of the wild-type receptor (Barak et al., 1997). If a mutant receptor under investigation displays loss of function, detection of the epitope-tagged version of the mutant protein at the cell surface indicates that the loss of function is due to the mutation. Failure of the tagged protein to intercalate into the membrane may not imply that the mutation results in loss of membrane expression, however, as the tag may impede membrane insertion. Binding properties of some ligands may be altered by addition of the tag, so the untagged analog should be used for binding experiments. This issue is particularly important for receptors with larger ligands such as peptides.

If antibody-labeling and epitope-tagging approaches are not feasible, the question of whether the receptor is expressed in the membrane may be answered via an intact cell binding assay with hydrophilic ligands. The hydrophilic βAR agonist [^{3}H] CGP-12177 does not permeate cells and therefore can only label cell surface receptors (Portenier et al., 1984). The D_2 dopamine receptor antagonists sulpiride and domperidone are also hydrophilic compared with ligands such as spiperone. If the B_{max} derived from a saturation binding experiment using a hydrophilic ligand is significantly lower than that derived using a lipophilic ligand, the receptor population may be mainly internal. Another method is to treat membranes with detergents such as Triton X-100 or digitonin. If such treatment increases the number of binding sites or restores binding that was lost in a mutant receptor, it is likely that the receptor is not expressed on the cell surface. Attributing differences in binding affinities to a mutation may be erroneous in this case as the differences may be due to differences in receptor localization.

2.B. Radioligand Binding and Data Analysis for Large Experiments

Analysis of receptor binding to wild-type and mutant receptors requires a saturable, stereoselective radioligand and a method for distinguishing between free and receptor-bound radioactivity. For a detailed review of receptor binding methods, see Bennett and Yamamura (1985). For receptors that bind small molecules, filtration is the most frequently used method for separating free from bound radioactivity. If the specific activity of the ligand is sufficiently high, scintillation proximity can be used (Nelson, 1987). This technique utilizes a scintillant complexed with a solid support coated with wheatgerm agglutinin (WGA). The membrane in the assay attaches to the WGA so that the energy from the bound radioligand activates the scintillant. Free ligand does not associate with the WGA and does not enter into close enough proximity to the scintillant to cause activation.

Regardless of the method for identifying bound radioligand, nonspecific binding of radioligand can occur. Assay tubes containing a receptor-saturating concentration of a competing unlabeled ligand should be included. If possible, the unlabeled compound should be structurally distinct from the radioligand so that radioligand binding to nonreceptor sites will not be displaced. The appropriate concentration of this compound should be determined by testing increasing concentrations in the presence of the radioligand to find the lowest concentration at which radioligand binding is minimized. The remaining level of radioligand binding in the presence of the unlabeled compound is subtracted from total binding during analysis.

Competition curves using selected agonists and antagonists allow for determination of changes in receptor affinity. IC_{50} values are only useful for comparison if the affinity for radioligand is not altered. It is critical to determine the affinity constant (K_d) for the radioligand for each mutant receptor. If the affinity of the radioligand differs, results of competition experiments should be reported using K_i values. The K_i can be calculated by the Cheng and Prusoff (1973) equation:

$$K_i = \frac{IC_{50}}{1 + [L]/K_d} \tag{1}$$

where [L] is the concentration of radioligand. A more precise K_i calculation can be obtained using the Munson and Rodbard (1988) equation:

$$K_i = \frac{IC_{50}}{1 + [L](y + 2)/[2K_d(y + 1) + y]} + K_d[y/(y + 2)] \tag{2}$$

where y is the bound/free ratio for the label in the absence of the competing drug. This ratio varies with the amount of protein used, and K_i values obtained using the two equations will diverge as the protein concentration is increased. Increasing the amount of protein in a binding assay results in an increase in the amount of ligand bound, which is only linear for low protein concentrations. Assays should not be run using protein concentrations that result in binding levels beyond the linear range.

The steps for mapping binding sites described in this chapter require generation of many mutant receptors and assessing binding affinities of many ligands on these receptors. Because of the large number of assays required to perform this analysis, a 96-well plate assay format is recommended. Most ligand binding assays are amenable to volumes under 500 μl, and filtration equipment is available for separating bound from free ligand in a 96-well format. For increased throughput, assays can exploit scintillation proximity technology as described above.

Data analysis of large volume experiments can also be streamlined by using a spreadsheet program such as Microsoft Excel, which can accept data files directly from most scintillation counters. Using the Solver tool in Excel, one can solve iterative nonlinear regression equations such as those used for IC_{50}, B_{max}, and K_d determinations (Rohatagi et al., 1995). For competition binding experiments in which the IC_{50} is desired, the following equation is used to define the theoretical curve:

$$TY = NS + \frac{(T - NS)}{1 + 10^{(X - \text{Log } IC_{50})}} \tag{3}$$

where TY is the theoretical response value, T and NS are the values for total and nonspecific binding, respectively, X is the logarithm of the competitor concentration, and IC_{50} is the competitor concentration at which one-half of specific radioligand binding is displaced. An estimate of the error between the TY and the actual data (AY) is determined by summing the squares of the differences between these two values:

$$SS = \Sigma(TY - AY)^2 Q \tag{4}$$

The Solver is employed to determine the nonlinear equation that best describes the experimental data. Solver is an add-in command in Excel that works by iteratively changing the values of specified cells and recalculating the value of a target cell. If the cell with the error term equation (SS) is the target cell, the log IC_{50} cell can be iteratively changed until the value of the error term is minimized. In addition to the IC_{50}, other parameters can be simultaneously incorporated into the error minimization such as T or NS.

The "goodness of fit" between the actual and theoretical data in nonlinear regression analysis is complex and can be estimated by determination of a correlation coefficient based on the following equation:

$$r^2 = \frac{SS}{(s_y)^2 (DF)} \tag{5}$$

where s_y is the standard deviation of the actual Y values at all values of X, and DF is degrees of freedom. DF is the number of actual Y values minus the number of parameters allowed to be iteratively changed by the Solver command. A spreadsheet-based analysis platform can significantly decrease the time required for competition binding data analysis.

2.C. Analysis of Receptor Function

When identifying sites that affect receptor binding affinity, it is often useful to determine whether the functional response of the mutant receptor is altered. For the class of GPCRs, functional parameters include GTPase activity, GTPγS binding, adenylyl cyclase stimulation or inhibition, arachidonic acid release, phosphoinositide turnover, and calcium mobilization. Assays should be selected that produce high-magnitude responses so that a mutation-induced diminished response can be easily differentiated from an abolished response. Adenylyl cyclase and calcium mobilization assays are preferable to GTPase assays because of higher signal-to-noise ratios.

Adenylyl cyclase activity can be measured either in whole cells or in membranes by quantifying the amount of cAMP produced during drug treatment. If a membrane preparation is used, ATP and pyruvate kinase must be added to the assay buffer for substrate generation. If a whole-cell assay is used, terminating the reaction by boiling lyses the cells and allows for detection of both intracellular and extracellular cAMP. As both fractions have been shown to respond to GPCR agonists, the signal-to-noise ratio is not compromised using this method (Egawa et al., 1988; Rosenberg and Dichter, 1989).

Reasonable detection throughput can be achieved using a 96-well radioimmunoassay system, many of which are commercially available. Scintillation proximity technology can be used to differentiate between antibody-bound and free ^{125}I-cAMP (Vagell et al., 1991). When measuring adenylyl cyclase stimulation, the diterpene forskolin can be used to stimulate cAMP production independent of receptor activation, providing a standard for interassay comparison. When measuring adenylyl cyclase inhibition, forskolin can be used to increase cAMP levels and observe receptor-mediated inhibition of the elevated baseline, which is easier to detect than inhibition of basal cAMP levels.

Another important cellular signal that can be effectively measured in transfected cell lines is phosphoinositide hydrolysis (Wilkinson, et al.). Measuring phosphoinositide turnover mediated by receptors coupled to Gq involves preloading cells with ^{3}H-inositol prior to stimulation and then performing agonist treatment in the presence of LiCl to inhibit IP_1 phosphatase. IP_1, IP_2, and IP_3 are then separated on Dowex resin and counted in a scintillation counter. Other Gq-mediated functional assays involve measurement of IP_3-induced calcium release from intracellular stores using fluorescent calcium-indicator dyes such as fura-2 (Matsuda et al., 1996).

3. MAPPING OF THE BIOGENIC AMINE BINDING SITE

The largest group of small molecule ligands can be loosely classified as biogenic amines (i.e., agonists for which a basic nitrogen is a critical binding element). These ligands include the catecholamines epinephrine, norepinephrine, and dopamine, as well as 5-hydroxytryptamine (5-HT), histamine, and acetylcholine (Fig. 3.1). These ligands activate subfamilies of GPCRs, each of which comprises several subtypes. The first biogenic amine binding site to be mapped was that of the β_2AR, which binds the catecholamines epinephrine and norepinephrine to activate the G protein Gs, stimulating adenylyl cyclase and leading to accumulation of intracellular cAMP (Dixon et al., 1986).

3.A. Scanning Deletion Mutagenesis

To map the general location of the binding site, scanning deletion mutagenesis was used to delete large regions of the receptor. Regions that could be deleted without affecting the binding parameters were judged not to be required for

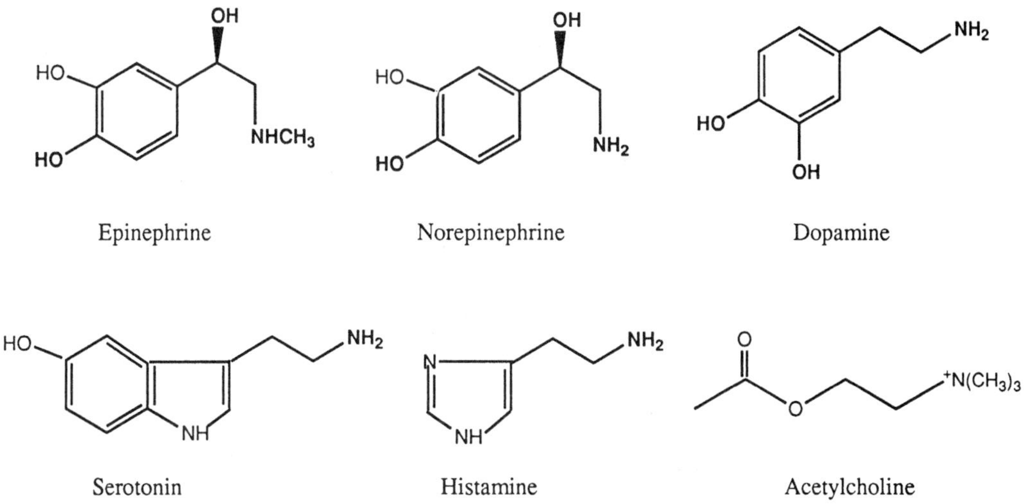

Figure 3.1. Endogenous ligands of biogenic amine receptors.

binding. This approach allows one to eliminate large domains from consideration when planning high-resolution two-dimensional mutagenesis experiments. However, it is important to remember that any loss of binding seen with such deletions is likely to result from a conformational disruption of the binding site, and no further interpretation of such data should be attempted. For the β_2AR, the deletion mutagenesis implicated the hydrophobic transmembrane core of the receptor in the interaction with the ligand (Dixon et al., 1987).

3.B. Construction of a Pharmacophore Map

For the β_2AR, a working map of the putative binding site could be derived from the wealth of studies that had explored the pharmacophore requirements for this receptor during the development of β-blockers and β-agonist drugs. As shown in Figure 3.2, the functional requirements at points of contact of the ligand were well-defined from these studies. Thus, it was relatively straightforward to infer potential contributions that the receptor might make to the binding interaction. For less well-studied receptors, the pharmacophore map would first have to be developed by chemical exploration of the structure–activity profile of analogs of the ligand. For the β_2AR, it was known that both agonists and antagonists require a protonatable nitrogen. Thus, it is reasonable to assume that the receptor should provide an acidic counterion for the amine. The β–OH group provides a chiral site of interaction for both agonists and antagonists, suggesting the presence of a hydrogen bond donor or acceptor in the receptor. The catechol ring is critical for agonist binding and activation, whereas antagonists tend to be phenoxypropanolamines with more hydrophobicity in the aromatic portion of the molecule. These data, taken together, can be used to construct a pharmacophore map of the binding pocket (Fig. 3.2).

A key assumption that can dramatically focus the experimental design of mutagenesis studies is the assumption that the agonist binding sites in different

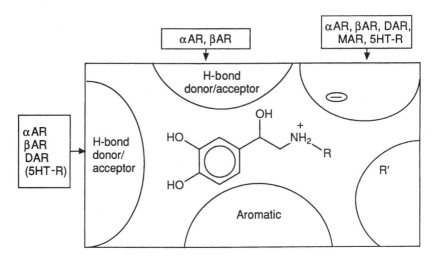

Figure 3.2. Map of ligand–receptor interactions in the binding site of the βAR. A catecholamine ligand is shown in a hypothetical binding site intercalated among the transmembrane helices of the receptor Each of the large semicircles represents a transmembrane helix of the receptor, inscribed with the type of binding interaction expected. Other GPCRs that would be expected to have similar interactions with their specific ligands are designated in boxes next to each helix. (Reproduced from Strader et al. [1989], with permission of the publisher.)

members of this receptor family are themselves structurally related. Thus, one would expect a particular pattern of conservation within the family of GPCRs for the residues providing the interactions with the endogenous agonist shown in Figure 3.2. For example, the putative counterion for the amine moiety might be expected to be conserved throughout the family of biogenic amine receptors, whereas the residues that interact with the catechol hydroxyl groups should be conserved among the catecholamine receptors but not among other biogenic amine receptors whose ligands do not contain the catechol functionality. This is a particular advantage in mapping small molecules that are the endogenous ligands for these receptors. Because the receptors did not evolve to bind synthetic small molecule antagonists, there is no reason to assume that the binding sites for such ligands would be conserved, and this shortcut cannot be applied beyond the endogenous agonists.

3.C. Specific Point Mutations

The suggestions derived from the pharmacophore map can then be followed up with specific point mutations. For example, to localize the putative counterion for the biogenic amine groups, each of the acidic residues in or near the transmembrane (TM) domain of the β_2AR was replaced by a neutral or basic residue to remove the negative charge (Strader et al., 1988b). Of these mutations, substitution of Asp[113] in TM3 abolished detectable ligand binding to the receptor, as assessed by an equilibrium binding assay. To detect lower affinity interactions, adenylyl cyclase activity was assayed as a measure of receptor signaling.

This experiment revealed that substitution of Asp[113] with glutamate or asparagine decreased the affinity of the receptor for the agonist by 100- and 10,000-fold, respectively, without affecting the maximum level of activation. These experiments demonstrated that the side chain of Asp[113] is important for the binding of agonists and antagonists to the β_2AR, but not for receptor activation. The data also suggested, but did not prove, that the loss of binding observed upon substitution of Asp[113] resulted from a loss of an ion pair connecting Asp[113] to the amine group in the ligand.

3.D. Two-Dimensional Mutagenesis

3.D.a. Plan A: Design of a Novel Binding Site. To obtain evidence for a specific ion pairing interaction between the ligand and the receptor, it is important to cause a predictable *gain* of function, because *loss* of function can result from either specific or nonspecific effects of the mutations. For the β_2AR, this was achieved by substitution of Asp[113] with a serine residue, thereby replacing an acidic side chain with one capable of serving as a hydrogen bond donor or acceptor (Strader et al., 1991). Catechol-containing nonamine analogs were then tested for their ability to activate the mutant receptor. The Ser[113] β_2AR was activated by catechol esters and ketones, compounds that could accept a hydrogen bond from the new serine side chain on the receptor. These compounds did not activate the wild-type receptor, consistent with the reduced ability of the aspartate side chain to serve as a hydrogen bond donor. These data provide strong evidence that there is a specific interaction linking the side chain of the residue at position 113 in the β_2AR to the functional group on the aliphatic end of the catechol-containing agonist. For the wild-type β_2AR, this would imply an ionic interaction between the carboxylate of Asp[113] and the amine moiety on the ligand.

3.D.b. Plan B: Additivity Analysis. Another less elegant, but frequently more achievable, approach to two-dimensional mutagenesis can be illustrated by experiments designed to pinpoint the sites of interaction of the catechol hydroxyl groups in the β_2AR. The original pharmacophore map (Fig. 3.2) had suggested hydrogen bonds linking the catechol groups of agonist ligands to the receptor. Substitution of potential hydrogen bonding residues in the receptor with alanine revealed two serine residues whose substitution resulted in a decrease in the affinity for catecholamine agonists but not for antagonists (Strader et al., 1988a). These serine residues were located in TM5, one helical turn apart at positions 204 and 207, consistent with the potential for forming two simultaneous hydrogen bonds with the agonist. To probe this possibility further, a two-dimensional approach was taken in which each of the serine residues was independently substituted with alanine. The effects of these mutations on the binding and activation by isoproterenol and by analogs in which either the *meta-* or the *para*-OH was replaced by H were examined. Removal of the –OH from either the receptor (by mutagenesis) or the ligand (by organic synthesis) reduced the affinity of the interaction by approximately 10-fold and reduced the efficacy by approximately 50%. The effects of substitution of Ser[204] with alanine were additive with the effects of substitution of the *para*-OH group of the ligand but not with substitution of the *meta*-OH group. The converse was true for substitution of Ser[207]. These data suggest that the catechol portion of the agonist interacts with the β_2AR through a specific hydrogen bond between the *meta*-OH group of the ligand and the side chain of Ser[204] in the receptor and a second specific hydrogen bond between the *para*-OH group of the ligand and Ser[207] in the receptor. These interactions appear to be critical for receptor activation, perhaps by positioning the catechol ring very precisely in the receptor binding pocket. Because the ligands are chemically accessible, the additivity analysis variant of two-dimensional mutagenesis is particularly applicable to mapping the binding site of GPCRs whose endogenous agonists are small molecules.

3.E. Conservation Pattern Analysis

As mentioned above, the observation that the ligands for the biogenic amine receptors fall into discrete structural classes would suggest that the receptor residues that interact with the various functional groups on the ligands should show specific patterns of conservation. Thus, Asp[113] might be expected to be conserved throughout the biogenic amine receptor family, whereas Ser[204] and Ser[207] should be specific for receptors that bind catecholamines. The presence of such a conservation pattern can provide further evidence to support the existence of a particular receptor–ligand interaction.

An aspartate residue is, in fact, conserved at the position in TM3 equivalent to Asp[113] in all biogenic amine receptors identified to date. This residue has been mutated in several other biogenic amine receptors (α-adrenergic, muscarinic, serotonin, dopamine, and histamine), always resulting in a marked decrease in ligand binding affinity (Wang et al., 1991, 1993; Fraser et al., 1989; Tomic et al., 1993; Gantz et al., 1992). The loss of function experiments have not been followed up with two-dimensional gain of function experiments in other receptors. However, given the results with the β_2AR and the pattern of conservation of the residue at this position, it seems reasonable to assume that the interaction is specific. Thus, homology mapping accompanied by limited point mutations can be used to extend observations from one receptor–ligand pair across a receptor family (Fig. 3.3).

While amino acid homology is a good preliminary indicator of ligand binding characteristics, the interactions between conserved amino acids and ligand moieties are not always identical. Conserved serine residues in catecholamine receptors reside at positions analogous to those of Ser[204] and Ser[207] in the β_2AR (Fig. 3.3). The conservation is not absolute, however: The serine analogous to Ser[204] is displaced by one position in some of the α_2 receptors. Experiments to determine the role of these residues in other catecholamine–receptor interactions have not been straightforward. While it is clear that these serine residues are important for catecholamine binding to other biogenic amine receptors, two-dimensional mutagenesis results have not always been readily interpretable for these other receptors. Additivity analysis has suggested an interaction between the Ser[204] analog and the *meta*-OH group for the α_{1A}- and α_{2A}-adrenergic receptors, but the data for the Ser[207] analog have been less clear (Hwa and Perez, 1996; Wang et al., 1991). These results may be explained by the fact that these relatively low-energy hydrogen bonds (estimated at 1.9 kcal for the β_2AR) can be substituted by other interactions with nearby (nonconserved) residues in the

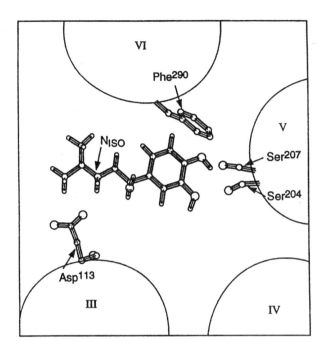

Receptor	Ligand	Asp^{113}	Ser$^{204, 207}$	Phe^{290}
Alpha-1	Norepinephrine	Asp	Ser, Ser	Phe
Alpha-2	Norepinephrine	Asp	Ser, Ser	Phe
Beta	Epinephrine	Asp	Ser, Ser	Phe
Dopamine	Dopamine	Asp	Ser, Ser	Phe
Muscarinic	Acetylcholine	Asp	Thr, Ala	Asn
Serotonin	Serotonin	Asp	Ser, Ala, Thr	Phe
Histamine	Histamine	Asp	Asp, Thr	Phe
Neurokinin	Substance P	Pro	Val, Tyr	His
Endothelin	Endothelin	Gln	Phe, Tyr	His
Bradykinin	Bradykinin	Ile	Asn, Gly	Thr

Figure 3.3. Proposed model for the ligand-binding site of the βAR viewed from the exterior face of the cell. The ligand isoproterenol is shown surrounded by TMIII–VI. The cationic amine of isoproterenol is indicated by N_{iso}. The proposed locations of the Asp^{113}, Phe^{290}, Ser^{203}, and Ser^{207} side chains are indicated. Examples of GPCR and their amino acids in the corresponding locations are listed. (Reproduced from Dixon et al. [1988], with permission of the publisher.)

mutant receptors. Alternatively, the catechol ring may fit slightly differently into the binding sites of the various catecholamine receptors.

The TM5 region appears to play a role in binding the heteroatoms in other biogenic amine ligands as well. In support of this hypothesis, Ser^{207}, but not

Ser[204], is conserved among the family of serotonin receptors, which bind the monohydroxylated agonist 5-HT. Point mutations of this residue in the serotonin receptor family suggest that this serine residue may play a role in agonist binding (Ho et al., 1992). However, two-dimensional mutagenesis will be needed to determine whether this serine represents the binding site for the 5-OH group, as would be predicted by homology analysis. In the muscarinic receptors, there is a conserved threonine residue adjacent to the position occupied by Ser[204] in the β_2AR, suggesting the potential for interactions with the ester oxygens of acetylcholine. Although not yet followed up with two-dimensional mutations to specify the interaction, it is notable that this threonine residue has been shown to be important for agonist binding to M_3 muscarinic receptors (Wess et al., 1992).

In the H_2 histamine receptor, the region of TM5 contains an aspartate and threonine residue located four positions apart from each other. When the H_2 receptor was first cloned, the authors discerned that histamine was the likely ligand by reasoning that the new receptor (1) must be a biogenic amine receptor because of the presence of an aspartate in TM3; (2) was not a catecholamine receptor because it lacked two serine residues in TM5; and (3) the imidazole group of histamine would, by analogy, be well-positioned to interact with the aspartate and threonine residues that were present in TM5 (Gantz et al., 1991). Subsequent mutagenesis work revealed that the aspartate and threonine are critical for histamine binding, although the specific molecular interactions have not yet been rigorously defined (Gantz et al., 1992). This receptor provides an excellent example of how homology patterns can be used to focus the mutagenesis studies on critical residues.

4. IDENTIFICATION OF SITES FOR ALLOSTERIC MODULATION

4.A. Assessment of Degree of Cooperativity

Chemicals that alter the binding of receptor ligands by simultaneous binding rather than direct displacement are said to exhibit noncompetitive or allosteric binding to the receptor. As has been demonstrated for other classes of proteins and enzymes, some GPCRs have been shown to have allosteric sites. Examples are binding of amiloride to α_2-adrenergic receptors, PD 81,723 to A_1-adenosine receptors, and gallamine to muscarinic receptors (Howard et al., 1987; Bruns et al., 1990; Stockton et al., 1983). The competitive site is defined as the site to which the endogenous agonist binds. Experiments that measure the degree of allosteric interaction should be used to interpret mutagenesis studies designed to identify amino acids that comprise noncompetitive binding sites.

In evaluating the effects of mutations on allosteric binding, two parameters must be investigated: the allosteric compound's affinity for the receptor and its degree of cooperativity with competitive agonists and antagonists (see Lazareno and Birdsall [1995] for analytical details). One way to define these parameters is to perform saturation binding experiments in the presence of different concentrations of the noncompetitive agent. The K_d obtained in the presence of the agent (K_{dA}) is divided by the K_d obtained in its absence to define the dose ratio, *DR*. Log (*DR*-1) is plotted against the log of the concentration of the agent. A curvilinear line indicates an allosteric interaction between the agent and the radioligand. A linear relationship indicates that either the interaction is competitive or the radioligand concentration used was not high enough to demonstrate noncooperativity. The value of the cooperativity factor, α, is obtained using the equation:

$$DR\text{-}1 = \frac{(\alpha - 1)[A]}{\alpha K_{dA} + [A]} \qquad (6)$$

where $[A]$ is the concentration of the allosteric ligand.

Another way to investigate cooperativity is to examine the dissociation rate of the radioligand in the presence of multiple concentrations of the agent. To perform these experiments, radioligand binding at equilibrium is disrupted by addition of a competitive ligand at various time points prior to termination of the assay. From the amount of residual binding at each time point, the dissociation rate of the radioligand is determined. A change in this rate induced by an agent is indicative of an allosteric interaction between the agent and the radioligand.

4.B. Mutations That Affect Allosteric Binding of Muscarinic Receptors

One of the first classes of receptors shown to possess a noncompetitive site was the muscarinic receptor family (Stockton et al., 1983). Lee et al. (1992) hypothesized that negatively charged amino acids in the TM regions of the M_1 receptor would be important for allosteric binding as most noncompetitive ligands at muscarinic receptors are highly positively charged. The TM regions were targeted based on a study in which trypsinization and purification of α_2-adrenergic receptors to remove extracellular and cytoplasmic domains did not affect allosteric binding by amiloride (Wilson et al., 1990). To investigate the contribution of aspartate residues to allosteric binding to M_1 receptors, three individual asparagine substitutions were made in TM2 (D71N) and TM3 (D99N and D122N). In each of these mutants, allosteric binding of gallamine was retained, evidenced by curvilinear Schild plots. However, the degree of cooperativity (α) was decreased only in the D71N and D99N mutants.

The aspartate in TM2 is highly conserved among GPCRs and has been shown to be important for receptor–G protein coupling (Chung et al., 1988) and regulation of agonist affinity by monovalent cations (Horstman et al., 1990). The loss of cooperativity of gallamine binding precipitated by this mutation indicates that small molecules like gallamine may interact with GPCRs in a similar manner to monovalent cations. In addition, the importance of Asp[79] in receptor activation is suggestive of an allosteric mechanism that involves disturbing the receptor–G protein interaction.

While the D99N mutation decreased the degree of cooperativity, it did not reduce the affinity of gallamine. However, this mutant receptor showed a decrease in the affinity for the competitive agonist carbachol (Fraser et al., 1989). The insensitivity of gallamine binding to this mutation conforms to the model of gallamine interacting with a site distinct from the agonist binding site.

The corresponding amino acids in the M_2 receptor, Asp[69] and Asp[97] but not Asp[120], were shown to be involved in allosteric binding of gallamine as well

(Leppik et al., 1994). In this study, a mutation in the second extracellular loop was also made, converting an acidic sequence unique to the M_2 receptor (EDGE) to the corresponding M_1 sequence (LAGQ). This mutation produced a fourfold increase in the degree of cooperativity of gallamine on ^{3}H-NMS binding, suggesting that for M_2 receptors the allosteric binding site involves residues outside of the hydrophobic core. Further delineation of these binding interactions with GPCRs will require two-dimensional mutagenesis experiments as described above for agonist ligands.

In summary, GPCRs whose endogenous agonists are small molecules provide a relatively simple system for exploring ligand binding sites. The chemical accessibility of the ligands allows the binding site to be mapped from the dual perspectives of the ligand and the receptor. In addition, the conservation of key aspects of the binding site throughout the family of biogenic amine receptors provides the framework with which the binding interactions can be viewed. Together, the family of small molecule receptors serves as a useful model for understanding the structure–function relationships of the extensive family of GPCRs.

ACKNOWLEDGMENT

The authors thank Dr. Robert Burrier for assistance in the area of computer-aided data analysis.

REFERENCES

Barak LS, Ferguson SS, Zhang J, Martenson C, Meyer T, Caron MG (1997): Internal trafficking and surface mobility of a functionally intact beta2-adrenergic receptor–Green Fluorescent Protein conjugate. Mol Pharmacol 51:177–184.

Bennett JP, Yamamura HI (1985): Neurotransmitter, hormone or drug receptor binding methods. In Enna SJ, Kuhar MJ, Yamamura HI (eds): Neurotransmitter Receptor Binding. New York: Raven Press, pp 61–90.

Bruns RF, Fergus JH, Coughenour LL, Courtland GG, Pugsley TA, Dodd JH, Tinney FJ (1990): Structure–activity relationships for enhancement of adenosine A_1 receptor binding by 2-amino-3-benzoylthiophenes. Mol Pharmacol 38:950–958.

Cheng Y-C, Prusoff WH (1973): Relationship between the inhibition constant (K_i) and the concentration of inhibitor which causes 50 percent inhibition (IC_{50}) of an enzymatic reaction. Biochem Pharmacol 22:3099.

Chung FZ, Wang CD, Potter PC, Venter JC, Fraser CM (1988): Site-directed mutagenesis and continuous expression of human beta-adrenergic receptors. Identification of a conserved aspartate residue involved in agonist binding and receptor activation. J Biol Chem 263:4052–4055.

Dixon RA, Kobilka BK, Strader DJ, Benovic JL, Dohlman HG, Frielle T, Bolanowski MA, Bennett CD, Rands E, Diehl RE, et al. (1986): Cloning of the gene and cDNA for mammalian beta-adrenergic receptor and homology with rhodopsin. Nature 321:75–79.

Dixon RA, Sigal IS, Candelore MR, Register RB, Scattergood W, Rands E, Strader CD (1987): Structural features required for ligand binding to the beta-adrenergic receptor. EMBO J 6:3269–3275.

Dixon RA, Sigal IS, Strader CD (1988): Structure–function analysis of the β-adrenergic receptor. Cold Spring Harb Symp Quant Biol 53(Pt. 1):487–497.

Egawa M, Hoebel BG, Stone EA (1988): Use of microdialysis to measure brain noradrenergic receptor function in vivo. Brain Res 458:303–308.

Fraser CM, Wang CD, Robinson DA, Gocayne JD, Venter JC (1989): Site-directed mutagenesis of ml muscarinic acetylcholine receptors: Conserved aspartic acids play important roles in receptor function. Mol Pharmacol 36:840–847.

Gantz I, DelValle J, Wang LD, Tashiro T, Munzert G, Guo YJ, Konda Y, Yamada T (1992): Molecular basis for the interaction of histamine with the histamine H_2 receptor. J Biol Chem 267:20840–20843.

Gantz I, Schaffer M, DelValle J, Logsdon C, Campbell V, Uhler M, Yamada T (1991): Molecular cloning of a gene encoding the histamine H_2 receptor. Proc Natl Acad Sci USA 88:429–433.

Ho BY, Karschin A, Branchek T, Davidson N, Lester HA (1992): The role of conserved aspartate and serine residues in ligand binding and in function of the 5-HT1A receptor: A site-directed mutation study. FEBS Lett 312:259–262.

Horstman DA, Brandon S, Wilson AL, Guyer CA, Cragoe EJ Jr, Limbird LE (1990): An aspartate conserved among G-protein receptors confers allosteric regulation of alpha 2-adrenergic receptors by sodium. J Biol Chem 265:21590–21595.

Howard MJ, Hughes RJ, Motulsky HJ, Mullen MD, Insel PA (1987): Interactions of amiloride with alpha- and beta-adrenergic receptors: Amiloride reveals an allosteric site on alpha 2-adrenergic receptors. Mol Pharmacol 32:53–58.

Hwa J, Perez DM (1996): The unique nature of the serine interactions for alpha 1-adrenergic receptor agonist binding and activation. J Biol Chem 27:6322–6327.

Lazareno S, Birdsall NJM (1995): Detection, quantification, and verification of allosteric interactions of agents with labeled and unlabeled ligands at G protein–coupled receptors: Interactions of strychnine and acetylcholine at muscarinic receptors. Mol Pharmacol 48:362–378.

Lee NH, Jingru H, El-Fakahany EE (1992): Modulation by certain conserved aspartate residues of the allosteric interaction of gallamine at the m1 muscarinic receptor. J Pharmacol Exp Ther 262:312–316.

Leppik RA, Miller RC, Eck M, Paquet J-L (1994): Role of acidic amino acids in the allosteric modulation by gallamine of antagonist binding at the m2 muscarinic acetylcholine receptor. Mol Pharmacol 45:983–990.

Liu J, Schöneberg T, van Rhee M, Wess J (1995): Mutational analysis of the relative orientation of transmembrane helices I and VII in G protein–coupled receptors. J Biol Chem 270:19532–19539.

Maneckjee R, Archer S, Zukin RS (1988): Characterization of a polyclonal antibody to the mu opioid receptor. J Neuroimmunol 17:199–208.

Matsuda S, Kusuoka H, Hashimoto K, Tsujimura E, Nishimura T (1996): The effects of proteins on $[Ca^{2+}]$ measurement: Different effects on fluorescent and NMR methods. Cell Calcium 20:425–430.

Munson PJ, Rodbard D (1988): An exact correction to the Cheng-Prusoff correction. J Recept Res 8:533.

Nelson N (1987): A novel method for the detection of receptors and membrane proteins by scintillation proximity radioassay. Anal Biochem 165:287–293.

Portenier M, Hertel C, Muller P, Staehelin M (1984): Some unique properties of CGP-12177. J Recept Res 4:108–111.

Rohatagi S, Hochhaus G, Mollmann H, Barth J, Derendorf H (1995): Pharmacokinetic interaction between endogenous cortisol and exogenous corticosteroids. Pharmazie 50:610–613.

Rosenberg PA, Dichter MA (1989): Extracellular cAMP accumulation and degradation in rat cerebral cortex in dissociated cell culture. J Neurosci 9:2654–2663.

Stockton JM, Birdsall NJ, Burgen AS, Hulme EC (1983): Modification of the binding properties of muscarinic receptors by gallamine. Mol Pharmacol 23:551–557.

Strader CD, Sigal IS, Dixon RAF (1989): Genetic approaches to the determination of structure–function relationships of G protein–coupled receptors. Trends Pharmacol Suppl 26–30.

Strader CD, Candelore MR, Hill WS, Sigal IS, Dixon RA (1988a): Identification of two serine residues involved in agonist activation of the beta-adrenergic receptor. J Biol Chem 264:13572–13578.

Strader CD, Gaffney T, Sugg EE, Candelore MR, Keys R, Patchett AA, Dixon RA (1991): Allele-specific activation of genetically engineered receptors. J Biol Chem 266:5–8.

Strader CD, Sigal IS, Candelore MR, Rands E, Hill WS, Dixon RA (1988b): Conserved aspartic acid residues 79 and 113 of the beta-adrenergic receptor have different roles in receptor function. J Biol Chem 263:10267–10271.

Strader CD, Sigal IS, Register RB, Candelore MR, Rands E, Dixon RAF (1987): Identification of residues required for ligand binding to the β-adrenergic receptor. Proc Natl Acad Sci USA 84:4384–4388.

Tomic M, Seeman P, George SR, O'Dowd BF (1993): Dopamine D1 receptor mutagenesis: Role of amino acids in agonist and antagonist binding. Biochem Biophys Res Commun 191:1020–1027.

Vagell ME, McGinnis MY, Possidente BP, Narasimhan VN, Lumia AR (1991): Olfactory bulbectomy increases basal suprachiasmatic cyclic AMP levels in male rats. Brain Res Bull 27:839–842.

Wang CD, Buck MA, Fraser CM (1991): Site-directed mutagenesis of alpha 2A-adrenergic receptors: Identification of amino acids involved in ligand binding and receptor activation by agonists. Mol Pharmacol 40:168–179.

Wang CD, Gallaher TK, Shih JC (1993): Site-directed mutagenesis of the serotonin 5-hydroxytryptamine2 receptor: Identification of amino acids necessary for ligand binding and receptor activation. Mol Pharmacol 43:931–940.

Weiss ER, Osawa S, Shi W, Dickerson CD (1994): Effects of carboxyl-terminal truncation on the stability and G protein–coupling activity of bovine rhodopsin. Biochemistry 33:7587–7593.

Wess J, Maggio R, Palmer JR, Vogel Z (1992): Role of conserved threonine and tyrosine residues in acetylcholine binding and muscarinic receptor activation. A study with m3 muscarinic receptor point mutants. J Biol Chem 267:19313–19319.

Wilkinson GF, Feniuk W, Humphrey PP (1997): Characterization of human recombinant somatostatin sst5 receptors mediating activation of phosphoinositide metabolism. Br J Pharmacol 121:91–96.

Wilson AL, Guyer CA, Cragoe EJ Jr, Limbird LE (1990): The hydrophobic tryptic core of the porcine alpha 2-adrenergic receptor retains allosteric modulation of binding by Na^+, H^+, and 5-amino–substituted amiloride analogs. J Biol Chem 265:17318–17322.

Zhang R, Tsai-Morris CH, Kitamura M, Buczko E, Dufau ML (1991): Changes in binding activity of luteinizing hormone receptors by site directed mutagenesis of potential glycosylation sites. Biochem Biophys Res Commun 181:804–808.

CHAPTER 4

MAPPING BINDING SITES FOR PEPTIDE G PROTEIN–COUPLED RECEPTORS: THE RECEPTOR FOR THYROTROPIN-RELEASING HORMONE

ROMAN OSMAN, ANNY-ODILE COLSON,
JEFFREY H. PERLMAN, LIISA J. LAAKKONEN,
and MARVIN C. GERSHENGORN

1	INTRODUCTION	60
2.	CONSTRUCTION OF MUTANT TRH-RECEPTORS AND THEIR CHARACTERIZATION	61
	A. Mutagenesis	61
	B. Transient Transfection	62
	C. Competition Binding Assays	63
	a. Cells in Monolayer	63
	b. Membrane Preparations	63
	c. Inositol Phosphate Assay	63
3.	MODELING THE TRANSMEMBRANE BUNDLE OF TRH-R	64
	A. Construction of a Set of Realistic Helices	65
	B. Forming the Templates	66
	C. Generic GPCR Model	66
	D. Construction of a TRH-R Model	67
4.	DETERMINATION OF INTRAMOLECULAR INTERACTIONS IN TRH-R	69
5.	DETERMINATION OF THE BINDING POCKET OF TRH-R	72
6.	MODELING THE TRANSMEMBRANE BINDING POCKET OF TRH-R	73
7.	MODELING THE EXTRACELLULAR LOOPS OF TRH-R	75
8.	BIOLOGICALLY ACTIVE CONFORMATION OF TRH	77
9.	PEPTIDE LIGANDS AND OTHER GPCRs	79
10.	CONCLUSIONS	80

Structure–Function Analysis of G Protein-Coupled Receptors, Edited by Jürgen Wess.
ISBN 0-471-25228-X   Copyright © 1999 Wiley-Liss, Inc.

I. INTRODUCTION

The purpose of this chapter is to describe methods used to characterize the binding sites for peptide ligands within the three-dimensional (3D) structure of GTP binding protein-coupled receptors (GPCRs). In particular, the goal is to identify specific amino acid residues within GPCRs that form intramolecular interactions that maintain the 3D structure of the receptor and others that inter-

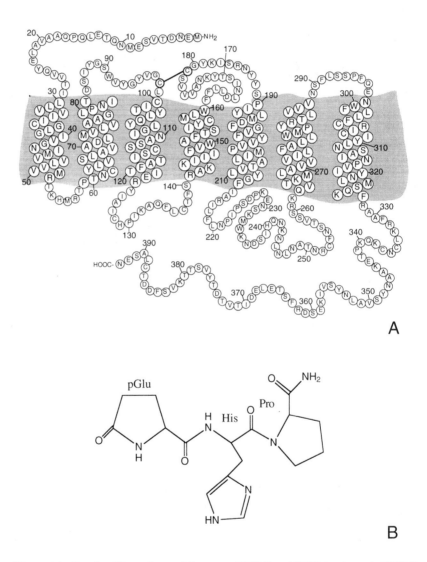

Figure 4.1. Putative 2D topology of the mouse TRH-R and TRH structure. **A:** TRH-R is an integral membrane protein with extracellular, transmembrane, and intracellular domains. The extracellular amino terminus and three extracellular loops are at the top of A. The seven α helices that span the plasma membrane are in the middle of A. The three intracellular loops and the intracellular carboxyl terminus are at the bottom of A. **B:** TRH: pyroglutamic acid–histidine–prolineamide.

act directly with the ligand. For the tripeptide thyrotropin-releasing hormone (TRH), the high affinity binding pocket appears to reside within the putative transmembrane (TM) bundle of the TRH receptor (TRH-R). Binding sites for other small ligands, such as neurotransmitters (see Chapter 3) and retinal (see Chapter 5), are found within the TM helices also. For larger peptide ligands, high affinity binding sites include residues that are present within the extracellular domains and residues located at the extracellular surfaces of TM helices. The high affinity binding determinants for large glycoprotein hormones, such as thyrotropin and luteinizing hormone, appear to involve the GPCR amino terminus primarily.

It is important to acknowledge that the 3D structure of no GPCR has been resolved at an atomic level. The only direct visualization of a GPCR is from two-dimensional (2D) projection maps of rhodopsin (see Chapter 12). In contrast, the structure of the seven TM-spanning protein bacteriorhodopsin, which is not a GPCR, has been resolved at an atomic level (Henderson et al., 1990). In the absence of direct structural information, a number of indirect methods have been employed to study the structures of GPCRs and their binding pockets. This chapter focuses on methods we have used to study the TRH-R (Gershengorn and Osman, 1996). Other methods and receptors are discussed at the end of the chapter.

To describe the methodology, examples are presented from the work we have carried out on the TRH-R. The amino acid sequence of TRH-R is presented as a putative 2D topology in Figure 4.1. TRH is a tripeptide: pyroglutamyl–histidine–prolineamide (Fig. 4.1). There are two important aspects to our approach to understanding the biology of the TRH-R in addition to the standard use of receptor mutagenesis. First, we have integrated experimental and computational approaches in an iterative fashion to model the 3D structure of TRH-R and to generate hypotheses regarding the structure–activity relationship of TRH-R. Second, to identify the TRH binding pocket, we have made substitutions in both TRH-R and the ligand, which has allowed us to address the critical issue of whether mutations in the receptor affect the maintenance of overall receptor conformation or, rather, directly affect specific interactions between TRH and its receptor.

2. CONSTRUCTION OF MUTANT TRH RECEPTORS AND THEIR CHARACTERIZATION

2.A. Mutagenesis

A polymerase chain reaction (PCR)–based method is usually used. If an endonuclease restriction site is conveniently located near a residue to be mutated, then one PCR reaction may be sufficient to introduce the mutation. The construction of Y106F TRH-R typifies this approach. Tyr^{106} is encoded by nucleotide bases 574–576. A unique $Nsi1$ restriction site is located at base pair (bp) 591 and a unique $BspH1$ site at bp 407. An antisense primer encoding the mutation was synthesized from bases 595 to 570, and a sense wild-type primer was synthesized from bases 400 to 420. Using standard PCR protocols, a 195 bp fragment was generated that was purified by agarose gel electrophoresis using glassmilk (Geneclean). The purified PCR fragment to be inserted into TRH-R was then digested with *Nsi*1 and *BspH*1. The eukaryotic expression plasmid encoding the mouse TRH-R, pCDM8mTRH-R, was also digested with *Nsi*1 and *BspH*1 and agarose gel purified. The PCR fragment and the digested vector fragment were co-precipitated in 50 μl ice-cold 7.5 M NH$_4$SO$_4$ and 340 μl ice-cold ethanol overnight at 4°C, centrifuged at 14,000g in the cold for 30 minutes, washed once with ice-cold 70% ethanol, and air dried. The pellet was then resuspended in 10 μl water and ligation performed with T4 DNA ligase overnight at 16°C.

If a restriction site is not conveniently located, then overlap PCR is used. An example of this approach is that used to construct C179A TRH-R. Cys[179] is encoded by bases 795 to 797. A unique *Nsi*1 restriction site is present at 591, and a unique *Sna*B1 restriction site is located at 1384. PCR product I was generated using a wild-type sense primer (bases 580 to 597) and an antisense 30-mer encoding a mutation for Cys[179]. PCR product II was generated using a sense 30-mer complementary to the amplimer encoding the mutation and an antisense wild-type primer (bases 1395 to 1376). Each PCR product was gel purified and then used in an overlap PCR reaction with the external primers to generate PCR product III. In the overlap reaction, low temperature (25°C) annealing is typically applied for the first 5 cycles, and then an annealing temperature 5°C below the melting temperature of that of the lower external primer is used for the next 25 cycles. PCR product III was then gel purified. Co-precipitation and ligation were carried out to subclone the PCR product into a vector. In this case, *Sna*B1 also cut in the vector, and, therefore, the PCR product was first subcloned into the cloning vector pBluescript containing wild-type TRH-R (pBSmTRH-R). The sequence encoding C179A TRH-R was then subcloned into pCDM8 using the unique restriction sites *Xho*1 (base 330) and *Not*1 (base 3500).

2.B. Transient Transfection

The mouse pituitary TRH-R sequence was subcloned into a eukaryotic expression vector such as pCDM8. African green monkey kidney COS-1 cells were transfected by the diethylaminoethyl (DEAE)–dextran method, which was modified from Cullen (1987). DNA is dissolved in water. DEAE–Dextran (Pharmacia) is prepared at 10 mg/ml in phosphate-buffered saline (PBS) and sterile filtered. On the day prior to transfection, 100-mm dishes are seeded with 1.5×10^6 cells in Dulbecco's modified Eagle's medium (DMEM) supplemented with 5% NuSerum (Life Technologies). A transfection cocktail of 3 ml per dish (containing usually 2 μg/ml DNA) is prepared from 2.85 ml Hanks' balanced salt solution (HBSS) with 10 mM HEPES, pH 7.4, and 0.15 ml DEAE–dextran final concentration 0.5 mg/ml). The transfection cocktail is vigorously vortexed immediately prior to use. The culture medium is aspirated from the dish, and the cells are washed twice with 10 ml warmed HBSS. The transfection cocktail is applied to the dish and incubated at 37°C for 30 minutes. Subsequently, 7 ml of DMEM containing 10% NuSerum and 80 μM chloroquine (chloroquine is prepared as 100× solution in HBSS and stored in the dark at 4°C up to 1 month) is added, and the incubation is continued at 37°C for 2.5 hours. The medium is then aspirated and replaced with 3 ml DMEM/10% NuSerum/10% dimethylsulfoxide for 2.5 minutes. After aspiration, 10 ml DMEM containing 10% NuSerum is added. On the next day, cells are split into 12- or 24-well plates at 100,000 or 50,000 cells/well with DMEM/5% NuSerum, and experiments are performed 1 or 2 days later.

2.C. Competition Binding Assays

[N-T-methylhistidine]TRH (MeTRH), which is ^{3}H labeled ([His-4-^{3}H, Pro-3,4-^{3}H]MeTRH, NEN Life Science Products), is commonly used as radioligand because the affinity and potency of MeTRH are 5–10-fold higher than that of TRH.

2.C.a. Cells in Monolayer. Wash cells grown in 12- or 24-well dishes once with 1 ml HBSS. Add 300 μl HBSS containing ^{3}H-MeTRH with or without 3 μl 100× unlabeled ligand. Incubate 1 to 3 hours, depending on incubation temperature and receptor affinity, until equilibrium binding is attained. After incubation, wash with 1 ml of ice-cold HBSS, pH 7.4 (three times on ice). After washing, add 1 ml 0.4 N NaOH and count 600 μl in a β counter.

2.C.b. Membrane Preparations. Two days after transfection, cells grown in 100-mm dishes are washed with 6 ml ice-cold HBSS/10 mM HEPES, pH 7.4 and incubated on ice for 10 minutes. The buffer is aspirated, and 6 ml ice-cold HBSS/10 mM EDTA is added, followed by an incubation on ice for 10 minutes. The cells are scraped with a rubber policeman and washed with 20 ml ice-cold PBS. Add 1.5 ml ice-cold 20 mM Tris-HCl (pH 7.6)/2 mM MgCl$_2$ (Tris-Mg buffer) for 10 minutes, and lyse the cells with 20 strokes of a Dounce (or similar) homogenizer. Centrifuge at 180g in the cold for 2 minutes to remove nuclei and unbroken cells. Collect the supernatant, and centrifuge again at low speed for 2 minutes. Transfer the supernatant to an Eppendorf tube and centrifuge in a microcentrifuge at 12,000 rpm for 15 minutes in the cold to collect membranes. Aspirate, resuspend the pellet in 1 ml cold Tris-Mg buffer, and centrifuge again at 12,000 rpm in the cold for 15 minutes. Aspirate and resuspend pellet in 1 ml cold Tris-Mg buffer. Distribute 50 μl of membrane suspension to glass tubes for incubation with ^{3}H-MeTRH and varying concentrations of unlabeled ligands. We have found that 1 100-mm dish provides sufficient material for 6–12 points in the case of a high affinity receptor and for 2–4 points in the case of lower affinity receptors.

2.C.c. Inositol Phosphate Assay. Incubate cells expressing TRH-Rs (in 12- or 24-well dishes) in growth medium containing 1 μCi myo-^{3}H-inositol/ml for at least 24 hours to radiolabel inositol-containing sugars and lipids. Wash cells three times with 1 ml HBSS, pH 7.4. Add 300 μl HBSS (pH 7.4)/10 mM LiCl. Li$^+$ inhibits inositol phosphatases and allows for accumulation of all inositol phosphates generated during the experimental incubation. After 5 minutes, add 3 μl 100× unlabeled ligand and incubate for 0.25 to 1 hour at 37°C. Terminate the incubation with 800 μl of methanol:water:concentrated HCl (100: 60:1). Transfer sample to a glass tube, and add 600 μl chloroform. Vortex and

centrifuge at low speed to separate phases. Inositol phosphates will be in the upper methanol–water phase and phosphoinositides in the lower chloroform phase. Transfer 200 μl from the lower phase to a scintillation vial, add 3 ml scintillation cocktail, and count in a β counter ("lipids"). Transfer 400 μl from the upper phase to a glass tube, and add 2 ml 5 mM inositol/0.1 M formic acid. Apply sample to a 1-ml Dowex anion-exchange column (AG 1×8 resin, Biorad) prewashed with 8 ml of 3 M ammonium formate/0.1 M formic acid and 8 ml water. Wash the column with 16 ml of 5 mM inositol/0.1 M formic acid to elute unphosphorylated inositol. Elute inositol phosphates with 3 ml of 1 M ammonium formate/0.1 M formic acid into scintillation vials, add 3 ml scintillation cocktail, and count in β counter ("inositol phosphates"). We routinely run 100 samples simultaneously.

An initial step in assessing the effect of an introduced mutation on TRH-R function is to study its effect on binding of ^{3}H-MeTRH. The first concentration chosen is generally 3 nM. If there is specific binding, a full competition binding assay is performed (saturation binding assays may also be performed). If no specific binding is observed with 3 nM ^{3}H-MeTRH, then the ^{3}H-MeTRH concentration is increased to 10 nM; concentrations higher than 10 nM result in unacceptably high levels of nonspecific binding. If no specific binding is detected with 10 nM ^{3}H-MeTRH, the mutant TRH-R is either poorly expressed or has a low affinity for the radioligand. An immunological approach is useful at this point to distinguish between these possibilities. Antibodies directed against epitopes in the extracellular domain of native receptors are useful but are usually difficult to generate. Another approach is to introduce an epitope for which an antibody is available within the extracellular domain of the receptor, usually within the amino terminus. Such epitope tags include FLAG, myc, and HA antigen. If such immunological tools are not available, as was the case for TRH-R (high affinity antibodies against the native TRH-R could not be generated, and epitope-tagged TRH-Rs were not recognized by their respective antibodies), a different approach must be employed. We used receptor signaling to attempt to distinguish whether a mutant TRH-R was poorly expressed or showed low ligand affinity. Dose–response curves for TRH-dependent stimulation of inositol phosphate formation were constructed. Due to its high solubility, doses as high as 1 mM TRH can be tested. An increase in TRH EC_{50} values indicates that the reason for the lack of radioligand binding is a reduction in ligand binding affinity. This method, however, does not distinguish between a receptor that is inactive and of low affinity versus a receptor that is poorly expressed on the cell surface. In either case, one may find no ligand binding and no activation. (More recently, we constructed TRH-Rs with extensions at their amino termini in which epitope tags are recognized by available antibodies.)

3. MODELING THE TRANSMEMBRANE BUNDLE OF TRH-R

A comprehensive analysis of GPCR sequences that addresses the position of conserved residues in an alignment of 200 sequences has been presented by Baldwin (1993). Based on the extent of lipid-exposed surface area at different "heights" of the protein, the analysis predicts the position, orientation, and tilting of helices in GPCRs, which agrees well with the rhodopsin footprint (Schertler et al., 1993). The proposed schematic structure of GPCRs has been shown to be consistent with a large number of mutagenesis studies, illustrating the essential validity of the proposed scheme (Baldwin, 1994). Systematic approaches to the construction of the TM domains of GPCRs have been reviewed (Ballesteros and Weinstein, 1995).

We have used Baldwin's analysis (1993) to construct a template for the helices of GPCRs that incorporates the general qualities of this superfamily of proteins. Such a template allows a systematic construction of GPCR models from realistic helices that are fitted to the template. In the past, models of many GPCRs have been constructed by homology to the crystal structure of bacteriorhodopsin. However, bacteriorhodopsin is not a GPCR, and the structure of its TM bundle is different from that of rhodopsin. Also, the sequence homology of many GPCRs with bacteriorhodopsin is low (Pardo et al., 1992). Our method does not depend on the assumption that certain sequences are homologous with a known structure. Rather, our method derives its universality from its generic template and is specific to a given sequence due to the construction of realistic, sequence-dependent helices. To test the method of template construction and of fitting the helices, we applied it to generate models of bacteriorhodopsin (whose structure is known) and of rhodopsin, for which a large collection of experimental data are available (see below). Utilizing this procedure, we have constructed a model of TRH-R that is being actively refined based on our experimental results (Perlman et al., 1994a,b, 1996; Laakkonen et al., 1996a).

3.A. Construction of a Set of Realistic Helices

The recognition that the TM bundle cannot be represented by ideal α helices derives from two observations. First, prolines in the TM helices are known to produce a kink because of the disruption of a hydrogen bond to the i-4 position. Second, the orientations of side chains in helices show special preferences that minimize steric clashes with the backbone. Thus, the construction of realistic helices takes advantage of the amino acid sequence both in the initial build up of the helix and in its final structure obtained by an energy minimization procedure.

In a first step, a helix of a given sequence is formed with representative side chain dihedrals. Then, the puckering of proline is set to *exo* because it is the dominant conformation in helices (Milner-White et al., 1992) and initial modeling of helices showed that the *endo* form caused sterical clashes between Pro C_δ and the backbone carbonyl three residues amino-terminal of it (X_{i-3}). Proline kinks are built in as defined by Sankararamakrishnan and Vishveshwara (1990) with the following backbone dihedrals (ϕ, ψ): $Pro_i(-57.2°, -43.9°)$, $X_{i-1}(-55.3° -50.6°)$, $X_{i-2}(-75.9°, -42.5°)$, $X_{i-3}(-68.9°, -37.9°)$. Bond angles that differ from standard values are $C_\alpha(X_{i-2}) -C(X_{i-2}) -N(X_{i-1}) = 119.0°$; $C(X_{i-2}) -N(X_{i-1}) -C_\alpha(X_{i-1}) = 122.2°$; and $N(X_{i-1}) -C_\alpha(X_{i-1}) -C(X_{i-1}) = 112.7°$. Finally, the constructed helices are energy minimized stepwise, first keeping the backbone frozen and then optimizing all degrees of freedom. This approach leads to stable helices and prevents the formation of inadvertent side chain–backbone H-bonds.

3.B. Forming the Templates

Baldwin's prediction of helix packing (1993) is expressed as three cross sections representing the intracellular, middle, and extracellular parts of a model GPCR (see Fig. 4 in Baldwin, 1993). The picture shows the positions of conserved residues in GPCRs and the extent of buried surface in each cross section. The superposition of the predicted slices matches well with the known rhodopsin footprint (Schertler et al., 1993). The cross sections were combined into a 3D model of helix axes in GPCRs by scaling to atomic distances and determining the vertical positions of the slices.

To derive a conversion factor between the 2D distances extracted from the cross sections and the 3D distances needed to construct the template, the cross sections (Baldwin, 1993) and the structure of bacteriorhodopsin (Henderson et al., 1990) were used. The center of helix 4 was defined as the x,y origin in every slice, and thus the z-direction coincides with the axis of helix 4. This agrees with the prediction that helix 4, which is the shortest of the TM helices, should be perpendicular to the membrane to span it entirely. Positions of helices in three layers (extracellular, midpoint, and intracellular) were defined.

To convert the cross sections presented in Baldwin's Figure 4 (1993) to molecular dimensions, the coordinates of bacteriorhodopsin (Henderson et al., 1990) were used to define a scaling factor. Interhelical distances, chosen to span the length and the width of the bundle (between helices 1–2, 1–5, 2–4, 3–5, 4–6, 6–7), were determined from the scanned data and from the bacteriorhodopsin structure. The distances in the crystal were measured between helical axes as determined by the program Dials_and_Windows (Sklenar et al., 1989; Swaminathan et al., 1990). The conversion factor 0.8913 ± 0.0059 cm/Å, measured between one-fifth, one-half, and four-fifths heights of the axis points in 2D, showed the smallest variance and was used to scale the data in the picture. The small variance of the conversion factor suggests that the picture is an accurate representation of bacteriorhodopsin structure.

3.C. Generic GPCR Model

The cross sections of the GPCR presented in Baldwin's analysis (1993) were scanned and measured similarly, and the data of helix axis positions were scaled by the conversion factor defined from bacteriorhodopsin data. A critical aspect of the GPCR model, which cannot be derived from the projected data presented by Baldwin (1993) or Schertler et al. (1993), is the vertical positions of the helices. We positioned the three planes 10 Å apart from each other because this is the average distance in bacteriorhodopsin from the middle to one-fifth or to four-fifths of the helix height. The centers of all helices could be placed at the same level, or the relative displacements from bacteriorhodopsin could be used. We chose to take the vertical displacements for our GPCR template from bacteriorhodopsin, but lowered helix 4 by 2.5 Å, as proposed by Tuffery et al. (1994). This is an arbitrary initial positioning, and the vertical displacements of the helices will be refined by experimental data on specific interhelical interactions. The completed GPCR template consists of 21 points, three per helix, and defines the helix axes in space. Realistic helices were fitted to the template. The helices were oriented according to the positions of conserved residues specified by Baldwin (1993). The conserved residues used for orientation were chosen from those in the middle height of the helices to minimize the effect of tilting on orientation.

The construction of receptor models on a template was implemented as a set of unix shell scripts and awk programs that use the program CHARMM (parameters 22b) (Brooks et al., 1983) for all coordinate manipulations and energy minimizations. The first script forms a helix of a given sequence with representative side chain dihedrals (see above). The second script combines the optimized helices into a bundle. It superimposes midpoints of helical axes obtained from Dials_and_Windows on those of the template and fits the axis points to the line that passes through the template. The third script rotates the helices around the lines fitted to their axes to angles defined for selected residues. Figure 4.2 shows a diagram of the stepwise procedure of fitting the helices to the template. To test the template and the construction method, we built a model of rhodopsin that agreed well with the structural aspects derived from biochemical and mutational experiments (Khorana, 1992). In particular, the important interaction in rhodopsin between Lys^{296} in helix 7 and Glu^{113} in helix 3 was observed.

3.D. Construction of a TRH-R Model

A molecular model of the TM bundle of TRH-R was constructed according to the procedure described above. No structural data on TRH-R are available, but there are inferences about the ligand binding site from mutational studies. These data were not used in construction of the model so that an unbiased starting structure for ligand docking was developed. Helix boundaries were defined from an alignment of 39 peptide receptors. Conserved residues were used (Baldwin, 1993) to align the sequences without insertions or deletions. The selected residues and the angles of their C_{α} with respect to the x-axis defined in the model are as follows (helix#–res#, angle): h1–Asn^{43}, $-31.09°$; h2–Asp^{71}, $113.15°$; h3–Ser^{112}, $144.51°$; h4–Trp^{150}, $158.56°$; h5–Pro^{203}, $-154.82°$; h6–Trp^{279}, $-64.71°$; h7–Asn^{316}, $-9.59°$. Sequences that contained Pro-X-Pro or X-Pro-Pro-X were not allowed inside the helices (MacArthur and Thornton, 1991), but prolines were allowed at amino termini. Lysines and arginines were allowed at intracellular termini of the helices (Ballesteros and Weinstein, 1992). The resulting helical boundaries are as follows: h1 (Leu^{30}–Arg^{52}); h2 (Pro^{59}–Thr^{84}); h3 (Cys^{100}–Arg^{123}); h4 (Arg^{141}–Trp^{160}); h5 (Pro^{190}–Phe^{213}); h6 (Gln^{263}–Val^{288}); h7 (Asn^{299}–Phe^{326}).

Helices formed from the chosen sequences were optimized, fitted to the GPCR template, and oriented according to the angles of the conserved residues (see Fig. 4.2). The entire bundle was optimized in CHARMM by 3,200 steps of adopted basis Newton-Raphson energy minimization. Several models of the environment represented as a continuum dielectric were tested. A constant dielectric of 1 or 4, or a distance-dependent dielectric function, did not change the positions of the helices significantly. It appears that the distance-dependent dielectric function served well in the optimization of the bundle.

The overall structure of the optimized model is similar to the template. In all minimization schemes, side chains moved to avoid close contacts and to

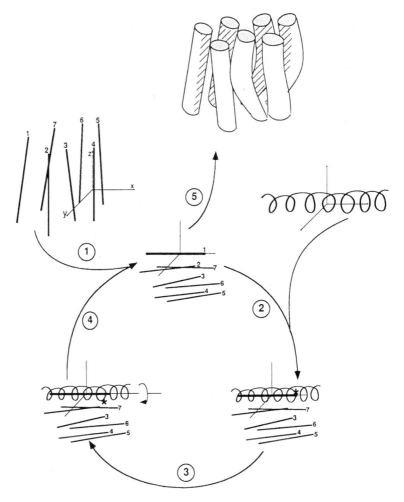

Figure 4.2. Construction of the computer-generated model of the transmembrane bundle of TRH-R. **1:** The template is oriented so that the guide of helix 1 coincides with the x-axis with the extracellular side in the positive direction. **2:** The optimized helix 1, oriented to lie on the x-axis with the extracellular side in the positive direction, is superimposed on the template. Their midpoints are made to coincide, and the line that fits best the axis points of helix 1 is matched to the line that passes through the guide for helix 1 in the template. **3:** Helix 1 is rotated around the x-axis so that the chosen conserved residue points in the desired direction. **4:** The entire template with helix 1 fitted is now oriented to position the guide for helix 2 on the x-axis. The cycle is repeated by adding (step 2), rotating the next helix (step 3), and reorienting the template (step 4) for the addition of each helix until the entire bundle is constructed. **5:** The finished bundle is rotated so that the guide of helix 4 coincides with the z-axis.

form new interactions. The tilts of the helices are similar to the template, and helices that are next to each other in the sequence are also close in space. In addition, helix 7 is close to the first three helices at all levels, and helix 3 comes close to helices 5, 6, and 7 at the bottom of the structure. In this arrangement helix 3 and helix 7 are surrounded by other helices, which is consistent with the high polar nature of their surfaces. The intracellular end of the helix bundle is more tightly packed than the extracellular one. However, the long axis of the bundle between helix 1 and helix 5 is approximately 30 Å at every level of the bundle.

The values of the interhelical angles between neighboring helices (closer than 13.5 Å) are in the range of 3.1° to 21.8° for parallel helices and −153.1° to 169.7° for antiparallel helices. This is in agreement with packing angles commonly seen in proteins (Chothia and Finkelstein, 1990). The angle between helix 4 and helix 5 is 179.5°, but the helices are not strictly antiparallel because helix 5 is strongly kinked. The angle measured between the extracellular halves of helix 4 and helix 5 is 10.4°, whereas the value between the intracellular portions is 1.9°. The latter seems to be less important because the distance between helix 4 and helix 5 at this level is 15.4 Å. In general, the variation in the interhelical angles shows that the receptor is not simply a collection of parallel and antiparallel helices.

A closer examination of the interhelical distances reveals an internal structure of the helices and highlights the difference between the real structure and the template. The template is a collection of straight lines that represent the helical axes, whereas the real helices are kinked. Helices 5, 6, and 7 are kinked at the prolines with axis kink angles of 166.2°, 164.1°, and 165.4°, respectively. The helices without prolines are approximately straight, with kink angles deviating minimally from straight lines: helix 1, 175.4°; helix 2, 172.8°; helix 3, 175.5°; helix 4, 178.0°. The internal structure of the helices can also be seen from the interhelical distances. Some distances, for example, between helix 2 and helix 5 and between helix 6 and helix 7, increase from the extracellular part to the middle and decrease again toward the intracellular side. Finally, it is worth noting that the orientation of the helices in the bundle did not change significantly during structure optimization. The analysis of the interhelical properties in the bundle indicates that the global structure of the TRH-R model is sound. Furthermore, stepwise heating to 300 K in 23 psec followed by molecular dynamics for 200 psec resulted in an energetically stable structure that maintained over 90% of the original helical hydrogen bonds. This model was tested experimentally and is in good agreement with results derived from mutational studies (see below).

4. DETERMINATION OF INTRAMOLECULAR INTERACTIONS IN TRH-R

The overall structures of all GPCRs are thought to be similar and consist of seven TM-spanning segments, an amino terminus, three intracellular loops (ICLs), three extracellular loops (ECLs), and a carboxy-terminal portion (Fig. 4.1). Thus, residues in the TRH-R that are conserved throughout the GPCR family may be considered as important in maintaining the conformation of TRH-R through interactions with other residues. Our model predicts that the TM region of TRH-R is arranged in a bundle with a central core that is maintained by interhelical interactions between several highly conserved amino acid residues. To provide evidence in support of this idea, we used an approach that compares the effects of double mutations of presumed interacting residues with the effects of single mutations of one of the two partner residues. The mutant TRH-Rs are characterized, and the intrinsic signaling activity, monitored as the level of inositol phosphate second-messenger formation stimulated by a maximally effective concentration of TRH, is estimated. The intrinsic signaling activity is used as an indication of the functional integrity of the mutant receptor, and it is assumed that there is a direct correlation between function and structure. In particular, substitutions that decrease signaling activity are thought to disrupt receptor conformation or prevent a conformational change that is essential for the formation of an activated receptor.

An example of this type of analysis is as follows (Perlman et al., 1997b). The highly conserved Asn^{43} in TM helix 1 (TM-1), Asp^{71} in TM-2, and Asn^{316} in TM-7 of TRH-R are predicted to interact with each other. These residues have been shown to be involved in controlling receptor function in several GPCRs. Specifically, our computer-generated model resulting from a 200 psec molecular dynamics simulation predicts that Asn^{43} interacts with Asp^{71}, which in turn contacts Asn^{316}, but that Asn^{43} and Asn^{316} do not interact directly. That is, Asp^{71} forms a bridge between Asn^{43} and Asn^{316} to constrain helices 1, 2, and 7 in a proximate arrangement of one to another. Each of these residues in the TRH-R was substituted by Ala. An Ala substitution is preferred as a first step in a mutational analysis because it eliminates all potential polar interactions and most hydrophobic interactions with minimum effects on the secondary structure of the α-helical TM domain. In contrast, a Gly substitution would remove all hydrophobic interactions but would potentially disrupt the α helix through increased flexibility. The maximal activity of N43A TRH-R was 37% that of wild-type TRH-R, and that of N316A TRH-R was 47% of wild-type TRH-R. There was no activity of D71A TRH-R, which, nevertheless, exhibited high binding affinity. The complete inactivity of D71A TRH-R was consistent with an additive effect of disrupting the bonds of Asp^{71} to both Asn^{43} and Asn^{316}. Importantly, the double-mutant N43A/N316A TRH-R was also completely inactive, which again was consistent with abolishing the bonds of Asp^{71} to each Asn. In addition, computer-generated models of the mutant D71A and of the double-mutant N43A/N316A TRH-Rs derived from 200 psec molecular dynamics simulations show similar conformational changes from native TRH-R and an excellent overlap of the two mutant receptors with a root mean square deviation between the C_α atoms of the helices of only 1.01 Å, while the root mean square deviation from wild type was 1.78 Å. Comparison of the structures of wild-type and D71A TRH-Rs shows that the helical bundle undergoes a significant change that is localized to the intracellular portion of the helices. The similarity of these changes for the mutant receptors and their difference from wild-type TRH-R suggests that a possible mechanism for loss of activity may be related to the rearrangement of helices. These rearrangements will induce conformational changes in the intracellular loops that are important for the interaction of the activated receptor with the G protein(s). The simulations therefore establish a proposed link between the disturbance of the helical bundle and the lack of coupling seen with the D71A and the N43A/N316A TRH-Rs.

To confirm these findings, polar substitutions were made to attempt to restore native interaction. The maximal stimulations of inositol phosphate formation with D71N, N316D, and D71N/N316D TRH-Rs were 52%, 98%, and 115% that of wild-type TRH-R. Thus, the additional substitution of Asn^{316} by Asp in D71N TRH-R, which re-establishes an Asp–Asn pair, results in restoration of activity, providing further evidence of an interaction between Asp^{71} and Asn^{316}.

A different type of analysis of complementary Ala mutations was used to show that there is a disulfide bond between Cys residues in ECL-1 and ECL-2 of TRH-R (Perlman et al., 1995). Disulfide bonds between highly conserved Cys residues in these ECLs have been shown to be present in several GPCRs and are thought necessary to maintain native receptor structure. In this analysis, an assumption is made that relative potencies of activation can be used to estimate relative affinities of binding in receptors of similar maximal activities (or efficacies) (Limbird, 1986). This assumption appears valid because we have shown that, in general, affinities parallel potencies in mutant TRH-Rs that signal with the same activity as wild-type TRH-Rs when both could be measured. The following analysis was performed on the two highly conserved Cys residues in ECL-1 (Cys^{98}) and ECL-2 (Cys^{179}) of TRH-R. C98A, C179A, and C98A/C179A TRH-Rs are maximally stimulated to form inositol phosphate second messengers to the same level as wild-type TRH-Rs but do not bind ^{3}H-MeTRH with high affinity. C98A TRH-R and C179A TRH-R exhibited EC_{50}s for inositol phosphate formation that were 4,400-fold and 640-fold higher than that of wild-type TRH-R, respectively. The EC_{50} for inositol phosphate formation of the double-mutant C98A/C179A TRH-R was only 5,600-fold higher than that of wild-type TRH-R. Additivity of effects in this case would be an increase in EC_{50} of 2.8 million-fold compared with wild-type TRH-R (4,400- $\times$ 640-fold). Therefore, the effect of mutating both Cys^{98} and Cys^{179} on the potency of activation is clearly not additive with the effect of the individual mutations. (We have found a number of TRH-Rs in which the effect of double mutations was additive, or more than additive, compared with the individual mutations [Gershengorn and Osman, 1996].) These findings indicate that Cys^{98} and Cys^{179} interact with each other.

The interpretation that Cys^{98} and Cys^{179} interact was corroborated by studying the effect of substituting Ser for Cys using the same approach that was employed for determining the interactions among Asn^{43}, Asp^{71}, and Asn^{316} (see above). Substituting Cys with Ser led to TRH-R mutants, C98S and C179S TRH-Rs, that exhibited decreased signal activities, whereas the double-mutant, C98S/C179S TRH-R exhibited signaling activity similar to wild-type TRH-R. The maximal signaling activities of C98S, C179S, and C98S/C179S TRH-Rs were 64%, 53%, and 97% that of wild-type TRH-R. The restoration of signaling activity in the double-mutant receptor to a level similar to wild-type TRH-R was most likely due to an interaction between the Ser residues at position 98 and position 179 because Ser residues can form a hydrogen bond that could substitute in a functional sense for the disulfide bond present in wild-type TRH-R. Further support for this interpretation came from experiments using the reducing agent dithiothreitol. In the presence of dithiothreitol, the potency of wild-type TRH-R was diminished whereas the potencies of C98A and C179A TRH-Rs were not. This disulfide bridge constituted one of the constraints employed in modeling the ECLs of the receptor (see below).

5. DETERMINATION OF THE BINDING POCKET OF TRH-R

While conserved residues appear to be critical for maintaining the common structure of GPCRs, it is reasonable to hypothesize that nonconserved residues are critical for the high affinity and specificity with which GPCRs bind their ligands. It is common to find residues whose mutation will result in loss of high affinity binding by a GPCR. Much more problematic is the determination of whether this affinity decrease is due to loss of a direct binding interaction with ligand or rather to an indirect effect on the binding pocket. An example of the latter would be C98A TRH-R (see above) in which a disulfide bond necessary for maintaining TRH-R in a high affinity conformation has been lost. The result is a distant effect on the binding pocket that lowers binding affinity. Cys^{98} does not directly interact with TRH. To address this issue, substitutions were made not only in TRH-R but in TRH as well. An initial direct interaction between the pyroGlu residue of TRH and Tyr^{106} in TM-3 was identified experimentally using the approach outlined below (Perlman et al., 1994b) and was used to begin construction of the computer-simulated model of the TRH/TRH-R complex (see below). The predictions of the model were then tested.

We have suggested that the binding pocket of TRH includes four residues that lie within the upper half of the transmembrane core. The four residues that were identified in our laboratory as part of the TRH binding pocket are Tyr^{106} and Asn^{110} in TM-3, Tyr^{282} in TM-6, and Arg^{306} in TM-7; they are predicted to form specific interactions with TRH (Fig. 4.3). As an example of how we test the predictions of our model of the binding pocket, we describe the analysis performed to determine whether there is an interaction between the His of TRH and Tyr^{282} in TM-6 (Perlman et al., 1996). We found that Y282A TRH-R does not bind ^{3}H-MeTRH with high affinity and that the EC_{50} of TRH for activation of Y282A TRH-R was 94,000-fold higher than that of wild-type TRH-R. Thus, we showed that Tyr^{282} was critical for binding. We tested various analogs of TRH for activation of wild-type and Y282A TRH-Rs. An analog of TRH in which His had been substituted by Val (Val^{2}TRH) was shown to exhibit an EC_{50} that was 960-fold higher than TRH at the wild-type TRH-R. Thus, we confirmed earlier work that His of TRH is important for binding. However, whereas the EC_{50} of Val^{2}TRH is 960-fold higher than that of TRH for wild-type TRH-R, it is only 24-fold higher than that of TRH at Y282A TRH-R. Thus, Y282A TRH-R has lost most of its ability to distinguish between His and Val at position 2 of TRH. This strongly suggests that Tyr^{282} recognizes, or binds, His of TRH.

More subtle receptor substitutions can be used to further define the nature of these interactions. As an example, Tyr^{282} was changed to Phe, which results in the retention of the aromatic ring and loss of the phenolic group. The TRH affinity of Y282F TRH-R was only ninefold lower than that of wild-type TRH-R, indicating, in conjunction with the Y282A TRH-R data, that it is the aromatic ring that is critical for high affinity binding of TRH. This was confirmed by constructing and testing Y282S and Y282N TRH-Rs, whose EC_{50}s were 57,000- and 170,000-fold higher, respectively, than wild-type TRH-R. Further insight was gained by comparing TRH and Val^{2}TRH binding by Y282F TRH-R; there was no loss of selectivity compared with wild-type TRH-R. Thus, we deduce that the

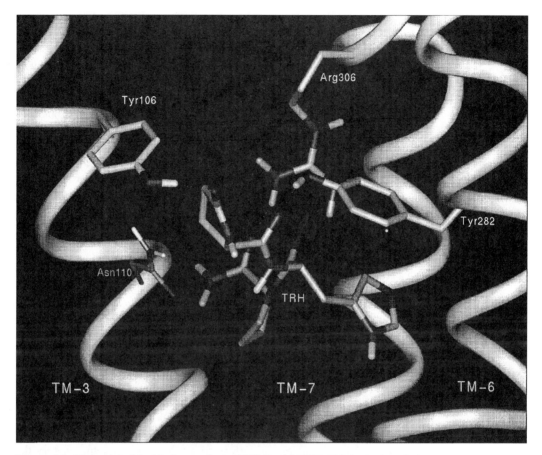

Figure 4.3. 3D model of the binding pocket of TRH-R with TRH. TM-1, -2, -4, and -5 have been omitted for clarity. The model is oriented so that the extracellular side is at the top. This model was constructed according to the guidelines described in the text.

ninefold increase in EC_{50} attributed to the phenolic group of Tyr^{282} is not due to an interaction with His of TRH. Similarly, refinements can be made by further alterations of the ligand. This method was used to confirm two other interactions between TRH-R and TRH: Asn^{110} with the ring N–H of the pyroGlu residue of TRH (Perlman et al., 1994a) and Arg^{306} with the terminal carboxamide of TRH (Perlman et al., 1996). The ability to define experimentally not only which residues of a GPCR are important for binding but also whether they are involved in direct binding of the ligand, and the manner in which they do so, is of critical importance in the characterization of a binding pocket.

6. MODELING THE TRANSMEMBRANE BINDING POCKET OF TRH-R

TRH was placed manually inside the helix bundle described above, and the side chains of residues in the ligand and the receptor were rotated manually to bring the interacting residues to close proximity. The pyroGlu in the ligand was placed to interact with Tyr^{106} (Perlman et al., 1994b) and Asn^{110} (Perlman et al.,

1994a) in the receptor. Tyr[282] in the receptor forms a stacking interaction with His of TRH, and Arg[306] forms H-bonds with the ProNH$_2$ and the backbone carbonyl of pyroGlu. Such an arrangement of TRH in the binding pocket should be considered as an initial structure only because the major interactions were designed manually and were not optimized by objective criteria. However, the design of an optimization method has to take into account the flexibility of TRH and the residues that define the binding pocket. To address the problem of the interaction of a flexible ligand, TRH, with a binding pocket in TRH-R that might possess its own conformational flexibility, we decided to utilize a novel technique that combines stochastic dynamics with intervening Monte Carlo steps, a mixed mode Monte Carlo/stochastic dynamics simulation (Guarnieri and Still, 1994; Guarnieri and Wilson, 1995). The critical aspect of this simulation method is its ability to explore simultaneously many thermally accessible conformational states of the receptor and the ligand. Attempts to enhance sampling have to take into account that intramolecular motions in large systems take place on a multiple time scale because bond stretching and bending occur on a femtosecond to picosecond time scale, whereas torsional rotations around flexible bonds occur on a nanosecond time scale. Thus, an efficient algorithm should be able to sample local wells as well as cross large energy barriers and yield a correct population distribution.

On the basis of the initial model constructed with standard energy minimization techniques, 15 mixed mode Monte Carlo/stochastic dynamics simulations were conducted to allow for extended sampling of the conformational states of the TRH/TRH-R complex (Laakkonen et al., 1996a). Because the entire system contains too many torsional angles to be varied in the Monte Carlo part of the mixed mode simulation, a specific portion of the system was defined to be a Monte Carlo–active zone. Thus, in the mixed mode simulation, the stochastic dynamics steps are performed on the entire system, whereas the Monte Carlo steps are performed only on the Monte Carlo–active zone. The active zone was chosen to include all nonpeptide torsional angles of TRH and all side-chain torsional angles of residues within TRH-R that had at least one atom within a sphere of 6 Å from any atom of the ligand (i.e., 46 residues with 97 free torsional angles). The simulation protocol consisted of heating the receptor–ligand complex to 600° K by steps of 50° K with 2 psec of stochastic dynamics at each temperature. All simulated annealings were then started at 600° K and cooled to 310° K in four steps of 100° K each. At each temperature, the system was simulated for 30 psec with a coupling constant of 0.4 psec. The overall time of each simulation was 120 psec. The elevated temperature (600° K) from which the simulated annealing protocol started allowed for significant fluctuations in the torsional angles of TRH and the residues of the Monte Carlo–active zone. This ensured good sampling of the conformational space of the complex. In addition, the multiple simulated annealing runs sampled the conformational space in an exhaustive way and provided a representative distribution of structures with successful conservation of the overall structure and maintenance of good helicity of the receptor.

To gain a molecular understanding of the selectivity of the receptor binding pocket for TRH, we calculated the energies of interaction between TRH and the individual residues in the TRH-R pocket. For this analysis, TRH was divided into four groups: three consisted of each side chain that included the C_α atom, and the fourth group consisted of the backbone atoms (i.e., two carbonyl groups and one N–H). This divided the TRH molecule in an analogous way to the experimental testing of ligand–receptor interactions. However, because the role of the backbone is more difficult to analyze experimentally, this partitioning served to predict the importance of the backbone for ligand–receptor interaction. The pairwise interaction energies of the four separate portions of TRH with the corresponding residues in the receptor provided a physicochemical basis for understanding the ligand–receptor complex. The interaction of pyroGlu with Tyr[106] showed a characteristic H-bonding pattern. However, a bimodal distribution of interaction energies was observed. This indicated that one of the populations was H-bonded whereas the other was not. Asp[195] was shown to compete with pyroGlu for the H-bond to Tyr[106]. Simulations in which Asp[195] was interacting with Arg[283], thus removing it from the vicinity of Tyr[106], resulted in a stable H-bond to pyroGlu. In all simulations, the imidazole of His showed a van der Waals attraction to Tyr[282] and a weak electrostatic repulsion from Arg[306]. This is consistent with the stacking of His and Tyr[282] observed in the simulations. The ProNH$_2$ had a strong and frequent H-bonding interaction with Arg[306]. These results are fully consistent with the mutational analysis of TRH analog binding (Gershengorn and Osman, 1996). The backbone carbonyls showed a frequent H-bonding interaction with the OH group of Tyr[282] and strong, often multiple, interactions with Arg[306]. The backbone of TRH inside the receptor exhibited an α-helical conformation, suggesting that the receptor, through its interaction with the ligand, provides the energy required for a conformational change in TRH from an extended to the folded form (see below).

7. MODELING THE EXTRACELLULAR LOOPS OF TRH-R

In spite of the rapidly increasing number of molecular models of GPCRs reported, only a few models have included ECLs (Dahl et al., 1991; Maloneyhuss and Lybrand, 1992; Findlay and Eliopoulos, 1990). However, these models did not critically address the difficulties involved in the construction of the loops in the absence of structural guidelines. Consequently, the construction of the ECLs was not described in detail, and the loops in the models were not tested experimentally.

The three putative ECLs of TRH-R shown in Figure 4.4 were constructed on the model of the TM bundle as four fragments: Asp[85] to Leu[99], Phe-[161] to Cys[179], Gly[180] to Ser[189], and Asn[289] to Glu[298], so that the disulfide bond between the conserved cysteines of ECL-1 and -2 (i.e., Cys[98] and Cys[179]; see above), was maintained throughout all simulations (Colson et al., 1998). The loops were then attached at one end to their respective target helix via a *trans* peptide bond. The backbone φ and ψ torsion angles of the loops were manually rotated so that the free end of each loop came into proximity with the other respective target helix. Adopted Basis Newton Raphson (ABNR) minimization was performed using CHARMM23 (Brooks et al., 1983) on the initial structure of the loops to establish a proper peptide bond between the carboxyl end of the loop and the amino terminus of the helix. Dihedral constraints were applied to the ω angles

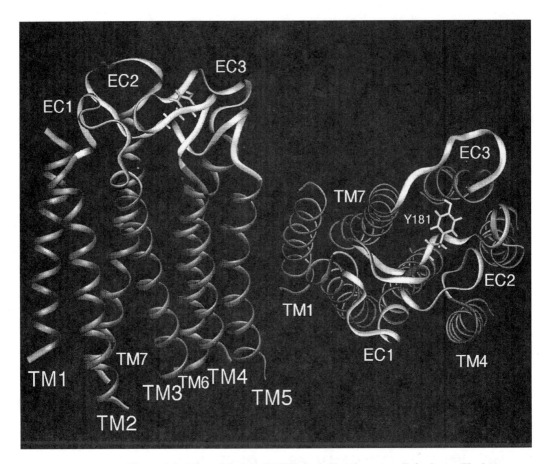

Figure 4.4. 3D model of the TRH-R, including the extracellular loops. The loops were constructed according to the guidelines described in the text. In the left panel, the model is oriented so that the extracellular side is at the top of the figure. A top view of this model is shown in the panel to the right.

of the loop–helix junctions to maintain the peptide bond in a *trans* orientation. Once appropriate peptide bond lengths were obtained at these junctions, the unconstrained loops were minimized for 1,000 steps, keeping the helical bundle frozen. The SHAKE algorithm was employed to fix all bonds to hydrogen atoms, and the environment was represented by a distance-dependent dielectric function.

For simulated annealings and molecular dynamics simulations, the minimized structure was heated to 1,500° K in 14 psec followed by 7 psec of constant temperature molecular dynamics simulation at 1,500° K. Fourteen structures were extracted from the trajectory at 1,500° K by sampling every 0.5 psec. Each structure was cooled down to 300° K in 60 psec and subsequently subjected to 100 psec of constant temperature molecular dynamics simulation at 300° K. Fourteen energy-minimized averaged structures were obtained over the stabilized portion of each trajectory and clustered according to their pairwise root mean square deviations into conformational families employing the program Xcluster (Shenkin and McDonald, 1994). At a level of 1.89 Å resolution, the cluster analysis reveals the existence of one major and six minor families. The major family is composed of seven members, suggesting that 50% of the structures have a common fold of the ECLs. Overall, ECL-1 is always positioned over helix 2 and helix 3 and packs against the first part (up to Cys[179]) of ECL-2. The first part of ECL-2 consistently starts with a short loop at its amino terminus and moves toward helix 1 or helix 7 to maintain the disulfide bridge at the carboxy terminus of ECL-1. The second part of ECL-2 lies below the first part and thus is closer to the helical bundle. It spans the area between helices 3, 5, 6, and 7. Consequently, it is positioned above the TM binding pocket. The backbone of ECL-3 is U shaped, but the side chains form close contacts with those of the second part of ECL-2 that occlude access to the TM binding pocket (Fig. 4.4).

The arrangement of the ECLs described above generates a cavity that includes Tyr[181], Arg[185], Asn[186], Tyr[187], Asn[289], Ser[290], and Phe[296]. The bottom of the cavity is formed by Lys[182], and movement of this residue exposes the TM binding pocket. Such an arrangement is consistent with the idea that TRH initially interacts with residues on the extracellular domain of the receptor and then moves into the TM binding pocket by inducing a conformational change that "opens" an entry channel. To test this hypothesis, mutations of selected residues were made (Perlman et al., 1997a). Binding and activation experiments revealed that mutation of Tyr[181] in ECL-2 (Fig. 4.4) lowers binding affinity and that Tyr[181] interacts with the pyroGlu moiety of TRH. Our model also predicts that the pyroGlu of TRH cannot simultaneously interact with residues in the TM helices and in the ECLs. Kinetic analyses of binding were consistent with the idea that there are initial interactions between TRH and the residues in the ECLs that form a putative entry channel into the TM binding pocket of TRH-R. These findings suggest a role for the ECLs in all GPCRs for small ligands that involves initially contacting the ligand and allowing entry into a TM binding pocket. This may be analogous to the interactions between larger peptide ligands and the ECLs of their GPCRs (see below).

8. BIOLOGICALLY ACTIVE CONFORMATION OF TRH

Our model of the TRH/TRH-R complex predicts not only the structure of the TM ligand binding pocket but also the bound conformation of the ligand. The conformation of free TRH has been determined in solution and by crystallography and consists of an extended peptide backbone with a preference for a *trans* peptide bond between His and Pro. The bound or active conformation of TRH has not been experimentally determined. The use of conformationally restricted analogs of TRH has proven to be of great benefit in addressing this issue. (A description of the chemical syntheses of these compounds is beyond the scope of this chapter. These analogs were synthesized by Dr. Kevin D. Moeller and colleagues of the Department of Chemistry, Washington University, St. Louis, Missouri [Laakkonen et al., 1996b; Li and Moeller, 1996].) To provide initial insight into the active structure of TRH, cyclohexylAla²TRH, in which cyclohexylAla is substituted for His, was synthesized. This substitution was needed to synthesize the restricted analogs. [Experiments are underway to obtain locked analogs with His at position 2 of TRH.] A methylene bridge connects the C_β of His to the C_δ of ProNH$_2$ and results in two stereoisomers, αCHTRH ([6S,9S,12S]-1-Aza-3-aminopyroglutamyl-4-cyclohexyl-9-carboxamide-2-oxo-bicyclo[4.3.0]non-2-ene) and βCHTRH ([6R,9S,12S]-1-Aza-3-aminopyroglutamyl-cyclohexyl-9-carboxamide-2-oxo-bicyclo[4.3.0]non-2-ene), that differ at the C_δ of the proline ring (Fig. 4.5). The unrestricted cyclohexylAla^{2}TRH had a 650-fold lower affinity than TRH, confirming the importance of the His of TRH in binding. Importantly, modeling of this analog showed that the position of pyroGlu relative to ProNH$_2$ was not different from TRH. Compared with unrestricted cyclohexylAla^{2}TRH, the affinity of αCHTRH was 45-fold *lower,* whereas, in contrast, the affinity of βCHTRH was 3.4-fold *higher.* The increased affinity of βCHTRH was observed despite the addition of a methylene bridge and a fused ring, which would be expected to inhibit binding through steric bulk. Thus, we conclude that the constraint in βCHTRH selects a preferred conformation for interaction with TRH-R. The levels of maximal stimulation of wild-type TRH-R signaling by cyclohexylAla^{2}TRH, βCHTRH, and αCHTRH were similar to that of TRH, and the changes in potency of these analogs paralleled the changes in affinity.

The conformations of αCHTRH and βCHTRH were investigated using the biased sample Monte Carlo method (Guarnieri and Wilson, 1995) and were then compared with that of TRH, which was simulated free in solution and in the binding pocket of TRH-R (see above). The structures of αCHTRH and βCHTRH were built in MacroModel (Mohamadi et al., 1990) and optimized with the AMBER force field (Weiner and Kollman, 1981; Weiner et al., 1984; McDonald and Still, 1992). The molecules were first subjected to Monte Carlo–simulated annealing (Kirkpatrick et al., 1983) in MacroModel with the Generalized Born/Surface Area solvation model (Still et al., 1990). Mean field population distributions of torsional angles collected from the simulated annealings were used to construct a mapping of the random numbers into tor-

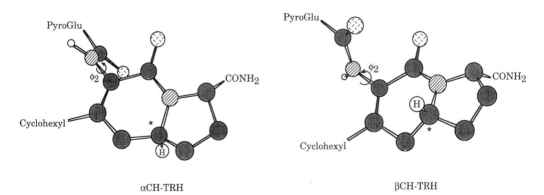

Figure 4.5. Structures of two conformationally restricted analogs of TRH - αCH-TRH and βCH-TRH. ·, Chiral carbon; , ▨ oxygen; , ▧ carbon; , ▨ nitrogen. PyroGlu, pyroglutamic acid. The major conformational difference between these analogs resides in the value of φ2.

sional populations (Conformational Memories) so that unpopulated areas are excluded from sampling (Guarnieri and Wilson, 1995). Each simulated annealing consisted of 16 cooling cycles that started at 800° K and decreased in 10 steps to 310° K, with a temperature protocol of Tn + 1 = 0.9·Tn. Fifty thousand Monte Carlo steps were computed at each temperature, resulting in a total of $8 \cdot 10^6$ steps for each simulated annealing. The variables in the simulations were the four free torsional angles. Two angles were changed simultaneously at each step by a random value between −180° and 180°. The bond lengths and bond angles were kept constant.

In the second part of the simulation, the Conformational Memories guided the collection of 1,000 structures from a Monte Carlo simulation of 500,000 steps at 310° K. The structures were clustered according to their pairwise root mean square differences with the use of the program Xcluster for structural comparisons (Shenkin and McDonald, 1994). Superposition of the models of TRH and βCHTRH shows that the prevalent conformations in solution are very different, not only in the orientation of their backbones but also in the positioning of their terminal carboxamides. The models of TRH in solution and bound to wild-type TRH-R are different also in that the bound TRH is not extended but rather takes on an α-helical turn. In contrast, the models of βCHTRH, which is the biologically more active conformer, in solution and TRH bound to TRH-R are superimposable. This strongly suggests that TRH in a model of the bound complex represents a biologically active conformation. The solution model of αCHTRH, the less biologically active conformer, and that of bound TRH do not superimpose well. Thus, this method can provide important independent verification of the modeling of a receptor–ligand complex and information as to whether the complex represents an active state.

The fact that bound TRH is not extended like its free form in solution raises the question whether the receptor recognizes TRH in its extended form and in the process of binding induces the conformational change or if it recognizes and binds the folded form, which may be a minor population in solution. One way of addressing this question is by estimating the energy difference between the extended and the folded forms. In an aqueous environment represented in MacroModel, the energy of the receptor-bound folded form is 52 kJ/mol higher than the extended form. Such an energetic difference would make the population of the folded form in aqueous solution too low to play an important role in recognition by the receptor. We conclude, therefore, that the receptor initially recognizes the extended form of TRH and subsequently induces the conformational change in the structure. Such a behavior would be consistent with the observation that the binding of TRH to its receptor exhibits biphasic kinetics (Hinkle and Kinsella, 1982).

9. PEPTIDE LIGANDS AND OTHER GPCRs

The approaches we have described above for the TRH-R and other methods have also been used to study the topology and binding pockets of other GPCRs. Proximity of TM-2 and TM-7 in rhodopsin (Rao et al., 1994) and in the gonadotropin-releasing hormone receptor (Zhou et al., 1994) were shown using site-specific mutations. Chimeric constructs have been useful in the study of the topology of GPCRs. Restoration of function by substitution of residues in chimeras have indicated proximity of TM-1 and TM-7 in the muscarinic (Liu et al., 1995) and adrenergic (Suryanarayana et al., 1992) receptors. The introduction of two His residues at the extracellular ends of TM-5 and TM-6 in the neurokinin NK_1 receptor converted an antagonist binding site into a high affinity site for Zn^{2+} ions, indicating proximity of TM-5 and TM-6 in this receptor (Elling et al., 1995). Introduction of two His residues at the cytoplasmic ends of TM-3 and TM-6 in rhodopsin inhibited activation and showed that these TMs are proximate (Sheikh et al., 1996). Cysteine substitutions into positions in the core of the GPCR bundle are predicted to be accessible to small, polar, sulfhydryl-specific reagents (see Chapter 2). Based on this prediction, the D2 dopamine receptor was studied with charged derivatives of methanethiosulfonate, and it was found that only mutation of a Cys in TM-3 could attenuate the inhibition of binding caused by methanethiosulfonate reagents. This allowed this Cys to be assigned to the binding crevice of the receptor (Javitch et al., 1994). Fluorescently labeled analogs, which can be quenched by hydrophobicity or protonation, have been used to characterize the environment of the binding pocket in the formyl peptide (Sklar et al., 1990) and the NK_1 receptors (Tota et al., 1994).

As noted above, peptides larger than TRH bind at least in part to the extracellular domains. The experimental and computational approaches we have outlined for the study of TRH-R are applicable independent of whether binding occurs in the TMs or the extracellular domains, or both. For example, a Glu in ECL-3 of the gonadotropin-releasing hormone receptor (GnRH-R) was shown to directly bind GnRH based on a comparison of GnRH analog affinities to both wild-type and mutant GnRH-Rs (Flanagan et al., 1994). The high affinity binding of glycoprotein hormones, such as luteinizing and thyroid-stimulating hormone, appears to be entirely attributable to interactions with the amino termini of their receptors. For example, the extracellular amino terminal domain alone of the luteinizing hormone receptor when expressed in human kidney 293 cells (in which it is not secreted but rather trapped) was found to allow high affinity binding (Xie et al., 1990). The localization of high affinity sites in the extracellular amino terminus raises the possibility of expressing the binding domain in a soluble form that would allow its structure to be resolved by crystallography or nuclear magnetic resonance spectroscopy. It is noteworthy, therefore, that the thyroid-stimulating hormone receptor is found on the cell surface as a heterodimer composed of the amino terminus and TM subunits linked by disulfide bond(s) and that there is spontaneous shedding of the amino terminal subunit by human thyroid cells as well as stably transfected cell lines (Couet et al., 1996).

10. CONCLUSIONS

Using an approach that integrates experimental and computational methods in an iterative fashion, we have modeled the 3D structure of the unliganded and TRH-occupied TRH-R and used these models to generate hypotheses regarding the structure–activity relationships of TRH and TRH-R. In particular, we have identified the TRH binding pocket within the TM bundle of TRH-R, delineated a conformation of TRH that appears to be biologically active, and uncovered a role for the ECLs of TRH-R in attracting TRH to the receptor and guiding TRH into the TM binding pocket. (The ECLs appear to be part of the binding pocket of GPCRs for larger peptides but may have this additional role also.)

Evidence in support of our model was acquired by studying the effects of TRH analogs and amino acid substitutions in TRH-R that has allowed us to address the critical issue of whether mutations in the receptor affect overall receptor structure or directly affect specific interactions with the ligand. This combined experimental and computational approach will be used in the future to delineate the molecular details of TRH-R activation and TRH-R coupling to G proteins.

REFERENCES

Baldwin JM (1993): The probable arrangement of the helices in G protein–coupled receptors. EMBO J 12:1693–1703.

Baldwin JM (1994): Structure and function of receptors coupled to G proteins. Curr Opin Cell Biol 6:180–190.

Ballesteros JA, Weinstein H (1992): Analysis and refinement of criteria for predicting structure and relative orientations of transmembranal helical domains. Biophys J 62:107–109.

Ballesteros JA, Weinstein H (1995): Integrated methods for the construction of three-dimensional models and computational probing of structure–function relations of G protein–coupled receptors. Methods Neurosci 25:366–428.

Brooks BR, Bruccoleri RE, Olafson BD, States DJ, Swaminathan S, Karplus M (1983): CHARMM: A program for macromolecular energy minimization and dynamics calculations. J Comp Chem 4:187–217.

Chothia C, Finkelstein AV (1990): The classification and origins of protein folding patterns. Annu Rev Biochem 59:1007–1039.

Colson AO, Perlman JH, Smolyar A, Gershengorn MC, Osman R (1998): Static and dynamic roles of extracellular loops in G-protein–coupled receptors: A mechanism for sequential binding of thyrotropin-releasing hormone to its receptor. Biophys J 74:1087–1100.

Couet J, Sar S, Jolivet A, Hai MTV, Milgrom E, Misrahi M (1996): Shedding of human thyrotropin receptor ectodomain—Involvement of a matrix metalloprotease. J Biol Chem 271:4545–4552.

Cullen BR (1987): Use of eukaryotic expression technology in the functional analysis of cloned genes. Methods Enzymol 152:684–704.

Dahl SG, Edvardsen O, Sylte I (1991): Molecular modeling of antipsychotic drugs and G protein coupled receptors. Therapie 46:453–459.

Elling CE, Nielsen SM, Schwartz TW (1995): Conversion of antagonist-binding site to metal-ion site in the tachykinin NK-1 receptor. Nature 374:74–77.

Findlay J, Eliopoulos E (1990): Three-dimensional modelling of G protein–linked receptors. TIPS 11:492–499.

Flanagan CA, Becker II, Davidson JS, Wakefield IK, Zhou W, Sealfon SC, Millar RP (1994): Glutamate 301 of the mouse gonadotropin-releasing hormone receptor confers specificity for arginine 8 of mammalian gonadotropin-releasing hormone. J Biol Chem 269:22636–22641.

Gershengorn MC, Osman R (1996): Molecular and cellular biology of thyrotropin-releasing hormone (TRH) receptors. Physiol Rev 76:175–191.

Guarnieri F, Still WC (1994): A rapidly convergent simulation method: Mixed Monte Carlo/Stochastic Dynamics. J Comp Chem 15:1302–1310.

Guarnieri F, Wilson SR (1995): Conformational memories and a simulated annealing program that learns: Application to LTB4. J Comp Chem 16:648–653.

Henderson R, Baldwin JM, Ceska TA, Zemlin F, Beckmann E, Downing KH (1990): Model for the structure of bacteriorhodopsin based on high-resolution electron cryo-microscopy. J Mol Biol 213:899–929.

Hinkle PM, Kinsella PA (1982): Rapid temperature-dependent transformation of the thyrotropin-releasing hormone–receptor complex in rat pituitary tumor cells. J Biol Chem 257:5462–5470.

Javitch JA, Li X, Kaback J, Karlin A (1994): A cysteine residue in the third membrane-spanning segment of the human D_2 dopamine receptor is exposed in the binding-site crevice. Proc Natl Acad Sci USA 91:10355–10359.

Khorana HG (1992): Rhodopsin, photoreceptor of the rod cell. An emerging pattern for structure and function. J Biol Chem 267:1–4.

Kirkpatrick S, Gelatt CD Jr, Vecchi MP (1983): Optimization by simulated annealing. Science 220:671–680.

Laakkonen L, Guarnieri F, Perlman JH, Gershengorn MC, Osman R (1996a): A refined model of the thyrotropin-releasing hormone (TRH) binding pocket. Novel mixed mode MonteCarlo/Stochastic Dynamics simulations of the complex between TRH and TRH receptor. Biochemistry 35:7651–7663.

Laakkonen L, Li WH, Perlman JH, Guarnieri F, Osman R, Moeller KD, Gershengorn MC (1996b): Restricted analogues provide evidence of a biologically active conformation of thyrotropin-releasing hormone. Mol Pharmacol 49:1092–1096.

Li W, Moeller KD (1996): Conformationally restricted TRH analogs: The compatibility of a 6,5-bicyclic lactam-based mimetic with binding to TRH-R. J Am Chem Soc 118:10106–10112.

Limbird LE (1986): Cell Surface Receptors: A Short Course on Theory and Methods. Boston: Martinus Nijhoff.

Liu J, Schöneberg T, Van Rhee M, Wess J (1995): Mutational analysis of the relative orientation of transmembrane helices I and VII in G protein–coupled receptors. J Biol Chem 270:19532–19539.

MacArthur MW, Thornton JM (1991): Influence of proline residues on protein conformation. J Mol Biol 218:397–412.

Maloneyhuss K, Lybrand TP (1992): Three-dimensional structure for the β_2 adrenergic receptor protein based on computer modeling studies. J Mol Biol 225:859–871.

McDonald QD, Still WC (1992): AMBER* Torsional parameters for the peptide backbone. Tetrahedron Lett 33:7743–7746.

Milner-White EJ, Bell LH, MacCallum PH (1992): Pyrrolidine ring puckering in cis- and trans-proline residues in proteins and polypeptides. Different puckers are favored in certain situations. J Mol Biol 228:725–734.

Mohamadi F, Richards NGJ, Guida WC, Liskamp R, Lipton M, Caufield C, Chang G, Hendrickson T, Still WC (1990): MacroModel—An integrated software system for modeling organic and bioorganic molecules using molecular mechanics. J Comp Chem 11:440–467.

Pardo L, Ballesteros JA, Osman R, Weinstein H (1992): On the use of the transmembrane domain of bacteriorhodopsin as a template for modeling the three-dimensional structure of guanine nucleotide–binding regulatory protein–coupled receptors. Proc Natl Acad Sci USA 89:4009–4012.

Perlman JH, Colson A-O, Jain R, Czyzewski B, Cohen LA, Osman R, Gershengorn MC (1997a): Role of the extracellular loops of the thyrotropin-releasing hormone receptor: Evidence for an initial interaction with thyrotropin-releasing hormone. Biochemistry 36:15670–15676.

Perlman JH, Colson A-O, Wang W, Bence K, Osman R, Gershengorn MC (1997b): Interactions between conserved residues in transmembrane helices 1, 2 and 7 of the thyrotropin-releasing hormone receptor. J Biol Chem 272:11937–11942.

Perlman JH, Laakkonen LJ, Guarnieri F, Osman R, Gershengorn MC (1996): A refined model of the thyrotropin-releasing hormone (TRH) receptor binding pocket. Experimental analysis and energy minimization of the complex between TRH and TRH receptor. Biochemistry 35:7643–7650.

Perlman JH, Laakkonen L, Osman R, Gershengorn MC (1994a): A model of the thyrotropin releasing hormone (TRH) receptor binding pocket. Evidence for a second direct interaction between transmembrane helix 3 and TRH. J Biol Chem 269:23383–23386.

Perlman JH, Thaw CN, Laakkonen L, Bowers CY, Osman R, Gershengorn MC (1994b): Hydrogen bonding interaction of thyrotropin-releasing hormone (TRH) with transmembrane tyrosine 106 of the TRH receptor. J Biol Chem 269:1610–1613.

Perlman JH, Wang W, Nussenzveig DR, Gershengorn MC (1995): A disulfide bond between conserved extracellular cysteines in the thyrotropin-releasing hormone receptor is critical for binding. J Biol Chem 270:24682–24685.

Rao VR, Cohen GB, Oprian DD (1994): Rhodopsin mutation G90D and a molecular mechanism for congenital night blindness. Nature 367:639–642.

Sankararamakrishnan R, Vishveshwara S (1990): Conformational studies on peptides with proline in the right-handed α-helical region. Biopolymers 30:287–298.

Schertler GFX, Villa C, Henderson R (1993): Projection structure of rhodopsin. Nature 362:770–772.

Sheikh SP, Zvyaga TA, Lichtarge O, Sakmar TP, Bourne HR (1996): Rhodopsin activation blocked by metal-ion–binding sites linking transmembrane helices C and F. Nature 383:347–350.

Shenkin PS, McDonald DQ (1994): Cluster analysis of molecular conformations. J Comp Chem 15:899–916.

Sklar LA, Fay SP, Seligmann BE, Freer RJ, Muthukumaraswamy N, Mueller H (1990): Fluorescence analysis of the size of a binding pocket of a peptide receptor at natural abundance. Biochemistry 29:313–316.

Sklenar H, Etchebest C, Lavery R (1989): Describing protein structure: A general algorithm yielding complete helicoidal parameters and a unique overall axis. Proteins 6:46–60.

Still WC, Tempczyk A, Hawley RC, Hendrickson T (1990): Semianalytical treatment of solvation for molecular mechanics and dynamics. J Am Chem Soc 112: 6127–6129.

Suryanarayana S, Von Zastrow M, Kobilka BK (1992): Identification of intramolecular interactions in adrenergic receptors. J Biol Chem 267:21991–21994.

Swaminathan S, Ravishanker G, Beveridge DL, Lavery R, Etchebest C, Sklenar H (1990): Conformational and helicoidal analysis of the molecular dynamics of proteins: "Curves." Dials and Windows for a 50 psec dynamic trajectory of BPTI. Proteins 8:179–193.

Tota MR, Daniel S, Sirotina A, Mazina KE, Fong TM, Longmore J, Strader CD (1994): Characterization of a fluorescent substance P analog. Biochemistry 33:13079–13086.

Tuffery P, Etchebest C, Popot J-L, Lavery R (1994): Prediction of the positioning of the seven transmembrane α-helices of bacteriorhodopsin. A molecular simulation study. J Mol Sol 236:1105–1122.

Weiner PK, Kollman PA (1981): AMBER: Assisted model building with energy refinement. A general program for modeling molecules and their interactions. J Comp Chem 2:287–303.

Weiner SJ, Kollman PA, Case DA, Singh UC, Ghio C, Alagona GA, Profeta SJ, Weiner P (1984): A new force field for molecular mechanical simulation of nucleic acids and proteins. J Am Chem Soc 106:765–784.

Xie Y-B, Wang H, Segaloff DL (1990): Extracellular domain of lutropin/choriogonadotropin receptor expressed in transfected cells binds choriogonadotropin with high affinity. J Biol Chem 265:21411–21414.

Zhou W, Flanagan C, Ballesteros JA, Konvicka K, Davidson JS, Weinstein, H, Millar RP, Sealfon SC (1994): A reciprocal mutation supports helix 2 and helix 7 proximity in the gonadotropin-releasing hormone receptor. Mol Pharmacol 45:165–170.

CHAPTER 5

LESSONS FROM RHODOPSIN

THOMAS P. SAKMAR and K. CHRISTOPHER MIN

1. INTRODUCTION   85
2. METHODS   87
     A. Preparation of Crude ROS Membranes   88
     B. Preparation of UW-ROS and Solubilized Rhodopsin   89
     C. Purification of Retinal Holotransducin and Transducin Subunits   89
     D. Transducin Activation Assays   91
     E. ROS Binding Assay   92
3. CHROMOPHORE–PROTEIN INTERACTIONS IN RHODOPSIN   93
4. MOLECULAR MECHANISM OF RHODOPSIN PHOTOACTIVATION   95
5. RHODOPSIN COUPLING TO TRANSDUCIN   97
6. CONSTITUTIVE ACTIVITY OF MUTANT OPSINS   98
7. STRUCTURAL MODELS OF RHODOPSIN   99

1. INTRODUCTION

Visual pigments comprise a large segment of the super-family of G protein–coupled receptors (GPCRs) (Sakmar, 1994, 1998). Since the molecular cloning of the gene for bovine rhodopsin in 1983 (Nathans and Hogness, 1983), a remarkable amount of information about structure–function relationships in GPCRs has been obtained using techniques of molecular biology. In the study of visual pigments in particular, site-directed mutagenesis has been employed to elucidate key structural elements, the opsin-shift mechanism, and the mechanism of receptor photoactivation.

One particular advantage of studying rhodopsin has been the opportunity to employ various spectroscopic methods in combination with site-directed mutagenesis. Optical spectroscopy and resonance Raman spectroscopy are possible because of the presence of the retinal chromophore, which is probed as a sensor of chromophore–protein interactions. Different spectroscopy techniques, such as Fourier-transform infrared (FTIR) and ultraviolet (UV)-visible difference spectroscopy, make use of the chromophore as an optical switch. Overexpression of

Structure–Function Analysis of G Protein-Coupled Receptors, Edited by Jürgen Wess.
ISBN 0-471-25228-X   Copyright © 1999 Wiley-Liss, Inc.

recombinant rhodopsin also allows a variety of biophysical methods to be used to address particular questions related to conformational changes and protein–protein interactions. Molecular models based on the two-dimensional projection structure of rhodopsin have also proven useful.

In summary, it is clear that information about rhodopsin structure and the molecular mechanism of rhodopsin photoactivation are relevant to understanding the molecular mechanism of signal transduction by GPCRs in general. This chapter focuses on recent technical and methodological advances in the study of recombinant rhodopsin that may be relevant to understanding the structure and function of GPCRs. Background information related to this chapter can be found in recent reviews (Fahmy and Sakmar, 1995; Sakmar, 1994, 1998).

Although they share many similarities with other GPCR types, there is significant specialization in visual pigments not found in other receptor families. In particular, pigments are made up of opsin apoprotein plus chromophore, 11-*cis*-retinal (Fig. 5.1). The chromophore is a cofactor and not a ligand in the classic sense because it is linked covalently via a protonated Schiff base bond to a specific lysine (Lys[296]) residue in the membrane-embedded domain of the protein. An important structural feature of the retinal chromophore in rhodopsin, in addition to its Schiff base linkage, is its extended conjugated polyene structure, which accounts for its visible absorption properties and allows for resonance structures (Rando, 1996). Rhodopsin has a broad visible absorption maximum (λ_{max}) at about 500 nm. Photoisomerization of the 11-*cis* to all-*trans* form of the retinylidene chromophore is the primary event in visual signal transduction, and it is the only light-dependent step. Upon photoisomerization of the chromophore, the pigment is converted to metarhodopsin II (MII) with a λ_{max} value of 380 nm. The MII intermediate is characterized by a deprotonated Schiff base chromophore linkage. MII is the active form of the recep-

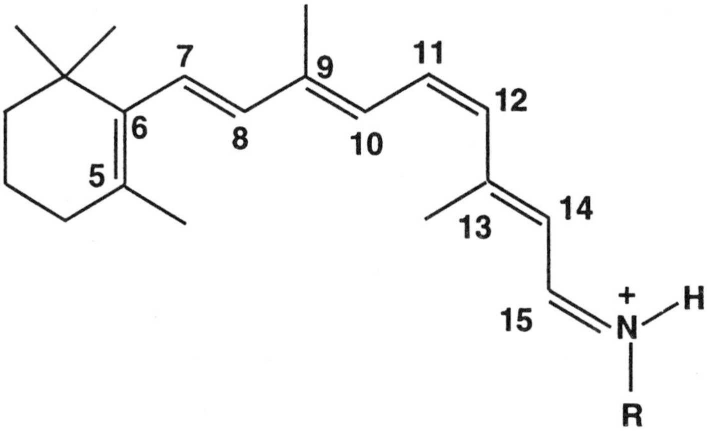

Figure 5.1. The chromophore of nearly all vertebrate visual pigments is 11-*cis*-retinylidene imine. The chromophore is covalently linked as a cofactor to Lys[296] on transmembrane (TM) helix 7 via a protonated Schiff base bond. Photoisomerization of the 11-*cis*-retinylidene chromophore to the all-*trans* form is the only light-dependent event in vision. The numbering of the carbons in the conjugated polyene system is given. R signifies opsin.

tor (R*), which catalyzes guanine–nucleotide exchange by the heterotrimeric G protein of the rod cell, transducin. In the case of the vertebrate visual system, GTP-bound transducin activates a cGMP phosphodiesterase, which lowers cGMP levels to close cGMP-gated cation channels in the plasma membrane of the rod cell. Light causes a graded hyperpolarization of the rod cell. The amplification, modulation, and regulation of the light response is of great physiological importance and has been discussed in detail elsewhere (Chabre, 1985; Stryer, 1991).

Bovine rhodopsin is the most extensively studied GPCR. A large amount of pigment ($\sim$0.5 mg) can be obtained from a single bovine retina by a sucrose density gradient centrifugation preparation of the rod outer segment (ROS) disc membranes. The pigment can be further purified by lectin affinity chromatography on concanavalin-A Sepharose resin. Rhodopsin is stable in solubilized form in a variety of detergents, including digitonin, dodecyl maltoside, and octyl glucoside. Rhodopsin was the first GPCR to be sequenced by amino acid sequencing (Ovchinnikov, 1982; Hargrave et al., 1983) and the first to be cloned (Nathans and Hogness, 1983, 1984). The cloning of the β_2-adrenergic receptor (Dixon et al., 1986) led to the identification of the structural homologies that now define the large family of GPCRs.

2. METHODS

The bovine retina has been the tissue of choice for the study of the biochemistry of vertebrate visual signal transduction due to the unique histology of the rod cell and to the fact that $\sim$95% of photoreceptor cells of the bovine retina are of the rod type. The relative ease with which ROSs, which contain all of the key components of the phototransduction system, can be isolated led to intense study by numerous investigators. More recent studies have been facilitated by the ability to express and to purify rhodopsin from transiently transfected mammalian cells in tissue culture (Oprian et al., 1987). Methods for the preparation of purified components of the bovine rod phototransduction system have been published, but over the course of our recent work a number of refinements were made, and several reported methods were combined to yield material appropriate for the biochemical reconstitutions required to assay recombinant receptors. This section presents methods for the preparation and biochemical assay of purified ROS membranes, holotransducin, and transducin α and $\beta\gamma$ subunits (Tα and T$\beta\gamma$).

To carry out reconstitutions with rhodopsin-containing membranes, a method was developed to remove peripherally associated proteins from crude ROS membranes. Several extractions of the membranes were performed to yield a preparation of urea-washed ROS (UW-ROS) that contained nearly pure rhodopsin in oriented phospholipid disc membrane bilayers. Assays were developed using these membranes to demonstrate functional binding of transducin to rhodopsin and to measure the increase in the rate of nucleotide exchange of purified transducin in the presence of light-activated rhodopsin. Transducin activation was measured by determining the rate of nucleotide uptake of a nonhydrolyzable analog of GTP, GTPγS. Time courses can be measured by removing aliquots of a reaction and applying them to nitrocellulose filters under vacuum. Another assay of rhodopsin-dependent transducin activation was developed that takes advantage of the previous observation that the intrinsic tryptophan fluorescence of the α subunit of transducin increases dramatically when GTP rather than GDP occupies its nucleotide binding pocket. This assay provides a real-time measurement of GTP uptake by transducin and allows the calculation of kinetic rate constants that define discrete steps in the activation pathway.

2.A. Preparation of Crude ROS Membranes

The following procedure is based on a combination of methods from a variety of sources (Papermaster and Dreyer, 1974; Papermaster, 1982; Hong and Hubbell, 1973; Fung et al., 1981; Ting et al., 1993). Frozen retinae were stored at $-80°C$ until use. Buffers used were homogenizing buffer (HB): 42% (w/w) sucrose, 65 mM NaCl, 0.2 mM $MgCl_2$, 5 mM HEPES, pH 7.5; dilution buffer (DB): 65 mM NaCl, 0.2 mM $MgCl_2$, 5 mM HEPES, pH 7.5; and isolation buffer (IB): 10 mM MOPS, pH 7.5, 60 mM KCl, 30 mM NaCl, 2 mM $MgCl_2$. All steps of the following procedure were carried out in a dark room illuminated only with dim red light (Kodak No. 1 safelight filters).

Vials of retinae (4×50 retinae/vial) were thawed in a water bath at $40°–50°C$. When the retinae were partially thawed, they were transferred to a 600-ml beaker. HB (300 ml) was supplemented with soybean trypsin inhibitory protein (TIP) to 30 mg/ml, aprotinin to 10 mg/ml, PMSF to 0.1 mM, pepstatin to 0.7 mg/ml, and leupeptin to 10 mg/ml. The protease inhibitors can be excluded if preparing UW-ROS, but are necessary to prevent proteolytic degradation of transducin and cGMP phosphodiesterase. HB (120 ml) was added to the retinae at room temperature, and they were stirred gently with a magnetic stirrer until completely thawed. The remaining steps were carried out on ice. The retinae were divided among 8 50-ml polypropylene Sorvall SS-34 centrifuge tubes. Each tube was vortexed for 1 minute at high speed. This step breaks the ROS from the inner segment and releases it from the retinal tissue. In this solution, the specific gravity of the ROS causes them to float. The crude retinal extract was centrifuged in an SS-34 rotor at 4,000 rpm for 4 minutes at 4°C. The supernatant fractions were transferred to a 500-ml Erlenmeyer flask sitting on ice. Each pellet was resuspended in 15 ml HB (ice cold) and spun again. The resulting supernatant fractions were pooled with those of the previous spin in the flask. DB (300 ml) was slowly added to the membrane suspension while carefully swirling. Under these conditions, the membranes can be pelleted by gentle centrifugation. The membranes were divided again and spun in an SS-34 rotor at 4,000 rpm for 4 minutes at 4°C. The resulting supernatant fractions were carefully removed by aspiration and discarded.

The membranes are only loosely pelleted at this point. The supernatant fraction appears quite red in color due mainly to the presence of hemoglobin and not to the loss of rhodopsin-containing membranes. Each of the pellets was resuspended in 5 ml of 38% (w/w) sucrose in IB (IB-Suc) supplemented with TIP to 30 mg/ml, PMSF to 0.1 mM, and DTT to 1 mM. The resuspended pellets were pooled, and the volume was brought up to 250 ml with IB-Suc. Under these conditions, the ROS membranes float, and contaminating membranes can be removed by centrifugation. The resuspended membranes were transferred to 10 50-ml SS-34 tubes and spun at 17,000 rpm for 15 minutes. The supernatant fractions were carefully transferred to a fresh 500-ml Erlenmeyer flask on ice, and 100 ml IB was added to the membrane suspension. The membrane suspension was transferred to six Beckman Ti-45 tubes and spun in an ultracentrifuge at 25,000 rpm for 15 minutes at 4°C. The supernatant fractions were removed by aspiration, and the membranes were resuspended in 120 ml IB, transferred to two Ti-45 tubes, and spun again. The pellets were resuspended again and spun as before. The ROS membranes can be stored overnight on ice until further processing of crude ROS to yield UW-ROS or transducin as described below.

2.B. Preparation of UW-ROS Membranes and Solubilized Rhodopsin

Crude bovine ROS membranes were prepared as described above. ROS membranes were then resuspended in wash buffer 1 (5 mM Tris-HCl, pH 7.5, 1 mM EDTA, 1 mM DTT, 5 mM NaF, 30 mM $AlCl_3$) and stored overnight at 4°C. At each step, 0.5 ml of wash buffer was used per retina. ROS membranes were collected by centrifugation at 100,000g for 15 minutes, resuspended in wash buffer 1, and spun again. The membranes were then washed sequentially with wash buffer 2 (20 mM Tris-HCl, pH 7.5, 1 mM EDTA, 1 mM DTT) three times, wash buffer 3 (200 mM Tris-HCl, pH 7.5, 20 mM $MgCl_2$, 1 mM DTT) once, wash buffer 4 (20 mM Tris-HCl, pH 7.5, 5 mM EDTA, 5 M urea) three times, and finally resuspended in assay buffer (10 mM PIPES, pH 7.5, 150 mM KCl, 4 mM Mg[OAc]$_2$, 0.1 mM EDTA, 1 mM DTT). The spectral ratio (A_{280}/A_{500}) of the UW-ROS is typically 1.8–2.0 measured after solubilizing the membranes in 1% (w/v) dodecyl maltoside. The concentration of rhodopsin was calculated using a molar extinction coefficient of 40,600 $cm^{-1} M^{-1}$. The yield from 200 retinae is typically 3–4 mmol. Illumination of rhodopsin was carried out using a 150-W light source and a 495-nm long-pass optical filter (Melles-Griot). A photobleaching-difference spectrum can be calculated by subtracting the UV-visible spectrum of rhodopsin in the dark from that after illumination (Fig. 5.2).

2.C. Purification of Retinal Holotransducin and Transducin Subunits

Holotransducin was purified essentially as described elsewhere (Fung et al., 1981) from crude ROS prepared as described above. The membranes were illuminated for 5 minutes using a 150-W projector lamp fitted with a 495-nm long-pass filter and were then extracted with hypotonic buffer (5 mM Tris-HCl, pH 7.5, 0.5 mM $MgCl_2$, 1 mM DTT, 0.1 mM PMSF) two additional times. The fourth centrifugation was carried out at 30,000 rpm for 15 minutes and the fifth at 31,000 rpm for 15 minutes. The supernatant fractions from these two steps were discarded. Transducin was released from the membranes by two extractions with hypotonic buffer supplemented with 100 μM GTP. Both centrifugation steps were carried out at 42,000 rpm for 15 minutes. The released

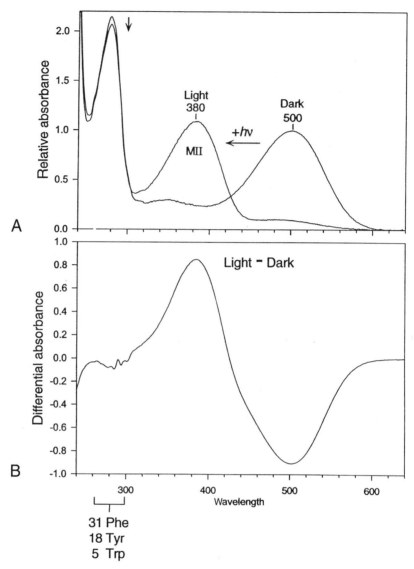

Figure 5.2. UV-visible spectroscopy of rhodopsin. **A:** UV-visible spectra in darkness and after illumination for purified rhodopsin in dodecyl maltoside (DM) detergent. Rhodopsin shows a characteristic broad visible absorbance with a λ_{max} value of 500 nm. The 280-nm peak represents the protein component. Upon illumination in DM, the pigment readily forms the metarhodopsin (MII) intermediate with a λ_{max} value of 380 nm. This spectral form of rhodopsin is the active conformation of the receptor (R*), which catalyzes guanine–nucleotide exchange by transducin. Identical results can be obtained with rhodopsin from bovine retinae purified by concanavalin-A lectin affinity chromatography or with recombinant rhodopsin expressed in COS cells and purified by immunoaffinity methods. **B:** Photobleaching-difference spectrum obtained by subtracting the dark spectrum from the light spectrum in A. In addition to the large difference peaks noted at 380 and 500 nm, a light-dependent change in the opsin component of the spectrum is noted in the 260–300 nm range. This change is due to changes in the environments of aromatic residues as discussed in the text.

holotransducin was filtered through a 0.45-mm filter and loaded directly onto a hexyl-agarose (12-atom spacer, Sigma) column (1.0 × 9 cm) equilibrated in hexyl-agarose column buffer (HCB) (10 mM MOPS, pH 7.5, 5 mM $MgCl_2$, 0.1 mM EDTA, 1 mM DTT, 0.1 mM PMSF) overnight using a peristaltic pump set to 7.5 ml per hour. The peristaltic pump was reset to 0.5 ml per minute, and the column was washed with 60 ml of HCB. After a second wash with 75 mM NaCl in HCB, transducin was eluted from the column with a step gradient of 300 mM NaCl in HCB. Peak fractions were pooled and dialyzed against glycerol storage buffer (40% glycerol [v/v] in HCB supplemented with 10 mM GDP). The total yield of purified transducin was typically 3–8 mg from 200 retinae. Purified transducin was stored at −20°C.

The subunits of transducin were prepared from purified holotransducin essentially as described (Shichi et al., 1984). A Blue Sepharose CL-6B (Pharmacia) column (1.5 × 9 cm) was equilibrated with Blue Sepharose column buffer (BSCB) (10 mM KH_2PO_4, pH 6.5, 2 mM $MgCl_2$, 0.1 mM EDTA, 1 mM DTT, 10% [w/v] glycerol) using a peristaltic pump set to 1.5 ml per minute. Purified transducin in glycerol storage buffer was diluted 20-fold in BSCB. Fractions (4.5 ml) were collected as the diluted transducin was loaded (fractions 1–9) onto the blue sepharose column and washed with 90 ml of BSCB (fractions 10–27). A 120-ml linear gradient from 0 to 2 M NaCl in BSCB was then applied to the column while collecting 2 ml fractions (fractions 28–72). Tβγ typically eluted early in the wash fractions, and Tα eluted in a broad peak in the middle of the NaCl gradient. Purified bovine transducin α (bTα) and transducin βγ (bTβγ) subunits were dialyzed against glycerol storage buffer and stored at −20°C.

2.D. Transducin Activation Assays

The filter-binding assay is based on the fact that transducin and its bound nucleotide are retained on nitrocellulose filters whereas free nucleotide passes through. The assay was carried out essentially as described elsewhere (Wessling-Resnick and Johnson, 1987; Min et al., 1993). Purified pigments were stored frozen at −20°C. Pigment concentrations were determined based on the A_{500} value immediately before each assay. The assay mixture generally consisted of purified pigment (5 or 10 nM), purified transducin (7 μM) and ^{35}S-GTPγS (20 μM) in 0.1 ml of assay buffer (50 mM Tris-Cl, pH 7.2, 100 mM NaCl, 4 mM $MgCl_2$, 1 mM DTT, and 0.01% dodecyl maltoside). Time course experiments were begun in darkness at 10°C by the addition of nucleotide. After a 2-minute incubation, the assay mixture was illuminated continuously with a 150-W projector lamp (Dolan-Jenner) affixed with a 495-nm long-pass filter. At 30-second intervals in darkness and 15-second intervals during illumination, 10-μl aliquots were removed and transferred to the nitrocellulose filters. The filters were washed and dried, and bound nucleotide was quantitated using a Phosphor Imager System (Molecular Dynamics, Inc.) or by liquid scintillation counting. A time course in darkness was performed to rule out pigment dark activity or constitutive activity by mutant apoprotein present in the assay mixture. Control experiments in the absence of transducin revealed that less than 0.03% of the nucleotide bound to the filter.

Spectrofluorimetric assays of transducin activation can also be carried out to determine precise kinetic rate constants (Fahmy and Sakmar, 1993). The active transducin α-subunit concentration in a preparation of holotransducin can be precisely determined by measuring rhodopsin-catalyzed nucleotide-induced fluorescence increase upon addition of different amounts of GTPγS (Fig. 5.3). The fluorescence assay employed was similar to one described elsewhere (Guy et al., 1990; Phillips and Cerione, 1988). Fluorescence was measured with a specially modified SPEX-Fluorolog II spectrofluorometer in signal/reference mode with excitation at 300 nm (2-nm band width) and emission at 345 nm (12-nm band width). Signal integration time was 2 seconds. The reaction mixture (1.6 ml) containing 10 mM Tris-Cl (pH 7.4), 100 mM NaCl, 2 mM MgCl$_2$, and 0.01% dodecyl maltoside, was cooled to 10°C and stirred continuously using a magnetic cuvette stirrer set at maximum speed. The addition of rhodopsin and nucleotide was done by injecting 50 μl of the appropriate solution into the cuvette with a gastight syringe kept at 10°C. The sample was continuously illuminated in the cuvette with 543-nm light from a HeNe-laser (Melles-Griot) connected to the sample compartment by a fiberoptic light guide. Stray light was efficiently blocked from reaching the detector by a double monochromator. Under conditions of the assay, the rate of photobleaching to form a 380-nm species characteristic of MII was complete in 15 seconds.

2.E. ROS Binding Assay

The biochemical basis of this assay is the observation by Kuhn (1980) that transducin binds tightly to illuminated ROS membranes, even under hypotonic buffer conditions, and can be released by the addition of GTP. Varying amounts

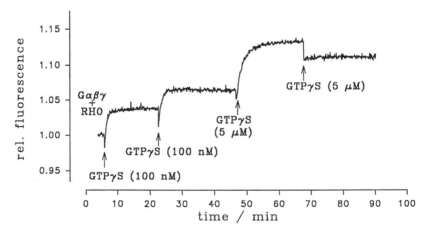

Figure 5.3. Spectrofluorometric titration of purified transducin with GTPγS in the presence of purified COS-cell rhodopsin. Intrinsic tryptophan fluorescence of the α subunit of transducin is measured. GTPγS binding by transducin, which is catalyzed by light-activated rhodopsin, causes a significant increase in fluorescence. The experiment is presented as a continuous time course with injections of GTPγS into the sample cuvette indicated by arrows. The experimental details of the assay are presented in the text.

of purified transducin and UW-ROS (5 mM rhodopsin) were mixed for 15 minutes at 4°C in the dark in hypotonic buffer (5 mM HEPES, pH 7.5, 2 mM MgCl$_2$, 1 mM DTT). Each tube was illuminated for 5 minutes on ice using a 150-W light source affixed with a 495-nm long-pass filter. To estimate the amount of GTPγS binding activity present at the start of the reaction, an aliquot was removed from each tube and assayed. Membranes from each tube were washed by sequential centrifugation at 100,000g for 10 minutes and resuspended in hypotonic buffer twice. The washed membranes were then assayed for GTPγS binding activity. GTPγS binding activity before and after washing of the membranes was determined by the addition of ^{35}S-GTPγS (20 μM) and incubation for 1 hour at room temperature before application of three aliquots from each reaction onto nitrocellulose filters. The filters were washed and air dried, and the amount of radioactivity bound to the filters was determined by use of a Phosphor Imager System (Molecular Dynamics).

3. CHROMOPHORE–PROTEIN INTERACTIONS IN RHODOPSIN

Even though different visual pigments employ the same retinylidene protonated Schiff base chromophore, their λ_{max} values span the visible spectrum from near UV at about 400 nm to far visible red at about 600 nm. Chromophore–protein interactions are responsible, directly or indirectly, for spectral tuning in visual pigments. Thus, differences in primary structure result in differences in spectral properties. The understanding of the molecular mechanism of spectral tuning has been advanced by the cloning of numerous visual pigments with different spectral properties. In many cases, primary structure alignments have led to testable hypotheses concerning the identities and mechanisms of key chromophore–protein interactions. For example, the cloning (Nathans et al., 1986) and *in vitro* spectral characterization (Oprian et al., 1991; Merbs and Nathans, 1992) of the human cone pigments has led to mutagenesis and Raman studies that have largely elucidated the mechanisms of green-red spectral tuning (Aseñjo et al., 1994; Kochendoerfer et al., 1997).

The chromophores in visual pigments lie within the helical bundle of the membrane-embedded domain of the receptor (Fig. 5.4). A number of studies have been carried out on bovine rhodopsin to investigate retinal–opsin interactions in the membrane-embedded domain of bovine rhodopsin. Several spectroscopic methods such as resonance Raman spectroscopy, FTIR-difference spectroscopy, and nuclear magnetic resonance spectroscopy have been employed (Birge, 1990; Siebert, 1995). Other approaches have included reconstitution of opsin apoprotein with synthetic retinal analogs (Honig et al., 1979) and photochemical crosslinking (Nakayama and Khorana, 1990; Zhang et al., 1994). Early work on the structure and function of recombinant bovine rhodopsin focusing on the use of techniques of molecular biology has been reviewed (Khorana, 1992; Nathans, 1992).

The environment of the Schiff base portion of the chromophore is a key protein–chromophore interaction in rhodopsin. Lys[296] and Glu[113] are the two amino acid residues that largely define the structure and function of the chromophore.

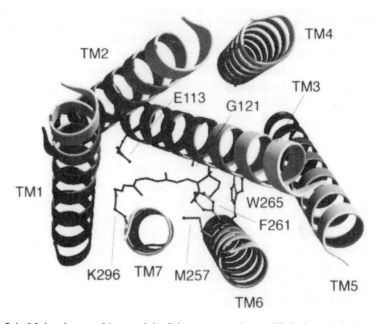

Figure 5.4. Molecular graphics model of the transmembrane (TM) domain bovine rhodopsin (Han et al., 1996b; Shieh et al., 1997). The molecule is viewed down the axis of TM helix 7 from above the plane of the disc membrane. TM helix 7 is thought to be roughly perpendicular to the membrane plane. TM helix 3 is tilted and buried within the bundle of TM helices. TM helices 3 and 6 largely define the retinal binding pocket, and key residues that interact with the chromophore include Glu^{113}, Gly^{121}, Phe^{261}, and Trp^{265}. The retinal lies roughly at the level of Gly^{121} and Phe^{261}. Near the extracellular end of TM helix 3, Glu^{113} serves as the counterion to the protonated Schiff base localized near C_{12} of the retinal. At the cytoplasmic end of TM helix 3 are three conserved residues (Glu^{134}, Arg^{135}, and Tyr^{136}, not shown), which are known to be involved in G protein binding. Movement of TM helices 3 and 6 in the plane of the membrane away from each other is thought to be required for receptor activation. An essential Cys^{110}–Cys^{187} disulfide bond at the extracellular surface of TM helix 3, which is highly conserved in GPCRs, might act as a pivot point for this motion ad restrict vertical translation of the helix 3 relative to the other helices. The model of the opsin is based in part on data from electron microscopy studies of reconstituted bovine and frog rhodopsin. The positions and assignments of the TM helices are according to the projection map of the protein and a primary structure analysis of the GPCR family. The rationale for positioning the chromophore has been described elsewhere (Shieh et al., 1997). (Reproduced from Han et al. [1996b], with permission of the publisher).

The Schiff base linkage of the chromophore to Lys^{296} is known to be protonated in the ground state of the receptor (Hargrave et al., 1982). Light-dependent deprotonation of the Schiff base is required for the formation of R* (Kibelbek et al., 1991; Longstaff et al., 1986; Zhukovsky et al., 1991). However, light can induce R* in the absence of a Schiff base chromophore linkage in certain opsin mutants lacking a lysine at position 296 (Zhukovsky et al., 1991). Glu^{113} in bovine rhodopsin serves as the counterion to the positive charge of the protonated Schiff base (Nathans, 1990; Sakmar et al., 1989; Zhukovsky and Oprian,

1989). Glu[113] is unprotonated and negatively charged in the ground state of rhodopsin (Fahmy et al., 1993). It becomes protonated upon light-dependent formation of MII and is the net proton acceptor for the Schiff base proton (Jäger et al., 1994). Glu[113] is located near C_{12} of the retinal polyene according to two-photon spectroscopy and nuclear magnetic resonance spectroscopy of retinal analogs and semiempirical quantum mechanical orbital calculations (Birge, 1990; Han et al., 1993; Han and Smith, 1995a,b). This location is supported by resonance Raman (Lin et al., 1992) and FTIR-difference spectroscopy studies (Fahmy et al., 1996) of recombinant mutant pigments lacking a carboxylic acid group at position 113.

The nature of the Glu[113]–protonated Schiff base interaction in the structure and function of rhodopsin has been elucidated indirectly by the discovery of constitutive activity among certain opsin mutants (Robinson et al., 1992; Rao and Oprian, 1996). Constitutive activity refers to the ability of an opsin to activate transducin in the absence of any chromophore. Generally, a mutation that disrupts the salt bridge between Glu[113] and Lys[296] in the opsin apoprotein leads to constitutive activity. For example, replacement of either Glu[113] or Lys[296] by a neutral amino acid results in a mutant opsin with constitutive activity. Other mutations such as G90D or A292E also result in constitutive activity, presumably because the introduction of the negatively charged residue affects the stability of the Glu[113]–Lys[296] salt bridge (Rao et al., 1994; Cohen et al., 1992). The mechanism of constitutive activity of opsins and the potential relevance of constitutive activity to visual diseases such as congenital night blindness have been recently reviewed (Rao and Oprian, 1996). The concept of constitutive activity as it applies to GPCRs in general is discussed in more detail below.

4. MOLECULAR MECHANISM OF RHODOPSIN PHOTOACTIVATION

Photoisomerization of the 11-*cis*-retinal chromophore induces the active receptor conformation R* (Bownds, 1967; Dratz and Hargrave, 1983; Oseroff and Callender, 1974). Recent studies have suggested that steric and electrostatic changes in the ligand binding pocket of rhodopsin may cause changes in the relative disposition of TM helices within the core of the receptor (Shieh et al., 1997). These changes may be responsible for transmitting a "signal" from the membrane-embedded domain to the cytoplasmic surface of the receptor. Tryptophan mutagenesis (Han et al., 1996a,b; Lin and Sakmar, 1996), mutagenesis of conserved amino acid residues on TM helices 3 and 6 (Han et al., 1996a,b), and the introduction of pairs of histidine residues at the cytoplasmic borders of TM helices to create sites for metal chelation (Sheikh et al., 1996) have recently provided insights regarding the functional role of specific helix–helix interactions in rhodopsin. A series of elegant studies using site-directed mutagenesis of rhodopsin in combination with spin-labeling and electron paramagnetic resonance spectroscopy has revealed that the molecular mechanism of receptor activation after chromophore isomerization involves outward rigid-body movements of TM helices 3 and 6 relative to the center of the helix bundle (Farrens et al., 1996) (see also Chapter 13). Because the arrangement of the seven TM helices is likely to be evolutionarily conserved among the family of GPCRs (Baldwin, 1993), the proposed motions of TM helices 3 and 6 may be a part of an activation mechanism shared among all GPCRs. In other receptor subtypes, ligand binding would have to be coupled to a change in the orientations of TM helices 3 and 6.

TM helix 3 of rhodopsin is known to be involved in chromophore–protein interactions and contains the retinal Schiff base counterion Glu[113]. Gly[121], which is strictly conserved in all visual pigments, is located near the middle of TM helix 3. The α carbon of Gly[121] was predicted to point toward the 9-methyl group of retinal based on a model of the chromophore orientation in the protein derived from solid-state nuclear magnetic resonance constraints (Han and Smith, 1995b). This model was tested using site-directed mutagenesis in combination with regeneration of mutant opsins with various chromophore analogs (Han et al., 1997a,b; 1998a). The results of these studies are consistent with the hypothesized orientation of the chromophore, which resides between Gly[121] and Phe[261] (Fig. 5.4). Furthermore, these functional interactions between TM helices 3 and 6 mediated by the chromophore appear to be crucial for receptor photoactivation, which is mediated by movement of these two helices.

Synthetic rhodopsin-derived peptides have been shown to compete with native rhodopsin for transducin binding (König et al., 1989). This has allowed the identification of the cytoplasmic loops 3–4 and 5–6 and a putative loop between the cytoplasmic termination of TM helix 7 and the palmitoylated Cys[322] and Cys[323] residues as transducin binding sites. Site-directed mutagenesis has further characterized groups of amino acids in these regions implicated in transducin binding and activation (Franke et al., 1990, 1992). Time-resolved (Farahbakhsh et al., 1993) and static electron paramagnetic resonance spectroscopy studies (Resek et al., 1993) on site-specific spin-labeled rhodopsin showed that the cytoplasmic ends of TM helices 3 and 7 undergo structural rearrangements in the vicinities of Cys[140] and Cys[316], respectively. These changes have been specifically assigned to the MII conformation. Cys[140] is close to the highly conserved Glu or Asp/Arg/Tyr triad at the cytoplasmic border of TM helix 3 (position 134–136 in bovine rhodopsin), which attracted attention in earlier studies because of its possible general importance for the function of GPCRs.

Replacement of Glu[134] by glutamine renders the photoactivated pigment about eightfold more efficient in activating transducin at alkaline pH than recombinant native rhodopsin (Fahmy and Sakmar, 1993). Therefore, it was suggested that Glu[134] is a good candidate for regulation of the transducin-binding region and may undergo a light-induced transition from an ionized to a protonated state (Fahmy and Sakmar, 1993; Cohen et al., 1993). Recent measurements of light-induced pH changes in the bulk-water phase monitored simultaneously with R* formation of the mutants E134Q and E134D showed the involvement of Glu[134] in proton uptake reactions (Arnis et al., 1994; Arnis and Hofmann, 1993). According to these results, it is likely that Glu[134] itself is a group that becomes protonated in MII. It has also been proposed that MII exists in two discrete conformations termed MII_a and MII_b, which are indistinguishable spectroscopically (Arnis and Hofmann, 1993). MII_b activates transducin and MII_a does not. It was proposed that one difference between these two forms of MII was the protonation state of Glu[134], where Glu[134] becomes protonated during the MII_a to MII_b transition.

The structural change detected by electron paramagnetic resonance spectroscopy may be directly related to protonation of Glu[134], which is expected to significantly alter the hydrogen-bonding properties of this amino acid. A rearrangement of neighboring hydrogen-bonding partners may then explain the conformational change. Indeed, electron paramagnetic resonance spectra of Glu[134] mutants showed that the receptor was in a partially active conformation (Kim et al., 1997). A possible functional importance of the relative positions of TM helices 3 and 4 of rhodopsin is also suggested on the basis of a conserved disulfide bond (Cys[110]–Cys[187]) between these helices on the extracellular receptor domain (Doi et al., 1990; Karnik et al., 1988; Karnik and Khorana, 1990), which stabilizes the MII conformation (Davidson et al., 1994). In terms of the chromophore, the suggested localization of light-induced sterical changes would be expected mostly to affect locations distal to the Schiff base linkage, which is situated at the interfaces of TM helices 2, 3, and 7 (Rao et al., 1994; Schertler et al., 1993; Baldwin, 1993).

5. RHODOPSIN COUPLING TO TRANSDUCIN

Perhaps the most extensively studied receptor–G protein interaction is that of bovine rhodopsin with transducin (Hofmann et al., 1995). Detailed biochemical and biophysical analysis of the R*–transducin interaction has been aided by mutagenesis of the cytoplasmic domain of bovine rhodopsin. Numerous rhodopsin mutants defective in the ability to activate transducin have been identified (Franke et al., 1992). Several of these mutant receptors were studied by flash photolysis (Franke et al., 1990), light scattering (Ernst et al., 1995), or proton uptake assays (Arnis et al., 1994). The salient result of these studies was that cytoplasmic loops 3–4 and 5–6 were involved in R*–transducin interaction. This finding was consistent with those from other approaches, including peptide competition studies (Arnis et al., 1994). Although relatively large segments of cytoplasmic loops are required for proper rhodopsin–transducin interaction (Franke et al., 1992), single amino acid substitutions within these domains can have dramatic effects on transducin activation. In addition, transducin binding and the activation of bound transducin were shown to be discrete steps involving different surface domains of the receptor (Franke et al., 1990). Mutant pigments with alterations of cytoplasmic loops 3–4 and 5–6 were characterized that formed spectrally normal MII-like photoproducts that bound transducin (Franke et al., 1990). However, the bound transducin was apparently defective in the release of GDP, which accounts for the block in pigment-catalyzed GTP uptake or GTPase activity by transducin observed when the mutant pigments were assayed (Ernst et al., 1995). These results were consistent with the idea that transducin binding and activation were discrete steps mechanistically, which could be uncoupled by specific amino acid substitutions on the cytoplasmic surface of the receptor.

The role of the carboxy-terminal tail of rhodopsin in the activation of transducin is less well defined. Although several biochemical studies have implicated the carboxy-terminal tail in transducin activation (Phillips et al., 1992; Takemoto et al., 1986), alanine-scanning mutagenesis failed to confirm that the proximal portion of the tail was required (Weiss et al., 1994; Osawa and Weiss, 1994). Additional work is needed to elucidate any role, subtle or otherwise, of the carboxy-terminal tail of rhodopsin in transducin activation. In addition, an important consideration will be to map the sites of specific transducin subunit interactions on the cytoplasmic surface of rhodopsin. This work may be facilitated by both the high-resolution crystal structure of heterotrimeric transducin (Lambright et al., 1996) and the ability to produce, using a baculovirus expression system, purified recombinant heterotrimeric transducin and to reconstitute it with recombinant rhodopsin.

6. CONSTITUTIVE ACTIVITY OF MUTANT OPSINS

Constitutive activity of GPCRs, defined as signaling in the absence of ligand, has been recognized as an important feature of these receptors since it was first reported to occur in mutant α_{1B}-adrenergic receptors (Cotecchia et al., 1990). The relationship between constitutive activity and ligand binding and activation can be described by the two-state model of GPCR function first proposed by Lefkowitz et al. (1993) and described quantitatively by Leff (1995). This model of GPCR function also seems to apply to rhodopsin (Han et al., 1998b).

Certain receptors also have intrinsically high levels of basal activity, which may be important for their physiological function (Tiberi and Caron, 1994; Cohen et al., 1997). Increases in signaling in the absence of ligand can also result from overexpression of either the receptor (Bond et al., 1995) or the G protein (Burstein et al., 1997). Rhodopsin, which is responsible for dim-light vision, has evolved a unique mechanism to minimize basal receptor activity and thus dark "noise." The chromophore 11-*cis*-retinal, which acts as a potent inverse agonist in rhodopsin, is covalently bound to the receptor to assure extremely low receptor signaling in the dark.

Constitutive activity of GPCRs has been implicated in the molecular pathophysiology of a number of human diseases. The mutant receptors responsible for these diseases are characterized by gain of function *in vivo* and *in vitro*. A partial list of these diseases includes familial male precocious puberty caused by constitutive activity of the luteinizing hormone receptor (Shenker et al., 1993), hyperfunctioning thyroid adenoma caused by constitutive activity of the thyrotropin receptor (Parma et al., 1993), hypocalcemia caused by constitutive activity of the Ca^{2+}-sensing receptor (Pollak et al., 1994), and Jansen-type metaphyseal chondrodysplasia caused by constitutive activity of the parathyroid hormone/parathyroid hormone–related peptide receptor (Schipani et al., 1995). Furthermore, Kaposi's sarcoma–associated herpes virus GPCR, homologous to the human interleukin-8 receptor, stimulates cellular proliferation by constitutively activating the phosphoinositide–inositol triphosphate–protein kinase C pathway and qualifies the receptor gene as a candidate viral oncogene (Arvanitakis et al., 1997).

Rhodopsin mutants that are constitutively active *in vitro* have also been found in visual diseases (Rao and Oprian, 1996). Congenital night blindness is caused by the rhodopsin mutation G90D or A292E (Dryja et al., 1993; Sieving et al., 1995), and forms of autosomal dominant retinitis pigmentosa are caused by the mutations K296E or K296M (Keen et al., 1991; Rim and Oprian, 1995). One of the differences between rhodopsin and other GPCRs is that the inverse agonist 11-*cis*-retinal, which is able to suppress the constitutive activity of the mutant opsins G90D (Rao et al., 1994) and A292E (Dryja et al., 1993), is present in cells containing the mutant receptors. The G90D and A292E mutant receptors exist, at least in part, in the inactive ligand-bound form rather than the constitutively active apo-receptor form. Interestingly, patients carrying the G90D mutation were shown to have an elevated absolute threshold for visual perception (Sieving et al., 1995). FTIR spectroscopy of expressed mutant pigment G90D also suggested a possible increased thermal barrier to light-dependent receptor activation (Fahmy et al., 1996). Rhodopsin mutants K296E and K296M, however, do not bind 11-*cis*-retinal due to the lack of the lysine for Schiff base attachment (Rim and Oprian, 1995; Robinson et al., 1992). Therefore, significant amounts of these mutant opsin apo-receptors may be present to activate transducin, independently of either light or ligand. Recently, retinal analogs were designed and synthesized that specifically inhibit constitutive activity of these opsin mutants (Yang et al, 1997).

Many of the activating mutations in GPCRs are located in TM helix 6 and in the cytoplasmic loop connecting TM helices 5 and 6 (van Sande et al., 1995; van Rhee and Jacobson, 1996; Porcellini et al., 1994). As discussed above, many activating mutations of rhodopsin disrupt a salt bridge between Glu^{113} in TM helix 3 and Lys^{296} in TM helix 7 (Rao and Oprian, 1996), which is not conserved in GPCRs in general. However, additional activating rhodopsin mutations have been reported on TM helix 6 as well at Met^{257} and Phe^{261} in the same region where activating mutations in many other GPCRs have been identified. In particular, Met^{257} may form an important and specific interhelical interaction that stabilizes the inactive receptor conformation by preventing TM helix 6 movement in the absence of all-*trans*-retinal. This stabilizing interaction may occur between Met^{257} and the NPXXY motif in TM helix 7, which is highly conserved in GPCRs. These results further support the idea that rhodopsin and other GPCRs share a common mechanism of receptor activation that involves specific changes in helix–helix interactions.

7. STRUCTURAL MODELS OF RHODOPSIN

Bacteriorhodopsin (BR), the light-driven proton pump of the halophilic bacterium *Halobacterium halobium,* contains seven TM helical segments and a retinylidene chromophore linked via a Schiff base to a specific lysine residue on TM helix 7. Although these similarities to rhodopsin are intriguing, there is no apparent primary structural homology among BR and visual pigments, including rhodopsin. In addition, BR does not couple to G proteins and has very short cytoplasmic loops compared with those of rhodopsin. Nevertheless, many of the same methodologies developed for the study of BR have also been applied to the study of rhodopsin. In addition, some of the paradigms involving the photochemistry and biophysics of membrane proteins involved in energy transduction or vectorial transport can be applied to both systems (Lanyi, 1995).

BR forms an ordered two-dimensional lattice in the "purple membrane" of *H. halobium,* which has made it possible to undertake high-resolution structural studies. The reconstructed three-dimensional structural model of BR (Henderson et al., 1990) and its subsequent refinements provided a template for the superimposition of the seven TM segments of GPCRs (Röper et al., 1994), including rhodopsin (Lin et al., 1992). Despite the recent reports of high-resolution three-dimensional structures for BR (Pebay-Peyroula et al., 1997; Kimura et al., 1997), the lack of a clear evolutionary relationship between BR and rhodopsin or other GPCRs makes modeling approaches based on BR structures somewhat speculative.

Recently, cryoelectron microscopy and image reconstruction were employed to determine the projection structures at about 9 Å resolution of bovine rhodopsin reconstituted into phospholipid bilayers (Schertler et al., 1993). Higher resolution structures of frog rhodopsin were subsequently reported (Schertler and Hargrave, 1995; Unger et al., 1997). Helix assignments and rotational orientations were proposed for the projection densities based on a comparison of a large number of GPCR amino acid sequences and on relevant biochemical and mutagenesis studies (Baldwin et al., 1997). The helix assignments for rhodopsin are consecutive at the extracellular membrane surface. However, different tilt angles of the helices as they traverse the membrane result in TM helix 3 being much more central in location at the cytoplasmic membrane surface.

Detailed molecular modeling based on these projection maps of rhodopsin has proven informative, and integrated approaches, which consider a variety of information in addition to sequence alignments, have been developed (Herzyk and Hubbard, 1995; Pogozheva et al., 1997). An example of one such model of rhodopsin is shown in Figure 5.4 (Shieh et al., 1997). Such models have been used extensively in ligand binding site analyses with the aim of "rationale drug design." In addition, when reasonable models are combined with experimental data on receptor activation, including studies of recombinant mutant pigments, a tentative assessment of intramolecular processes that may be of functional importance during activation of rhodopsin, and of other GPCRs, might be attempted either empirically or by molecular dynamic simulations (Gether et al., 1997; Colson et al., 1998).

It should be noted that models based on the projection maps of rhodopsin generally only consider the membrane-embedded domain of the receptor. However, TM helices can be connected by loop structures predicted by using secondary structure prediction algorithms. Also, some three-dimensional structural information concerning the cytoplasmic surface of rhodopsin has been inferred by solving the solution structures of peptides corresponding to putative cytoplasmic domains. Multidimensional nuclear magnetic resonance (NOSEY) spectroscopy and circular dichroism were employed to study peptides corresponding to the third cytoplasmic loop (loop 5–6) and the carboxy-terminal tail of bovine rhodopsin (Yeagle et al., 1995, 1997). The loop 5–6 peptide seemed to fold to form a stable structure that could be docked onto the ends of the appropriate TM helices identified in the projection models (see Chapter 16).

ACKNOWLEDGMENTS

T.P.S. is an Associate Investigator of the Howard Hughes Medical Institute. Support was also provided by the Allene Reuss Memorial Trust. K.C.M. was supported in part by NIH MSTP grant GM 07739. We thank our collaborators on studies of visual phototransduction over the past 6 years: H. Bourne, K. Fahmy, S.G. Graber, K.P. Hofmann, D.S. Kliger, R.A. Mathies, K. Nakanishi, H. Ostrer, F. Siebert, and S.O. Smith.

REFERENCES

Arnis S, Fahmy K, Hofmann KP, Sakmar TP (1994): A conserved carboxylic acid group medicates light-dependent proton uptake and signaling by rhodopsin. J Biol Chem 269:23879–23881.

Arnis, S, Hofmann, KP (1993): Two different forms of metarhodopsin II: Schiff base de-protonation precedes proton uptake and signaling state. Proc Natl Acad Sci USA 90:7849–7853.

Arvanitakis L, Geras-Raaka E, Varma A, Gershengorn MC, Cesarman E (1997): Human herpes virus KSHV encodes a constitutively active G-protein–coupled receptor linked to cell proliferation. Nature 385:347–350.

Aseñjo AB, Rim J, Oprian DD (1994): Molecular determinants of human red/green color discrimination. Neuron 12:1131–1138.

Baldwin JM (1993): The probable arrangement of the helices in G protein–coupled receptors. EMBO J 12:1693–1703.

Baldwin JM, Schertler GF, Unger VM (1997): An alpha-carbon template for the transmembrane helices in the rhodopsin family of G-protein–coupled receptors. J Mol Biol 272:144–164.

Birge RR (1990): Nature of the primary photochemical events in rhodopsin and bacteriorhodopsin. Biochim Biophys Acta 1016:293–327.

Bond RA, Leff P, Johnson TD, Milano CA, Rockman HA, McMinn TR, Apparsundaram S, Hyek MF, Kenakin TP, Allen LF, Lefkowitz RJ (1995): Physiological effects of inverse agonists in transgenic mice with myocardial overexpression of the β_2-adrenoceptor. Nature 374:272–276.

Bownds D (1967): Site of attachment of retinal in rhodopsin. Nature 216:1178–1181.

Burstein ES, Spalding TA, Brann MR (1997): Pharmacology of muscarinic receptor subtypes constitutively activated by G proteins. Mol Pharmacol 51:312–319.

Chabre M (1985): Trigger and amplification mechanisms in visual phototransduction. Annu Rev Biophys Biophys Chem 14:331–360.

Cohen DP, Thaw CN, Varma A, Gershengorn MC, Nussenzveig DR (1997): Human calcitonin receptors exhibit agonist-independent (constitutive) signaling activity. Endocrinology 138:1400–1405.

Cohen GB, Oprian DD, Robinson PR (1992): Mechanism of activation and inactivation of opsin: Role of Glu113 and Lys296. Biochemistry 31:12592–12601.

Cohen GB, Yang T, Robinson PR, Oprian DD (1993): Constitutive activation of opsin: Influence of charge at position 134 and size at position 296. Biochemistry 32: 6111–6115.

Colson A-O, Perlman JH, Smolyar A, Gershengorn MC, Osman R (1998): Static and dynamic roles of extracellular loops in G-protein–coupled receptors: A mechanism of sequential binding of thyrotropin-releasing hormone to its receptor. Biophys J 74:1087–1100.

Cotecchia S, Exum S, Caron MG, Lefkowitz RJ (1990): Regions of the α_1-adrenergic receptor involved in coupling to phosphatidylinositol hydrolysis and enhanced sensitivity of biological function. Proc Natl Acad Sci USA 87:2896–2900.

Davidson FF, Loewen PC, Khorana HG (1994): Structure and function in rhodopsin: Replacement by alanine of cysteine residues 110 and 187, components of a conserved disulfide bond in rhodopsin, affects the light-activated metarhodopsin II state. Proc Natl Acad Sci USA 91:4029–4033.

Dixon RAF, Kobilka BK, Strader DJ, Benovic JL, Dohlman HG, Frielle T, Bolanowski MA, Bennett CD, Rands E, Diehl RE, Mumford RA, Slater EE, Sigal IS, Caron MG, Lefkowitz RJ, Strader CD (1986): Cloning of the gene and cDNA for mammalian β-adrenergic receptor and homology with rhodopsin. Nature 321:75–79.

Doi T, Molday RS, Khorana HG (1990): Role of the intradiscal domain in rhodopsin assembly and function. Proc Natl Acad Sci USA 87:4991–4995.

Dratz EA, Hargrave PA (1983): The structure of rhodopsin and the rod outer segment disk membrane. Trends Biochem Sci 8:128–131.

Dryja TP, Berson EL, Rao VR, Oprian DD (1993): Heterozygous missense mutation in the rhodopsin gene as a cause of congenital stationary night blindness. Nature Genet 4:280–283.

Ernst OP, Hofmann KP, Sakmar TP (1995): Characterization of rhodopsin mutants that bind transducin but fail to induce GTP nucleotide uptake. Classification of mutant pigments by fluorescence, nucleotide release, and flash-induced light-scattering assays. J Biol Chem 270:10580–10586.

Fahmy K, Jäger F, Beck M, Zvyaga TA, Sakmar TP, Siebert F (1993): Protonation states of membrane-embedded carboxylic acid groups in rhodopsin and metarhodopsin II: A Fourier-transform infrared spectroscopy study of site-directed mutants. Proc Natl Acad Sci USA 90:10206–10210.

Fahmy K, Sakmar, TP (1993): Regulation of the rhodopsin–transducin interaction by a highly conserved carboxylic acid group. Biochemistry 32:7229–7236.

Fahmy K, Sakmar, TP (1995): The photoactivated state of rhodopsin and how it can form. Biophys Chem 56:171–181.

Fahmy K, Zvyaga TA, Sakmar TP, Siebert F (1996): Spectroscopic evidence for altered chromophore–protein interactions in low-temperature photoproducts of the visual pigments responsible for congenital night blindness. Biochemistry 35:15065–15073.

Farahbakhsh ZT, Hideg K, Hubbell WL (1993): Photoactivated conformational changes in rhodopsin: A time-resolved spin label study. Science 262:1416–1419.

Farrens DL, Altenbach C, Yang K, Hubbell WL, Khorana HG (1996): Requirement of rigid-body motion of transmembrane helices for light activation of rhodopsin. Science 264:768–770.

Franke RR, König B, Sakmar TP, Khorana HG, Hofmann KP (1990): Rhodopsin mutants that bind but fail to activate transducin. Science 250:123–125.

Franke RR, Sakmar TP, Graham RM, Khorana HG (1992): Structure and function in rhodopsin. Studies of the interaction between the rhodopsin cytoplasmic domain and transducin. J Biol Chem 267:14767–14774.

Fung BK-K, Hurley JB, Stryer L (1981): Flow of information in the light-triggered cyclic nucleotide cascade of vision. Proc Natl Acad Sci USA 78:152–156.

Gether U, Lin S, Ghanouni P, Ballesteros JA, Weinstein H, Kobilka BK (1997): Agonists induce conformational changes in transmembrane domains III and VI of the β_2 adrenoceptor. EMBO J 16:6737–6747.

Guy PM, Koland JG, Cerione RA (1990): Rhodopsin-stimulated activation–deactivation cycle of transducin: Kinetics of the intrinsic fluorescence response of the α subunit. Biochemistry 29:6954–6964.

Han M, DeDecker BS, Smith SO (1993): Localization of the retinal protonated Schiff base counterion in rhodopsin. Biophys J 65:899–906.

Han M, Groesbeek M, Sakmar TP, Smith SO (1997a): The C_9-methyl group of retinal interacts with glycine 121 in rhodopsin. Proc Natl Acad Sci USA 94:13442–13447.

Han M, Groesbeek M, Smith SO, Sakmar TP (1998a): The role of the C_9-methyl group in rhodopsin activation: Characterization of mutant opsins with the artificial chromophore 11-*cis*-9-demethyl-retinal. Biochemistry 37:538–545.

Han M, Lin SW, Smith SO, Sakmar TP (1996a): The effects of amino acid replacements of glycine 121 on transmembrane helix 3 of rhodopsin. J Biol Chem 271:32330–32336.

Han M, Lin SW, Minkova M, Smith SO, Sakmar TP (1996b): Functional interaction of transmembrane helices 3 and 6 in rhodopsin. Replacement of phenylalanine 261 by alanine causes reversion of phenotypes of glycine 121 replacement mutants. J Biol Chem 271:32337–32342.

Han M, Lou J, Nakanishi K, Sakmar TP, Smith SO (1997b): Partial agonist activity of 11-*cis*-retinal in rhodopsin mutants. J Biol Chem 272:23081–23085.

Han M, Smith SO (1995a): High-resolution structural studies of the retinal–Glu113 interaction in rhodopsin. Biophys Chem 56:23–29.

Han M, Smith SO (1995b): NMR constraints on the location of the retinal chromophore in rhodopsin and bathorhodopsin. Biochemistry 34:1425–1432.

Han M, Smith SO, Sakmar TP (1998b): Constitutive activation of rhodopsin by mutation of methionine 257 on transmembrane helix 6. Biochemistry 37:8253–8261.

Hargrave PA, Downds D, Wang JK, McDowell JH (1982): Retinyl peptide isolation and characterization. Methods Enzymol 81:211–215.

Hargrave PA, McDowell JH, Curtis DR, Wang JK, Juszczak E, Fong S-L, Mohana Rao JK, Argos P (1983): The structure of bovine rhodopsin. Biophys Struct Mech 9:235–244.

Henderson R, Baldwin JM, Ceska TA, Zemlin F, Beckmann E, Downing KH (1990): Model for the structure of bacteriorhodopsin based on high-resolution electron cryomicroscopy. J Mol Biol 213:899–929.

Herzyk P, Hubbard RE (1995): Automated method for modeling seven-helix transmembrane receptors from experimental data. Biophys J 69:2419–2442.

Hofmann KP, Jäger S, Ernst OP (1995): Structure and function of activated rhodopsin. Israel J Chem 35:339–355.

Hong K, Hubbell WL (1973): Lipid requirements for rhodopsin regenerability. Biochemistry 12:4517–4523.

Honig B, Kinur U, Nakanishi K, Balogh-Nair V, Gawinowicz MA, Arnaboldi M, Motto MG (1979): An external point-charge model for wavelength regulation in visual pigments. J Am Chem Soc 101:7084–7086.

Jäger F, Fahmy K, Sakmar TP, Siebert F (1994): Identification of glutamic acid 113 as the Schiff base proton acceptor in the metarhodopsin II photointermediate of rhodopsin. Biochemistry 33:10878–10882.

Karnik SS, Khorana HG (1990): Assembly of functional rhodopsin requires a disulfide bond between cysteine residues 110 and 187. J Biol Chem 265:17520–17524.

Karnik SS, Sakmar TP, Chen H-B, Khorana HG (1988): Cysteine residues 110 and 187 are essential for the formation of correct structure in bovine rhodopsin. Proc Natl Acad Sci USA 85:8459–8463.

Keen TJ, Inglehearn CF, Lester DH, Bashir R, Jay M, Bird AC, Jay B, Bhattacharya SS (1991): Autosomal dominant retinitis pigmentosa: Four new mutations in rhodopsin, one of them in the retinal attachment site. Genomics 11:199–205.

Khorana HG (1992): Rhodopsin, photoreceptor of the rod cell. An emerging pattern for structure and function. J Biol Chem 267:1–4.

Kibelbek J, Mitchell DC, Beach JM, Litman BJ (1991): Functional equivalence of metarhodopsin II and the G_t-activating form of photolyzed bovine rhodopsin. Biochemistry 30:6761–6768.

Kim JM, Altenbach C, Thurmond RL, Khorana HG, Hubbell WL (1997): Structure and function in rhodopsin: Rhodopsin mutants with a neutral amino acid at E134 have a partially activated conformation in the dark state. Proc Natl Acad Sci USA 94: 14273–14278.

Kimura Y, Vassylyev DG, Miyazawa A, Kidera A, Matsushima M, Mitsuoka K, Murata K, Hirai T, Fujiyoshi Y (1997): Surface of bacteriorhodopsin revealed by high-resolution electron crystallography. Nature 389:206–211.

Kochendoerfer GG, Wang Z, Oprian DD, Mathies RA (1997): Resonance Raman examination of the wavelength regulation mechanism in human visual pigments. Biochemistry 36:6577–6587.

König B, Arendt A, McDowell JH, Kahlert M, Hargrave PA, Hofmann KP (1989): Three cytoplasmic loops of rhodopsin interact with transducin. Proc Natl Acad Sci USA 86:6878–6882.

Kuhn H (1980): Light- and GTP-regulated interaction of GTPase and other proteins with bovine photoreceptor membranes. Nature 283:587–589.

Lambright DG, Sondek J, Bohm A, Skiba NP, Hamm HE, Sigler PE (1996): The 2.0 Å crystal structure of a heterotrimeric G protein. Nature 379:311–319.

Lanyi JK (1995): Bacteriorhodopsin as a model for proton pumps. Nature 375:461–463.

Leff P (1995): The two-state model of receptor activation. Trends Pharmacol Sci 16:89–97.

Lefkowitz RJ, Cotecchia S, Samama P, Costa T (1993): Constitutive activity of receptors coupled to guanine nucleotide regulatory proteins. Trends Pharmacol Sci 14:303–307.

Lin SW, Sakmar TP (1996): Specific tryptophan UV-absorbance changes are probes of the transition of rhodopsin to its active state. Biochemistry 35:11149–11159.

Lin SW, Sakmar TP, Franke RR, Khorana HG, Mathies RA (1992): Resonance Raman microprobe spectroscopy of rhodopsin mutants: Effect of substitutions in the third transmembrane helix. Biochemistry 31:5105–5111.

Longstaff C, Calhoon RD, Rando RR (1986): Deprotonation of the Schiff base of rhodopsin is obligate in the activation of the G protein. Proc Natl Acad Sci USA 83:4209–4213.

Merbs SL, Nathans J (1992): Absorption spectra of human cone pigments. Nature 356:433–435.

Min KC, Zvyaga TA, Cypess AM, Sakmar TP (1993): Characterization of mutant rhodopsins responsible for autosomal dominant retinitis pigmentosa. Mutations on the cytoplasmic surface affect transducin activation. J Biol Chem 268:9400–9404.

Nakayama TA, Khorana HG (1990): Orientation of retinal in bovine rhodopsin determined by cross-linking using a photoactivatable analog of 11-*cis*-retinal. J Biol Chem 265:15762–15769.

Nathans J (1990): Determinants of visual pigment absorbance: Role of charged amino acids in the putative transmembrane segments. Biochemistry 29:937–942.

Nathans J (1992): Rhodopsin: Structure, function, and genetics. Biochemistry 31:4923–4930.

Nathans J, Hogness DS (1983): Isolation, sequence analysis, and intron–exon arrangement of the gene encoding bovine rhodopsin. Cell 34:807–814.

Nathans J, Hogness DS (1984): Isolation and nucleotide sequence of the gene encoding human rhodopsin. Proc Natl Acad Sci USA 81:4851–4855.

Nathans J, Thomas D, Hogness DS (1986): Molecular genetics of human color vision: The genes encoding blue, green, and red pigments. Science 232:193–202.

Oprian DD, Aseñjo AB, Lee N, Pelletier SL (1991): Design, chemical synthesis, and expression of genes for the three human color vision pigments. Biochemistry 30:11367–11372.

Oprian DD, Molday RS, Kaufman RJ, Khorana HG (1987): Expression of a synthetic bovine rhodopsin gene in monkey kidney cells. Proc Natl Acad Sci USA 84:8874–8878.

Osawa S, Weiss ER (1994): The carboxyl terminus of bovine rhodopsin is not required for G protein activation. Mol Pharmacol 46:1036–1040.

Oseroff AR, Callender RH (1974): Resonance Raman spectroscopy of rhodopsin in retinal disk membranes. Biochemistry 13:4243–4248.

Ovchinnikov YA (1982): Rhodopsin and bacteriorhodopsin: Structure–function relationships. FEBS Lett 148:179–191.

Papermaster DS (1982): Preparation of retinal rod outer segments. Methods Enzymol 81:48–52.

Papermaster DS, Dreyer WJ (1974): Rhodopsin content in the outer segment membranes of bovine and frog retinal rods. Biochemistry 13:2438–2444.

Parma J, Duprez L, Van Sende J, Cochauz P, Gervy C, Mockel J, Dumont J, Vassart G (1993): Somatic mutations in the thyrotropin receptor gene cause hyperfunctioning thyroid adenomas. Nature 365:649–651.

Pebay-Peyroula E, Rummel G, Rosenbusch JP, Landau EM (1997): X-ray structure of bacteriorhodopsin at 2.5 Angstroms from microcrystals grown in lipidic cubic phases. Science 277:1676–1681.

Phillips WJ, Cerione RA (1988): The intrinsic fluorescence of the α subunit of transducin. Measurement of receptor-dependent guanine nucleotide exchange. J Biol Chem 263:15498–15505.

Phillips WJ, Wong SC, Cerione RA (1992): Rhodopsin/transducin interactions. II. Influence of the transducin–βγ subunit complex on the coupling of the transducin–α subunit to rhodopsin. J Biol Chem 267:17040–17046.

Pogozheva ID, Lomize AL, Mosberg HI (1997): The transmembrane 7-alpha-bundle of rhodopsin: Distance geometry calculations with hydrogen bonding constraints. Biophys J 72:1963–1985.

Pollak MR, Brown EM, Estep HL, McLaine PN, Kifor O, Park J, Hebert SC, Seidman CE, Seidman JG (1994): Autosomal dominant hypocalcemia caused by a Ca^{2+}-sensing receptor gene mutation. Nature Genet 8:303–307.

Porcellini A, Ciullo I, Laviola L, Amabile G, Fenzi G, Avvedimento VE (1994): Novel mutations of thyrotropin receptor gene in thyroid hyperfunctioning adenomas. J Clin Endocrinol Metab 79:657–661.

Rando RR (1996): Polyenes and vision. Chem Biol 3:255–262.

Rao VR, Cohen GB, Oprian DD (1994): Rhodopsin mutation G90D and a molecular mechanism for congenital night blindness. Nature 367:639–642.

Rao VR, Oprian DD (1996): Activating mutations of rhodopsin and other G protein–coupled receptors. Annu Rev Biophys Biomol Struct 25:287–314.

Resek JF, Farahbakhsh ZT, Hubbell WL, Khorana HG (1993): Formation of the meta II photointermediate is accompanied by conformational changes in the cytoplasmic surface of rhodopsin. Biochemistry 32:12025–12032.

Rim J, Oprian DD (1995): Constitutive activation of opsin: Interaction of mutants with rhodopsin kinase and arrestin. Biochemistry 34:11938–11945.

Robinson PR, Cohen GB, Zhukovsky EA, Oprian DD (1992): Constitutively active mutants of rhodopsin. Neuron 9:719–725.

Röper D, Jacoby E, Krüger F, Engels M, Grötzinger J, Wollmer A, Strassburger W (1994): Modeling of G-protein coupled receptors with bacteriorhodopsin as a template. A novel approach based on interaction energy differences. J Receptor Res 14:167–186.

Sakmar TP (1994): Opsins. In Peroutka SJ (ed): Handbook of Receptors and Channels: G Protein Coupled Receptors. Boca Raton, FL: CRC Press, pp 257–277.

Sakmar TP (1998): Rhodopsin: A prototypical G protein–coupled receptor. Progr Nucl Acid Res Mol Biol 59:1–34.

Sakmar TP, Franke RR, Khorana HG (1989): Glutamic acid 113 serves as the retinylidene Schiff base counterion in bovine rhodopsin. Proc Natl Acad Sci USA 86:8309–8313.

Schertler GFX, Hargrave PA (1995): Projection structure of frog rhodopsin in two crystal forms. Proc Natl Acad Sci USA 92:11578–11582.

Schertler GFX, Villa C, Henderson R (1993): Projection structure of rhodopsin. Nature 362:770–772.

Schipani E, Kruse K, Jüppner H (1995): A constitutively active mutant PTH–PTHrP receptor in Jansen-type metaphyseal chondrodysplasia. Science 268:98–99.

Sheikh S, Zvyaga TA, Lichtarge O, Sakmar TP, Boume HR (1996): Rhodopsin activation blocked by metal-ion–binding sites linking transmembrane helices C and F. Nature 383:347–350.

Shenker A, Laue L, Kosugi S, Merendino JJ Jr, Minegishi T, Cutler GB Jr (1993): A constitutively activating mutation of the luteinizing hormone receptor in familial male precocious puberty. Nature 365:652–654.

Shichi H, Yamamoto K, Somers RL (1984): GTP binding protein: Properties and lack of activation by phosphorylated rhodopsin. Vision Res 24:1523–1531.

Siebert F (1995): Infrared spectroscopy applied to biochemical and biological problems. Methods Enzymol 246:501–526.

Shieh T, Han M, Sakmar TP, Smith SO (1997): The steric trigger in rhodopsin activation. J Mol Biol 269:373–384.

Sieving PA, Richards JE, Naarendorp F, Bingham EL, Scott K, Alpern M (1995): Darklight: Model for nightblindness from the human rhodopsin Gly-90 to Asp mutation. Proc Natl Acad Sci USA 92:880–884.

Stryer L (1991): Visual excitation and recovery. J Biol Chem 266:10711–10714.

Takemoto DJ, Morrison D, Davis LC, Takemoto LJ (1986): C-terminal peptides of rhodopsin. Determination of the optimum sequence for recognition of retinal transducin. Biochem J 235:309–312.

Tiberi M, Caron MG (1994): High agonist-independent activity is a distinguishing feature of the dopamine D1B receptor subtype. J Biol Chem 269:27925–27931.

Ting TD, Goldin SB, Ho Y-K (1993): Purification and characterization of bovine transducin and its subunits. In Iyengar A (ed): Methods in Neurosciences, Photoreceptor Cells, vol 15. New York: Academic Press, pp 180–195.

Unger VM, Hargrave PA, Baldwin JM, Schertler GFX (1997): Arrangement of rhodopsin transmembrane α-helices. Nature 389:203–206.

van Rhee AM, Jacobson KA (1996): Molecular architecture of G protein–coupled receptors. Drug Dev Res 37:1–38.

van Sande J, Parma J, Tonacchera M, Swillens S, Dumont J, Vassart G (1995): Somatic and germline mutations of the TSH receptor gene in thyroid diseases. J Clin Endocrinol Metab 80:2577–2585.

Weiss ER, Osawa S, Shi W, Dickerson CD (1994): Effects of carboxyl-terminal truncation on the stability and G protein–coupling activity of bovine rhodopsin. Biochemistry 33:7587–7593.

Wessling-Resnick M, Johnson GL (1987): Transducin interactions with rhodopsin: Evidence for positive cooperative behavior. J Biol Chem 262:12444–12447.

Yang T, Snider BB, Oprian DD (1997): Synthesis and characterization of a novel retinylamine analog inhibitor of constitutively active rhodopsin mutants found in patients with autosomal dominant retinitis pigmentosa. Proc Natl Acad Sci USA 94: 13559–13564.

Yeagle PL, Alderfer JL, Albert AD (1995): Structure of the third cytoplasmic loop of bovine rhodopsin. Biochemistry 34:14621–14625.

Yeagle PL, Alderfer JL, Albert AD (1997): Three-dimensional structure of the cytoplasmic face of the G protein receptor rhodopsin. Biochemistry 36:9649–9654.

Zhang H, Lerro KA, Yamamoto T, Lien TH, Sastry L, Gawinowicz MA, Nakanishi K (1994): The location of the chromophore in rhodopsin: A photoaffinity study. J Am Chem Soc 116:10165–10173.

Zhukovsky EA, Oprian DD (1989): Effect of carboxylic acid side chains on the absorption maximum of visual pigments. Science 246:928–930.

Zhukovsky EA, Robinson PR, Oprian DD (1991): Transducin activation by rhodopsin without a covalent bond to the 11-*cis*-retinal chromophore. Science 251:558–560.

CHAPTER 6

SPLIT RECEPTORS AS TOOLS FOR ANALYZING G PROTEIN–COUPLED RECEPTOR STRUCTURE

MASAHIRO KONO and DANIEL D. OPRIAN

1. INTRODUCTION   109
2. BACKGROUND   111
   A Disulfide Crosslinking   111
   B. Split Receptors   111
   C. Disulfide Crosslinking of Split Receptors   112
3. METHODS   112
   A. Nomenclature   112
   B. Design and Synthesis of Split Receptor Genes   112
   C. Strategy for Split Sites   113
   D. Expression and Reconstitution of Split Rhodopsins   113
   E. Oxidative Crosslinking   114
4. EXPERIMENTAL CONSIDERATIONS   117
5. CURRENT WORK AND FUTURE DIRECTIONS   117

1. INTRODUCTION

We have developed a method for probing G protein–coupled receptor (GPCR) structure. Tertiary contacts between helices are identified by engineering disulfide crosslinks between Cys residues introduced at two sites in a "split receptor" construct (Yu et al., 1995). By split receptors, we mean that our GPCR is no longer a single polypeptide but rather comprised of two complementary halves of the protein forming a functional two-fragment GPCR (see below). If a single cysteine on one fragment of a split receptor forms a disulfide bond with a single cysteine on the other fragment, the protein will have the same mobility on a nonreducing SDS-PAGE gel as that of the native full-length protein. In contrast, if a disulfide bond does not bridge the two fragments, then the

Structure–Function Analysis of G Protein-Coupled Receptors, Edited by Jürgen Wess.
ISBN 0-471-25228-X   Copyright © 1999 Wiley-Liss, Inc.

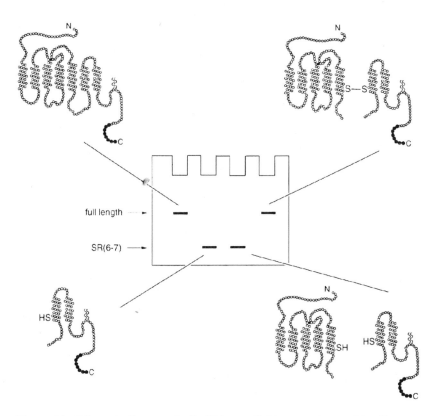

Figure 6.1. Schematic diagram of a disulfide crosslinking experiment with split receptors. In the Western blots described in this chapter, we used the rho-1D4 monoclonal antibody, which is specific for the carboxy-terminal eight amino acid residues of rhodopsin (indicated as filled circles) (Molday and MacKenzie, 1983). SDS-PAGE is performed under nonreducing conditions. When the two halves of the split receptor are not linked by a disulfide bond, the protein appears on a Western blot probed with the 1D4 antibody as a band migrating with the same mobility as that of the carboxy-terminal fragment alone (lanes 2 and 3). When the two halves of the split receptor are linked by a disulfide bond, the protein appears as a band migrating with the same mobility as that of the full-length protein (lanes 1 and 4).

protein will appear as a band with a higher mobility than that of the full-length protein (Figure 6.1). Site-specific mutagenesis can be used to place individual Cys residues anywhere in each split receptor fragment, and if two Cys residues form a disulfide bond then the two sites can be deemed as likely to be in tertiary contact. In this manner the tertiary structure of a GPCR can be mapped.

In this chapter, we describe how we have set up this system using split rhodopsins. Rhodopsin is a prototypical member of the GPCR family of proteins. The best three-dimensional structure for a GPCR is from electron cryomicroscopy studies of rhodopsin (Schertler et al., 1993; Unger and Schertler, 1995; Unger et al., 1997). Furthermore, rhodopsin has several properties that lend itself to structural and functional studies, including a distinctive color for a properly folded protein (absorption max at 500 nm), a very convenient and fast activation switch (light), spectral shifts corresponding to intermediate states during the photoreaction, assays for measuring its ability to activate G protein, and its ability to be expressed at relatively high levels. These aspects of rhodopsin are discussed in much more detail in other chapters in this volume. Because of these characteristics, it is a very good candidate for determining tertiary contacts through disulfide crosslinking.

2. BACKGROUND

2.A. Disulfide Crosslinking

The idea of systematically engineering disulfide crosslinks between transmembrane helices and probing for them via gel mobility shifts on Western blots is not new. The technique was pioneered by Falke and Koshland (1987) to map the helix–helix contact sites between subunits of the homodimeric chemotactic aspartate receptor. In this approach, amino acid residues are individually replaced with Cys in the transmembrane region. The proteins are then exposed to a chemical oxidant to crosslink juxtaposed cysteines, and the samples are examined for disulfide crosslinked dimers by electrophoretic mobility shifts on nonreducing SDS gels. This method has been used to characterize the transmembrane domain of the bacterial aspartate (Falke and Koshland, 1987; Lynch and Koshland, 1991), Tar (Pakula and Simon, 1992), and Trg (Lee et al., 1994) chemotactic receptors. Because all of these receptors are homodimers, when a disulfide bond formed, a single cysteine residue on one helix crosslinked with a cysteine at the same location on the other subunit, forming a dimer that was easily detected by electrophoresis. The problem with studying GPCR structure using this method is that the interhelical contacts that are of primary interest are in the tertiary and not quaternary structure of the protein.

2.B. Split Receptors

To circumvent this problem, we decided to use split GPCRs. Again, the idea of making split receptors, itself, is not new. Reconstituting fully functional two-fragment membrane proteins was pioneered by Khorana and coworkers, who purified bacteriorhodopsin fragments and refolded a functional proton pump by mixing together complementary protein fragments (Huang et al., 1981). Subsequently, several other membrane proteins were shown to form functional proteins from two or more fragments, including ß$_2$-adrenergic receptor (Kobilka et al., 1988), the sodium channel (Stühmer et al., 1989), adenylylcyclase (Tang et al., 1991), muscarinic acetylcholine receptors (Maggio et al., 1993; Schöneberg et al., 1995), lactose permease (Bibi and Kaback, 1990; Wrubel et al., 1990, 1994; Zen et al., 1994), the yeast a-factor transporter (Berkower and Michaelis, 1991), the Shaker potassium channel (Naranjo et al., 1997), and rhodopsin (Ridge et al., 1995, 1996; Yu et al., 1995). Generally, these split proteins were made by coexpressing separate plasmids containing genes encoding protein fragments, allowing the cells to synthesize and fold the protein.

2.C. Disulfide Crosslinking of Split Receptors

Our initial study demonstrated the feasibility to crosslink split receptors via disulfide bonds (Yu et al., 1995). We describe this method in more detail in this chapter. We note that it has recently been used to identify tertiary contacts in the lactose permease (Wu and Kaback, 1996).

3. METHODS

3.A. Nomenclature

Our nomenclature for the split rhodopsin mutants is adapted from that used for split mutants of the β_2-adrenergic receptor, as described by Kobilka et al. (1988). Thus SR(1–3) corresponds to the amino-terminal split receptor fragment containing the first three transmembrane helices. SR(4–7) corresponds to the carboxy-terminal split receptor fragment containing the last four helices. SR(1–3/4–7) corresponds to the split receptor complex assembled from SR(1–3) and SR(4–7) (Figure 6.2). Point mutations are indicated by beginning with the name of the fragment, followed by a colon, followed by the single letter code for the wild-type amino acid and amino acid number, followed by the single letter code for the new amino acid. For example, the double mutant in which Val^{204} is changed to Cys in SR(1–5) and Phe^{276} is changed to Cys in SR(6–7) is designated as SR(1–5:V204C/6–7:F276C).

3.B. Design and Synthesis of Split Receptor Genes

We have prepared three split receptor constructs (Yu et al., 1995): SR(1–3/4–7), SR(1–4/5–7), and SR(1–5/6–7) (Figure 6. 2). Ridge et al. (1995, 1996) have constructed functional rhodopsins with similar (and in two cases identical) split sites. The genes for each split fragment were made as an *Eco*RI–*Not*I cassette

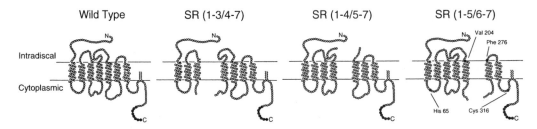

Figure 6.2. Schematic representation of the structure of the wild-type and split rhodopsins described in this chapter. The nomenclature is adopted from Kobilka et al. (1988) and described in detail in the text. The carboxy-terminal eight amino acids in each structure are highlighted by filled circles to indicate the location of the epitope recognized by the rho-1D4 monoclonal antibody (Molday and MacKenzie, 1983). SR(1–3/4–7) is split between Pro^{142} and Met^{143} in the second cytoplasmic loop separating transmembrane segments 3 and 4. SR(1–4/5–7) is split between Pro^{194} and His^{195} in the second intradiscal loop separating transmembrane segments 4 and 5. SR(1–5/6–7) is split between Ser^{240} and Ala^{241} in the third cytoplasmic loop separating transmembrane segments 5 and 6.

(Yu et al., 1995) and derived from a synthetic gene for bovine rhodopsin cloned into a pMT3 expression vector (Ferretti et al., 1986; Oprian et al., 1987; Franke et al., 1988).

For amino-terminal fragments, the sequence was exactly the same as in full-length rhodopsin until the split site where a stop codon was introduced followed by a *Not*I restriction site. For carboxy-terminal fragments, the 3′ sequence was exactly the same as in full-length rhodopsin, and the 5′ end was constructed with an *Eco*RI restriction site, followed by a Kozack consensus sequence, followed by an initiating Met codon (unless the first residue of the split site is already a Met), followed by the sequence of the remainder of the protein fragment.

3.C. Strategy for Split Sites

There are several factors that need to be considered in deciding where a membrane protein should be split. Ideally, we wanted to position split sites in the interhelical loop regions where the effects on folding and function are anticipated to be minimal. Rhodopsin can be cleaved by thermolysin at 240/241 in the third cytoplasmic loop and by chymotrypsin at 146/147 in the second cytoplasmic loop and still remain functional. We chose the former site to create our SR(1–5/6–7) construct. We chose a site near the chymotryptic digest site between Pro[142] and Met[143] to generate SR(1–3/4–7) because the carboxy-terminal fragment SR(4–7) did not require an additional initiating Met since the native Met at that location could be used. SR(1–4/5–7) was split between Pro[194] and His[195]. This site had been shown by Khorana and coworkers (Doi et al., 1990) to be in the middle of a 12 amino acid residue stretch in the second extracellular loop (residues 191–202) that could tolerate point mutations and deletions. In contrast, the native sequence on either end of this region seemed to be required for a properly expressed and folded protein (Doi et al., 1990).

3.D. Expression and Reconstitution of Split Rhodopsins

The split opsin gene was transiently transfected into COS cells as described previously (Oprian et al., 1987; Oprian, 1993; Yu et al., 1995). Briefly, confluent plates (10 cm) of COS cells were cotransfected with 2 µg of each plasmid per plate using the DEAE/dextran and DMSO shock method as detailed elsewhere (Oprian et al., 1987; Oprian, 1993; Yu et al., 1995). Cells were harvested on the third day after transfection.

Split rhodopsin was reconstituted and immunopurified with the 1D4 antibody, which recognizes the last eight amino acid residues of the native carboxy terminus of rhodopsin, in the same manner as originally described for full-length rhodopsin (Molday and MacKenzie, 1983; Oprian et al., 1987; Oprian, 1993). The cells were incubated with 20 µM 11-*cis*-retinal in PBS (pH 7), solubilized in 1% dodecyl maltoside (DM) in PBS, bound to a 1D4–Sepharose 4B affinity matrix, washed with 0.1% DM in PBS, and eluted from the matrix with a peptide corresponding to the last 18 amino acid residues of rhodopsin. Typically, about 2 µg rhodopsin can be purified per plate in this manner.

When the 1D4-immunopurified split rhodopsins are subjected to nonreducing SDS-PAGE and probed on Western blots for the amino-terminal fragments, only individual fragments, SR(1–3), SR(1–4), and SR(1–5), not crosslinked full-length rhodopsin-sized proteins, are detected; when probed with an antibody directed against the carboxy-terminal fragment, only the individual split receptor fragments SR(4–7), SR (5–7), and SR(6–7) are detected (Ridge et al., 1995; Yu et al., 1995). While this result was expected for SR(1–4/5–7) and SR(1–5/6–7), it was somewhat surprising for SR(1–3/4–7) since the two halves of this mutant should be bridged on the intradiscal side by the native Cys^{110}–Cys^{187} disulfide bond (Kamik and Khorana, 1990). In subsequent studies we showed that if SR(1–3/4–7) is denatured in the presence of N-ethylmaleimide (NEM) or if Cys^{185} is changed to Ser, the protein migrates with the same mobility as that of the full-length rhodopsin on nonreducing SDS gels (Kono et al., 1998). These results suggest that the native disulfide bond exists in SR(1–3/4–7), but upon denaturation the thiolate of Cys^{185} reacts with Cys^{187}, releasing Cys^{110} in a disulfide bond exchange reaction. When NEM is included in the gel-loading buffer, Cys^{185} reacts rapidly with NEM, preventing it from participating in an exchange reaction with the native disulfide bond. Similarly, mutation of Cys^{185} to Ser eliminates the thiol and the possibility of the disulfide exchange reaction.

All three split rhodopsins displayed wild-type–like functional properties. They produced wild-type–like pigments with absorption maxima at 500 nm, which shifted to 380 nm after exposure to light, typical of the active Meta II intermediate (Ridge et al., 1995; Yu et al., 1995). SR(1–4/5–7) was expressed at a similar level as full-length wild-type rhodopsin, as determined by the optical density of purified samples at 500 nm; SR(1–3/4–7) was expressed at about 30%, and SR(1–5/6–7) was expressed at about 50% of wild-type levels, respectively.

All three split rhodopsins were active as determined by the ability to activate transducin in a light-dependent manner (Ridge et al., 1995; Yu et al., 1995). SR(1–3/4–7) and SR(1–4/5–7) had wild-type levels of specific activity, whereas SR(1–5/6–7) had about 25% the activity of wild-type. The low activity of SR(1–5/6–7) was expected since a similar split receptor mutant of the ß$_2$-adrenergic receptor was found to have only 25% the activity of the wild-type receptor (Kobilka et al., 1988).

3.E. Oxidative Crosslinking

Disulfide crosslinks typically were induced in the split receptor constructs using a copper phenanthroline oxidant according to the procedure described by Lee et al. (1994). A 5× stock solution of copper phenanthroline was prepared containing 15 mM $CuSO_4$, 45 mM phenanthroline, and 10% glycerol in 50 mM sodium phosphate buffer, pH7. Four microliters of this copper solution was mixed with 1–10 µl of purified rhodopsin such that the final reaction volume was 20 µl. The reactions were allowed to proceed under dim red light at 37°C until terminated by addition of (final concentrations) 12.5 mM NEM and 12.5 mM EDTA in Laemmli gel load buffer (60 mM Tris buffer [pH 6.8] containing 2% [w/v] SDS, 6% [w/v] sucrose, and 0.005% [w/v] bromophenol blue). When desired, a final concentration of 5% (v/v) ß-mercaptoethanol or 50 mM DTT was added to the denatured copper-treated sample. Specifically, 20 µl of the gel load/quench buffer was added to each sample to stop the reaction and was comprised of the following: 8 µl gel load buffer consisting of 300 mM Tris, 10% SDS, 30% sucrose, and 0.025% bromophenol blue, 4 µl 125 mM NEM (freshly prepared in water), 1 µl 500 mM EDTA, and 7 µl water. Two microliters of mercaptoethanol was added after quenching, if necessary.

Samples were then subjected to SDS-PAGE on 12%–15% polyacrylamide gels using a BioRad mini gel apparatus and analyzed on Western blots for crosslinking of fragments using the antirhodopsin antibody 1D4 as the primary antibody for the carboxy-terminal fragments and an alkaline phosphatase-conjugated antimouse IgG as the secondary one (Promega). Proteins were then visualized by reaction with nitroblue tetrazolium (NBT) and 5-bromo-4-chloro-3-indolyl-1-phosphate (BCIP) according to directions from Promega.

In preliminary studies to show the feasibility of engineering disulfides between juxtaposed Cys residues, we selected residues 200, 204, and 276, which had been shown by Elling et al. (1995) to form a zinc binding site when substituted for His at the corresponding sites in the tachykinin NK_1 receptor. As shown in Figure 6.3, SR(1–5:V204C) was crosslinked to SR(6–7: F276C) only when both mutations were present and after oxidation in the presence of copper phenanthroline. Disulfide bond formation is confirmed by the observation that the electrophoretic mobility of the protein increases to that of the free carboxy-

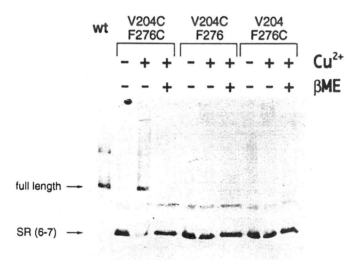

Figure 6.3. Western blot analysis of crosslinked SR(1–5/6–7) mutants based on an engineered zinc binding site in the tachykinin NK_1 receptor (Elling et al., 1995; Yu et al., 1995). From left to right, lanes contain wild-type (wt) full-length rhodopsin, SR(1–5:V204C/6–7:F276C), SR(1–5:V204C/6–7), and SR(1–5/6–7:F276C) as indicated above the Western blot. For each split rhodopsin mutant, there is a set of three lanes: before treatment with copper (–) after treatment with copper (+), and after treatment with copper followed by treatment with 5% (v/v) mercaptoethanol (+ ßME). The copper-treated samples were incubated in 3 mM copper phenanthroline for 20 minutes at 37°C as described in the text. The positions of full-length rhodopsin and the SR(6–7) fragment are indicated by arrows. (Reproduced from Yu et al. [1995], with permission of the publisher.)

terminal fragment, SR(6–7), after treatment of the full-length protein with ß-mercaptoethanol. The existence of a disulfide crosslink between residues 204 and 276 is illustrated by the fact that the single Cys mutants SR(1–5:V204C/6–7) and SR(1–5/6–7:F276C) never comigrated with the full-length protein even after copper treatment. Both single mutants and the double mutant prior to oxidation displayed normal spectral and activity properties (Yu et al., 1995). Cys residues at positions 200 and 276 also formed a disulfide bond, but this mutant did not form a pigment and formed a disulfide bond without oxidant (not shown). These disulfide crosslinking results using split receptor fragments not only demonstrate the great potential of the method but also suggest that different GPCRs share a common structure. The ability to engineer disulfide bonds between nearby residues in rhodopsin at positions juxtaposed in NK_1 receptors (as determined by an engineered zinc binding site) suggests that helix–helix packing is similar in these two receptors.

We also tested the split receptor mutants for formation of a disulfide bond between residues 65 and 316. Yang et al. (1996) have shown by site-directed spin-labeling experiments that these residues are within 10 Å from each other and that a Cys at both of these positions resulted in spontaneous formation of a disulfide bond. As shown in Figure 6.4, Cys^{65} also forms a disulfide crosslink to Cys^{316} in the split rhodopsin mutant SR(1–5:H65C/6–7). The full-length protein band appeared even in the absence of treatment with copper. When the na-

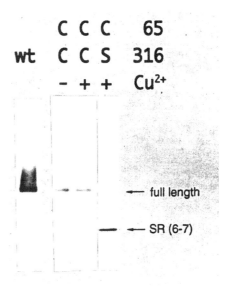

Figure 6.4. Western blot analysis of crosslinked SR(1–5/6–7) mutants based on spin labeling and disulfide crosslinking studies on full-length rhodopsin by Yang et al. (1996). From left to right, lanes contain wild-type (wt) full-length rhodopsin, SR(1–5:H65C/6–7), SR(1–5:H65C/6–7), and SR(1–5:H65C/6–7:C316S). The SR(1–5/6–7) mutants are indicated by the residue at positions 65 and 316. Residue 316 is a native Cys. Full-length split rhodopsin for SR(1–5:H65C/6–7) appears independent of copper treatment. The positions of full-length rhodopsin and the SR(6–7) fragment are indicated by arrows.

tive Cys[316] was mutated to Ser, the crosslinked product was no longer seen on Western blots.

4. EXPERIMENTAL CONSIDERATIONS

The primary advantage of engineering disulfide bonds in split receptors is its simplicity. It requires basic equipment available in most modern biology laboratories, and the results are clear—there is either a full-length receptor band or there is not, indicating the presence or absence of a disulfide bond. Although it is also possible to engineer Cys residues into full-length proteins, it is much more laborious to assay for the presence of disulfide bonds. In addition, the number of mutants needed for a systematic scan between two helices makes experiments using full-length proteins even more labor intensive. To completely scan a pair of helices of 20 residues each, only 40 single Cys mutants would have to be constructed in the case of split receptors (20 Cys mutants per split receptor fragment); 1,600 double mutants would have to be constructed to conduct the same experiment with a full-length protein.

One must, however, be cautious when interpreting disulfide crosslinking results. A positive result (crosslink) does not necessarily indicate that the two residues are in tertiary contact to each other in the native, folded state. The crosslink could result from trapping sparsely populated conformational states that arise from thermal fluctuations in the global structure of the protein. When a disulfide bond forms in this case, an unfavored conformation will accumulate from the irreversible nature of the covalent disulfide that forms. Also, disulfide crosslink formation in the presence of copper phenanthroline has been reported for cysteines thought to be as far as 10Å away from each other (Pakula and Simon, 1992), a distance too far for cysteines in disulfide bonds. Partial crosslinks are also problematic and likely also to arise from irreversible trapping of different conformations of the protein. The lack of a crosslink should also be interpreted with caution. The two residues may be close but not in a proper orientation or microenvironment to support formation of a disulfide bond. In these instances, it is important to consider the kinetics of disulfide formation as well as data from cross-linking of other residues in the region revealing periodicity consistent with a known secondary structure.

5. CURRENT WORK AND FUTURE DIRECTIONS

The use of split receptors in studying the structure of membrane proteins has great potential. Using disulfide crosslinking of split receptors, we are focusing on limited scans between helices to begin mapping the tertiary structure of rhodopsin. In addition, we have found crosslinks that depend on the activation state of rhodopsin. In other words, the crosslinking pattern of the dark inactive state can differ from the active state of the protein. These results will enable us to compare and contrast the tertiary structures of both states and gain a better understanding of specific changes involved in GPCRs' ability to activate G proteins.

ACKNOWLEDGMENTS

We thank Timothy D. McKee for reading the manuscript and providing helpful comments.

REFERENCES

Berkower C, Michaelis S (1991): Mutational analysis of the yeast a-factor transporter STE6, a member of the ATP binding cassette (ABC) protein superfamily. EMBO J 10:3777–3785.

Bibi E, Kaback HR (1990): In vivo expression of the lac Y gene in two segments leads to functional lac permease. Proc Natl Acad Sci USA 87:4325–4329.

Doi T, Molday RS, Khorana HG (1990): Role of the intradiscal domain in rhodopsin assembly and function. Proc Natl Acad Sci USA 87:4991–4995.

Elling CE, Nielsen SM, Schwartz, TW (1995): Conversion of antagonist-binding site to metal-ion site in the tachykinin NK-1 receptor. Nature 374:74–77.

Falke JJ, Koshland DE Jr (1987): Global flexibility in a sensory receptor: A site-directed cross-linking approach. Science 237:1596–1600.

Ferretti L, Karnik SS, Khorana HG, Nassal M, Oprian DD (1986): Total synthesis of a gene for bovine rhodopsin. Proc Natl Acad Sci USA 83:599–603.

Franke RR, Sakmar TP, Oprian DD, Khorana HG (1988): A single amino acid substitution in rhodopsin (lysine 248 → leucine) prevents activation of transducin. J Biol Chem 263:2119–2122.

Huang K-S, Bayley H, Liao M-J, London E, Khorana HG (1981): Refolding of an integral membrane protein—Denaturation, renaturation, and reconstitution of intact bacteriorhodopsin and two proteolytic fragments. J Biol Chem 256:3802–3809.

Karnik SS, Khorana HG (1990): Assembly of functional rhodopsin requires a disulfide bond between cysteine residues 110 and 187. J Biol Chem 265:17520–17524.

Kobilka BK, Kobilka TS, Daniel K, Regan JW, Caron MG, Lefkowitz RJ (1988): Chimeric α_2-, β_2-adrenergic receptors: Delineation of domains involved in effector coupling and ligand binding specificity. Science 240:1310–1316.

Kono M, Yu H, Oprian DD (1998): Disulfide bond exchange in rhodopsin. Biochemistry 37:1302–1305.

Lee GF, Burrows GG, Lebert MR, Dutton DP, Hazelbauer GL (1994): Deducing the organization of a transmembrane domain by disulfide cross-linking. J Biol Chem 269: 29920–29927.

Lynch BA, Koshland DE Jr (1991): Disulfide cross-linking studies of the transmembrane regions of the aspartate sensory receptor of *Escherichia coli*. Proc Natl Acad Sci USA 88:10402–10406.

Maggio R, Vogel Z, Wess J (1993): Reconstitution of functional muscarinic receptors by co-expression of amino- and carboxyl-terminal receptor fragments. FEBS Lett 319: 195–200.

Molday RS, MacKenzie D (1983): Monoclonal antibodies to rhodopsin: Characterization, cross-reactivity, and application as structural probes. Biochemistry 22:653–660.

Naranjo D, Kolmakova-Partensky L, Miller C (1997): Expression of split Shaker K channels in *Xenopus* oocytes. Biophys J 72:A11.

Oprian DD (1993): Expression of opsin genes in COS cells. Methods Neurosci 15: 301–306.

Oprian DD, Molday RS, Kaufman RJ, and Khorana HG (1987): Expression of a synthetic bovine rhodopsin gene in monkey kidney cells. Proc Natl Acad Sci USA 84:8874–8878.

Pakula AA, Simon, MI (1992): Determination of transmembrane protein structure by disulfide cross-linking: The *Escherichia coli* Tar receptor. Proc Natl Acad Sci USA 89:4144–4148.

Ridge KD, Lee SSJ, Abdulaev NG (1996): Examining rhodopsin folding and assembly through expression of polypeptide fragments. J Biol Chem 271:7860–7867.

Ridge KD, Lee SSJ, Yao LL (1995): *In vivo* assembly of rhodopsin from expressed polypeptide fragments. Proc Natl Acad Sci USA 92:3204–3208.

Schertler GFX, Villa C, Henderson R (1993): Projection structure of rhodopsin. Nature 362:770–772.

Schöneberg T, Liu J, Wess, J (1995): Plasma membrane localization and functional rescue of truncated forms of a G protein–coupled receptor. J Biol Chem 270:18000–18006.

Stühmer W, Conti F, Suzuki H, Wang X, Noda M, Yahagi N, Kubo H, Numa S (1989): Structural parts involved in activation and inactivation of the sodium channel. Nature 339:597–603.

Tang W-J, Krupinski J, Gilman AG (1991): Expression and characterization of calmodulin-activated (type I) adenylylcyclase. J Biol Chem 266:8595–8603.

Unger VM, Hargrave PA, Baldwin JM, Schertler, GFX (1997): Arrangement of rhodopsin transmembrane α-helices. Nature 389:203–206.

Unger VM, Schertler GFX (1995): Low resolution structure of bovine rhodopsin determined by electron cryo-microscopy. Biophys J 68:1776–1786.

Wrubel W, Stochaj U, Ehring R (1994): Construction and in vivo analysis of new split lactose permeases. FEBS Lett 349:433–438.

Wrubel W, Stochaj U, Sonnewald U, Theres C, Ehring R (1990): Reconstitution of an active lactose carrier *in vivo* by simultaneous synthesis of two complementary protein fragments. J Bacteriol 172:5374–5381.

Wu J, Kaback HR (1996): A general method for determining helix packing in membrane proteins *in situ*: Helices I and II are close to helix VII in the lactose permease of *Escherichia coli*. Proc Natl Acad Sci USA 93:14498–14502.

Yang K, Farrens DL, Altenbach C, Farahbakhsh ZT, Hubbell WL, Khorana HG (1996): Structure and function in rhodopsin. Cysteines 65 and 316 are in proximity in a rhodopsin mutant as indicated by disulfide formation and interactions between attached spin labels. Biochemistry 35:14040–14046.

Yu H, Kono M, McKee TD, Oprian DD (1995): A general method for mapping tertiary contacts between amino acid residues in membrane-embedded proteins. Biochemistry 34:14963–14969.

Zen KH, McKenna E, Bibi E, Hardy D, Kaback HR (1994): Expression of lactose permease in contiguous fragments as a probe for membrane-spanning domains. Biochemistry 33:8198–8206.

CHAPTER 7

METAL–IONS AS ATOMIC SCALE PROBES OF G PROTEIN–COUPLED RECEPTOR STRUCTURE

JOHN A. SCHETZ and DAVID R. SIBLEY

1. METAL CATIONS AS ATOMIC SCALE PROBES OF PROTEIN
   STRUCTURE   122
2. THEORETICAL ASPECTS OF METAL CATION
   COMPLEXATION BEHAVIOR   122
   A. Electrostatics Explain 1A and 2A Metal Cation Stability
      and Selectivity Sequences for Their Anionic Sites:
      The Eisenman Series   122
   B. Ionization Potential Partly Explains First Row d-Transition
      Stability and Selectivity Sequences for Their Anionic Sites:
      The Irving-Williams Series   124
   C. Favorable Complexation Entropies Are a Function of
      Inverse Ionic Radius, Increasing Valence, and the Number
      of Coordinating Groups   125
   D. Other Factors Affecting Metal Complexation Behavior   125
3. INDUCTIVE VERSUS DEDUCTIVE APPROACHES TO
   STUDYING METAL–PROTEIN INTERACTIONS   127
4. EXPERIMENTAL PROCEDURES FOR DETERMINATION
   AND ANALYSIS OF METAL INTERACTIONS WITH A GPCR   128
   A. Screening for Metal Binding Properties   128
   B. Pharmacological Relevance of Metal Binding as Determined
      by Dose-Dependence, Saturability, and Reversibility   130
   C. Thermodynamics of Metal Binding   130
   D. Null Pharmacological Analyses of the Molecular Mechanisms
      of Metal–Protein Interactions   133
   E. Kinetic Analysis of Metal Binding   134
5. TARGETING PUTATIVE METAL COORDINATION SITE
   RESIDUES BY SUPERFAMILY SEQUENCE COMPARISONS
   AND SEQUENCE SUBTRACTIONS   137

Structure–Function Analysis of G Protein-Coupled Receptors, Edited by Jürgen Wess.
ISBN 0-471-25228-X   Copyright © 1999 Wiley-Liss, Inc.

1. METAL CATIONS AS ATOMIC SCALE PROBES OF PROTEIN STRUCTURE

Common methodologies for probing protein structure include chemical modification and mutational analyses. Here metal cations are introduced as another system for probing the molecular structural changes associated with G protein–coupled receptor (GPCR) protein function. Perhaps the most attractive feature of metal cations as probes is their atomic scale size. The dimensions of free metal cations are comparable to the side chain of a single amino acid as well as to some common secondary structural features of proteins (Table 7.1). In much the same manner that the limit of the resolving power of the light microscope is restricted to the smallest wavelength of visible light, the limit of the resolving power of any probe is dictated by its dimensions relative to the dimensions of the structures being measured. For this reason, metal cations should be able to provide highly resolved information concerning protein molecular structure under dynamic conditions. Additionally, the periodic properties of metal cations can be exploited to study structure–activity relationships of the probe with respect to the probed site. While the focus of this chapter is on the use of metal cations as atomic scale probes in their own right, it is important to keep in mind that metal probes can be used in conjunction with other common probing systems (De Biasi et al., 1993).

2. THEORETICAL ASPECTS OF METAL CATION COMPLEXATION BEHAVIOR

2.A. Electrostatics Explain 1A and 2A Metal Cation Stability and Selectivity Sequences for Their Anionic Sites: The Eisenman Series

The theoretical concepts for describing the 1A metal cation selectivity sequence for an anionic ligating site were put forth by Eisenman (1962) and Eigen and Winkler (1970). Eisenman rigorously enumerated these concepts and demonstrated that electrostatics explain the "concave upward" alkali (1A) metal cation selectivity sequences for a variety of biophysical and biochemical systems, including the interactions of metal cations with nonaqueous polar solvents and counterions, with electrode or lipid bilayer barriers, and with chelators, carriers, pores, and a variety of voltage-sensitive cation channels (Eisenman et al., 1976; Eisenman and Horn, 1983; Hille, 1992). Truesdell and Christ (1967) realized that the principles underpinning the so-called Eisenman series for monovalent alkali cations could be applied equally as well to divalent alkaline earth (2A) cation selectivity sequences. Remarkably, the Eisenman series adequately explains not only the selectivity preferences for 1A and 2A metal cations, but also the overall stability series of these metal cations for an anionic coordination site. An important consequence of the Eisenman series is that the order of a particular selectivity sequence changes in a predictable fashion that is governed by the magnitude (strong or weak) of the field strength of the anionic site. For cations bearing the same charge, it is the effective anionic field strength that determines both the selectivity sequence and the stability of

TABLE 7.1. Some Common Bond Strengths and Dimensions of Protein Structures, Metal Cations, Nonmetal Elements and Chemical Bonds

Classification	Dimensions	Strength (kcal/mol)	Distance Factor
Chemical bond lengths (Å)[a]			
Covalent	1.5	90 (36–125)	—
Reinforced ionic	—	(10)	—
Ionic	2.5	3 (5)	$1/r$
Hydrogen	3.0	1 (1–7)	—
Ion dipole	—	(1–7)	$1/r^2$
Dipole–dipole	—	(1–7)	$1/r^3$
Van der Waals	3.5	0.1 (0.5–1)	$1/r^4 - 1/r^6$
Protein structure lengths (Å)[b]			
β-Strand (untwisted/twisted)	6.4–7.6		
α-Helix	5.4		
Turn (3_{10} helix)	6.0		
Molar volumes (cm³/mol)[c]			
Amino acids (grouped in quartiles)			
1: Gly, Ala, Ser, Asp	36.3, 52.6, 54.9, 68.4		
2: Cys, Thr, Asn, Pro, Glv, Val	70.8, 71.2, 72.6, 73.6, 84.7, 85.3		
3: Gln, His, Met, Leu, Iso	88.9, 91.9, 97.7, 101.8, 101.8		
4: Lys, Arg, Phe, Tyr, Trp	105.1, 109.1, 113.9, 116.2, 136.7		
Atoms			
Carbon	9.9		
Hydrogen	3.1		
Nitrogen	1.5		
Oxygen	2.3		
Sulfur	15.5		
Van der Waals diameters (Å)[d]			
Liquid water	2.8		
Elemental cations			
Al^{3+}-Cs^+	1–3.4		
Atomic groupings			
Methyl	4.0		
Aromatic carbon	3.7		
Carbonyl carbon	3.0		
Amines	3.0		
Keto, ether, hydroxyl oxygen	2.8		
Free or ether sulfur	3.7		

[a]Distance factors and bond strengths in parentheses are from Zimmerman and Feldman (1981). The remaining bond strength and bond length data are from Alberts et al. (1989). A dashed line means no numerical value was given. The term *distance factor* refers to the distances over which these forces are effective.
[b]The length data from Gennis (1989) refer to longest lengths.
[c]The molar volume values for the amino acids are from Eisenberg and Crothers (1979) and include the volume of the amide backbones. The volume contribution of the side chain alone can be calculated by subtracting the volume of the amide backbone as volume of amide backbone = volume of glycine − the volume of hydrogen (33.2 = 36.3 − 3.1). The volume of a single amino acid (or side chain) can be calculated by dividing the molar volume by Avogadro's number (6.0221×10^{23} particles/mol).
[d]Data are from Eisenberg and Crothers (1979) and Weast (1985).

the metal–anionic site complex. Typically, at anionic sites of low field strength monovalent 1A cations are preferred, and as the field strength becomes stronger divalent 2A cations are preferred. Concomitantly, the overall stability of a metal–anionic site complex increases with increasing anionic field strength for both monovalent and divalent cations. While the distance separating coordinating anionic sites influences overall anionic field strength, it does not alter the selectivity pattern among cations bearing the same charge. Rather, coordination site separation is tantamount to the selectivity preferences between monovalent and divalent metals (Truesdell and Christ, 1967; Eisenman et al., 1976). In general, a coordination site separation of less than 4 Å favors complexation of 2A metals, while site separations above 5 Å favor 1A metals.

2.B. Ionization Potential Partly Explains First Row d-Transition Stability and Selectivity Sequences for Their Anionic Sites: The Irving-Williams Series

The Eisenman series explains alkali and alkaline earth metal cation complexing behavior, but fails to adequately explain complexation behavior for other metals (e.g., d-transition). This is due in part to 1A and 2A metal cations having spherical, unfilled outer shell orbitals while the d-transition metals have nonspherical, unfilled inner shell orbitals. Although limited in scope, the Irving-Williams series describes the order of stability of complexes formed by first row d-transition metals whose inverse ionic radii and second ionization potentials increase monotonically from manganese to copper.

In its original form, Irving and Williams (1953) represented the order of their stability sequence as Mn < Fe < Co < Ni < Cu > Zn. They emphasize the characteristic drop in stability for zinc following a maximum at copper, but in actuality zinc stabilities generally decrease below those of nickel too. Remarkably, the Irving-Williams series explains the complexing behavior of these divalent cations for a wide variety of chelating agents. The stability of the Irving-Williams metal complexes is a function of ionization potential, ionic radius, and to some extent the nature of the ligating group. For example, the stability of the metal complexes of the Irving-Williams series increases monotonically with cumulative metal ionization potential. Importantly, the relative magnitude of the metal-complex stabilities relies somewhat on the types of functional groups that form the coordination site, while the stability pattern for a particular ligand usually does not. Coordination sites composed of nitrogen ligands complex the Irving-Williams series metals more strongly than those with oxygen ligands in the general order nitrogen type > nitrogen–oxygen type > oxygen type. The only exception to this is for manganese, whose stability pattern is reversed (Irving and Williams, 1953). If sulfur–oxygen and sulfur-nitrogen type ligand groups are included in the coordinating site, then some selectivity is achieved for pseudo-noble-gas configuration (PNGC) cations (Vallee and Coleman, 1964). With sulfur-ligating groups, the stability of all metal complexes increases and the selectivity for manganese through copper follows the Irving-Williams series. However, the stability of zinc complexes are now equivalent to or greater than those of nickel. Cadmium forms lower stability complexes than zinc when coordinating sites are derived from either oxygen or nitrogen groups, but this order reverses when sulfur is added as a ligating group. The ligating group preferences for mercury are the same as for copper, but mercury complexes are even more stable (Vallee and Coleman, 1964).

2.C. Favorable Complexation Entropies Are a Function of Inverse Ionic Radius, Increasing Valence, and the Number of Coordinating Groups

Since neither the Eisenman series nor the Irving-Williams series directly takes entropy effects into account, one might expect that entropic contributions to metal complexation either follow the order of free energy changes or are negligibly small in comparison to enthalpy contributions. In general, this will be the case for monovalent cations and for metal complexes with low coordination numbers. On the other hand, entropy effects can become considerable for higher valence metals, especially for the smaller cations and higher coordination number complexes[1] (Fig. 7.1). Strikingly, the relationship between favorable complexation entropy versus coordination number and inverse ionic radius applies for a much wider range of cations than originally envisioned by Martell (1960) and is also clearly a function of valence. Chelation of iron in its divalent versus its trivalent state demonstrates this contribution of valence to entropy stabilization effects.

Constraining the ligating groups so that they form a ring or cyclized structure often greatly increases metal complex stability relative to their uncyclized forms (Martell, 1960). Within a cyclized coordination site, increasing the number of ligating groups also tends to increase metal-complex stability. This is due to an increase in overall anionic strength, but only to the point that there is minimal steric interference between ligating groups. Increasing the diameter of the chelator ring in an effort to reduce steric hindrance as additional ligating groups are added simultaneously decreases the anionic field strength. This balance between coordination number and anionic field strength has been systematically demonstrated using homologs of EDTA as the chelator, and maximal stability is reached when five to six ligating groups form the metal coordination site (Schwartzenbach and Ackermann, 1948; Martell, 1960). In some cases, the number of ligating groups in the coordination site can affect selectivity, which is due to the preference of metals to assume certain geometries in the coordinated state. Because these coordination geometry preferences are not absolute and are frequently small in comparison to the other factors governing selectivity, they cannot be employed as singular predictors of selectivity.

2.D. Other Factors Affecting Metal Complexation Behavior

Perhaps the most general property of metal cations is the tendency for metal-complex stabilities to increase as the cumulative ionization potential of the test metal increases. Most striking is that this trend holds for a broad range of metal cations, including alkali, alkaline earth, 3B, first row d-transition, PNGC, and

---

[1]Given that $\Delta G = \Delta H - T\Delta S$, an entropy change of, for example, only 5 cal/K $\cdot$mol at room temperature (298°K) contributes -1.5 kcal/mol to the favorable free energy change. Because $\Delta G = RT\ln K$, such an overall change in free energy would translate to a greater than 10-fold change in the stability or "affinity" constant of the metal complex.

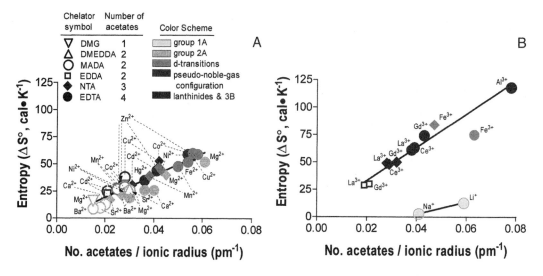

Figure 7.1. The stability of metal complexes is a function of complexation entropy. **A, B:** $\Delta S°$ means that the entropy measurements were taken at standard state, i.e., 1 atm and 298°K. The number of acetates refers to the number of negatively charged groups at the coordination site of the chelator, and the radius refers to the ionic radius of the metal cation. The arrowheads on some of the dashed lines refer to data points that are obscured by other data points (i.e., underneath). Studies with EDTA analogs have demonstrated that favorable entropy contributions can substantially increase the stability of metal complexes when the metals are small and the coordination site has multiple ligating groups. In addition, the magnitude of the favorable entropy change increases as the valence of the metal cation increases. DMEDDA, dimethylethylenediaminediacetate; DMG, dimethylglycinate; MADA, methylaminediacetate; EDDA, ethylenediiminodiacetate; NTA, nitrilotiracetate; EDTA, ethylenediaminetetraacetate. Metal complex entropy values are from Martell (1960) and Martell and Smith (1974). Ionic radius values are from Weast (1985). (See color plates.)

lanthinide series cations. Thus, the ionization potential of the metal in its aqueous valence state can be used as a guide to anticipate the overall metal-complex stability series for a given coordination site. This correlation is not necessarily expected to hold if metal complexation is strongly influenced by other factors, such as accessibility to the coordination site due to physical or kinetic barriers (e.g., a coordination site located at the end of a narrow cavity). In limited cases, selectivity may arise when the "coordination site" is associated with a barrier system that has a multi-ion coordination site. For example, the barium selectivity of certain zeolite electrodes can be enhanced by doping the zeolite with barium (Truesdell and Christ, 1967), and the selectivity of voltage-dependent calcium channels can be explained by assuming that multiple calcium ions interact with the channel pore (Hess and Tsein, 1984; Hille, 1992).

Functional groups other than electron-donating nitrogens, oxygens, and sulfurs can also influence complexing stability and selectivity. In general, the weakest coordinating groups are ether oxygens, which are about the strength of hydroxylalkyl groups with no proton displacement. Somewhat stronger complexes are formed (up to fivefold) when methoxy and hydroxyl groups contribute to the coordinating site. Still stronger metal complexes are formed by sulfur ethers. While the stabilizing contribution of these groups are much weaker than for the carboxylate, amine, and negative sulfur groups, they can have important and sometimes dramatic effects on metal selectivity when arranged in the appropriate orientation. For example, alkoxides and phenoxides that are positioned ortho to a phenolic (or other electron-donating groups) have very high affinities for trivalent and tetravalent metals (Martell, 1960). As was noted for the Irving-Williams series, the type of ligating groups that form the coordination site can affect metal-complex stabilities. In general, both alkali series form the strongest complexes with oxyanions, the d-transition metals complex strongly with nitrogen and sulfur groups, and the PNGC metals, in particular, form superstable complexes with sulfur (O'Sullivan, 1982).

3. INDUCTIVE VERSUS DEDUCTIVE APPROACHES TO STUDYING METAL–PROTEIN INTERACTIONS

Both inductive and deductive approaches have evolved for probing protein structure and function with metal cations. For an inductive approach, one would first construct a model using the available structural data for a given protein. Predictions are then made on the basis of the model, outcomes are tested empirically, and the model is adjusted accordingly. By progressively testing various structural features in this manner, a general picture of protein structure and function emerges. Studies with the neurokinin-1 receptor have shown, for example, that substitution of amino acids identified as docking residues for the antagonist CP96,345 with histidyl residues (imidazole nitrogen ring) produces a mutant receptor that is functionally antagonized by zinc and resistant to CP96,345 (Elling et al., 1995). The agonist properties of the mutant neurokinin-1 receptors are not affected by these histidine substitutions, suggesting that the mutations did not cause any gross structural changes in the receptor. Engineering of metal binding sites is especially useful for testing GPCR models of helical orientation and connectivity (Sheikh et al., 1996; Elling et al., 1997). As the topic of induced metal binding properties in GPCR proteins has been reviewed recently (Elling et al., 1997), our focus will be on deductive approaches to probing protein structure with metal cations.

A deductive approach for probing metal–protein interactions takes advantage of endogenous metal binding sites on proteins. An advantage of this approach is that it does not rely on model building at the onset, and the pool of potentially interesting metal–protein interactions is established by screening the property of interest (e.g., drug binding or G protein coupling). However, the challenge of deductive methods is locating the coordinating site for the metal cation on the protein. Often it is useful to narrow the number of candidate amino acids by detailed pharmacological analysis of molecular mechanisms and comparing amino acid sequences of metal-sensitive and metal-insensitive GPCR superfamily members. In certain cases, the effective metals can be grouped according to some periodic property and thus provide insight into the number, strength, and types of ligating groups that might form the anionic site.

For example, sodium sensitivity is a property of some GPCRs that results from the binding of sodium cations to an anionic site on the receptor protein. In the case of D2 dopamine receptors, millimolar concentrations of sodium ions, and to a lesser extent lithium and potassium ions, enhance the binding of certain substituted benzamide antagonists. This would correspond to an anionic site of relatively low field strength (corresponding to Eisenman series number X) and low overall stability (corresponding to weak binding affinity). Interestingly, D2 dopamine receptor mutagenesis studies demonstrate that a single carboxyl group (the side chain of aspartate at position 80) is largely responsible for this sodium selective effect on substituted benzamide antagonist binding (Neve et al., 1991). Indeed, all of our theoretical expectations are met in that sodium binds to a relatively weak anionic site on dopamine receptors that is monovalent and formed by an oxyanion.

Importantly, the sodium effect is not a direct competition between sodium and substituted benzamide antagonists for a common site, rather, these sites appear to be allosterically coupled (Neve et al., 1990). This suggests that the theoretical concepts as well as the biological substrates responsible for 1A metal cation selectivity for an anionic site are analogous, even if the metal acts at an allosteric site. This is critical, because even a metal cation that binds directly to a GPCR ligand binding site is likely to have some allosteric properties. The reason for this is that even small ligands are likely to have more than one docking point, and the biggest metal cations are still much smaller than the smallest endogenous ligands (e.g., catecholamines). Thus, measurements of GPCR–metal interactions are, by nature, likely to be indirect.

4. EXPERIMENTAL PROCEDURES FOR DETERMINATION AND ANALYSIS OF METAL INTERACTIONS WITH A GPCR

4.A. Screening for Metal Binding Properties

The most practical approach to screening the effects of metals on GPCRs is to measure changes in radioligand binding via rapid filtration techniques. These assays are cost-effective and can be performed under a variety of conditions using standard laboratory equipment, and high affinity radioligands are available for a large variety of GPCRs. The critical variables to consider when screening for metal effects with radioligand binding include the source of receptors and the choice of radioligand, metal salts, and buffer conditions. It is highly desirable to screen metal effects on cloned GPCRs expressed in cell lines. A major concern is that metals can influence the stability of receptors derived from endogenous tissue sources (Oliveira et al., 1983; Scheuhammer and Cherian, 1985; Braestrup and Anderson, 1987), and often tissues contain a variety of GPCRs, making it difficult to distinguish multiple receptor populations from a negative heterotropic cooperative effect.

These potential problems can be overcome by expressing high levels of the GPCR of interest in a cell line that does not contain other receptors that bind the primary radioligand with high affinity. Although the properties that we choose to measure will to some extent determine what effects we will find, from a practical standpoint it is initially desirable to use a radiolabeled antagonist as the ligand for the primary site. In contrast to most agonists, the binding of antagonists to GPCRs is much less likely to be influenced by the particular state of the receptor, such as the extent of G protein coupling. Furthermore, endogenous metals, such as sodium and magnesium, can affect agonist affinity states, which will complicate interpretation of the results of the other metals that are being screened. Similarly, the buffer system should not contain large concentrations of metals other than the test metal. One reason for this is that even if antagonists are used the possibility remains that the addition of metals commonly added to the binding buffer (e.g., calcium, sodium, potassium, and magnesium) will in some manner compete with the test metals. If possible, chloride salts of the test metals should be used when screening for metal effects. A variety of ultrapure, relatively inexpensive metal salts are commercially available, and virtually all of the metals of interest form stable water-soluble chloride salts. Furthermore, chloride is the least reactive halide, and, if only chloride salts of metals are used in a cation screen, then it is possible to control for any potential counteranion effects. Consequently, the interactions of different metals can be compared directly with minimal interference from the counteranion. A more detailed anion control might include screening a variety of salts (e.g., $ZnCl_2$, $ZnSO_4$, and ZnAcetate) for dose equivalency. In addition, it is advisable to test for pure charge effects by screening a positively charged, nonmetal replacement electrolyte, like N-methyl-D-glucamine or choline.

The principles described for metal cation screening of GPCRs are illustrated with data collected for ^{3}H-antagonist binding to clonal cell lines expressing either the rat dopamine D3 receptor or the rat adenosine A_1 receptor. Chinese hamster ovary (CHO) cells were selected as cell background because untransfected CHO cells do not specifically bind the D3 antagonist ^{3}H-methylspiperone or the A_1 antagonist ^{3}H-DPCPX (1,3-dipropyl-8-cyclopenylxanthine), and nonspecific binding is low. Membranes were prepared from stable CHO cell lines expressing high levels of either receptor (0.5–2 pmol/mg membrane protein) using standard techniques (Schetz and Sibley, 1997). Binding assays were conducted in 50 mM Tris, pH 7.4, at 23°C at a single high concentration of radioligand (0.5–1 nM). Nonspecific binding was determined in the presence of saturating concentrations of (+)-butaclamol for the dopamine receptors and adenosine for the A_1 receptor, respectively. Membranes (51 ± 9 μg membrane protein/ml) and assay components were equilibrated at 23°C for 75 minutes, then rapidly washed with 3 × 3.0 ml of ice-cold 50 mM Tris, pH 7.4, at 0°C through 0.3% polyethylenimine-treated Whatman GF/C glass filters. Filters were mixed in 3.5 ml Cytoscint, and the radioactivity was counted on a scintillation counter with a 48% efficiency. All metals were screened as their chloride salts at a single high concentration, with the exception of silver, which was in the form of silver nitrate.[2] Stock solutions for most metals were made up in either water or, in a few cases, 5 mN HCl and then diluted to a final assay concentration of 5 mM. Mercury, gold, and silver were less soluble and were instead tested at a final assay concentration of 50–500 μM. When the effects of

[2]$AgNO_3$ is much more water soluble than AgCl. In addition, some metals, like zinc and mercury, are much more soluble in dilute acid solutions.

metals on ^{3}H-antagonist binding to D3 and A_1 receptors are grouped for comparison, large differences in metal sensitivities are observed for aluminum, zinc, gold, and cadmium (Fig. 7.2).

4.B. Pharmacological Relevance of Metal Binding as Determined by Dose Dependence, Saturability, and Reversibility

For a metal effect on drug binding to be considered pharmacologically relevant, the dose-dependence, saturability and reversibility of the effect must first be established. The dose dependence and saturability of a metal effect can be determined by competition curve analysis, and the shapes of metal competition curves can provide clues as to possible binding mechanisms. This point is illustrated by copper and cadmium inhibition of ^{3}H-methylspiperone binding to D2-like dopamine receptor subtypes (Fig. 7.3A,B). Both copper and cadmium inhibit all D2-like dopamine receptors in a monophasic, dose-dependent, and saturable manner, but the receptor subtype-selective differences for copper correspond to a difference in binding affinity, while for cadmium the differences are in the pseudo-Hill slopes (n_H).

Because the effects of metal–protein interactions are being tested with equilibrium binding methods, it is important to verify the reversibility of metal effects. Here metals have a distinct advantage over most other methods because relatively inert, high affinity chelators are available for most metals (O'Sullivan, 1982; Martell and Smith, 1974). Furthermore, these metal chelators do not have to be highly selective because the binding buffer contains only the test metal. EDTA reversal of zinc inhibition of ^{3}H-SCH23390 binding to cloned D1B dopamine receptors is provided as an example (Fig. 7.3C). A typical metal reversal experiment includes three groups run in parallel: one group with chelator only, a second group with metal only, and a third group with both metal and chelator. First, an IC_{50} concentration of zinc is "pre-"equilibrated with D1B receptors (i.e., in the absence of radioligand and chelator to allow the metal to interact with the receptor unimpeded). Next, both radioligand binding and metal chelation reactions are initiated by addition of ^{3}H-SCH23390 and EDTA (10-fold in excess). Subsequent saturation isotherm analysis provides direct evidence that the effect of zinc is completely reversed by chelation (Fig. 7.3C), and therefore it is appropriate to further analyze the macroscopic and microscopic interactions of zinc with dopamine receptors using standard binding techniques.

4.C. Thermodynamics of Metal Binding

Thermodynamic measures are critical for determining the overall energetics and driving forces of a reversible reaction, which include changes in free energy (ΔG), entropy (ΔS), and enthalpy (ΔH). For a metal binding reaction, the thermodynamic parameters are derived by measuring metal binding affinity as a function of temperature using the equation, $G = -RT \times ln(1/K_i)$, where $R = 0.00199$ kcal/mol·K. First, the free energy changes of the metal binding reaction are evaluated using standard competition curve analysis performed at a variety of different temperatures (e.g., 0°, 10°, 23°, 37°C). As K_i values are derived from the Cheng and Prussoff (1973) equation, it is necessary to deter-

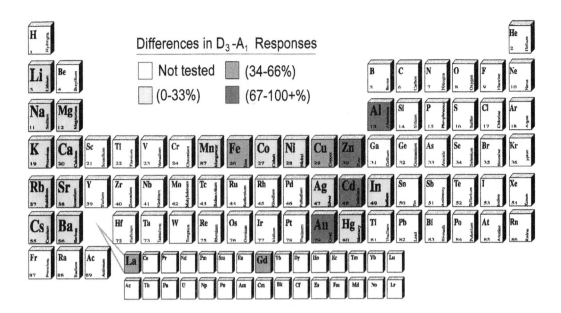

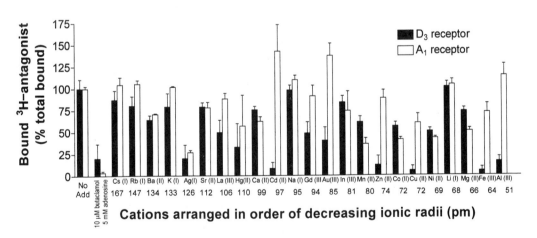

Figure 7.2. Screening for the effects of a single high concentration of metal cations on ^{3}H-antagonist binding to GPCRs. The effects of a single high concentration (0.05–5 mM) of metals on the binding of the dopamine D3 and adenosine A_1 receptors were screened using ^{3}H-methylspiperone (945 ± 350 pM) and ^{3}H-DPCPX (1,3-dipropyl-8-cyclopenylxanthine, 703 ± 260 pM), respectively. The pattern of metal sensitivity of ^{3}H-antagonist binding to dopamine D3 receptors is representative of the metal sensitivities of all cloned rat dopamine receptors (Schetz and Sibley, 1997, and unpublished results). The equilibrium binding data shown in the histogram were grouped in thirds according to the relative magnitude of metal sensitivities to D3 verses adenosine A_1 receptors. These relative differences in metal sensitivities are shown schematically as a function of their position in the periodic table. (See color plates.)

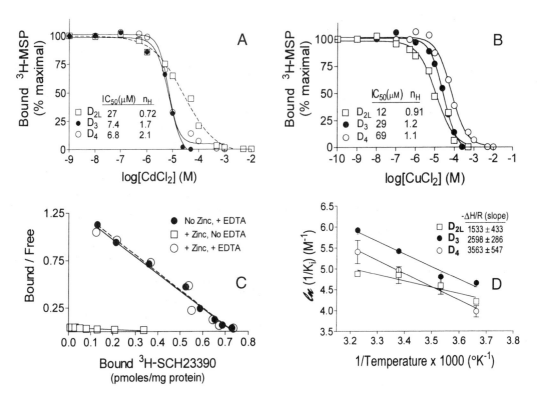

Figure 7.3. Dose dependence, saturability, reversibility, and thermodynamics of metal binding to dopamine receptors. **A,B:** Copper and cadmium competition curves of ^{3}H-methylspiperone (^{3}H-MSP) binding to D2-like dopamine receptor subtypes (n = 3). Standard error bars have been omitted for clarity. The corresponding K_i ±s.d. values for copper and cadmium are 0.4 ± 0.2 and 2.4 ± 2.8 μM for D2L, 14 ± 17 and 2.5 ± 1.9 μM for D3, and 41 ± 15 and 4.7 ± 2.6 μM for D4. Competition assays were performed using 471 ± 85 pM, 1,006 ± 531 pM and 404 ± 116 pM ^{3}H-methylspiperone for D2L, D3, and D4 receptors, respectively. **C:** Representative Rosenthal plots of EDTA reversal of zinc inhibition of ^{3}H-SCH23390 saturation isotherm binding to D1B dopamine receptors. Membranes were first pretreated with or without 260 μM ZnCl$_2$ for 50 minutes, after which time the binding assay was initiated by addition of radioligand and 2.5 mM EDTA. Reactants were equilibrated for an additional 80 minutes before harvesting by rapid filtration. Averaged K_d and B_{max} values for each group are 58 ± 5 pM and 0.64 ± 0.14 pmol/mg protein for −zinc/+EDTA (solid circles), 1,234 ± 294 pM and 0.36 ± 0.11 pmol/mg protein of +zinc/−EDTA (open squares), and 55 ± 1 pM and 0.62 ± 0.16 pmol/mg protein for +zinc/+EDTA (open circles) (n = 2). **D:** Van't Hoff plots of the thermodynamics of zinc binding to D2-like dopamine receptor subtypes. ^{3}H-methylspiperone saturation isotherms and zinc competition curves were obtained for each receptor subtype at four different temperatures, 0°, 10°, 23°, and 37°C. The calculated free energy, enthalpy, and entropy values of zinc binding for D2L, D3, and D4 receptors are ΔG = −1.66 ± 0.09, −2.59 ± 0.47, and −2.61 ± 0.3 kcal/mol; ΔH = 3.05 ± 0.86, 5.17 ± 0.11, and 7.09 ± 1.09 kcal/mol; and ΔS = 1.73 ± 0.13, 3.55 ± 0.19, and 3.16 ± 0.19 cal/K·mol, respectively (n = 2). Competition curves were performed using 300 ± 120 pM, 790 ± 50 pM, and 1,100 ± 10 pM ^{3}H-methylspiperone for D2L, D3, and D4, respectively.

mine the values for both the IC_{50} of the metal and the K_d of the radioligand at each temperature. Since $ln(1/K_i) = -\Delta H/RT + \Delta S/R$, the enthalpy changes are calculated from the slope of a van't Hoff plot, $-\Delta H/R =$ slope (Fig. 7.3D). Once ΔG and ΔH are known, the corresponding changes in entropy are derived from the second law of thermodynamics by solving for ΔS in the equation, $\Delta G = \Delta H - T\Delta S$.

Often changes in enthalpy are interpreted as corresponding to conformational changes in a protein and entropy effects are interpreted as a reordering of water molecules; however, these interpretations are limited to simple (one step) reactions. The reason for this is that thermodynamic measures are macroscopic changes representing the sum of all reaction events at equilibrium. Despite this, changes in the entropy and enthalpy of metal binding can still provide information on the dominant forces governing metal–protein complexation, and the magnitude of favorable free energy changes may provide insight into the type(s) of chemical bonding interactions that might account for metal–GPCR complexation. If we assume a one-step model for the binding of zinc to D2-like dopamine receptors, then both the magnitude of the free energy changes and the fact that the reaction is entropy driven ($-\Delta G$, $-\Delta S$, and $+\Delta H$) suggest that zinc is essentially chelated by dopamine receptors, possibly resulting in the reordering of hydrogen bonds.

4.D. Null Pharmacological Analyses of the Molecular Mechanisms of Metal–Protein Interactions

For the reasons outlined above, the effects of metals on radioligand binding to GPCRs are likely to be allosteric. These allosteric effects can be effectively analyzed with null pharmacological methods (Ehlert, 1988). The basic idea underlying null methods is to compare the extent to which the data deviate from a model of a perfectly competitive (ideal) inhibitor and then to quantify this deviation or plateau as allosteric cooperativity. The two most useful approaches for measuring metal–GPCR interactions with radioligand binding are variations of the Schild plot called Schild-type plots.

First there is the competition curve style Schild-type plot. As for any competition curve, inhibition of radioligand binding at a single fixed dose is measured as a function of metal cation (inhibitor) dose. The unique feature of Schild analysis is that the competition curves are repeated at three or more different fixed concentrations of radioligand simultaneously. The critical information lies in the direction and magnitude of the shifts in IC_{50} values at the different radioligand concentrations. The comparative nature of Schild analysis requires that we establish a model of purely competitive inhibition for our radioligand binding system. In the form $K_i = IC_{50}/(1 + [radioligand]/K_d)$, the Cheng and Prusoff (1973) equation models the dose–response behavior expected for a purely competitive binding interaction. The different outcomes of the Cheng-Prusoff equation can also help to narrow the range of possible binding mechanisms. If, for example, the IC_{50} values for a test metal were to decrease as the fixed concentration of radioligand was increased (a leftward shift), then this would provide strong evidence for an uncompetitive mechanism of inhibition whereby the metal and radioligand would form a complex prior to binding the receptor site (Cheng and Prusoff, 1973). Alternatively, if a metal were acting as a noncompetitive inhibitor of radioligand binding, then in theory the IC_{50} values should be independent of the concentration of radioligand. In practice, though, it may be difficult to distinguish between a noncompetitive inhibition and an allosteric competitive inhibition[3] with low cooperativity, as is the case for zinc inhibition of ^{3}H-methylspiperone binding to D2L and D4 receptors (Fig. 7.4B,C). The final possible outcome is that a metal binding interaction is characterized as competitive allosteric with a large degree of cooperativity, as in the case for D1A receptors (Fig. 7.4A). Notice that as the degree of cooperativity increases the data more closely approximate a purely competitive interaction. While informative, the competition style Schild-type plots may not distinguish between subtle but distinct binding mechanisms. For this, a second type of Schild analysis must be employed using the dose–response range we established for the Schild-type competition curves.

Saturation isotherm style Schild-type plots represent the second type of Schild analysis. Saturation isotherms of radioligand binding are obtained in the usual manner except that several isotherms are performed simultaneously in the presence of different doses of metal (inhibitor). The degree of allosteric cooperativity can be directly visualized as the plateau region (α) on the dose-ratio (K_d[zinc]/K_d[no zinc]) curve plotted as log (Dose-Ratio-1) verses log[metal] (Fig. 7.4D–F). Note that the value for α equals 1 for D4 receptors, indicating lack of cooperativity. This is consistent with the finding that zinc is a noncompetitive inhibitor of ^{3}H-methylspiperone binding to D4 receptors; zinc primarily reduces D4 receptor density (unpublished observation). The saturation isotherm style Schild-type plots for D1A and D2L receptors clearly show that for both subtypes zinc acts as a competitive allosteric modulator with differing degrees of cooperativity (Fig. 7.4D,E). Thus, cooperativity measures can distinguish purely competitive interactions from highly cooperative allosteric competitive interactions and weakly cooperative allosteric competitive interactions from purely noncompetitive interactions. When testing for allosteric interactions it is critical to employ a large enough range of either radioligand or metal (inhibitor) so that any highly cooperative allosteric interactions can be observed. In the case of competition style analysis, concentrations near and far beyond the K_d for the radioligand should be employed. For saturation isotherm style analysis, the entire effective metal (inhibitor) concentration range should be utilized as far as the limits of experimental detection permit. Especially at these higher metal concentrations, it is critical that buffering conditions are adequate because a sharp change in pH can produce an anomalous plateau in a saturation isotherm style Schild-type plot (Kenakin and Beek, 1982).

4.E. Kinetic Analysis of Metal Binding

Often it is desirable to complement equilibrium measurements with kinetic analyses because the equilibrium affinity constant is related to the kinetic rate constants of association and dissociation. Because for many binding reactions

---

[3]Note that competitive allosteric inhibition can also be described as negative heterotropic cooperativity.

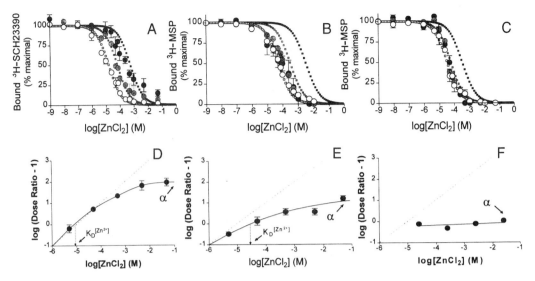

Figure 7.4. Molecular mechanisms: null pharmacological analyses of zinc inhibition of ^{3}H-antagonist binding to D1A, D2L, and D4 dopamine receptors. **A–C:** Competition curve style Schild-type analyses: dose–response inhibition of ^{3}H-antagonist binding to dopamine receptors by zinc as a function of increasingly higher concentrations of ^{3}H-antagonist. Zinc inhibition curves of ^{3}H-antagonist binding to each dopamine receptor subtype were generated at three different concentrations of ^{3}H-antagonist: 37 ± 9, 322 ± 93, and $3{,}187 \pm 890$ pM for D1A (A); 25 ± 0.3, 263 ± 8, and $2{,}795 \pm 35$ pM for D2L (B); and 52 ± 11, 540 ± 99 and $5{,}610 \pm 778$ pM for D4 receptors (C) (n = 3). The open, gray, and solid symbols correspond to binding from the lowest to highest concentration radioligand, respectively. The large circles correspond to the experimentally measured values. The small squares correspond to the expected theoretical binding values if zinc inhibition of ^{3}H-antagonist were purely competitive, i.e., if it obeyed the equation $IC_{50}^2 = IC_{50}^1 \times \{1 + [(\text{radioligand}^2)/K_d]\} / \{1 \pm [(\text{radioligand}^1)/K_d]\}$, and had a pseudo Hill slope $n_H = 1.00$. The theoretical and experimental IC_{50} values for zinc inhibition from lowest to highest ^{3}H-antagonist concentrations are D1A, 13, 50, and 421 μM and 13, 46, 167 μM; D2L, 55, 325, and 3,200 μM and 55, 130, 109 μM; and D4, 34, 61, and 350 μM and 34, 41, and 47 μM (s.d. $\leq$ 20%). **D–F:** Saturation isotherm style Schild-type plot analysis of zinc inhibition of ^{3}H-antagonist binding to D1A (D), D2L (E), and D4 (F) dopamine receptors, respectively. These saturation isotherm style Schild-type plots illustrate the relative K_d shifts [log(dose ratio $-$ 1)] versus log[ZnCl$_2$]. The gray dashed lines transecting each plot are theoretically expected for a perfectly competitive inhibitor. Zinc binds these dopamine receptor subtypes with varying degrees of cooperativity: zinc binding is highly cooperative ($\alpha \cong 100$) for D1A, weakly cooperative ($\alpha \cong 10$) for D2L, and not cooperative ($\alpha \cong 1$) for D4 receptors. When binding is cooperative, the theoretical $K_d^{[Zn^{2+}]}$ for zinc at the "unoccupied" (no radiolabeled antagonist bound) receptor can be estimated as the x-intercept value on Schild plots yielding values of approximately 9 and 40 μM for D1A and D2L, respectively. (D, E, reprinted from Lippincott-Raven Publishers.)

the rate of association is invariant (i.e., diffusion controlled), most of the mechanistic information can be obtained from measuring changes in the dissociation kinetics. A perfectly competitive binding interaction obeys the law of mass action, $K_{affinity} = K_{association}/K_{dissociation}$ (Weiland and Molinoff, 1981). For allosteric

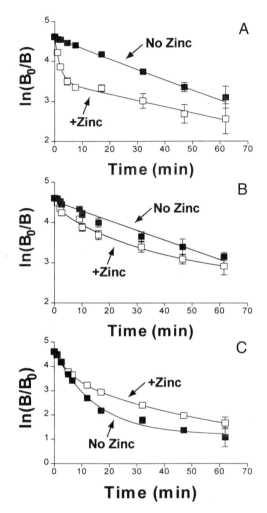

Figure 7.5. Molecular mechanisms: effect of zinc on the rate of ^{3}H-antagonist dissociation from D1A **(A)**, D2L **(B)**, and D4 **(C)** dopamine receptors. The concentrations of radioligand were 250 ± 23 pM ^{3}H-SCH23390 for the D1A receptor and 668 ± 15 pM and $1,034 \pm 94$ pM ^{3}H-methylspiperone for D2L and D4 receptors, respectively. The dissociation rate data were transformed to fit the function $ln(B/B_0)$ versus time. Dissociation rates were measured by first equilibrating dopamine receptors with radioligand at 37°C followed by addition of excess nonlabeled antagonist (2–3 μM) without significant dilution either in the absence (black squares) or presence (white squares) of 5–10 mM zinc chloride. The calculated k_{off} rates for D1A receptors are $k_1 = 0.030 \pm 0.012$ min^{-1} in the absence and $k_1 = 0.024 \pm 0.024$ min^{-1} and $k_2 = 0.75 \pm 0.29$ min^{-1} in the presence of zinc. For D2L receptors the rates are 0.056 ± 0.023 min^{-1} in the absence and $k_1 = 0.38 \pm 0.16$ min^{-1} and $k_2 = 0.029 \pm 0.011$ min^{-1} in the presence of zinc. The k_{off} rates for D4 receptors are $k_1 = 0.23 \pm 0.017$ min^{-1} and $k_2 = 0.029 \pm 0.004$ min^{-1} in the absence, and $k_1 = 0.23 \pm 0.028$ min^{-1} and $k_2 = 0.027 \pm 0.003$ min^{-1} in the presence of zinc (n = 3). Although the calculated dissociation rates for D4 in the presence and absence of zinc are not significantly different, the dissociation in the presence of zinc appears slower because the proportion of receptors undergoing the faster dissociation is about 10%–15% smaller than in the absence of zinc.

interactions, this equation may not necessarily hold, but the equilibrium affinity and kinetic rate constants will still be proportionally related. For this reason, any allosteric effect on the equilibrium constant of the primary radioligand should be reflected in its dissociation rate and vice versa. The subtype-selective mechanistic effects of zinc on D1A, D2L, and D4 dopamine receptors illustrate this point. At equilibrium, zinc inhibits ^{3}H-antagonist binding to D1A and D2L dopamine receptors by an allosteric mechanism, while at D4 receptors the mechanism of inhibition is noncompetitive (Fig. 7.4). Similarly, in the presence of zinc, the rate of ^{3}H-antagonist dissociation is accelerated for D1A and D2L receptors but not for D4 receptors (Fig. 7.5). In other words, a decrease in the equilibrium affinity constant is analogous and proportional to an acceleration in the kinetic dissociation rate. In general, changes in dissociation rate kinetics will better illustrate highly cooperative allosterism, while weakly cooperative allosterism is mostly easily observed with saturation isotherm style Schild-type analysis.

Measuring the approach to equilibrium with kinetic methods provides additional details concerning possible binding reaction mechanisms. For example, purely competitive and noncompetitive inhibitors of radioligand binding would not change the existing dissociation rate(s) of the radioligand. A purely cooperative binding interaction would accelerate radioligand dissociation, but would not change the number of steps in the reaction. In the case of D1A and D2L dopamine receptors, zinc accelerates ^{3}H-antagonist dissociation by inducing an additional accelerated phase to the dissociation while leaving the initial dissociation phase intact. This provides perhaps the most compelling evidence that zinc and ^{3}H-antagonist bind to competitively coupled, yet physically distinct, sites on D1A and D2L dopamine receptor proteins (i.e., zinc is an allosteric modulator of ^{3}H-antagonist binding). It follows then that the binding of zinc to D1A and D2L dopamine receptors induces a destabilizing transition state in the receptor protein. This is in contrast to the effect of zinc on D4 receptors, which is to completely occlude ^{3}H-antagonist binding. Remarkably, the macroscopic mechanisms (a favorable change in entropy) are the same for all D2-like dopamine receptors even though the molecular mechanisms of zinc inhibition are unique. It is the allosteric nature of zinc modulation that permits this apparent dichotomy in macroscopic and microscopic properties.

5. TARGETING POTENTIAL COORDINATION SITE RESIDUES BY SUPERFAMILY SEQUENCE COMPARISONS AND SEQUENCE SUBTRACTIONS

The potential pool of amino acids that are likely to form part of the metal coordination site can be drastically narrowed by exploiting the metal sensitivities of receptor isoforms (Satin et al., 1992). Recall that during the mass metal screening (Fig. 7.3), dopamine receptors and the adenosine A_1 receptor were found to have markedly different sensitivities to four of the metals, including zinc. By comparing the amino acid sequences of the zinc-sensitive dopamine receptors with the zinc-insensitive adenosine receptor, one can begin to correlate the metal binding properties with structural features of their protein receptors. This is achieved by first aligning the amino acid sequences of all five of the zinc-sensitive dopamine receptors to one another and then retaining only those amino acids that are conserved. Those residues that are conserved in all dopamine receptors except for D4 receptors are also distinguished because the molecular mechanisms of zinc inhibition of the D4 receptor were different from all other dopamine receptors. This minimal dopamine receptor framework is further refined by comparing it to the corresponding sequence for the zinc-insensitive adenosine receptor. The resulting amino acid differences between the zinc-sensitive and zinc-insensitive GPCRs thus provide a minimal GPCR metal binding site framework.

Mapping the location of the zinc binding site on dopamine receptors is especially challenging for two reasons. First, the sequences for both the zinc-sensitive and zinc-insensitive GPCRs are moderately conserved, which results in about two dozen amino acids remaining in the minimal framework. Second, the most effective metals do not fit neatly into either of the established metal series; rather, it is PNGC and smaller trivalent metals that most effectively inhibit antagonist binding to dopamine receptors. These factors suggest that other physicochemical properties must be considered to explain zinc selectivity. Significantly, zinc binding involves a relatively small change in free energy that is entropy driven. A plausible explanation would be that the coordination site contains more than one ligating group, and at least one of these is presumed to be a sulfur group. Those amino acids that are suspected of participating in the metal–protein coordination site can now be tested with complementary methods, such as chemical modification and mutational analysis.

REFERENCES

Alberts B, Bray D, Lewis J, Raff M, Roberts K, Watson JD (1989): Molecular Biology of the Cell, 2nd ed. New York: Garland Publishing, Inc., p 89.

Braestrup C, Anderson PH (1987): Effects of heavy metal cations and other sulfhydryl reagents on brain dopamine D1 receptors: Evidence for involvement of a thiol group in the conformation of the active site. J Neurochem 48:1667–1672.

Cheng Y-C, Prusoff WH (1973): Relationship between the inhibition constant (K_i) and the concentration of inhibitor which causes 50 per cent inhibition (IC_{50}) of an enzymatic reaction. Biochem Pharmacol 22:3099–3108.

De Biasi M, Drewe JA, Kirsch GE, Brown AM (1993): Histidine substitution identifies a surface position and confers Cs^+ selectivity on a K^+ pore. Biophys J 65:1235–1242.

Ehlert FJ (1988): Estimation of the affinities of allosteric ligands using radioligand binding and pharmacological null methods. Mol Pharmacol 33:187–194.

Eigen M, Winkler R (1970): Alkali-ion carriers: Dynamics and selectivity. In Schmidt FO (ed): The Neurosciences, vol II. New York: Rockefeller University Press, pp 685–696.

Eisenberg D, Crothers D (1979): Physical Chemistry With Applications to the Life Sciences. Menlo Park: The Benjamin/Cummings Publishing Company, Inc., p 508.

Eisenman G (1962): Cation selective glass electrodes and their mode of operation. Biophys J 2(suppl 2):259–323.

Eisenman G, Horn R (1983): Ionic selectivity revisited: The role of kinetic and equilibrium processes in ion permeation through channels. J Membrane Biol 76:197–225.

Eisenman G, Krasne S, Ciani S (1976): Further studies on ion selectivity. In Clark L, Lubbers D, Silver I, Simon W, Kessler M (eds): Ion and Enzyme Electrodes in Medicine and Biology. Berlin: Urban and Schwartzenberg, pp 3–22.

Elling CE, Moller S, Nielsen M, Schwartz TW (1995): Conversion of antagonist-binding site to metal-ion site in the tachykinin NK-1 receptor. Nature 374:74–77.

Elling CE, Thirstrup K, Nielsen SM, Hjorth SA, Schwartz TW (1997): Metal-ion sites as structural and functional probes of helix–helix interactions in 7TM receptors. Ann NY Acad Sci 814:142–151.

Gennis RB (1989): Biomembranes molecular structure and function. In Cantor CR (ed): Springer Advance Texts in Chemistry. New York: Springer-Verlag.

Hess P, Tsien RW (1984): Mechanism of ion permeation through calcium channels. Nature 309:453–456.

Hille B (1992): Ionic Channels of Excitable Membranes. Sunderland: Sinauer Associates, Inc.

Irving H, Williams RJP (1953): The stability of transition-metal complexes. J Chem Soc 637:3192–3210.

Kenakin TP, Beek D (1982): A quantitative analysis of histamine H2-receptor–mediated relaxation of rabbit trachea. J Pharmacol Exp Ther 220:353–357.

Martell A (1960): The relationship of chemical structure to metal-binding action. In Seven MJ, Johnson LA (eds): The 1959 Proceedings of a Symposium Sponsored by Hahnemann Medical College and Hospital Philadelphia, Metal Binding in Medicine. Philadelphia: J.B. Lippincott Company, pp. 1–18.

Martell AE, Smith RM (1974): Critical Stability Constants, vol 1, Amino Acids. New York: Plenum Press.

Neve KA, Cox BA, Henningsen RA, Spanoyannis A, Neve RL (1991): Pivotal role for aspartate-80 in the regulation of dopamine D2 receptor affinity for drugs and inhibition of adenylyl cyclase. Mol Pharmacol 39:733–739.

Neve KA, Henningsen RA, Kinzie JM, De Paulis T, Schmidt DE, Kessler RM, Janowsky A (1990): Sodium-dependent isomerization of dopamine D-2 receptors characterized using [^{125}I]epidepride, a high-affinity substituted benzamide ligand. J Pharmacol Exp Ther 25:1108–1116.

Oliveira CR, Wajda I, Lajtha A, Carvalho AP (1983): Effects of cations and temperature on the binding of [^{3}H]spiperone to sheep caudate nucleus. Biochem Pharmacol 32:417–422.

O'Sullivan JW (1982): 17. Stability constants of metal complexes. In Dawson RMC (ed): Data for Biochemical Research, 2nd ed. New York: Oxford University Press, pp 423–434.

Satin J, Kyle JW, Chen M, Bell P, Cribbs L, Fozzard HA, Rogart RB (1992): A mutant of TTX-resistant cardiac sodium channels with TTX-sensitive properties. Science 256:1202–1205.

Schetz JA, Sibley DR (1997): Zinc allosterically modulates antagonist binding to cloned D_{1A} and D_{2L} dopamine receptors. J Neurochem 68:1990–1997.

Scheuhammer AM, Cherian MG (1985): Effects of heavy metal cations, sulfhydryl reagents and other chemical agents on striatal D_2 dopamine receptors. Biochem Pharmcol 34:3405–3413.

Schwartzenbach G, Ackermann H (1948): Komplexone XII. Die Homologen der Äthylendiammin-tetraessigsäure und irhe Erdalkalikomplexe. Helv Chim Acta 31:1029–1048.

Sheikh SP, Zvyaga TA, Lichtarge O, Sakmar TP, Bourne HR (1996): Rhodopsin activation blocked by metal-ion-binding sites linking transmembrane helices C and F. Nature 383:347–350.

Truesdell AH, Christ CL (1967): Glass electrodes for calcium and other divalent cations. In Eisneman G (ed): Glass Electrodes for Hydrogen and Other Cations. New York: Marcel Dekker, pp 293–321.

Vallee BL, Coleman JE (1964): Metal coordination and enzyme action. In Florkin M, Stolz EH (eds): Comprehensive Biochemistry: Enzymes General Considerations, vol 12. Amsterdam: Elsevier Publishing Co., pp 165–236.

Weiland GA, Molinoff PB (1981): Qualitative analysis of drug–receptor interactions. I. Determination of kinetic and equilibrium properties. Life Sci 21:313–330.

Weast RC (1985): In Weast RC, Astle MJ, Beyer WH (eds): CRC Handbook of Chemistry and Physics: A Ready Reference Book of Chemical and Physical Data, 66th ed. Boca Raton: CRC Press, Inc., pp E-67, E-74, F-164.

Zimmerman JJ, Feldman S (1981): Physical–chemical properties and biological activity. In Foye WO (ed): Principles of Medicinal Chemistry, 2nd ed. Philadelphia: Lea Febiger, pp 11–51.

CHAPTER 8

GENETIC APPROACHES FOR STUDYING THE STRUCTURE AND FUNCTION OF G PROTEIN–COUPLED RECEPTORS IN YEAST

CHRISTINE M. SOMMERS and MARK E. DUMONT

1. INTRODUCTION   142
2. THE YEAST PHEROMONE RESPONSE PATHWAY   142
3. REASONS TO STUDY GPCRs IN YEAST   144
   A. Screening Libraries of Random Receptor Mutations   144
   B. Screening Libraries of Random G Protein and Effector Mutations   145
   C. Screening for Ligands That Can Interact With a Given GPCR   145
   D. Screening for Proteins That Mediate or Regulate Cellular Signaling Pathways   146
   E. Use of the Yeast Two-Hybrid System   146
4. ASSAYS OF THE YEAST PHEROMONE RESPONSE PATHWAY   146
   A. Mating Efficiency   147
      a. Quantitative Mating Assay   147
      b. Selection for Yeast With Increased Mating Efficiency   148
   B. Expression of Pheromone-Responsive Reporter Genes   149
      a. Detection of *FUS1–lacZ* Expression on Plates   149
      b. Liquid Assay for *FUS1–lacZ* Expression   150
   C. Growth Arrest   151
5. CREATION OF LIBRARIES OF MUTANT RECEPTORS   152
   A. Oligonucleotide-Directed Random Mutagenesis   153
   B. Mutagenesis Using PCR and Plasmid Recombination   156
   C. Verification of Mutant Isolates   158
6. BIOCHEMICAL ANALYSES OF GPCRs IN YEAST   159
7. THE SECOND-SITE SUPPRESSOR APPROACH TO STUDYING GPCR STRUCTURE AND FUNCTION   159

Structure–Function Analysis of G Protein-Coupled Receptors, Edited by Jürgen Wess.
ISBN 0-471-25228-X   Copyright © 1999 Wiley-Liss, Inc.

I. INTRODUCTION

The diverse family of proteins classified as G protein–coupled receptors (GPCRs) includes two membrane proteins of the baker's yeast *Saccharomyces cerevisiae* that serve as receptors for mating pheromones. The availability of a variety of genetic screens and selections both for and against function of these receptors, together with rapid and efficient techniques for making genetic alterations in yeast, make it possible to study these receptors in ways that would be difficult in mammalian systems. Furthermore, the similarities of the pheromone receptors to other GPCRs, and the interchangeability of components of the GPCR signaling pathway between yeast and mammals, allow extension of the results and some of the techniques to mammalian signaling. We summarize below the types of questions that have been addressed using these analyses and details of some of the critical experimental procedures. Particular reference is made to recent procedures used in our laboratory for the isolation of intragenic second-site suppressors in a yeast GPCR. For information on standard procedures of yeast genetics, cell biology, and molecular biology, see Rose et al. (1990), Guthrie and Fink (1991), and, in particular, the contribution by Sprague (1991) on the mating pathway.

2. THE YEAST PHEROMONE RESPONSE PATHWAY

The haploid phase of *S. cerevisiae* consists of cells of two different mating types, *Mat**a*** and *Matα*, that fuse to form the organism's diploid zygote phase. *Mat**a*** cells secrete a-factor, a 12 residue peptide with an attached farnesyl group. *Matα* cells secrete α factor, a 13 residue peptide with the amino acid sequence WHWLQLKPGQPMY. a-factor and α factor bind to receptors encoded by the *STE3* and *STE2* genes that are expressed on the surface of *Matα* and *Mat**a*** cells, respectively. Binding of pheromone induces a set of physiological changes in the cell, including cell cycle arrest, changes in cell shape, and increased expression of a number of genes involved in cell fusion (for review, see Blumer and Thorner, 1991; Dohlman et al., 1991; Kurjan, 1992; Bardwell et al., 1994).

Like other GPCRs, the yeast pheromone receptors have an amino acid sequence containing seven hydrophobic segments presumed to form transmembrane α helices. In addition, signal transduction initiated by the pheromone receptors proceeds via a set of proteins with strong similarities to mammalian heterotrimeric G proteins (Table 8.1). Although the a-factor and α-factor receptors have no discernible sequence similarity, they couple to the same G protein. *S. cerevisiae* also contains a second Gα-like protein, encoded by the *GPA2* gene, that may couple to a third yeast GPCR involved in modulating growth and in induction of the pseudohyphal phase of these yeast (Yun et al., 1997).

The physiological responses to the presence of pheromone secreted by cells of the opposite mating type involve the following steps with similarities to other known GPCR-linked signaling pathways: (1) binding of pheromone to the receptor; (2) interaction of the activated receptor with the G protein trimer, leading to release of a molecule of GDP bound to the G protein α subunit;

TABLE 8.1. Relevant Genes of the *S. cerevisiae* Pheromone Response Pathway

Gene Name	Role in Signaling
STE2	α-Factor receptor
STE3	a-Factor receptor
STE4	G protein β subunit
STE5	"Scaffold protein"; possible effector of G protein
STE12	Transcription factor inducing expression of pheromone-responsive genes
STE18	G protein γ subunit
GPA1	G protein α subunit
BAR1	Protease that degrades α-factor
SST2	RGS protein stimulating adaptation to pheromone
FAR1	Interacts with CLN2 to mediate pheromone-induced cell cycle arrest

(3) binding of GTP to the Gα subunit, leading to dissociation of this subunit from the β and γ subunits; (4) activation of downstream effectors and a MAP kinase-type protein phosphorylation cascade triggered by release of the G protein β and γ subunits. The *GPA1*-encoded Gα subunit appears to play only an inhibitory role in signaling, thus deletion of the *GPA1* gene leaves free β and γ subunits, resulting in constitutive activation of the pathway. The most likely candidate for the effector immediately downstream of the G protein is the product of the *STE5* gene [Whiteway et al., 1995]; (5) phosphorylation of a transcription factor encoded by the *STE12* gene, leading to increased transcription of mating-related genes; and (6) adaptation to the pheromone stimulus. The mechanisms thought to be involved in adaptation include degradation of ligand by the product of the *BAR1* gene, phosphorylation of receptors, endocytosis of receptors, and stimulation of the GTPase activity of Gα by the product of the *SST2* gene.

The similarity between the yeast pheromone response pathway and GPCR-mediated signaling in mammalian systems is exemplified by the interchangeability of components between the systems. A number of mammalian GPCRs, including the human M_1 and rat M_5 muscarinic acetylcholine receptors, human dopamine receptors, and bovine visual opsin (Payette et al., 1990; Huang et al., 1992; Sander et al., 1994a,b; Mollaaghababa et al., 1996), have been expressed and shown to bind ligand in membranes of *S. cerevisiae*. In addition, several receptors, including the human β_2-adrenergic receptors, rat somatostatin receptors, the rat A_{2A} adenosine receptor, the human growth hormone-releasing hormone receptor, and a human receptor for lysophosphatidic acid (King et al., 1990; Price et al., 1995, 1996; Kajkowski et al., 1997; Erickson et al., 1998) are capable of coupling to the yeast signaling pathway so as to activate the pheromone response upon binding their non-pheromone ligands. In the case of the β_2-adrenergic receptor, such coupling only occurs when a mammalian Gsα subunit is coexpressed in yeast with the heterologous GPCR. However, mammalian somatostatin and adenosine receptors are capable of coupling to the endogenous *GPA1* gene product (Price et al., 1995, 1996). Expression in yeast of intact mammalian Gα subunits, as well as certain chimeras between mammalian and yeast Gα subunits, can complement the constitutive signaling and growth arrest associated with deletion of the endogenous *GPA1*-encoded Gα subunit (Kang et al., 1990), but neither the mammalian α subunits nor the chimeras can participate in pheromone-induced signal transduction. However, coupling between the yeast α-factor receptor and a chimeric yeast–rat Gαs subunit has been reported in the unusual circumstance where a chimeric yeast–mammalian Gα is fused directly to the receptor (Medici et al., 1997).

3. REASONS TO STUDY GPCRs IN YEAST

The ease of genetic manipulation of yeast combined with the availability of phenotypic screens and selections involving the pheromone response make possible approaches to studying the structure and function of G proteins that would be difficult or impossible in mammalian systems. Furthermore, the high efficiency of homologous genetic recombination in yeast makes it a straightforward and rapid procedure to delete or replace chromosomal genes to create a genetic background appropriate for genetic screens. The recent completion of the yeast genome project means that the sequences of all genetic loci are readily accessible. The presence of only two to three GPCRs, two G protein α subunits, and one each of β and γ subunits, all of which can be easily removed by standard genetic procedures, makes it possible to study even low-level signaling events in the absence of any background from endogenous receptors or G protein subunits. These factors allow the following experimental approaches.

3.A. Screening Libraries of Random Receptor Mutations

Site-directed mutagenesis has been widely used as a tool for studying GPCR-linked signaling pathways in mammalian systems, but most such studies examine the effects of particular amino acid substitutions on function. In only a handful of cases has it been possible to screen random mutations affecting components of mammalian signaling systems (Lerner, 1994; Burstein et al., 1995; Hill-Eubanks et al., 1996). In contrast, the availability of high-efficiency protocols for transformation of yeast cells and a battery of genetic screens and selections for mutations affecting mating pathways make it possible to create and screen large libraries of random mutations. Mutations leading to defective, hypersensitive, constitutively active, and dominant-negative pheromone receptors have been identified from libraries of random mutations (Boone et al., 1993; Weiner et al., 1993; Clark et al., 1994; Stefan and Blumer, 1994; Konopka et al., 1996; Sommers and Dumont, 1997; Leavitt LM, Macaluso CR, Kim KS, and Dumont ME, unpublished results). Random screening has been used to identify pheromone receptors with altered ligand specificities, as in the case of mutant α-factor receptors from *S. cerevisiae* that have gained the ability to respond to pheromone from a related yeast species, *Saccharomyces kluyvei* (Marsh, 1992). The application of this type of genetic analysis has also allowed the identification of intragenic second-site suppressors in the α-factor receptor (Sommers and Dumont, 1997; see also below). Such suppressors are mutations in one region of a gene that suppress the effects of a second mutation in a different region of the same gene. The compensatory effects of such interacting pairs of mutations can be used to identify helix–helix contacts in receptors and other membrane proteins. While the random screens of receptor mutations published to date have focused on endogenous yeast proteins, it is anticipated that similar techniques will soon be applied to mammalian receptors that can couple to the pheromone response pathway.

In addition to allowing identification of particular examples of certain types of mutations, screening of random libraries in yeast makes possible recovery of a complete collection of all single base mutations giving rise to a desired phenotype. Even the analysis of incomplete collections of mutations allows calculation of the frequency with which a particular type of mutation will occur in the overall population.

3.B. Screening Libraries of Random G Protein and Effector Mutations

The transformation and screening techniques that have been applied to receptors can also be applied to G proteins and downstream effectors. Thus, mutations of the yeast G protein subunits that affect coupling to receptor (Kallal and Kurjan, 1997) and other steps of the pheromone response (Stone and Reed 1990; Leberer et al., 1992; Whiteway et al., 1994) have been uncovered from random mutant libraries. Similarly, single mutations and suppressing pairs of mutations that affect interactions among the subunits and G protein–effector coupling have been identified from random screens (Whiteway et al., 1994). Information from these mutations has aided in interpretation of the recently published x-ray crystallographic structures of trimeric G proteins (Lambright et al., 1996; Wall et al., 1995). The ability to screen for receptor–G protein interactions in a yeast cell that lacks any other endogenous G proteins may provide a way of determining the basis for the understanding the specificity with which GPCRs signal via certain G proteins.

3.C. Screening for Ligands That Can Interact With a Given GPCR

The ability to quickly and reliably identify yeast colonies expressing a receptor that can respond to a ligand is already being employed in identifying new agonists and antagonists for known receptors. Most of these studies are focusing on mammalian receptors expressed in yeast. Using colorimetric or growth-based assays, it is possible to use high-throughput robotic-based systems to test yeast cultures for responses to a large bank of chemical compounds for the identification of lead compounds (Bass et al., 1996). A novel genetic approach to the identification of receptor ligands was recently applied to the endogenous yeast α-factor receptor. Libraries of genes encoding randomly mutagenized α-factor peptides were transformed into a yeast strain expressing the α-factor receptor. Peptides exhibiting agonist and antagonist activity were identified using screens for receptor function in colonies expressing the library components

(Manfredi et al., 1996). The use of the yeast two-hybrid system (see below) for identifying ligands that can interact with a soluble fragment of a GPCR has also been reported (Kajkowski et al., 1997).

3.D. Screening for Proteins That Mediate or Regulate Cellular Signaling Pathways

Yeast can be used for expression cloning of proteins that affect receptor function. For instance, the yeast *SST2* protein is an example of an RGS protein that accelerates the GTPase activity of heterotrimeric G proteins (Dohlman et al., 1996). This gene is a normal component of the desensitization mechanism in the yeast pheromone response pathway that was uncovered in screens of random mutations affecting pheromone signaling. *SST2* mutations can be complemented by mammalian RGS proteins (Druey et al., 1996). Thus, the use of yeast genetics may allow discovery of new modulators of signaling, either in the endogenous signaling pathway or through expression cloning of genes from mammalian systems. The completion of the yeast genome project means that all the endogenous yeast genes related to the signaling pathway are known, even if they have not yet been implicated in signaling.

3.E. Use of the Yeast Two-Hybrid System

The yeast two-hybrid system is a method for detecting protein–protein interactions that has been widely applied to endogenous yeast proteins as well as proteins from other organisms (for review, see Fields and Sternglanz, 1994). Because this system is based on the assembly of interacting proteins into functional transcriptional activators, it requires that the interacting proteins be imported into the nucleus. This precludes use of the two-hybrid system for studying interactions among membrane proteins. Nonetheless, there have been recent successes in the use of the two-hybrid system to detect interactions among soluble fragments of membrane proteins. For example, an unexpected interaction has been detected between carboxy-terminal tail fragments of α_{2A}- and α_{2B}-adrenergic receptors and the eukaryotic initiation factor 2B, a cytoplasmic guanine nucleotide exchange protein (Klein et al., 1997). An interaction has been detected between the product of the yeast *GPA2* gene, encoding a Gα-like protein, and predicted cytoplasmic regions of a protein that resembles a GPCR (Yun et al., 1997). An interaction between an extracellular amino-terminal fragment of the growth hormone-releasing hormone receptor and its hormone ligand can be detected using the two-hybrid system (Kajkowski et al., 1997).

4. ASSAYS OF THE YEAST PHEROMONE RESPONSE PATHWAY

The diversity of responses of haploid yeast cells to pheromone binding has provided numerous ways of detecting activation of the pathway. In the course of decades of research on the mechanisms of mating, these assays have been extensively developed not only as a way of quantitating mating, but also as a way of selecting and screening for interesting mutations. The fact that these techniques are indicative of signaling function, and not just of association between signaling components, makes them particularly useful. We present below three assays that can be used for both assaying the pheromone response and screening for mutants. Two additional assays that have been used primarily for quantitation are projection formation (detection of changes in cell shape by microscopic examination; see Sprague, 1991) and agglutination (detection of the aggregation of cells of opposite mating types through measurement of light-scattering changes; see Moore, 1983).

It is often desirable to conduct assays of the pheromone response using *bar1*⁻ strains to minimize the effects of different cell densities and culture conditions on pheromone levels and to avoid selection of *bar*1 mutations. However, even in *bar1*⁻ strains, for reasons that are not yet completely understood, not all assays necessarily measure the same fractional level of response to a given dose of pheromone (Moore, 1983).

We routinely conduct assays of the pheromone response in triplicate using three independent yeast transformants. This prevents false conclusions that could arise from variations in plasmid copy number, variations in site of integration (for integrated constructs), or inadvertent use of strains containing spontaneous mutations in the receptor or other components of the pheromone response pathway.

4.A. Mating Efficiency

The end product of the pheromone response pathway is the fusion of two haploid cells to form a diploid. Under optimum conditions, haploid yeast cells can be induced to mate with cells of the opposite mating type with nearly 100% efficiency. In cells lacking a major component of the pathway, such as a pheromone receptor, this efficiency can be reduced by a factor of more than 10^6, providing a wide dynamic range for assaying signaling. Mating efficiency can usually be assayed using two haploid strains with different auxotrophic markers. When two such strains mate to form a diploid, genetic complementation of the two markers allows growth on selective media that will not support growth of either of the initial haploids. A ratio of mated cells to input cells can be obtained by plating a mixture of the haploid cells to be tested with an excess of haploid cells of the opposite mating type on plates that only support growth of the diploid cells and comparing the number of colonies with the number obtained on plates that support growth of the diploids and one of the haploid strains. In some genetic backgrounds, if mating efficiency is high and there is a significant difference in the growth rates of diploids and haploids, mating efficiency can also be estimated from the numbers of colonies of different sizes on a single plate containing medium that supports growth of both the diploids and the haploids being tested.

4.A.a. Quantitative Mating Assay. This assay is performed in our laboratory as follows (see Sommers and Dumont, 1997; based on an earlier description by Sprague, 1991):

1. Grow cultures of the strain to be assayed and a strain of the opposite mating type with a complementary auxotrophic marker to an OD_{600} of 1–2.
2. In a 1.5-ml microcentrifuge tube, mix 1×10^6 cells of the strain to be tested and 1×10^7 cells of the strain of the opposite mating type. An OD_{600} of 1 corresponds to approximately 2×10^7 cells/ml.
3. Collect this mixture of cells on a sterile filter (Millipore HAWP, 0.45-μm pore size, autoclaved). Processing of many samples can be achieved by pipetting into the end of small filtration manifolds connected at the other end to a vacuum line.
4. Remove the filters from the manifold, place them on the surface of a petri dish containing YPD medium (Rose et al., 1990), and allow to grow at 30°C for 3 hours.
5. Place the filters in 1.5-ml tubes containing 0.5 ml of SD-10 medium (Rose et al., 1990), sonicate in a bath-type sonicator for 10 seconds, vortex for 10 seconds, and plate directly or diluted on selective medium.

4.A.b. Selection for Yeast With Increased Mating Efficiency. Because of the wide variation in mating efficiencies of strains with various defects in the pheromone response pathway, it is also possible to use mating as a selection for gain-of-function mutations as follows:

1. Plate as many as several thousand colonies of a mutagenized strain to be screened (see below) on petri plates containing selective medium and allow to grow for several days.
2. Prepare lawns of cells of a yeast strain that can mate to the strain to be tested. The cells used for the lawn should be of the opposite mating type from the strain to be tested and should have a complementary auxotrophic marker. Lawns are prepared by spreading cells evenly on a plate containing YPD, allowing growth overnight at 30°C, and then replica plating onto a second YPD plate.
3. Replica plate the colonies to be tested onto the freshly replica-plated lawns of the opposite mating type.
4. Culture at 30°C for 6–7 hours (time depends on the particular strains used).
5. Replica plate to medium selective for diploid cells.

Background levels of mating in deficient strains can limit the diversity of the mutant collections that can be successfully screened because the colonies arising from mating by such strains can outnumber the colonies resulting from gain-of-function mutations. However, if the number of background mating events is not overwhelming, colonies arising from the low-frequency background events can be distinguished from genuine mutational events. This is possible because the background mating events give rise to single isolated colonies on a plate containing medium selective for diploids, whereas mutational events that convert a whole colony to a higher mating efficiency can give rise to multiple closely packed colonies. We find that the particular type of velveteen fabric used for replica plating can greatly affect the yield and selectivity of this selection. Lawns are prepared with synthetic velveteens that have a high yield of transfer. Plates containing individual colonies are replica plated with a short nap cotton velveteen that transfers less efficiently but maintains well-separated colonies.

By using a recessive resistance to a drug such as canavanine or cyclohex-imide, it is also possible to select for nonmating cells. In this case, the haploid strain to be subjected to the selection would contain a gene for resistance to the drug, while the haploid strain of the opposite mating type and any diploids resulting from mating would be sensitive.

4.B. Expression of Pheromone-Responsive Reporter Genes

Activation of the pheromone response pathway leads to increased transcription of numerous genes involved in preparing cells to mate, including the *FUS1* gene, which encodes a cell surface protein (MccAffrey et al., 1987). Thus, one of the most useful assays for activation of the pheromone response pathway is to measure expression of *FUS1*, which can increase 50-fold in response to pheromone binding to receptor. Expression can be monitored using Northern blotting to quantitate *FUS1* RNA levels (Sprague, 1991), but is more commonly detected by fusing the 5' region of the *FUS1* gene to a reporter gene such as the *Escherichia coli* lacZ gene, encoding β-galactosidase, or to the yeast *HIS3* gene, conferring histidine prototrophy. Fusion to lacZ allows detection of signaling-competent cells as blue colonies on solid culture medium containing the colorimetric β-galactosidase substrate X-gal (5-bromo-4-chloro-3-indolyl-β-D-galactoside). In addition, quantitative assays of *FUS1–lacZ* expression are readily performed as liquid assays using o-nitrophenyl-β-D-galactoside (ONPG) as a colorimetric substrate. Both of these techniques are described in more detail below. Expression of *FUS1–HIS3* fusion genes is detected by monitoring cell growth, either as colonies on culture plates or, quantitatively, by measuring the turbidity of liquid cultures. The sensitivity of assays of *FUS1–HIS3* expression can be adjusted by adding various concentrations of aminotriazole, a competitive inhibitor of the *HIS3* enzyme (Stevenson et al., 1992).

4.B.a. Detection of FUS1–lacZ Expression on Plates.
Strains to be tested are first cultured on plates containing a selective or rich medium. These initial cultures are replica plated to plates containing X-gal. Because β-galactosidase exhibits maximum activity near neutral pH, it is important that these plates be buffered (for formulation of this medium, see Rose et al., 1990). Purified synthetic α-factor (available from several commercial sources; at least 100 μl of water containing 6 ng α-factor per plate for a typical *bar1⁻* strain or 20 ng α-factor for a *BAR1⁺* strain) is spread on the plates and allowed to dry shortly before use. a-Factor is usually purified from cultures of *Mat***a** cells (Sprague, 1991). Replica plating onto the medium used for the assay reduces the variation caused by different growth rates of different strains in the presence of

pheromone and ensures that a large number of cells will be present because most yeast strains do not grow very well at neutral pH.

Assays of *FUS1–lacZ* expression on plates can be confounded by variations in cell growth rate reflecting cell cycle arrest caused by signaling. However, the plate assays are extremely sensitive to low-level expression of *FUS1–lacZ*. If allowed to grow long enough, even strains that are completely deficient for pheromone signaling will eventually turn blue.

4.B.b. Liquid Assay for FUS1–lacZ Expression. This assay can provide reliable quantitation of the induction of *FUS1* expression triggered by the pheromone response pathway and allow determination of the EC_{50} for ligand and the magnitude of the response at saturation:

1. Grow overnight cultures in selective or rich medium to an OD_{600} approximately equal to 1. Assays are performed in triplicate, using three independent yeast transformants.
2. Dilute the cultures to an OD_{600} of 0.04 in YPD medium and allow to grow to an OD_{600} of at least 0.1.
3. Add α-factor to desired concentrations, depending on the genetic background and particular receptor to be tested. The α-factor is diluted into solutions containing 5 μg/ml cytochrome *c* as a carrier to prevent loss of peptide at low concentrations. Allow cells to grow for 2 hours. Culture tubes can be maintained in a tray of water at 30°C during dilution and transfer steps.
4. Culture tubes are transferred to an ice bath, and the OD_{600} of representative cultures is determined. Approximately 2.5×10^6 cells are transferred to 1.5-ml microcentrifuge tubes centrifuged for 6 minutes in a microcentrifuge and resuspended in 500 μl Z buffer (Rose et al., 1990; 16.1 g $Na_2HPO4 \cdot 7H_2O$, 5.5 g $NaH_2PO_4 \cdot H_2O$, 0.75 g KCl, 0.246 g $MgSO_4 \cdot 7H_2O$ in a total volume of 1 liter, pH adjusted to 7.0) containing freshly added β-mercaptoethanol (0.27 ml/liter) and SDS (to 0.004% w/v final concentration). Enough of the original culture is set aside to allow determination of the exact OD_{600} at a later time, following freezing, if necessary.
5. Chloroform (30 μl) is added; the cells are vortexed and then incubated at 28°C for 5 minutes.
6. The assay is started by the addition to each tube of 200 μl of Z buffer containing freshly added β-mercaptoethanol (0.27 ml/liter) and 4 mg/ml ONPG. The samples are incubated at 28°C for an appropriate amount of time (usually 20 minutes for a normal $STE2^+$ strain). The reaction is terminated by addition of 500 μl of 1 M Na_2CO_3, then centrifuged 5 minutes in a microcentrifuge. The OD_{420} of the supernatant is determined and the relative β-galactosidase activity is calculated as

$$\text{Activity} = \frac{OD_{420}}{(OD_{600} \text{ of culture}) \times (\text{time of assay}) \times (\text{volume of culture used})}$$

4.C. Growth Arrest

Induction of the pheromone response leads to growth arrest in the G_1 phase of the cell cycle. This phenomenon serves as the basis for both a genetic selection and a quantitative assay of the pheromone response. The selection consists simply of identifying cells with robust growth in the presence of high concentrations of pheromone. Because of adaptation processes, cells that have not mated can eventually resume growth in the presence of a- or α-factor, but colonies of nonresponding cells form larger colonies on plates containing pheromone.

The quantitative assay based on growth arrest, commonly referred to as the *halo assay*, involves measuring the size of a halo of arrested cells surrounding aliquots of pheromone spotted onto a lawn of the cells to be tested (Sprague, 1991). As performed in our laboratory, cells to be tested are cultured to stationary phase. A total of 3×10^6 cells brought up to a volume of 1.5 ml with sterile water are mixed with 2 ml of melted soft agar (consisting of the appropriate SD dropout medium containing 1% bacto-agar) that had been preincubated at 55°C. The mixture is poured onto a prewarmed (37°C) plate of the same culture medium (but containing 2% agar) and spread evenly by tilting the plate. After the plate cools, a total of 3 µl of water containing varying amounts of mating factor (10–300 ng of α factor for a *Mata bar1⁻* strain) is pipetted at an array of points on the plate. When the diameter of the circular region of growth inhibition surrounding the site of pheromone addition is plotted against the log of the mass of pheromone added, a straight line is usually obtained. Strains with different EC_{50}s for pheromone generally give rise to lines with similar slopes but different y-intercepts.

This assay is most useful for identifying strains expressing receptors with different EC_{50}s for pheromone or strains that fail to respond to pheromone at all. The halos of strains expressing receptors with similar EC_{50}s but different maximal responses at saturating ligand concentrations are sometimes difficult to distinguish (Leavitt LM, Macaluso CR, Kim KS, and Dumont ME, unpublished results).

In some strains it is desirable to uncouple the pheromone response pathway from cell cycle arrest. For example, strains expressing constitutively active receptors, or strains where the signaling response is constitutively activated because of lack of G protein α subunit, can go into permanent cell cycle arrest, resulting in lethality. There are several ways to study signaling mutations that could result in permanent growth arrest: (1) Deletion of the *FAR1* gene. The product of this gene interacts with the G_1 cyclin encoded by *CLN2* to promote pheromone-dependent cell cycle arrest. Upon deletion of the *FAR1* gene, yeast strains are still capable of *STE12*-mediated transcriptional activation of pheromone-responsive genes, but they exhibit only a slight pheromone-dependent growth defect and mate with reduced efficiency (Price et al., 1995). (2) Maintenance of strains as *Mata/Matα* diploids, which do not express key components of the mating apparatus and do not respond to pheromone. The effects of mutant alleles on the mating pathway are determined by sporulating the diploid strains and studying the potentially lethal effects on the resulting haploid spores (Kang et al., 1990). (3) Plasmid shuffle. Cells containing a recessive mutation that causes constitutive signaling can be propagated by maintaining a normal copy of a gene on a plasmid (Medici et al., 1997). To study the effects on cells of the presence of a single mutant allele, the plasmid is cured from the strain either by removing the need to maintain an auxotrophic marker or by using a drug to select for cells that have lost the marker, such as 5-fluoroorotic acid (FOA) for loss of the *URA3* gene. (4) Mutation of downstream components. A temperature-sensitive *ste5* mutation has been used for this purpose. At elevated temperatures, this strain fails to signal in response to stimuli or additional mutants that activate the pheromone response pathway. At low temperatures, signaling is restored, and the effects of the stimuli or mutations can be tested (Konopka et al., 1996).

5. CREATION OF LIBRARIES OF MUTANT RECEPTORS

To screen or select for interesting mutations affecting particular signaling molecules, it is necessary to design a strategy for mutagenizing the relevant genes. Conventional techniques for *in vivo* mutagenesis of yeast using radiation or chemical mutagens can lead to the introduction of mutations into any gene in the chromosome. This is a useful outcome if the purpose of the experiment is to identify new genes affecting signaling. However, this type of *in vivo* mutagenesis is not an efficient way to obtain a large library of mutations in a particular gene of interest for several reasons: (1) The yield of mutations in the gene of interest can be low; (2) the spectrum of base substitutions obtained is limited; and (3) genetic back-crosses are required to verify that the phenotype results from mutation of the specific gene of interest.

A convenient way of targeting mutations to a particular gene of interest is to express the gene on a plasmid in a host strain from which the chromosomal copy of the gene has been deleted. This ensures that the only copy of the gene in the cell is the plasmid-borne one. Mutagenesis of the plasmid can then be conducted *in vitro* and the library of mutated genes introduced into yeast cells by transformation. This greatly reduces (but does not eliminate, see below) the likelihood that mutant phenotypes result from mutations in genes other than the one being mutagenized. It also greatly facilitates cloning, verifying, and sequencing the mutations, and, with the proper choice of plasmid, can also make it possible to use the same system to study the effects of site-directed mutations.

A standard method for *in vitro* random mutagenesis of a plasmid has been the treatment of the plasmid with hydroxylamine (see Rose et al., 1990). However, this method does not allow for targeting of mutations to a particular region of a plasmid unless additional steps of subcloning from one plasmid to another are used. It is generally desirable to minimize the number of cloning steps in the creation of libraries both for convenience and to maintain the maximum diversity of mutants. Mutations in the promoter region of a gene or even in the sequences involved in plasmid maintenance in yeast can give rise to phenotypes that are sometimes difficult to distinguish from genuine missense or nonsense mutations. In addition, the mutational spectrum of hydroxylamine treatment is limited. However, this approach was successfully used to identify constitutively activating mutations in the α-factor receptor (Konopka et al., 1996). In this study, 12 alleles exhibiting the desired phenotype were recovered, and all 12 were found to contain a mutation of proline 258 to leucine in the sixth predicted transmembrane segment, even though other constitutively active alleles exist (Clark et al., 1994; Sommers CM, Martin NP, and Dumont ME, unpublished results).

5.A. Oligonucleotide-Directed Random Mutagenesis

Oligonucleotide-directed mutagenesis provides a rapid and convenient way of targeting random mutations to a restricted region of a gene. Random mutagenesis is accomplished by conducting the mutagenesis with oligonucleotides that contain a small percentage of each of three incorrect nucleotides at each position in the chain. This is usually accomplished in automated oligonucleotide synthesizers by using a fifth bottle containing a mixture of all four phosphoramidites (Hinkle et al., 1990; Hutchison et al., 1991; technical information supplied by Applied Biosystems) at a low concentration and introducing a 50:50 mix of contents of this bottle with the correct phosphoramidite for the base at that position in the gene. Thus, the region that will be mutagenized is determined by the choice of oligonucleotides, and is restricted to 60 or so nucleotides by the current limitations of automated oligonucleotide synthesis. A major advantage of this approach over chemical mutagenesis is the generality of the spectrum of mutants produced, since substitutions of each of the incorrect bases at each position in the sequence are equally likely. Gel purification of synthetic oligonucleotides is recommended to reduce byproducts that could introduce deletions.

A variety of techniques for introducing the oligonucleotide-encoded mutations into the gene of interest are available:

1. Ligation into a vector replacing the region between unique restriction sites. The usefulness of this approach is restricted by the requirement for unique restriction sites flanking the mutagenized region, by the requirement to synthesize two strands or to use a polymerase to create a second strand, and by the need to carry out ligation reactions, which could restrict the diversity of the resulting library.

2. Polymerase chain reaction (PCR) amplification using the mixture of mutagenic oligonucleotides as one of the primers. This approach requires either the presence of a unique restriction site within one of the primer sequences, making it possible to ligate the PCR product into restriction-cut vector, or the use of a two-step procedure to incorporate the mutagenized products into the gene of interest (Chen and Przbyla, 1994). Caution must be exercised to ensure that any mutant phenotypes uncovered are actually due to mutations in the region of interest and are not caused by spurious mutations elsewhere in the gene introduced by error-prone polymerases used for PCR. Such mutations can be difficult to detect without sequencing the entire amplified region.

3. Conventional techniques of site-directed mutagenesis using single- or double-stranded DNA. These approaches allow the introduction of mutations into any region of a gene, with no need for restriction sites at particular positions. In particular, the single-stranded DNA procedure of Kunkel et al. (1987) has the advantages of simplicity, high efficiency, and minimal requirement for *in vitro* DNA replication (thereby minimizing the possibility of spurious mutations outside the targeted region). The technique relies on the use of a mutant strain of *E. coli* that can incorporate uracil into single-stranded DNA. Following hybridization of the mutagenic primer and synthesis of the second strand that does not contain uracil, the first strand is degraded during transformation into normal *E. coli* strains, providing a selection for the mutagenized strand of DNA. Genes to be mutagenized can be propagated on phagemids, which can be easily manipulated as double-stranded plasmids with large DNA inserts, but which can also produce single-stranded DNA when transfected into *E. coli* in the presence of a helper phage. Detailed protocols for isolation of single-stranded phagemid DNA and for performing the mutagenic reactions, have been described elsewhere (Kunkel et al., 1987; Sambrook et al., 1989; Hutchison et al., 1991).

4. Direct introduction of mutagenic oligonucleotides into yeast cells. This technique has been most successfully applied for introducing gain-of-function mutations into genes for which there exists a strong screen or selection (Yamamoto et al., 1992). The potential of this technique for large-scale random mutagenesis remains unproven; however, the approach has the potential to allow for extremely rapid creation of mutants.

The number of possible single base substitutions in a 50 residue oligonucleotide is 150. However, to be reasonably assured of obtaining every possible component of such a library it is necessary to screen a much larger number of transformants because of the following factors:

1. A library that contains an average of one substitution per oligonucleotide will contain, on average, one-third single base substitutions. The remainder of components will be divided approximately equally between oligonucleotides with no substitutions and oligonucleotides with more than one substitution (Hinkle et al., 1990). Although they are more difficult to analyze, components carrying multiple mutations can significantly increase coverage of the library, especially in cases where most substitutions are not detrimental to gene function (Sommers and Dumont, 1997).

2. Techniques for oligonucleotide-directed mutagenesis have varying efficiencies. It is often, but not always, possible to achieve more than 80% efficiency using the procedure of Kunkel et al. (1987), but a more conservative estimate would be 50%.

3. Not all components are present at equal abundance in a random library. This makes it necessary to collect at least several times the total number of expected clones in the library to have a reasonable likelihood of

recovering any individual component (Finkel et al., 1994). Thus, for a 50 residue oligonucleotide, it would be necessary to screen about 3,000 colonies ($150 \times 3 \times 2 \times 3$) to be reasonably sure to recover every interesting mutation. This number of colonies is easily achieved by the procedure described below, making it possible in a single experiment to conduct a nearly complete screen of a mutational library based on an oligonucleotide long enough to encompass the length of a GPCR transmembrane segment or a small hydrophilic loop (Sommers and Dumont, 1997).

For identification of single amino acid substitutions with particular effects on protein function, it is usually desirable to introduce an average of one base substitution per oligonucleotide. This is accomplished by introducing incorrect bases into each position at an abundance that is equal to the reciprocal of the number of nucleotides in the oligonucleotide. Thus, for a total length of 50 nucleotides, one would introduce each of the mixed bases at an abundance of $1/3 \times 1/50$ or 0.7%. Although oligonucleotide-directed mutagenesis using incorrect bases introduced in this way would be expected to produce only missense and nonsense mutations, we have, in some cases, recovered significant numbers of frameshifts resulting from short deletions. These may arise from unnatural chemical groups that are introduced during oligonucleotide synthesis and excised upon transformation into E. coli (see Hecker and Rill, 1998). It is advisable to subject 20–30 clones that have not been genetically selected to DNA sequencing to determine whether deletions or other abnormalities occur at high frequency.

The general outline of the procedure for random oligonucleotide-directed mutagenesis in yeast is as follows:

1. Clone the gene of interest into a phagemid, and culture in the presence of helper phage. Prepare uracil-containing single-stranded DNA as described by Kunkel et al. (1987) except for the use of two polyethylene glycol/NaCl precipitations to remove endogenous priming activity.

2. Synthesize a mutagenic oligonucleotide containing a low percentage of incorrect base at each position, as described above.

3. Use the oligonucleotide for site-directed mutagenesis, as described by Kunkel et al. (1987). Use only a severalfold excess of oligonucleotide over template to prevent preferential annealing of certain components of the oligonucleotide library to template.

4. Transform the mutagenesis reactions into competent E. coli so as to obtain the number of colonies required for complete sampling of the library. Conventional CaCl$_2$-based transformation protocols may not provide adequate transformation frequencies; however, electroporation or higher efficiency chemical transformation protocols (Sambrook et al., 1989) are adequate.

5. Add 5 ml of YT medium (Sambrook et al., 1989) to the plates containing the transformed E. coli colonies and resuspend the colonies. Grow this liquid culture for several generations (so as to allow only minimal loss of

library diversity), and isolate the plasmid library using a standard DNA miniprep procedure.

6. Transform the plasmid DNA into yeast using the lithium acetate procedure described by Gietz and Woods (1994), but omitting the carrier DNA, which we find can sometimes increase the background level of spontaneous chromosomal mutations to a level that can be detected in subsequent screens.

7. Screen yeast transformants for signaling by the techniques described above.

5.B. Mutagenesis Using PCR and Plasmid Recombination

In many cases, it is undesirable to restrict the region to be mutagenized to the 60 or so nucleotides that can be altered with a single oligonucleotide in the procedures described above. For example, there are many circumstances where one would wish to screen an entire gene for a particular type of mutation such as a second-site suppressor. However, even when an entire gene is to be randomly mutagenized, there are advantages to conducting the mutagenesis in sections. This facilitates sequencing of the mutant alleles and can give a rapid indication of the diversity of the mutations recovered, even before they are sequenced.

For mutagenesis of regions smaller than entire genes, but larger than can be altered with a single oligonucleotide, it is convenient to use misincorporation of nucleotides by *Taq* DNA polymerase during PCR. The normal moderate error rate of *Taq* DNA polymerase can be increased by elevating concentrations of Mg^{2+}, Mn^{2+}, and two deoxynucleoside triphosphates to the point where as many as 2% of bases in the final products of a multi-cycle PCR reaction are the result of misincorporation. Because PCR amplifies double-stranded DNA, misincorporation of only two nucleotides can lead to efficient substitution of all three incorrect bases at any given position (Cadwell and Joyce, 1994; Sommers CM, Martin NP, and Dumont ME, unpublished results).

Although it is possible to conduct PCR reactions so as to be able to ligate the mutagenized product as a substitute for a particular region of the targeted gene, a simpler and more general procedure is to co-transform the linear PCR products with a linearized plasmid, allowing the efficient homologous recombination systems of yeast to combine the two linear DNAs into a circular plasmid (Fig. 8.1). This procedure works because linear DNAs transform yeast poorly compared with circular plasmids. It was originally conducted with gapped plasmids from which a region of the sequence between two restriction sites had been excised (Muhlrad et al., 1992; Staples and Diekman, 1993). However, if it is possible to easily distinguish the desired mutant alleles from alleles resulting from recircularization of the starting plasmid, gapping of the plasmid can be replaced by simple linearization at a single restriction site (Brenner et al., 1994; Sommers and Dumont, 1997). The use of single restriction sites has the distinct advantage that the region to be mutagenized can be selected simply by the choice of primers for the PCR reaction as long as the region to be amplified overlaps both ends of the linearized plasmid by about 50 nucleotides. The num-

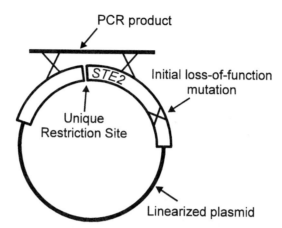

Figure 8.1. Creation of a random mutant library by PCR under error-prone conditions followed by recombination with linearized plasmid. The PCR product is shown overlapping the position of a unique restriction site in the *STE2* gene. The "X" indicates the site of an initial loss-of-function mutation for which second-site intragenic suppressors might be sought. The plasmid is linearized at the unique restriction site and co-transformed into yeast along with the PCR product. Homologous recombination at the indicated positions results in replacing the original wild-type sequences with mutagenized DNA.

ber of yeast colonies resulting from recircularization of linearized plasmid in the absence of PCR product is only about 10% of the number of colonies obtained in the presence of PCR product, indicating that 90% of colonies result from recombination events (Sommers and Dumont, 1997). Furthermore, in contrast to many other techniques for generating mutational libraries, the transformants resulting from recombination are each independent clones resulting from the recombination of a unique PCR product with plasmid. The only potential for nonrandom redundancy in the library arises from mutational events that occur in early cycles of the PCR reactions. These can be minimized by using high concentrations of PCR template.

To facilitate the use of PCR for mutagenesis of the *STE2* gene, we have introduced restriction sites at intervals of about 250 nucleotides throughout the transmembrane regions. This span between sites is small enough to allow fairly complete screening of single base substitutions in each library and to facilitate sequencing of mutant alleles but large enough to allow mutagenesis of the seven transmembrane segments of *STE2* in four or five separate reactions. If there is an average of a single base substitution per PCR product, and assuming 90% of colonies arise from recombination with PCR products, less than 10,000 colonies should be required for nearly complete screening (see above). At a mutational rate of 2%, each isolate will have an average of five mutations; thus, it may be desirable to adjust the concentration of the limiting nucleotides to decrease the average number of mutations per segment. The presence of multiple mutations in each mutant allele complicates the determination of the particular base substitution causing the mutant phenotype, but allows for more complete coverage of the library in a smaller number of colonies because most mutations appear to have minimal effects on receptor function (Sommers and Dumont, 1997).

The procedure for PCR mutagenesis is as follows:

1. PCR amplify a linearized template encoding the target gene in the presence of 7 mM $MgCl_2$, 50 mM KCl, 10 mM Tris (pH 8.3), 1 mM dCTP, 1 mM dTTP, 0.2 mM dATP, 0.2 mM dGTP, and 0.5mM $MnCl_2$. The $MnCl_2$ should be maintained as a $10\times$ stock separate from the other concentrated PCR components. A 100-μl reaction should also include 20 fmol of template DNA, 30 pmol of each primer, and 5 units of *Taq* DNA polymerase (Cadwell and Joyce, 1994). In screening for functional alleles, such as second-site suppressors, it is a good idea to use a template containing a loss-of-function mutation outside the region being amplified. This eliminates the possibility that colonies with functional receptors could arise by transformation with unamplified template. The PCR reactions should run for 30 cycles of 94°C for 1 minute, 45°C (depending on the primers) for 1 minute, and 72°C for 1 minute.

2. Check the PCR products by agarose gel electrophoresis. However, it is not usually necessary to purify or precipitate the amplified DNA before transformation.

3. Linearize the plasmid to be recombined with the PCR products by digesting with the appropriate restriction enzyme. Extract with phenol/chloroform/isoamyl alcohol (24:24:1) to inactivate the enzyme, and ethanol precipitate.

4. Co-transform into yeast (Gietz and Woods, 1994) several micrograms each of PCR product and linearized plasmid. Screen yeast transformants for signaling (see above).

5.C. Verification of Mutant Isolates

If a colony displaying the desired phenotype is detected among a plasmid library of random mutations, the next step is to recover the plasmid in order to sequence the mutation and verify that the mutant phenotype is caused by a mutation in the targeted region of the gene of interest. Plasmids are isolated by culturing the strain to stationary phase, centrifuging the cells, and vortexing the pellet with glass beads in the presence of detergents and phenol/chloroform/isoamyl alcohol (Rose et al., 1990). To obtain adequate amounts of the relevant plasmid, the aqueous phase is transformed into *E. coli* and the plasmid is re-isolated using a miniprep procedure. To make sure that the phenotype is plasmid linked, it is advisable to confirm that the mutant phenotype is obtained following retransformation of the plasmid into the original host strain used for the screen. In a screen for rare second-site suppressors in *STE2* as many as half the strains that regained function appeared to result from chromosomal rather than plasmid-linked mutations. Surprisingly, many of the chromosomal mutations appeared to be dominant, since they were also recovered in a *Mata*/*Mata* diploid background.

Once the mutation has been localized to the plasmid, the mutated region can be sequenced using double-stranded techniques with primers internal to the gene. If multiple mutations are present, it is often desirable to re-create individual mutations by site-directed mutagenesis as described above, but using a single defined oligonucleotide. This procedure also provides verification that the mutant phenotype results from sequence alterations in the regions of the gene targeted for mutagenesis and not from spontaneous mutations elsewhere on the plasmid. If sufficient clones are recovered from a genetic screen for rare phenotypes, the identification of important amino acid substitutions is facilitated by the repeated recovery of the same mutation from multiple independent clones. However, if the mutagenesis was performed by PCR, there is a possibility that a neutral mutation that arises in an early cycle of PCR could be transmitted to a significant fraction of the final products.

6. BIOCHEMICAL ANALYSES OF GPCRs IN YEAST

Many of the same biochemical and molecular biological techniques that have been applied to mammalian systems can also be applied to GPCRs expressed in yeast. Binding of the pheromone agonist, as well as several known antagonists, to endogenous yeast receptors and to mammalian receptors expressed in yeast has been quantitated (Jenness et al., 1986; Raths et al., 1988), and effects of G protein coupling on receptor affinity have been detected (Blumer and Thorner, 1990). Ligand has been crosslinked to receptor (Blumer et al., 1988). Proteolysis has been used to monitor ligand-induced changes in the conformation of the receptor and for studying subcellular trafficking of receptors (Bükü-soglu and Jenness, 1996; Schandel and Jenness, 1994). However, proteolysis of whole cells is inefficient because of the presence of the cell wall, and some enzymes used to disrupt the cell wall lead to degradation of membrane proteins. Portions of the α factor have been produced as synthetic peptides and studied by nuclear magnetic resonance and circular dichroism (Reddy et al., 1994). Recent success in purifying the yeast α-factor receptor and reconstituting it into membranes (David et al., 1997) opens up the possibility of many additional biochemical and biophysical studies.

7. THE SECOND-SITE SUPPRESSOR APPROACH TO STUDYING GPCR STRUCTURE AND FUNCTION

The usefulness of random screening techniques in yeast is illustrated by the recent isolation of loss-of-function mutations and a second-site intragenic suppressor of these mutations in our laboratory. The initial screen for loss of signaling function was conducted by oligonucleotide-directed mutagenesis using a small percentage of incorrect bases, as described above, using a 45 base oligonucleotide spanning the hydrophobic region of the third transmembrane segment of the α-factor receptor. Screening focused on the identification of partially defective alleles because most completely nonfunctional alleles were found to consist of deletions and nonsense mutations, demonstrating that amino acid substitutions causing complete loss of function occur even less frequently than nonsense mutations. The screen resulted in a collection of 26 plasmids with single base changes comprising only 12 different mutations. The extremely nonconservative nature of the mutations recovered, together with a statistical analysis of the results of the screen, indicate that the majority of possible mutations in this region, including many nonconservative substitutions, have little or no effect on receptor function. This is surprising in view of recent evidence from electron crystallography and modeling suggesting that the third transmembrane segments of GPCRs are the most buried in the structure, implying that they participate in more protein–protein interactions in the membrane than other transmembrane segments (Baldwin et al., 1997).

Second-site suppressors of the initial loss-of-function mutations were created using PCR mutagenesis, as described above, starting with the selected defective alleles in the third transmembrane segment. The mutagenized regions encompassed all six remaining transmembrane segments. Suppressor alleles were identified using a selection for increased mating and a screen for *FUS1–lacZ* induction. Both procedures yielded some of the same substitutions. Three second-site suppressors were identified (see Fig. 8.2). One of these, $Tyr^{266} \rightarrow Cys$, in the sixth transmembrane segment, was allele specific, only suppressing an initial $Glu^{143} \rightarrow Lys$ mutation. Furthermore, the $Tyr^{266} \rightarrow Cys$ mutation, when studied as an individual mutation, was defective for signaling. The allele specificity of this $Tyr^{266} \rightarrow Cys$ suppressor, combined with the fact that the combination of two defective alleles leads to a functional receptor, is indicative of the existence of specific protein–protein contacts in the receptor. Such contact between the third and sixth transmembrane segments is consistent with recent models for helix packing in the GPCR family (Farrens et al., 1996; Han et al., 1996; Sheikh et al., 1996; Baldwin et al., 1997). The existence of such contacts may be confirmed using biochemical techniques, such as disulfide crosslinking (Chervitz et al., 1995; Farrens et al., 1996; Yang et al., 1996). Receptors with cysteine residues introduced at the sites of both mutations in the α-factor receptor (Glu^{143} and Tyr^{266}) are currently being examined.

Two intragenic second-site suppressors, $Arg^{58} \rightarrow Gly$ and $Met^{218} \rightarrow Thr$, were recovered multiple times, indicating that the total number of potential suppressors of the particular initial mutations studied is not very high. Upon recreating the individual mutations and mutation pairs, it was discovered that these two suppressors are not allele specific; they were capable of suppressing multiple initial loss-of function mutations. Furthermore, when studied as individual mutations, they exhibited near-normal signaling. These characteristics indicate that these mutations are "global" suppressors. They stabilize the protein structure or function in such a way as to be able to compensate for defects at multiple locations in the receptor.

The ability to readily screen for loss-of-function and gain-of-function mutations as illustrated in this example raises the possibility of a combined genetic and biochemical approach to mapping the interactions between transmembrane helices in pheromone receptors and other GPCRs expressed in yeast. Potential interacting residues would be identified using genetics, and the existence of direct interactions would be tested by disulfide crosslinking. There are some technical difficulties to be overcome, as most previous disulfide crosslinking studies of GPCRs have been conducted at extramembrane loops that may be more exposed to oxidants than groups buried in the membrane. Furthermore, intramolecular crosslinks appear to cause only minimal

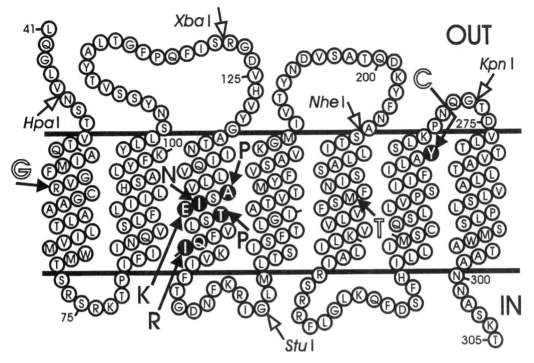

Figure 8.2. Schematic diagram of the transmembrane regions of the α-factor receptor encoded by the yeast *STE*2 gene. Filled-in circles indicate the positions of loss-of-function mutations that can be suppressed by intragenic second-site suppressors. The substitutions causing loss of function are indicated by filled letters. Amino acid substitutions isolated as suppressors are indicated as unfilled letters. The substitution Tyr[266]→Cys suppresses only the loss-of-function mutation Glu[143]→Lys. The substitution Met[218]→Thr suppresses Glu[143]→Lys and Thr[144]→Pro and Ala[140]→Pro but not Ile[142]→Asn or Ile[150]→Arg. The substitution Gly[58]→Arg suppresses all the indicated loss-of-function mutations in the third transmembrane segment. Loss of function due to the substitution Tyr[266]→Cys is suppressed by Glu[143]→Lys. The positions of unique restriction sites used for PCR mutagenesis are indicated. The extracellular amino-terminal and the intracellular carboxy-terminal tails have been omitted.

effects on the mobility of GPCRs on SDS gels; thus analysis of crosslinking would most likely be conducted on receptors expressed as separate fragments, an approach that seems promising in the case of the α-factor receptor (Martin et al., 1999). The use of genetics for identifying potential interacting residues may provide a more efficient way of identifying potential crosslinking partners than the wholesale approach of cysteine scanning mutagenesis, which can require the production of an extremely large number of mutants for analyzing polytopic membrane proteins with more than a few transmembrane segments. The identification of interacting residues via the combination of genetics and biochemical crosslinking may serve as a useful complement to the many ongoing attempts to model the helix–helix packing of GPCRs and other transmembrane proteins for which functional screens can be developed in yeast.

ACKNOWLEDGMENTS

We thank the members of our laboratory, Karen Kim, Negin Martin, LuAnn Leavitt, and Ping Yi, for comments and insights helpful in the preparation of this chapter. In addition, we acknowledge discussions with Beth Grayhack, Eric Phizicky, and Fred Sherman that contributed significantly to the development of our understanding of mating and mutational approaches in yeast, and we thank Beth Grayhack for her comments on this manuscript. We are also grateful for funding from the American Heart Association (grant 95010260) and the American Cancer Society (grant VM169) that made possible various aspects of this research.

REFERENCES

Baldwin JM, Schertler GFX, Unger VM (1997): An alpha-carbon template for the transmembrane helices in the rhodopsin family of G-protein–coupled receptors. J Mol Biol 272:144–164.

Bardwell L, Cook JG, Inouye CJ, Thorner J (1994): Signal propagation and regulation in the mating pheromone response pathway of the yeast *Saccharomyces cerevisiae.* Dev Biol 166:363–379.

Bass RT, Buckwalter BL, Patel BP, Pausch MH, Price LA, Strnad J, Hadcock JR (1996): Identification and characterization of novel somatostatin antagonists. Mol Pharmacol 50:709–715.

Blumer KJ, Reneke JE, Thorner J (1988): The *STE2* gene product is the ligand-binding component of the α-factor receptor of *Saccharomyces cerevisiae.* J Biol Chem 263:10836–10842.

Blumer KJ, Thorner J (1990): β and γ subunits of a yeast guanine nucleotide-binding protein are not essential for membrane association of the α subunit but are required for receptor coupling. Proc Natl Acad Sci USA 87:4363–4367.

Blumer KJ, Thorner J (1991): Receptor G-protein signaling in yeast. Annu Rev Physiol 53:37–57.

Boone C, Davis NG, Sprague GF (1993): Mutations that alter the third cytoplasmic loop of the a-factor receptor lead to a constitutive and hypersensitive phenotype. Proc Natl Acad Sci USA 90:9921–9925.

Brenner C, Bevan A, Fuller RS (1994): Biochemical and genetic methods for analyzing specificity and activity of a precursor-processing enzyme: Yeast kex2 protease, kexin. Methods Enzymol 244:152–167.

Burstein ES, Spalding TA, Hill-Eubanks D, Brann M (1995): Structure–function of muscarinic receptor coupling to G proteins: Random saturation mutagenesis identifies a critical determinant of receptor affinity for G proteins. J Biol Chem 270: 3141–3146.

Büküsoglu G, Jenness DD (1996): Agonist-specific conformational changes in the yeast α-factor pheromone receptor. Mol Cell Biol 16:4818–4823.

Cadwell RC, Joyce GF (1994): Mutagenic PCR. PCR Methods Appl 3:S136–S140.

Chen B, Przybyla AE (1994): An efficient site-directed mutagenesis method based on PCR. BioTechniques 17:657–659.

Chervitz SA, Lin CM, Falke JJ (1995): Transmembrane signaling by the aspartate receptor: Engineered disulfides reveal static regions of the subunit interface. Biochemistry 34:9722–9733.

Clark CD, Palzkill P, Botstein D (1994): Systematic mutagenesis of the yeast mating pheromone receptor third intracellular loop. J Biol Chem 269:8831–8841.

David NE, Gee M, Andersen B, Naider F, Thorner J, Stevens RC (1997): Expression and purification of the *Saccharomyces cerevisiae* α-factor receptor (Ste2p), a 7-transmembrane-segment G protein coupled receptor. J Biol Chem 272:15553–15561.

Dohlman HG, Song JP, Ma D, Corchesne WE, Thorner J (1996): Sst2, a negative regulator of pheromone signaling in the yeast *Saccharomyces cerevisiae:* Expression, localization and physical association with Gpa1 (the G-protein α subunit). Mol Cell Biol 16:5194–5209.

Dohlman HG, Thorner J, Caron MG, Lefkowitz RJ (1991): Model systems for the study of seven-transmembrane-segment receptors. Annu Rev Biochem 60:653–688.

Druey KM, Blumer KJ, Kang VH, Kehri JH (1996): Inhibition of G protein–mediated MAP kinase activation by a new mammalian gene family. Nature 379:742–746.

Erickson JR, Wu JJ, Goddard JG, Tigyi G, Kawanishi K, Tomei LD, Kiefer MC (1998): Edg-2/Vzg-1 couples to the yeast pheromone response pathway selectively in response to lysophosphatidic acid. J Biol Chem 273:1506–1510.

Farrens DL, Altenbach C, Yang K, Hubbell W, Khorana G (1996): Requirement for rigid body motion of transmembrane helices of light activation of rhodopsin. Science 274:768–770.

Fields S, Sternglanz R (1994): The two-hybrid system: An assay for protein–protein interactions. Trends Genet 10:286–291.

Finkel RA, Bent S, Schardl CL (1994): Determining the probability of obtaining a desired clone in an amplified or shuttle library. Biotechniques 16:580–582.

Gietz RD, Woods RA (1994): High efficiency transformation in yeast. In Johnston JA (ed): Molecular Genetics of Yeast: Practical Approaches. New York: Oxford University Press, pp 124–134.

Guthrie C, Fink GR (1991): Guide to Yeast Genetics and Molecular Biology. San Diego: Academic Press.

Han M, Lin SW, Smith SO, Sakmar TP (1996): Functional interaction of transmembrane helices 3 and 6 in rhodopsin. Replacement of phenylalanine 261 by alanine causes reversion of a glycine 121 replacement mutant. J Biol Chem 271:32337–32342.

Hecker KH, Rill RL (1998): Error analysis of chemically synthesized polynucleotides. BioTechniques 24:256–260.

Hill-Eubanks D, Burstein ES, Spalding TA, Brauner-Osborne H, Brann M (1996): Structure of a G-protein–coupling domain of a muscarinic receptor predicted by saturation mutagenesis. J Biol Chem 271:3058–3065.

Hinkle PC, Hinkle PV, Kaback HR (1990): Information content of amino acid residues in putative helix VIII of the *lac* permease from *Escherichia coli.* Biochemistry 29:110989–10994.

Huang H-J, Liao C-F, Yang B-C, Kuo T-T (1992): Functional expression of the rat M5 muscarinic receptor in yeast. Biochem Biophys Res Commun 182:1180–1186.

Hutchison CA, Swanstrom R, Loeb DD (1991): Complete mutagenesis of protein coding domains. Methods Enzymol 202:356–390.

Jenness DD, Burkholder AC, Hartwell LH (1986): Binding of α-factor to *Saccharomyces cerevisiae:* Dissociation constant and number of binding sites. Mol Cell Biol 6:318–320.

Kajkowski EM, Price LA, Pausch MH, Young KH, Ozenberger BA (1997): Investigation of growth hormone releasing hormone receptor structure and activity using yeast expression technologies. J Recept Signal Transduct Res 17:293–303.

Kallal L, Kurjan J (1997): Analysis of the receptor binding domain of Gpa1p the G_α subunit involved in the yeast pheromone response pathway. Mol Cell Biol 17:2897–2907.

Kang Y-S, Kane J, Kurjan J, Stadel JM, Tipper DJ (1990): Effects of expression of mammalian $G\alpha$ proteins on the yeast pheromone response signal transduction pathway. Mol Cell Biol 10:2582–2590.

King K, Dohlman HG, Thorner J, Caron MG, Lefkowitz RJ (1990): Control of yeast mating signal transduction by a mammalian β_2-adrenergic receptor and $G_s\alpha$ subunit. Science 250:121–123.

Klein U, Ramirez MT, Kobilka BK, von Zastrow M (1997): A novel interaction between adrenergic receptors and the α-subunit of eukaryotic initiation factor 2B. J Biol Chem 272:19099–19102.

Konopka JB, Margarit SM, Dube P (1996): Mutation of Pro-258 in transmembrane domain 6 constitutively activates the G protein–coupled α-factor receptor. Proc Natl Acad Sci USA 93:6764–6769.

Kunkel TA, Roberts JD, Zakour RA (1987): Rapid and efficient site-specific mutagenesis without phenotypic selection. Methods Enzymol 154:488–492.

Kurjan J (1992): Pheromone response in yeast. Annu Rev Biochem 61:1097–1129.

Lambright DG, Sondek J, Bohm A, Skiba NP, Hamm HE, Sigler PB (1996): The 2.0 Å structure of a heterotrimeric G protein. Nature 379:311–319.

Leberer E, Dignard D, Hougan L, Thomas DY, Whiteway MS (1992): Dominant-negative mutations of a yeast G-protein β subunit identify two functional regions involved in pheromone signaling. EMBO J 11:4805–4813.

Lerner M (1994): Tools for investigating functional interactions between ligands and G protein coupled receptors. Trends Neurosci 17:142–146.

Manfredi JP, Klein C, Herrero JJ, Byrd DR, Truehart J, Wielser WT, Fowlkes DM, Broach JR (1996): Yeast α mating factor structure–activity relationship derived from genetically selected peptide agonists and antagonists of Ste2p. Mol Cell Biol 16: 4700–4709.

Marsh L (1992): Substitutions in the hydrophobic core of the α-factor receptor of *Saccharomyces cerevisiae* permit response to *Saccharomyces kluveri* α-factor and to antagonist. Mol Cell Biol 12:3959–3966.

Martin WP, Leavitt LM, Sommers CM, Dumont ME (1999). Assembly of G protein–coupled receptors from fragments: Identification of functional receptors with discontinuities in each of the loops connecting transmembrane segments. Biochemistry 38:682–695.

MccAffrey G, Clay FJ, Kelsay K, Sprague G (1987): Identification and regulation of a gene required for cell fusion during mating of the yeast *Saccharomyces cerevisiae*. Mol Cell Biol 7:2680–2690.

Medici R, Bianchi E, Di Segni G, Tocchini-Valentini GP (1997): Efficient signal transduction by a chimeric yeast–mammalian G protein α subunit Gpa1-Gsα covalently fused to the yeast receptor Ste2. EMBO J 16:7241–7249.

Mollaaghababa R, Davidson FF, Kaiser C, Khorana G (1996): Structure and function in rhodopsin: Expression of functional mammalian opsin in *Saccharomyces cerevisiae*. Proc Natl Acad Sci USA 93:11482–11486.

Moore SA (1983): Comparison of dose–response curves for α-factor–induced cell division arrest, agglutination, and projection formation of yeast cells. J Biol Chem 258:13849–13856.

Muhlrad D, Hunter R, Parker R (1992): A rapid method for localized mutagenesis of yeast genes. Yeast 8:79–82.

Payette P, Gossard F, Whiteway M, Dennis M (1990): Expression and pharmacological characterization of the human M1 muscarinic receptor in *Saccharomyces cerevisiae.* FEBS Lett 266:21–25.

Price LA, Kajkowski EM, Hadcock JR, Ozenberger BA, Pausch MH (1995): Functional coupling of a mammalian somatostatin receptor to the yeast pheromone response pathway. Mol Cell Biol 15:6188–6195.

Price LA, Strnad J, Pausch MH, Hadcock JR (1996): Pharmacological characterization of the Rat A_{2A} adenosine receptor functionally coupled to the yeast pheromone response pathway. Mol Pharmacol 50:829–837.

Raths SK, Naider F, Becker JM (1988): Peptide analogues compete with the binding of α-factor to its receptor in *Saccharomyces cerevisiae.* J Biol Chem 263:17333–17341.

Reddy AP, Tallon MA, Becker JM, Naider F (1994): Biophysical studies on fragments of the α-factor receptor protein. Biopolymers 34:679–689.

Rose MD, Winston F, Hieter P (1990): Methods in Yeast Genetics. Cold Spring Harbor, NY: Cold Spring Harbor Laboratory.

Sambrook J, Fritsch EF, Maniatis T (1989): Molecular Cloning: A Laboratory Manual, 2nd ed. Cold Spring Harbor, NY: Cold Spring Harbor Laboratory.

Sander P, Grünewald S, Bach M, Haase W, Reiländer H, Michel H (1994a): Heterologous expression of the human D_{2S} dopamine receptor in protease-deficient *Saccharomyces cerevisiae* strains. Eur J Biochem 226:697–705.

Sander P, Grünewald S, Maul G, Reiländer H, Michel H (1994b): Constitutive expression of the human D_{2S} dopamine receptor in the unicellular yeast *Saccharomyces cerevisiae.* Biochim Biophys Acta 1193:255–262.

Schandel KA, Jenness DD (1994): Direct evidence for ligand-induced internalization of the yeast α-factor pheromone receptor. Mol Cell Biol 14:7245–7255.

Sheikh SP, Zvyaga TA, Lichtarge O, Sakmar TP, Bourne H (1996): Rhodopsin activation blocked by metal–ion-binding sites linking transmembrane helices C and F. Nature 383:347–350.

Sommers CM, Dumont ME (1997): Genetic interactions among the transmembrane segments of the G protein coupled receptor encoded by the yeast *STE2* gene. J Mol Biol 266:559–575.

Sprague G (1991): Assay of yeast mating reaction. Methods Enzymol 194:77–93.

Staples RR, Diekman CL (1993): Generation of temperature-sensitive *cbp1* strains of *Saccharomyces cerevisiae* by PCR mutagenesis and *in vivo* recombination: Characteristics of the mutant strains imply that *CBP1* is involved in stabilization and processing of cytochrome *b* pre-mRNA. Genetics135:981–991.

Stefan CJ, Blumer KJ (1994): The third cytoplasmic loop of a yeast G-protein coupled receptor controls pathway activation, ligand discrimination, and receptor internalization. Mol Cell Biol 14:3339–3349.

Stevenson BJ, Rhodes N, Errede B, Sprague G (1992): Constitutive mutants of the protein kinase *STE11* activate the yeast pheromone response pathway in the absence of G protein. Genes Dev 6:1293–1304.

Stone DE, Reed SI (1990): G protein mutations that alter the pheromone response. Mol Cell Biol 10:4439–4446.

Wall MA, Coleman DE, Lee E, Iñiguez-LLuhl JA, Posner B, Gilman A, Sprang SR (1995): The structure of the G protein heterotrimer $G_{i\alpha1}\beta_1\gamma_2$. Cell 83:1047–1058.

Weiner JL, Guttierez-Steil C, Blumer KJ (1993): Disruption of receptor–G protein coupling in yeast promotes the function of an *SST2*-dependent adaptation pathway. J Biol Chem 268:8070–8077.

Whiteway MS, Clark K, Leberer E, Dignard D, Thomas DY (1994): Genetic identification of residues involved in association of α and β G-protein subunits. Mol Cell Biol 14:3223–3229.

Whiteway MS, Wu C, Leeuw T, Clark K, Fourest-Lieuvin A, Thomas DY, Leberer E (1995): Association of the yeast pheromone response G protein beta gamma subunits with the MAP kinase scaffold Ste5p. Science 269:1572–1575.

Yamamoto T, Moerschell RP, Wakem LP, Ferguson D, Sherman F (1992): Parameters affecting the frequencies of transformation and co-transformation with synthetic oligonucleotides in yeast. Yeast 8:935–948.

Yang K, Farrens DL, Altenbach C, Farahbakhsh ZT, Hubbell WL, Khorana HG (1996): Structure and function in rhodopsin. Cysteines 65 and 316 are in proximity in a rhodopsin mutant as indicated by disulfide formation and interactions between attached spin labels. Biochemistry 35:14040–14046.

Yun C-W, Tamaki H, Nakayama R, Yamamoto K, Kumagai H (1997): G-protein coupled receptor from yeast *Saccharomyces cerevisiae*. Biochem Biophys Res Commun 240: 287–292.

CHAPTER 9

CONSTITUTIVELY ACTIVE RECEPTOR MUTANTS AS PROBES FOR STUDYING THE MECHANISMS UNDERLYING G PROTEIN–COUPLED RECEPTOR ACTIVATION

SUSANNA COTECCHIA, FRANCESCA FANELLI, ALEXANDER SCHEER, and PIER G. DE BENEDETTI

1. INTRODUCTION   168
2. COMBINED COMPUTATIONAL AND EXPERIMENTAL
   MUTAGENESIS OF THE α_{1b}-AR   169
   A. Methodological Approaches to Receptor Modeling   169
   B. The Search for **R** and **R*** by MD Analysis of Constitutively
      Active α_{1b}-AR Mutants   170
   C. The Mechanistic Role of the Conserved E/DRY Sequence
      in Receptor Activation   171
   D. The Upgraded α_{1b}-AR Model   174
   E. The Concerted Movement of the Helices and Intracellular
      Loops in Receptor Activation   174
   F. Comparison Between Mutation- and Agonist-Induced
      Activation of the Receptor   178
   G. The Potential Common Role of Conserved Amino Acids
      in GPCR Activation   179
3. CONCLUSIONS AND FUTURE OUTLOOK   180

Structure–Function Analysis of G Protein-Coupled Receptors, Edited by Jürgen Wess.
ISBN 0-471-25228-X   Copyright © 1999 Wiley-Liss, Inc.

I. INTRODUCTION

Mutational analysis of several G protein–coupled receptors (GPCRs) has revealed that the extracellular regions and the transmembrane (TM) helices contribute to the formation of the ligand binding site, whereas the amino acid sequences of the intracellular (i) loops appear to mediate receptor–G protein coupling and to interact with regulatory proteins (Khorana, 1992; Savarese and Fraser, 1992; Wess, 1997; Bourne, 1997). Yet very little is known about how binding of the extracellular signals triggers receptor activation (i.e., the "conformational switch" of the receptor leading to productive receptor–G protein coupling).

The first evidence that point mutations in a GPCR could trigger receptor activation was described for the α_{1b}-adrenergic receptor (AR) (Cotecchia et al., 1990). Remarkably, all possible amino acid substitutions of A293 in the carboxy-terminal portion of the i3 loop of the α_{1b}-AR induced variable levels of constitutive (agonist-independent) activity (Kjelsberg et al., 1992). Similar mutations were also found to increase the constitutive activity of the β_2-AR and α_2-AR (Ren et al., 1993; Samama et al., 1993). A detailed analysis of the properties of constitutively active AR mutants was instrumental in proposing the "allosteric ternary complex" model to describe GPCR behavior (Samama et al., 1993). This model introduces an explicit isomerization constant regulating the equilibrium of GPCR between at least two interconvertible allosteric states, **R** (inactive or ground state) and **R*** (active state). It is assumed that only **R*** can effectively interact with the G protein. In the absence of the agonist, **R** predominates possibly because a structural constraint might prevent sequences of the intracellular loops to interact with the G proteins. On the other hand, agonists can trigger the equilibrium toward **R***, thus favoring its stabilization. The position of the equilibrium between **R** and **R*** might be responsible for the various levels of basal (agonist-independent) activity observed for different wild-type GPCRs as it has been described for two dopamine receptor subtypes (Tiberi and Caron, 1994). In addition, constitutively activating mutations might trigger the formation of **R***, releasing the structural constraint that keeps GPCRs inactive. These findings led to the hypothesis that activating mutations mimic, at least to some extent, the conformational change triggered by agonist binding a GPCR.

Spontaneously occurring activating mutations have been discovered for several GPCRs and are responsible for several human diseases. In addition, several constitutively activating mutations have been identified in site-directed mutagenesis studies of different GPCRs (reviewed by Scheer and Cotecchia, 1997). A number of activating mutations seem to cluster in the carboxy-terminal portion of the i3 loop and in helix 6 of GPCRs as initially reported for the adrenergic receptors. However, constitutive activation of GPCRs can be triggered by mutations distributed in different domains, including the TM helices, extracellular as well as intracellular regions.

The identification of receptor sites susceptible to constitutive activation might contribute to elucidate the structural basis of the "conformational switch" underlying GPCR activation. The most convincing example is found in the rhodopsin system in which an important contribution to our understanding of receptor activation came from the investigation of a constitutively active opsin mutant occurring in a severe form of autosomal dominant retinitis pigmentosa (Robinson et al., 1992). Mutation of K296 in helix 7 is predicted to disrupt a salt bridge between helices 3 and 7 that contributes to keep the opsin inactive, thus resulting in its constitutive activation.

Although site-directed mutagenesis and biophysical studies on different GPCRs (described in other chapters of this volume) provided many insights into the structure–function relationships of these proteins, a consistent structural description of the molecular changes underlying the conversion of **R** to **R*** is still lacking. The definitions of **R** and **R*** refer to macroscopic entities behaving as binary inactive/active switches defined according to the thermodynamic principles of allosteric transition (Wyman and Gill, 1990). On the other hand, each of two or several possible functional receptor macroscopic forms may correspond to large collections of microscopic states of the protein. Thus, the challenging question is which pattern of microscopic configurations predominates in each functional state of the receptor and which concerted intramolecular motion allows these proteins to behave as binary switches. Even if a highly resolved structure of a GPCR was immediately available, the description of the mechanisms of activation in terms of both structure and dynamics would remain to be elucidated.

In this chapter, we illustrate a strategy to explore the potential molecular changes correlated with the transition of **R** to **R*** using the Gq-coupled α_{1b}-AR as a model system.

2. COMBINED COMPUTATIONAL AND EXPERIMENTAL MUTAGENESIS OF THE α_{1b}-AR

2.A. Methodological Approaches to Receptor Modeling

To probe structure–function relationships of GPCRs, several useful molecular models of these membrane proteins were built using different methods (Cronet et al., 1993; Fanelli et al., 1995a; Findlay and Eliopoulos, 1990; MaloneyHuss and Lybrand, 1992; Trumpp-Kallmeyer et al., 1992). To investigate the potential intramolecular motions underlying different functional states of the receptor, three-dimensional model building of receptor structure was combined with computational simulation of receptor dynamics. The strategy usually chosen in these studies is to compare the dynamic behavior of the receptor molecule in the free, antagonist-bound, and agonist-bound forms (Fanelli et al., 1995a,b; Luo et al., 1994; Zhang and Weinstein, 1993).

However, two drawbacks limit the strategy based on docking of functionally different ligands (agonist and antagonist). One is that the receptor states explored in these simulations depend on the configuration of the residues predicted to interact with the ligand; such residues, however, still remain ill-defined for many GPCRs. The second drawback is that this approach does not help clarify how GPCRs become active even in the absence of agonist.

The approach proposed here is a combination of computer-simulated (using Molecular Dynamics [MD]) and experimental mutagenesis of the α_{1b}-AR to investigate the structural–dynamic features characterizing the inactive and active receptor states. The α_{1b}-AR model was built using an iterative *ab initio* procedure started as a comparative MD study on the helix bundle of seven GPCRs (α_{1b}-AR, β_2-AR, α_2-AR, dopamine D2, 5-HT1A, muscarinic M_1, and bovine rhodopsin not complexed with retinal) (Fanelli et al., 1995a). In the first model (model I), the initial arrangement of the helices was based on the structural constraints inferred from the analysis of $\sim$200 GPCR sequences (Baldwin, 1993). Model I was progressively upgraded by adding the intracellular and extracellular domains of the α_{1b}-AR (models II and III) (Scheer et al., 1996; Fanelli et al., 1998). The analysis of the MD trajectories was used to compare the structural–dynamic features of functionally different receptor mutants with those of the wild-type α_{1b}-AR and to predict key residues the mutation of which would either constitutively activate or inactivate the receptor.

Molecular modeling was achieved by means of the molecular graphics package QUANTA 96 (Molecular Simulations, Waltham, MA). Minimization and Molecular Dynamics simulations were performed by means of CHARMM (Brooks et al., 1983). The methodology employed to build the receptor models and to simulate their dynamics has been described in detail in previous papers (Fanelli et al., 1995a, 1998; Scheer et al., 1996).

2.B. The Search for R and R* by MD Analysis of Constitutively Active α_{1b}-AR Mutants

The comparative MD study of the helical bundle of different GPCRs highlighted the structural–functional role of conserved polar amino acids (Fanelli et al., 1995a,b). Among these, N63 (helix 1), D91 (helix 2), N344, and Y348 (helix 7) form a transmembrane "polar pocket" consistent with other modeling studies (Oliveira et al., 1994). This prompted us to explore the potential molecular changes occurring at the level of the "polar pocket" in all 19 constitutively active α_{1b}-AR mutants carrying mutations of A293 (Kjelsberg et al., 1992). Our hypothesis was that comparing the structural–dynamic features of the wild-type α_{1b}-AR with those of receptor mutants showing different degrees of constitutive activity would help identify some of the structural changes correlated with the transition from **R** to **R*** (in the absence of agonist).

A new model of the α_{1b}-AR (model II) was built as previously described (Scheer et al., 1996) by adding all three extracellular loops, the i1 and i2 loops, the amino-terminal and the carboxy-terminal portions of i3, and the amino-terminal portion of the carboxy-terminal tail (carboxy tail).

The structural–dynamic features characteristic for the wild-type and the active forms of the α_{1b}-AR primarily involved the interaction pattern of R143 contained in the highly conserved E/DRY sequence lying at the cytosolic end of helix 3 (Fig. 9.1). In the minimized average structure of the wild-type receptor, R143 was directed toward helix 2, interacting with the conserved D91 (helix 2) of the "polar pocket." Moreover, several cationic amino acids on i3 (i.e., R288 and K291) were found buried with respect to the cytosol. Activating mutations of A293 resulted in minimized average structures showing almost progressive shifts of R143 out of the "polar pocket" and cytosolic exposure of R288 and K291. In our model, A293 lies within the carboxy-terminal portion of i3 with its side chain directed toward several residues of the amino-terminal α-helical portion of the same loop. Any alternative side chains introduced by mutations change the inter-

HELIX 1	A I S V G L V L G A F I L F A I V G N I L V I L S V e e e e t t t t t t t t t t t t t t t t t t t i i	70
i1	A C N R H L R T P T N L L L T T T T L L i L	81
HELIX 2	Y F I V N L A I A D L L L S T V L P F S A T L E V i i i i t t t t t t t t t t ' t t t t t e e e e	107
e1	L G Y W V L G R I F C L L L T T T T L L L L	118
HELIX 3	D I W A A V D V L C C T A S I S L C A I S I D R Y I G J e e e e t t t t t t t t t t t t t t t t t i i i i i j	147
i2	R Y S L Q Y P T L V T R R L T T T T L L L T T T L L	160
HELIX 4	K A I L A L L S V W V L S T V I S I G P L L G W K i i t t t t t t t t t t t t t t t t e e e e e	185
e2	E P A P N D D K E C G V T E E L L L L L L L L L L L L T T L L L	201
HELIX 5	F Y A L F S S L G S F Y I P L A V I L V M Y C R V Y ' J A K R T T K e e e e t t t t t t t t t t t t t t t i i i i i i i i i i	235
i3	N L E A G V M K E M S N S K E L T L R I H S K N F H E D T L S S T K A K G H N P R S S I A V K L F L L L L L L H H H H H H H H H H H H H H L T T T L T T T T L L L L L L L L L L T T T L L T T L	284
HELIX 6	K F S R E K K A A K T L G I V V G M F I L C W L P F F ' A L P L G S i i i i i i i i i i i i i t t t t t t t t t t t t t t t e e e e	318
e3	L F S T L K P P D L L L L L L L L	327
HELIX 7	A V F K V V F W L G Y F N S C L N P I I Y P C S S K E K R A F M R I L G e e e e t t t t t t t t t t t t t t t t t i i i i i i i i i i i i i	364

Figure 9.1. Amino acids included in the theoretical model of the α_{1b}-AR. The model includes the seven helices, the three extracellular loops (e1, e2, and e3), and the three intracellular loops (i1, i2, and i3). For the intracellular (i) and the extracellular (e) receptor domains, the secondary structure (H, helix; T, turn; L, loop) assigned in the input arrangement is indicated. For the helices, the predicted topology of amino acids (e, extracellular; t, transmembrane; i, intracellular) is indicated.

helical interaction pattern and consequently promote rigid body motions of helices 5 and 6 that propagate to the whole helix bundle, inducing a detachment of R143 from D91 (helix 2). Remarkably, the extent of the arginine shift seemed related to the rank order of constitutive activation induced by mutations of A293.

Based on the features highlighted by this comparative MD analysis (Scheer et al., 1996), we assumed that the average structure of the wild-type α_{1b}-AR could represent the inactive or ground state (**R**) of the receptor. On the other hand, the similarities observed for the constitutively active receptor mutants (i.e., the shift of R143 out of the "polar pocket" and the exposure of several cationic amino acids toward the cytosol) could be characteristic for the active forms (**R***) of the α_{1b}-AR.

2.C. The Mechanistic Role of the Conserved E/DRY Sequence in Receptor Activation

To search for structural–dynamic perturbations other than mutations that could allow the α_{1b}-AR to become active, we referred to recent findings in the rhodopsin system. In this system, a key event in forming **R*** appears to be light-induced proton uptake by rhodopsin occurring after Schiff base deprotonation (Arnis et al., 1994; Fahmy and Sakmar, 1993). One of at least two anionic residues that undergo proton uptake appears to be glutamate E134, which is part of the highly conserved E/DRY motif at the cytosolic side of helix 3. Thus, we investigated whether the transition from **R** to **R*** for the wild-type α_{1b}-AR would also depend on protonation of D142 (corresponding to E134 in rhodopsin).

Because the results discussed above refer to the average structure of the α_{1b}-AR carrying all ionizable residues in their charged form, we simulated the wild-type α_{1b}-AR structure with D142 in its protonated (neutral) form and compared it with the previous structures. The resulting average minimized structure of the protonated α_{1b}-AR shared several distinct features with the constitutively active mutants simulated previously. Most importantly, the interaction pattern of R143 and the cytoplasmic exposure of several cationic amino acids closely resembled those of the constitutively active mutants.

To challenge the predictive power of the theoretical model, we simulated the mutation of D142 into alanine. The alanine residue neutralizes the negative charge at position 142 while introducing only a few degrees of freedom due to side-chain torsion angles. The finding that the average structure resulting from this mutation mimicked that of constitutively active receptors carrying mutations of A293 led to the correct prediction that mutation of D142 into alanine could constitutively activate the α_{1b}-AR (Scheer et al., 1996). These results are in striking agreement with recent findings by Cohen et al. (1993) showing that mutating the homologous E134 of rhodopsin into glutamine can cause light-independent activation.

The role of the aspartate of the E/DRY sequence in the α_{1b}-AR was further investigated by introducing all possible natural amino acids at D142 (Scheer et al., 1997). Because all receptor mutants displayed various levels of agonist-independent activity, we were able to quantitatively analyze the relationship between the structural–dynamic features of the receptor and the extent of constitutive activity. A linear relationship was found between the hydrophobicity of the residue present at position 142 and the extent of constitutive activity induced by the mutation. According to the definition of hydrophobicity as empirical descriptor (obtained as mean value of free energy transfer from a polar to a nonpolar environment), the mechanism triggering receptor activation may involve the translocation of D142 from the cytosol to a less polar environment.

The analysis of receptors carrying mutations of D142 pointed again to the important functional roles of D91 (helix 2) and R143. When D142 is in the anionic form, the interaction of D91 and R143 is highly favored, maintaining the arginine and several other amino acids of i2 and i3 in a buried state with respect to the cytosol. In contrast, protonation of D142 or constitutively activating mutations at this position, although not structurally identical, all share the same main structural effect consisting of the shift of R143 out of the "polar pocket." We found that the higher the extent of a constitutive activity of the mutant, the deeper the burying of the residue at position 142 and the larger the distance between D91 and R143 in the average mutant structures.

These findings led to the hypotheses that (1) a network of hydrogen-bonding interactions between residues of the "polar pocket" (N63 in helix 1, D91 in helix 2, N344 and Y348 in helix 7) and R143 contributes to stabilize the α_{1b}-AR in its inactive state, (2) the hydrophobic/hydrophilic character of D142 is regulated by protonation/deprotonation, and (3) the increased hydrophobicity of D142 promotes internalization of its side chain, which induces changes in the interaction pattern of D142 with residues on helix 2, helix 4, and i2, leading to the shift of R143 out of the "polar pocket."

Thus, the aspartate of the E/DRY sequence may constitute a fundamental switch of α_{1b}-AR activation through protonation/deprotonation of its side chain.

Our findings suggested that the main role of R143 of the conserved E/DRY motif is to mediate receptor activation by either directly interacting with the G protein heterotrimer or by allowing several amino acids in i2 and i3 (i.e., R288 and K291) to attain the proper configuration for the formation of a G protein docking site.

This hypothesis is supported by the observation that different mutations of R143 can almost entirely inactivate the α_{1b}-AR (Fig. 9.2) (Scheer et al., 1996). These results are in agreement with recent studies in which mutations of the conserved arginine in several GPCRs, including rhodopsin (Acharya and Karnik, 1996), the M_1 and M_2 AchR (Zhu et al., 1994) and the V2 vasopressin receptor (Rosenthal et al., 1993), produced loss of receptor–G protein coupling.

A crucial role of R143 in receptor activation is further supported by the finding that the α_{1b}-AR double receptor mutants D142A/R143A and A293E/R143A did not display any constitutive or agonist-induced activity (Fig. 9.2). Thus, the mutation of R143 into alanine was able to abolish the constitutive activity induced by mutations of D142 or A293. In agreement with the experimental findings, the computer-simulated structures did not display the structural–dynamic features shown by the active mutants D142A and A293E (results not shown). Thus, R143 seems to represent an important structural mediator of the constitutive activation induced by the two different mutations.

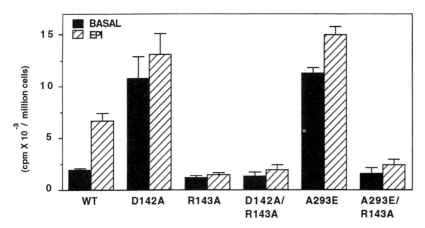

Figure 9.2. Inositol phosphate response mediated by the wild-type and different α_{1b}-AR mutants. Inositol phosphates (IP) were measured in COS-7 cells expressing the wild-type α_{1b}-AR (WT) or different single (D142A, A293, and R143A) and double (D142/R143A, A293E/R143A) receptor mutants as previously described (Scheer et al., 1996). Receptor numbers were 800, 310, 410, 200, 650, and 880 fmol/10^6 cells for WT, D142A, R143A, D142/R143A, A293, and A293E/R143A, respectively. IP responses were measured in the absence (BASAL) or presence of 10^{-4} M epinephrine (EPI).

2.D. The Upgraded α_{1b}-AR Model

To gain new insights into the mechanism of receptor activation and to further challenge the computational procedure employed, the α_{1b}-AR model was modified and completed including new experimental information available on GPCRs (Fanelli et al., 1998). In particular, the structural information derived from electron micrographs of two-dimensional fragrhodopsin crystals (Unger et al., 1997) were used to upgrade the α_{1b}-AR model. The tilt of the helices in the α_{1b}-AR input arrangement obtained previously was slightly modified according to the tilt angles of the seven TM helices estimated from the map of frog rhodopsin (Unger et al., 1997). In addition, the i3 loop of the α_{1b}-AR (model II) was completed by adding residues 236–284.

Figure 9.1 shows the TM (t), the extracellular (e), and the intracellular (i) domains included in α_{1b}-AR model III, together with the secondary structures assigned to the input receptor model. A large conformational space was explored introducing different perturbations onto the input arrangement (i.e., translations and rotations of the helices and loops as well as modifications of the side-chain torsion angles). The wild-type α_{1b}-AR input structure selected among a very large number of input arrangements tested was used to produce the input structures of the receptor mutants. The procedure employed to select the starting arrangement also included the testing of different combinations of distance constraints between the backbone oxygen atoms of residue i and the backbone nitrogen atoms of residue i + 4, excluding prolines. A comparative analysis was performed on the wild-type receptor and several mutant structures averaged over the last 100 psec of the MD simulation time and minimized (Fanelli et al., 1998). Several similarities as well as some differences were found between the α_{1b}-AR structures and the α-carbon atom model recently proposed by Baldwin et al. (1997) as template for the TM helices of the rhodopsin family of GPCRs.

2.E. The Concerted Movement of the Helices and Intracellular Loops in Receptor Activation

The upgraded α_{1b}-AR input structure (model III) allowed us to define new structural–dynamic features of the active and inactive states of the receptor. In addition to the D142A and A293E mutants, the comparative analysis of the constitutively active receptors was extended to the C128F receptor carrying an activating mutation in helix 3 (Perez et al., 1996). Investigations of inactive receptor mutants focused on receptors carrying mutations of R143 or D91 into alanine.

In the average minimized structure of the wild-type α_{1b}-AR in its ground state (Fig. 9.3a), conserved polar amino acids are involved in the following interactions: (a) N63 (helix 1) interacts with D91 (helix 2); (b) D91 (helix 2) is also engaged in charge-reinforced H-bonding interactions with R143 (helix 3) and N344 (helix 7); (c) D142 (helix 3) is involved in charge-reinforced H-bonding interactions with R159 (carboxy-terminal portion of i2); (d) R143 makes charge-reinforced H-bonding interactions with D91 (helix 2) and van der Waals interactions with Y348 (helix 7); (e) Y144 (DRY sequence) is di-

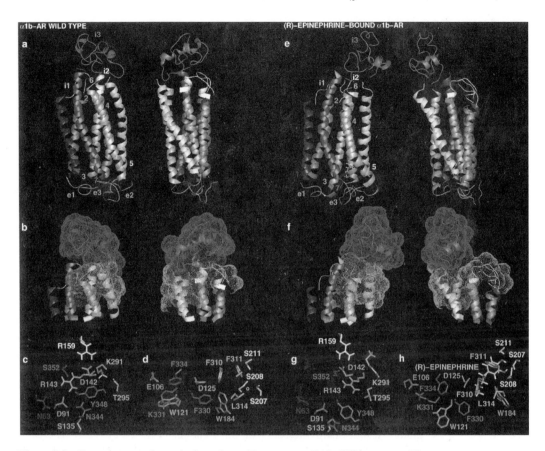

Figure 9.3. Ground state and agonist-bound α₁ᵦ-AR structures. **Left:** Wild-type α₁ᵦ-AR (ground state). **Right:** R(−)-epinephrine-bound α₁ᵦ-AR. All views are in a direction perpendicular to the helix main axis. **a, e:** Two views of the average minimized structure. The seven helices are colored in blue (1), orange (2), green (3), pink (4), yellow (5), sky blue (6), and violet (7), respectively. The i1, i2, and i3 loops are colored in green, white, and cyclamen, respectively; whereas e1, e2, and e3 are colored in carnation, respectively. **b, f:** Two views of the cytosolic half of the average minimized structure. The solvent-accessible surface computed on i2, i3, and the cytosolic extensions of helices 5 and 6 are displayed. **c, g:** Some amino acids (colored according to their location) in the environment of D91 (helix 2), D142, and R143 of the E/DRY. **d, h:** Some amino acids in the environment of D125 (helix 3) and S207 (helix 5). (See color plates.)

rected towards helix 5, (f) W307 (helix 6) makes H bonds with S132 (helix 3) and van der Waals contact with N340 (helix 7); and (g) N340 (helix 7) forms van der Waals interactions with W307 (helix 6).

Moreover, D125 (helix 3), conserved in all cationic neurotransmitter receptors, is surrounded by a cluster of aromatic amino acids F310 [helix 6], F330 [helix 7] and F334 [helix 7]) (Fig. 9.3).

Mutation of D142 into alanine (Fig. 9.4a) results in changes of interaction patterns involving this conserved residue. In particular, replacement of D142 by alanine breaks the coulombic interaction with R159 and triggers the "internalization" of the residue into a water-shielded environment where van der Waals interactions with aromatic residues on helix 2 constitute a major stabilizing factor. Following this perturbation, helices 2, 3, and 4 undergo large rotational and vertical movements that propagate to the other four helices, thus promoting a rearrangement of the interaction patterns involving several conserved polar amino acids.

As shown in Figure 9.4a, the following interactions represent some of the structural features characterizing the constitutively active D142A receptor mutant: (a) N63 (helix 1) retains the interaction found in the wild-type receptor ground state structure; (b) D91 makes H-bonding interactions with S135 (helix 3) and N344 (helix 7); (c) R143 forms H bonds with S352 (helix 7); (d) Y144 interacts with R225 (helix 5); (e) W307 contacts I133 (helix 3); (f) N340 (helix 7) interacts with N344; and (g) Y348 (helix 7) is directed toward helix 6 and is engaged in dispersion interactions with R143. D125 also faces helix 6, surrounded by the same aromatic cluster as in the wild-type receptor.

Interestingly, some of the interaction patterns observed in the D142A structure are largely conserved in the average minimized structures of two other constitutively active mutants, A293E and C128F (results not shown). However, the helix movements leading to the average interaction patterns shared by the D142A, A293E, and C128F mutants are different.

In the wild-type α_{1b}-AR structure, A293 in the carboxy-terminal portion of i3 is oriented toward several residues within the α-helical amino-terminal portion of the same loop. C128 in the extracellular half of helix 3 is surrounded by residues of helices 2 and 7. In the constitutively active mutant A293E, the glutamate side chain interacts with Y227 and K231 (in the cytosolic extension of helix 5) and consequently promotes, through a complex series of intramolecular interactions, rigid body movements of helices 5 and 6 that propagate to other helices. On the other hand, in the constitutively active C128F mutant the phenylalanine introduced by mutation is engaged in new interactions involving F310 (helix 6) and F334 (helix 7), thus promoting motions of helix 3 that can propagate to other helices (results not shown).

Interestingly, in all three constitutively active mutants (D142A, A293E, and C128F), the motions of the helices, although differently triggered, induce the projection toward the cytosol of an α-helical segment in i3 involving amino acids 242–255 (Fig. 9.4b). The three receptor structures seem to share the opening of a site formed by i2, the cytosolic extension of helix 5, and amino acids 242–258 in i3 characterized by a large solvent-accessible surface. The electrostatic potential corresponding to this surface protruding out from the membrane is positive and can complement the negative isopotential surface of G protein (Gq) domains hypothesized to form the receptor–G protein interface (results not shown).

Our model supports the hypothesis that the transmembrane domains of the receptor play an essential mechanistic role in allowing the flexible cytosolic loops to interact effectively with the G protein heterotrimer. In particular, our analysis suggests that amino acids 242–255 in i3 might play a crucial role in the interaction of α_{1b}-AR with the G protein. These findings are consistent with our previous experimental results (Cotecchia et al., 1990) showing that substitution

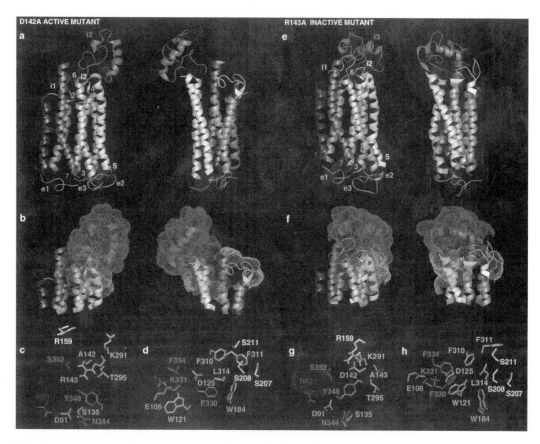

Figure 9.4. Active and inactive α_{1b}-AR mutant structures. **Left:** Constitutively active D142A mutant. **Right:** Inactive R143A mutant. Other details are as in Figure 9.3. (See color plates.)

of amino acids 252–259 in i3 of the α_{1b}-AR almost completely abolished receptor-mediated increases of inositol phosphate production.

In agreement with previous results (Scheer et al., 1997), model III also indicates that protonation of D142 promotes rigid body motions of TM helices that confer to the α_{1b}-AR the structural–dynamic features characteristic of the constitutively active mutants (results not shown). These results strengthen our hypothesis that the functional equilibrium between the **R** and **R*** states of the α_{1b}-AR depends, at least in part, on the equilibrium between the protonated and the deprotonated forms of D142.

On the contrary, in the wild-type receptor (carrying D142 in the anionic form) (Fig. 9.3a), as well as in the inactive mutants D91N (results not shown) and R143A (Fig. 9.4e), the cationic amino acids R288 and K291 are buried with respect to the cytosol, and amino acids 242–255 are oriented differently compared with the active structures. Furthermore, due to the different arrangement of the helices and loops, the i2 loop and the cytosolic extension of helix 5 are more buried with respect to the cytosol compared with the active structures.

The inactive receptor mutants, however, differ from the ground state of the wild-type α_{1b}-AR in that they cannot be activated by agonists. In the D91N mutant, the asparagine side chain is neutral, whereas in the wild-type receptor the aspartate is in its anionic form. Mutational evidence suggests that D91 (helix 2) remains deprotonated (anionic form) during receptor activation. This is supported by experimental findings that mutation of D91 in the α_{1b}-AR (Scheer et al., 1996) and of the homologous aspartate in other GPCRs can profoundly impair receptor–G protein coupling (Savarese and Fraser, 1992).

In the D91N mutant (Fig. 9.4e), R143 is directed toward helix 2 in a fashion that it feels the electrostatic field generated by D142 (which is involved in a salt bridge with R159). It is conceivable that in receptor forms that can be activated by agonists, R143 is attracted by D91 more than by D142, whereas the opposite might occur in inactive receptors. Moreover, due to movements of helix 3, a novel constraining interaction occurs in the D91N and R143A mutants involving D125 and K331 at the extracellular side of the helix bundle.

2.F. Comparison Between Mutation and Agonist-Induced Activation of the Receptor

An important question is whether the active states of the α_{1b}-AR induced by activating mutations share structural similarities with the receptor conformation induced by agonist binding. Thus, the MD trajectories of the constitutively active receptor mutants were compared with those resulting from simulations of the agonist–α_{1b}-AR complex.

R(−)-epinephrine was docked into the input structure of the wild-type α_{1b}-AR carrying the deprotonated D142 (Fig. 9.3e). Previous experimental findings indicated that S207 (helix 5) plays a crucial role in catecholamine binding to the α_{1b}-AR (Cavalli et al., 1996). Thus, MD simulations of the (−)-epinephrine–α_{1b}-AR complex were started by testing different combinations of distance constraints between S207 (helix 5) and the catecholic oxygen atoms of the ligand. Finally, we selected the minimized average structure resulting from the simulation in which S207 is constrained to act as H-bonding donor and acceptor for the *meta*- and *para*-hydroxyl groups, respectively, of the ligand. This pattern allows the cationic nitrogen atom of (−)-epinephrine to form strong charge-reinforced H-bonding interactions with D125 (helix 3). This model is in agreement with our previous experimental findings showing that D125 is essential for both agonist and antagonist binding to the α_{1b}-AR (Cavalli et al., 1996).

According to our model, the interactions between the catecholic hydroxyl groups of the ligand and S207 (helix 5), its protonated nitrogen atom and D125 (helix 3) allow the agonist to bridge helices 3 and 5. Furthermore, helix 6 is also involved in agonist binding due to van der Waals attractive interactions between F311 (helix 6) and the aromatic ring of the ligand. S208 and S211 (helix 5), which are involved in agonist-induced activation of the receptor (Cavalli et al., 1996), interact with residues on helices 6 and 4, respectively. These interactions might contribute to propagate the agonist-induced structural changes from the binding site to helices 6 and 4.

Comparison of the minimized average structures of the free and (−)-epinephrine-bound forms of the wild-type α_{1b}-AR (Fig. 9.3) shows that the agonist induces a strong structural rearrangement propagating from the extracellular to the intracellular side. In particular, the structural modifications of the binding site following agonist docking involve a rotational movement of helix 5 that induces the shift of S207 from an interhelical position (facing helix 4) toward the core of the helix bundle. The movements occurring in the ligand binding domain propagate through the helix bundle, inducing a change of the arrangement of the amino acids of the "polar pocket" near the cytosol. The main effect of this rearrangement consists of breaking the charge-reinforced H-bonding interaction between D91 and R143, promoting the shift of R143 out from the "polar pocket," and translocating positively charged amino acids toward the cytosol. Interestingly, the rigid body movements of the helices allow the opening of a site between the i2 and i3 loops characterized by a large solvent-accessible surface area (Fig. 9.3).

Importantly, this investigation indicates that there is a structural link between the agonist binding site and the receptor domains involved in G protein coupling. This is further supported by our findings showing that activating mutations (Fig. 9.4a) as well as protonation of D142 (results not shown) induce modifications of the receptor binding site. Such changes might underlie the increased binding affinity of agonists observed for several constitutively active receptor mutants (Scheer et al., 1996). However, the structural determinants involved in the increased binding affinity need to be further investigated.

Notwithstanding the similarities highlighted above, the agonist-induced active structure also displayed clear differences from the structures of the constitutively active mutants (i.e., D142A, A293E, and C128F). These findings are consistent with recent results from site-directed spin labeling experiments on rhodopsin that suggest that the constitutively activating E134Q mutation (E134 corresponds to D142 in the α_{1b}-AR) induces helix motions different from those caused by photoactivation (Kim et al., 1997). In fact, the E134Q mutant did not display the dramatic increase in distance between helices C (3) and F (6) observed upon photoactivation of wild-type rhodospin (Farrens et al., 1996). Interestingly, a marked separation of the cytosolic ends of helices 3 and 6 was also not observed in the structures of the constitutively active α_{1b}-AR mutants D142A and A293E.

2.G. The Potential Common Role of Conserved Amino Acids in GPCR Activation

The combined computer-simulated and experimental mutagenesis studies of the α_{1b}-AR indicated that several conserved amino acids (D91, S132, R143, W307, N344, and Y348) might drive helix movements resulting in rearrangement of the cytosolic domains and the exposure of a site with docking complementary for the G protein. We suggest that the structural roles played by these conserved amino acids might be shared by other members of the rhodopsin family of GPCRs. This is supported by findings that mutations of some homologous conserved amino acids induce loss of function in several GPCRs (Savarese and Fraser, 1992).

Experimental evidence seems to indicate that, whereas inactivating mutations of GPCRs involve conserved amino acids, activating mutations usually occur at structurally crucial interhelical positions with a low degree of conservation. One exception are mutations that involve the negatively charged amino acid of the E/DRY sequence for which, despite its high degree of conservation, activating mutations have been reported (Cohen et al., 1993; Scheer et al., 1996).

Our finding that neutralization of D142 in the α_{1b}-AR causes constitutive activity suggests that this residue has to become protonated to allow receptor activation. Because of the high degree of conservation of the E/DRY motif, we suggest that the equilibrium between the two prototropic forms of its negatively charged residue might affect the equilibrium between the ground and active states of several GPCRs. One might also predict that the conformational switch of several GPCRs induced by an agonist could be coupled to a change of the pKa of the conserved E/D residue by changing its environment and thereby favoring protonation.

So far, activating mutations of the negatively charged residue of the E/DRY motif have been reported only for rhodopsin and the α_{1b}-AR. However, this observation does not invalidate the hypothesis that this residue might become protonated in the active states of other GPCRs. On the one hand, the tendency of the E/DRY sequence to become protonated may vary among GPCRs due to different structural features of its environment. On the other hand, receptor responses other than constitutive activity including agonist-induced activation of GPCRs might also be associated with the protonated state of the E/DRY sequence. Interestingly, mutation of the homologous aspartate in the gonadotropin-releasing hormone receptor into asparagine enhanced agonist-induced receptor responses (Arora et al., 1997). In the muscarinic M_1 receptor (Lu et al., 1997), some mutations of the homologous aspartate dramatically decreased receptor expression. However, in the case of those receptor mutants that were expressed, the efficiency of agonist-induced receptor-mediated responses was similar if not better than with the wild-type receptor.

3. CONCLUSIONS AND FUTURE OUTLOOK

In this chapter, we described the iterative approach employed to build an activation model of the α_{1b}-AR by means of an *ab initio* procedure. The models are limited due to the fact that they are based on simplifications and assumptions. However, when the models are built to handle the available experimental information and are used in a comparative context (as we have done), they can be used to rationalize complex experimental findings and to provide suggestions for new experiments.

ACKNOWLEDGMENTS

We acknowledge the financial support of the Fonds National Suisse pour la Recherche Scientifique (31-51043.97), the European community (BMH4-cig7-2152), the CNR, and Ministero dell'Università e della Ricerca Scientifica (funds 40%). We are grateful to the CICAIA (University of Modena) for its technical support and for the use of its computer facilities.

REFERENCES

Acharya S, Karnik SS (1996): Modulation of GDP release from transducin by the conserved Glu[134]–Arg[135] sequence in rhodopsin. J Biol Chem 271:25406–25411.

Arnis S, Fahmy K, Hofmann KP, Sakmar TP (1994): A conserved carboxylic acid group mediates light-dependent proton uptake and signaling by rhodopsin. J Biol Chem 269:23879–23881.

Arora KK, Cheng Z, Catt KJ (1997): Mutations of the conserved DRS motif in the second intracellular loop of the gonadotropin-releasing hormone receptor affect expression, activation, and internalization. Mol Endocrinol 11:1203–1212.

Baldwin JM (1993): The probable arrangement of the helices in G protein–coupled receptors. EMBO J 12:1693–1703.

Baldwin JM, Schertler GF, Unger VM (1997): An alpha-carbon template for the transmembrane helices in the rhodopsin family of G-protein–coupled receptors. J Mol Biol 272:144–164.

Bourne HR (1997): How receptors talk to trimeric G proteins. Curr Opin Cell Biol 9: 134–142.

Brooks BR, Bruccoleri RE, Olafson BD, States DJ, Swaminathan S, Karplus M (1983): CHARMM: A program for macromolecular energy minimization and dynamics calculations. J Comput Chem 4:187–217.

Cavalli A, Fanelli F, Taddei C, De Benedetti PG, Cotecchia S (1996): Amino acids of the α_{1b}-adrenergic receptor involved in agonist binding—Differences in docking catecholamines to receptor subtypes. FEBS Lett 399:9–13.

Cohen GB, Yang T, Robinson PR, Oprian DD (1993): Constitutive activation of opsin: Influence of charge at position 134 and size at position 296. Biochemistry 32: 6111–6115.

Cotecchia S, Exum S, Caron MG, Lefkowitz RJ (1990): Regions of the α_1-adrenergic receptor involved in coupling to phosphatidylinositol hydrolysis and enhanced sensitivity of biological function. Proc Natl Acad Sci USA 87:2896–2900.

Cotecchia S, Ostrowski J, Kjelsberg MA, Caron MG, Lefkowitz RJ (1992): Discrete amino acid sequences of the α_1-adrenergic receptor determine the selectivity of coupling to phosphatidylinositol hydrolysis. J Biol Chem 267:1633–1639.

Cronet P, Sander C, Vriend G (1993): Modeling of transmembrane seven helix bundles. Prot Eng 6:59–64.

Fahmy K, Sakmar TP (1993): Regulation of the rhodopsin–transducin interaction by a highly conserved carboxylic acid group. Biochemistry 32:7229–7236.

Fanelli F, Menziani MC, Cocchi M, De Benedetti PG (1995a): Comparative molecular dynamics study on the seven-helix bundle arrangement of G-protein coupled receptors. J Mol Struct (Theochem) 333:49–69.

Fanelli F, Menziani MC, De Benedetti PG (1995b): Computer simulations of signal transduction mechanism in α_{1b}-adrenergic and m3-muscarinic receptors. Prot Eng 8:557–564.

Fanelli F, Menziani C, Scheer A, Cotecchia S, De Benedetti PG (1998): *Ab initio* modeling and molecular dynamics simulation of the α_{1b}-adrenergic receptor activation. Methods companion. Methods Enzymol 14:302–317.

Farrens DL, Altenbach C, Yang K, Hubbell WL, Khorana HG (1996): Requirement of rigid-body motion of transmembrane helices for light activation of rhodopsin. Science 274:768–770.

Findlay J, Eliopoulos E (1990): Three-dimensional modelling of G protein–linked receptors [published erratum appears in Trends Pharmacol Sci 1991 Mar;12(3):81]. Trends Pharmacol Sci 11:492–499.

Khorana HG (1992): Rhodopsin, photoreceptor of the rod cell. An emerging pattern for structure and function. J Biol Chem 267:1–4.

Kim JM, Altenbach C, Thurmond RL, Khorana HG, Hubbell WL (1997): Structure and function in rhodopsin: Rhodopsin mutants with a neutral amino acid at E134 have a partially activated conformation in the dark state. Proc Natl Acad Sci USA 94: 14273–14278.

Kjelsberg MA, Cotecchia S, Ostrowski J, Caron MG, Lefkowitz RJ (1992): Constitutive activation of the α_{1b}-adrenergic receptor by all amino acid substitutions at a single site. J Biol Chem 267:1430–1433.

Lu ZL, Curtis CA, Jones, PG, Pavia J, Hulme EC (1997): The role of the aspartate–arginine–tyrosine triad in the m1 muscarinic receptor—Mutations of aspartate 122 and tyrosine 124 decrease receptor expression but do not abolish signaling. Mol Pharmacol 51:234–241.

Luo X, Zhang D, Weinstein H (1994): Ligand-induced domain motion in the activation mechanism of a G-protein–coupled receptor. Prot Eng 7:1441–1448.

MaloneyHuss K, Lybrand TP (1992): Three-dimensional structure for the β_2-adrenergic receptor protein based on computer modeling studies. J Mol Biol 225:859–871.

Oliveira L, Paiva AC, Sander C, Vriend G (1994): A common step for signal transduction in G protein–coupled receptors. Trends Pharmacol Sci 15:170–172.

Perez DM, Hwa J, Gaivin R, Mathur M, Brown F, Graham RM (1996): Constitutive activation of a single effector pathway: Evidence for multiple activation states of a G protein–coupled receptor. Mol Pharmacol 49:112–122.

Ren Q, Kurose H, Lefkowitz RJ, Cotecchia S (1993): Constitutively active mutants of the α_2-adrenergic receptor [published erratum appears in J Biol Chem 1994 Jan 14;269(2):1566]. J Biol Chem 268:16483–16487.

Robinson PR, Cohen GB, Zhukovsky EA, Oprian DD (1992): Constitutively active mutants of rhodopsin. Neuron 9:719–725.

Rosenthal W, Antaramian A, Gilbert S, Birnbaumer M (1993): Nephrogenic diabetes insipidus. A V2 vasopressin receptor unable to stimulate adenylyl cyclase. J Biol Chem 268:13030–13033.

Samama P, Cotecchia S, Costa T, Lefkowitz RJ (1993): A mutation-induced activated state of the β_2-adrenergic receptor. Extending the ternary complex model. J Biol Chem 268:4625–4636.

Savarese TM, Fraser CM (1992): *In vitro* mutagenesis and the search for structure–function relationships among G protein–coupled receptors. Biochem J 283:1–19.

Scheer A, Cotecchia S (1997): Constitutively active G protein–coupled receptors: Potential mechanisms of receptor activation. J Recept Signal Transduct Res 17:57–73.

Scheer A, Fanelli F, Costa T, De Benedetti PG, Cotecchia S (1996): Constitutively active mutants of the α_{1b}-adrenergic receptor: Roles of highly conserved polar amino acids in receptor activation. EMBO J 15:3566–3578.

Scheer A, Fanelli, F, Costa T, De Benedetti PG, Cotecchia S (1997): The activation process of the α_{1b}-adrenergic receptor: Potential role of protonation and hydrophobicity of a highly conserved aspartate. Proc Natl Acad Sci USA 94:808–813.

Tiberi M, Caron, MG (1994): High agonist-independent activity is a distinguishing feature of the dopamine D1B receptor subtype. J Biol Chem 269:27925–27931.

Trumpp-Kallmeyer S, Hoflack J, Bruinvels A, Hibert M (1992): Modeling of G-protein–coupled receptors: Application to dopamine, adrenaline, serotonin, acetylcholine, and mammalian opsin receptors. J Med Chem 35:3448–3462.

Unger VM, Hargrave PA, Baldwin JM, Schertler GF (1997): Arrangement of rhodopsin transmembrane alpha-helices. Nature 389:203–206.

Wess J (1997): G-protein–coupled receptors: Molecular mechanisms involved in receptor activation and selectivity of G-protein recognition. FASEB J 11:346–354.

Wyman J, Gill SJ (eds) (1990): Functional Chemistry of Biological Macromolecules. Mill Valley, CA: University Science Books.

Zhang D, Weinstein H (1993): Signal transduction by a 5-HT2 receptor: A mechanistic hypothesis from molecular dynamics simulations of the three-dimensional model of the receptor complexed to ligands. J Med Chem 36:934–938.

Zhu SZ, Wang SZ, Hu JR, Elfakahany EE (1994): An arginine residue conserved in most G-protein coupled receptors is essential for the function of the m1 muscarinic receptor. Mol Pharmacol 45:517–523.

β_2-ADRENOCEPTOR–Gsα FUSION PROTEIN AS A MODEL FOR THE ANALYSIS OF RECEPTOR–G PROTEIN COUPLING

ROLAND SEIFERT and BRIAN K. KOBILKA

1. INTRODUCTION	186
A. The β_2-Adrenoceptor–Gs System as a Model for the Analysis of Receptor–G Protein Interaction	186
B. Methods for the Analysis of β_2-Adrenoceptor–Gs Interaction	186
C. Receptor–G Protein Fusion Proteins as a Novel Approach To Study Receptor–G Protein Interaction	188
D. Use of Sf9 Cells To Express and Study β_2-ARGsα Fusion Proteins	188
2. MATERIALS AND METHODS	189
A. Materials	189
B. Receptor Binding Assay	190
C. GTPase Assay	190
D. GTPγS Binding Assay	190
E. Adenylyl Cyclase Assay	191
F. Miscellaneous	191
3. RESULTS AND DISCUSSION	191
A. General Properties of β_2-ARGsα Fusion Protein	191
B. Agonist Competition Studies	192
C. GTPase Studies	195
D. GTPγS Binding Studies	196
E. Adenylyl Cyclase Studies	198
F. The Importance of Stoichiometry and Relative Orientation With Respect to Each Other and of Signaling Components for Their Interaction	199
G. Uses of Fusion Proteins	200
4. SUMMARY	200

Structure–Function Analysis of G Protein-Coupled Receptors, Edited by Jürgen Wess.
ISBN 0-471-25228-X   Copyright © 1999 Wiley-Liss, Inc.

I. INTRODUCTION

I.A. The β₂-Adrenoceptor–Gs System as a Model
for the Analysis of Receptor–G Protein Interaction

The β_2-adrenoceptor (β_2-AR) presumably represents the most extensively studied G protein–coupled receptor (GPCR) and is considered as a prototypical receptor for the entire GPCR family (Ostrowski et al., 1992; Strader et al., 1994; Hein and Kobilka, 1995). In most studies, the synthetic catecholamine $(-)$-isoproterenol (ISO) and not the endogenous ligand epinephrine is used as an agonist for the β_2-AR. The availability of numerous agonists with different intrinsic activities, antagonists, and inverse agonists and the clinical importance of the β_2-AR contribute to the popularity of the β_2-AR as a GPCR model. The β_2-AR couples to the G protein Gs, which activates adenylyl cyclase (AC) (Gilman, 1987; Hein and Kobilka, 1995). AC generates cAMP from ATP. cAMP activates protein kinase A, which phosphorylates target proteins and thereby alters cell function.

Gs consists of a specific α subunit (Gsα) and a $\beta\gamma$ complex, which may differ in composition ($\beta_x\gamma_y$) (Gilman 1987; Neer, 1995). Figure 10.1 illustrates the cyclic interaction of the β_2-AR and Gs. In the resting state, Gsα is GDP liganded (Gilman, 1987). When an agonist binds to the β_2-AR, the receptor undergoes a distinct conformational change (Gether et al., 1997), which renders the β_2-AR capable of releasing GDP from Gsα. GDP release is the rate-limiting step of G protein activation (Gilman, 1987). Following GDP release, a high-affinity ternary complex, consisting of agonist, β_2-AR, and guanine nucleotide-free Gsα is formed (De Lean et al., 1980; Seifert et al., 1998b,c). In the next step, GTP binds to Gsα. GTP-liganded Gsα uncouples from the β_2-AR, a process that can be monitored by disruption of high-affinity agonist binding (De Lean et al., 1980; Seifert et al. 1998b,c). Another consequence of GTP-binding to Gsα is that the interaction of Gsα with the $\beta\gamma$ complex is destabilized so that free GTP-liganded Gsα is generated. GTP-liganded Gsα then activates AC (Gilman, 1987). G protein deactivation is achieved by the GTPase activity of Gsα, which hydrolyzes GTP to GDP and P_i (Cassel and Selinger, 1976). GDP-liganded Gsα and the $\beta\gamma$ complex reassociate, thus completing one G protein activation–deactivation cycle.

Not only the agonist-occupied β_2-AR but even the "empty" receptor can activate Gs, a property that is referred to as *constitutive activity* (Samama et al., 1994; Chidiac et al., 1994; Gether et al., 1995; Seifert et al., 1998b,c). Constitutive β_2-AR activity is uncovered by the inhibitory effects of inverse agonists on basal Gs activity. The prototypical inverse agonist for the β_2-AR is [erythro-DL-1(7-methylindan-4-yloxy)-3-isopropylaminobutan-2-ol] (ICI 118,551) (Samama et al., 1994; Seifert et al., 1998b,c).

I.B. Methods for the Analysis of β₂-Adrenoceptor–Gs Interaction

Commonly, β-AR–Gs interactions are monitored indirectly by measurement of AC activity (method 4, Fig. 10.1) (e.g., Chidiac et al., 1994; Samama et al., 1994; Gether et al., 1995). The AC assay takes advantage of the signal amplifi-

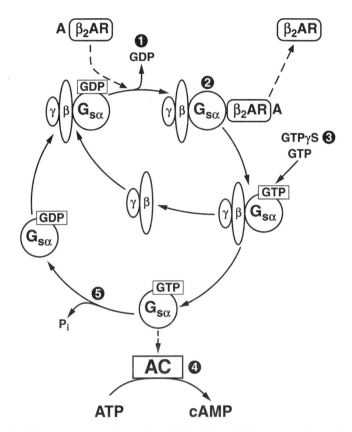

Figure 10.1. Interactions of the β₂-AR with Gs. The individual steps of the G protein activation–deactivation cycle are described in the text. The circled numbers indicate the methods available for studying β₂-AR–Gs interaction. **1:** GDP release assay. **2:** High-affinity agonist binding assay (ternary complex formation). **3:** GTPγS binding assay. **4:** AC assay. **5:** GTPase assay. A, agonist.

cation at the effector system level and monitors cAMP accumulation. A disadvantage of the AC assay is that it is difficult to control for the impact of AC availability on signaling efficiency (Alousi et al., 1991; MacEwan et al., 1996).

Unfortunately, studies aimed at analyzing β₂-AR–Gsα interaction at the G protein level are difficult to perform. Specifically, receptor-regulated GDP release (method 1, Fig. 10.1) can be studied only in selected systems expressing β-ARs (Cassel and Selinger, 1978). The extent of GDP release from Gsα may be too small to be detected above the background of more rapidly GDP/GTP-exchanging G proteins such as Gi proteins.

High-affinity agonist binding (method 2, Fig. 10.1) is a well-established method to monitor β₂-AR–Gs interaction, but ternary complex formation can only be studied in those cell lines and expression systems that express sufficiently high concentrations of Gs (De Lean et al., 1980; Samama et al., 1993; Bertin et al., 1994). Accumulation of high-affinity ternary complexes is monitored at equilibrium in the absence of guanine nucleotides (i.e., under conditions under which GTP binding to the G protein cannot occur). GTP and the

GTPase-resistant GTP analog guanosine $5'\text{-}O\text{-}(3\text{-thiotriphosphate})$ (GTPγS) inhibit high-affinity agonist binding.

GTPγS binds to G protein α subunits with high affinity (Gilman, 1987; Wieland and Jakobs, 1994). However, detection of β-AR–stimulated GTPγS binding to Gsα (method 3, Fig. 10.1) is also hampered by the high abundance of pertussis toxin-sensitive G proteins and may require their functional elimination by specific treatments before Gsα regulation can be studied (Wieland and Jakobs, 1994). The GTPγS binding assay monitors an abortive G protein activation–deactivation cycle because GTPγS is not hydrolyzed and binds very tightly to the α subunit.

Because in most membrane systems Gi proteins and nonspecific nucleotidases are abundant, detection of β-AR–stimulated GTP hydrolysis (method 5, Fig. 10.1) is also difficult (Cassel and Selinger, 1976). The steady-state GTPase activity monitors the outcome of multiple G protein activation–deactivation cycles.

In reconstituted systems with highly purified β-AR and Gs, agonist-stimulated GTPγS binding and GTP hydrolysis can be monitored more easily than in cell membranes because the background caused by other G proteins and nucleotidases is reduced (Brandt and Ross, 1986; Cerione et al., 1984). However, because of the difficulties in obtaining purified receptor and Gs in sufficient quantity, purity, and stability, this approach is not generally available for extensive biochemical and pharmacological studies.

I.C. Receptor–G Protein Fusion Proteins as a Novel Approach To Study Receptor–G Protein Interaction

The fusion protein technique was introduced by Bertin et al. (1994). The fusion protein was created by ligating the DNA for the β_2-AR to the DNA for Gsα, resulting in the generation of β_2-ARGsα DNA. β_2-ARGsα was expressed in S49 cyc$^-$ lymphoma cells, which do not endogenously express Gsα. In the fusion protein, there is a defined 1:1 stoichiometry between GPCR and G protein, while in S49 lymphoma wild-type cells and physiological tissues, Gsα is expressed at about a 100-fold higher concentration than β_2-AR (Ransnäs and Insel, 1988; Kawai and Arinze, 1991). β_2-ARGsα reconstituted high-affinity agonist binding as efficiently as the nonfused Gsα at the β_2-AR in S49 lymphoma wild-type cells. Of particular interest, β_2-ARGsα was more efficient than the nonfused system at promoting agonist stimulation of AC. Unfortunately, the expression level of β_2-ARGsα achievable in S49 cyc$^-$ cells is relatively low (<1 pmol/mg) so that the analysis of receptor–G protein coupling in terms of GTPγS binding and GTP hydrolysis is still difficult to perform.

I.D. Use of Sf9 Cells To Express and Study β_2-ARGsα Fusion Proteins

To establish a model in which β_2-ARGsα could be expressed at higher levels than in S49 cyc$^-$ lymphoma cells, we chose the *Spodoptera frugiperda* insect cell (Sf9)/baculovirus expression system. In Sf9 cells, the β_2-AR can readily be expressed at levels of up to 20–40 pmol/mg (Chidiac et al., 1994; Gether et al.,

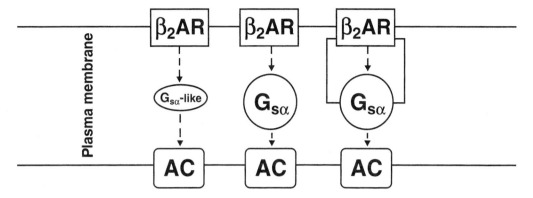

Figure 10.2. Analysis of β_2-AR–Gs interaction in Sf9 insect cells. Signaling *via* the β_2-AR in Sf9 membranes was studied in three different settings. First, we studied coupling of the β_2-AR to the endogenous Gsα-like G protein of the insect cells. Second, we co-expressed the β_2-AR with mammalian Gsα. Third, we fused the β_2-AR to mammalian Gsα.

1995). Additionally, Sf9 cells have already been proven to be suitable for reconstitution of GPCRs with cognate G proteins (Butkerait et al., 1995; Grünewald et al., 1996; Barr et al., 1997; Clawges et al., 1997). Importantly as well, Sf9 cells can be cultured in large scale so that sufficient material even for extensive studies can readily be obtained (Gether et al., 1995). We studied β_2-AR–mediated signal transduction in three settings (Fig. 10.2). First, we expressed the β_2-AR alone. In this situation, the β_2-AR can couple to endogenous Gsα-like G proteins of insect cells to activate AC (Kleymann et al., 1993; Chidiac et al., 1994; Gether et al., 1995). In the second setting, we co-expressed the β_2-AR together with mammlian Gsα. Third, we studied the properties of a fusion protein of the β_2-AR with mammalian Gsα.

2. MATERIALS AND METHODS

2.A. Materials

[γ-^{32}P]GTP (6,000 Ci/mmol), [α-^{32}P]ATP (3,000 Ci/mmol), and ^{35}S-GTPγS (1,000–1,500 Ci/mmol) were from NEN-DuPont (Boston, MA). ^{3}H-dihydroalprenolol (^{3}H-DHA) (85–90 Ci/mmol) was from Amersham (Arlington Heights, IL). All unlabeled nucleotides were obtained from Boehringer Mannheim (Mannheim, Germany). ICI 118,551 was from RBI (Natick, MA). ISO, ($\pm$)-alprenolol, creatine phosphate, creatine kinase, mono(cyclohexylammonium) phosphoenolpyruvate, pyruvate kinase, myokinase, and alumina for column chromatography (super 1, type WN-6, neutral) were from Sigma (St. Louis, MO). Glass fiber filters (GF/C) were from Schleicher and Schuell (Dassel, Germany). Single-use cellulose filter fast-flow columns for the AC assay were from E & K Scientific Products (Campbell, CA). For receptor binding, GTPγS binding, and AC assays, a buffer referred to as *binding buffer* (75 mM Tris/HCl, 12.5 mM MgCl$_2$, and 1 mM EDTA, pH 7.4) was used.

2.B. Receptor Binding Assay

Before experiments, membranes were pelleted by a 15-minute centrifugation at 4°C and 15,000g and resuspended in binding buffer. The purpose of the washing procedure is to remove as far as possible endogenous GTP and GDP, which could interfere with high-affinity agonist binding. ISO at different concentrations (50 μl) was added to tubes, followed by the addition of 400 μl of binding buffer with or without GTPγS (final conc., 10 μM). Thereafter, suspended membranes (25 μl per tube, 15–40 μg of protein, depending on the expression level) were added. Reactions were initiated by the addition of ^{3}H-DHA (25 μl per tube; final conc., 1 nM). Tubes were mixed and incubated for 75–90 minutes at 25°C under shaking at 200–250 rpm. Bound ^{3}H-DHA was separated from free radioligand by filtration through GF/C filters, followed by three washes with 2 ml binding buffer (4°C). Radioactivity was determined by liquid scintillation counting.

2.C. GTPase Assay

β_2-AR ligands (10 μl) were added to tubes. Subsequently, 50 μl of a reaction mixture containing (final concentrations) 0.1 μM GTP, 1.0 mM $MgCl_2$, 0.1 mM EDTA, 0.1 mM ATP, 1 mM adenylyl imidodiphosphate, 5 mM creatine phosphate, 40 μg of creatine kinase, and 0.2% (mass/vol.) bovine serum albumin in 50 mM Tris/HCl, pH 7.4, were added. Suspended membranes (10 μg of protein in 20 μl) were then added to the tubes. Assay tubes were incubated for 3 minutes at 25°C before the addition of 20 μl [γ-^{32}P]GTP (0.1–0.5 μCi/tube). To assess nonenzymatic degradation of [γ-^{32}P]GTP, an excess of unlabeled GTP (1 mM) was added to control tubes. Usually, the amount of ^{32}P-P$_i$ released under these conditions was <1% of the total amount of radioactivity added. Reactions were conducted for 20 minutes at 25°C and were terminated by the addition of 900 μl of an ice-cold slurry consisting of 5% (mass/vol) activated charcoal and 50 mM NaH_2PO_4, pH 2.0. Charcoal-quenched reaction mixtures were centrifuged for 15 minutes at room temperature at 15.000g. Seven hundred microliters of supernatant was removed, and ^{32}P-P$_i$ was determined by liquid scintillation counting.

2.D. GTPγS Binding Assay

GDP at different concentrations (50 μl) was added to tubes, followed by the addition of 50 μl of β_2-AR ligand and 300 μl of binding buffer supplemented with (final concentration) 0.05% (mass/vol) bovine serum albumin. Subsequently, suspended membranes (15 μg of protein in 50 μl) were added. Reactions were initiated by the addition of 50 μl of ^{35}S-GTPγS (0.25 μCi/tube; final conc., 1.0 nM). Tubes were mixed and incubated for 40 minutes at 25°C under shaking at 200–250 rpm. Nonspecific binding was determined in the presence of 10 μM GTPγS and was <0.2% of total binding. Bound ^{35}S-GTPγS was separated from free ^{35}S-GTPγS by filtration through GF/C filters, followed by three washes with 2 ml binding buffer (4°C). Radioactivity was determined by liquid scintillation counting.

2.E. Adenylyl Cyclase Assay

GTP or distilled water (5 μl each) was added to tubes, followed by the addition of β₂-AR ligand (5 μl). Subsequently, 20 μl of Sf9 membranes suspended in binding buffer (50–75 μg of protein for membranes expressing β₂-AR; 15–20 μg of protein for other types of membranes) were added. Assay tubes were incubated for 3 minutes at 37°C before the addition of reaction mixture (20 μl) containing (final concentrations) 40 μM [α-³²P]ATP (2.5–3.0 μCi/tube), 2.7 mM mono(cyclohexyl)ammonium phosphoenolpyruvate, 0.125 IU of pyruvate kinase, 1 IU of myokinase, and 0.1 mM cAMP. Reactions were conducted for 20 minutes and were terminated by addition of 20 μl 2.2 N HCl. Denatured membrane protein was pelleted by a 3-minute centrifugation at room temperature and 15,000g. Sixty-five microliters of supernatant was applied onto chromatography columns filled with 1.3 g of neutral alumina. cAMP was eluted into 20-ml scintillation counting tubes by adding 4 ml of 0.1 M ammoinum acetate, pH 7.0, into the column reservoir (Alvarez and Daniels, 1990). ³²P-cAMP was determined by liquid scintillation counting. Blank values were generally <0.01% of total amount of [α-³²P]ATP added.

2.F. Miscellaneous

Protocols for the construction of β₂-AR–Gsα DNA, culture of Sf9 cells, membrane preparation, and immunoblotting are described by Seifert et al. (1998b,c). Protein was determined using the Bio-Rad DC protein assay (Bio-Rad). Data were analyzed by nonlinear regression, using the Prism program (GraphPad).

3. RESULTS AND DISCUSSION

3.A. General Properties of β₂-ARGsα Fusion Protein

We fused the DNA coding for the human β₂-AR with the DNA coding for the long splice variant of rat Gsα in a two-step overlap-extension PCR procedure (Seifert et al., 1998b,c). At the amino terminus, the β₂-AR bears the FLAG epitope, which is recognized by the M₁ antibody (Fig. 10.3). At the carboxy terminus, the β₂-AR contains a hexahistidine tag. We cloned the fusion protein DNA into the baculovirus vector pVL 1392. Sf9 cells were infected with recombinant baculoviruses, and membranes were prepared.

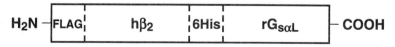

Figure 10.3. Schematic structure of the β₂-ARGsα fusion protein. The DNA coding for the human β₂-AR (hβ₂) was fused to the DNA coding for the long splice variant of rat Gsα (rG$_{sαL}$) in a two-step overlap-extension PCR protocol. At the receptor amino terminus, the β₂-AR is tagged with the FLAG epitope, and at the carboxy terminus the β₂-AR bears a hexahistidine tag (6His). (Modified from Seifert et al. [1998b], with permission of the publisher.)

As had been shown previously (Gether et al., 1995), the apparent molecular mass of the β_2-AR expressed in Sf9 cells is 52 kD (Fig. 10.4). The apparent molecular mass of the long splice variant of Gsα on SDS-polyacrylamide gels is 52 kD as well (Graziano et al., 1987). Therefore, the molecular mass of the β_2-AR–Gsα fusion protein would be expected to be 104 kD. Indeed, both the M_1 antibody (Fig. 10.4) and the anti-Gsα Ig (Fig. 10.5) reacted with a 104 kD protein in Sf9 membranes. No degradation products of β_2-ARGsα fusion protein were detected. The endogenous Gsα-like G protein did not react with the anti-Gsα Ig used in our study (Mumby and Gilman, 1991).

Because of the defined 1:1 stoichiometry of GPCR to G protein α subunit in the β_2-AR–Gsα fusion protein (Fig. 10.3), ^{3}H-DHA saturation binding determines both β_2-AR and Gsα expression levels. Membranes expressing β_2-AR and β_2-AR–Gsα bound ^{3}H-DHA in a monophasic and saturable manner with K_d values of 0.36 ± 0.01 nM and 0.36 ± 0.03 nM, respectively. Depending on the particular cell culture conditions chosen (Seifert et al., 1998b), the B_{max} values of ^{3}H-DHA binding varied between 0.2–11.8 pmol/mg for the β_2-AR and 0.6–12.9 pmol/mg for β_2-ARGsα.

3.B. Agonist Competition Studies

In Sf9 membranes expressing the β_2-AR alone at a level of 5.0–6.1 pmol/mg, ISO inhibited DHA binding according to a monophasic function with low affinity (K_i, 192 ± 22 nM), and GTPγS had virtually no effect on the ISO-

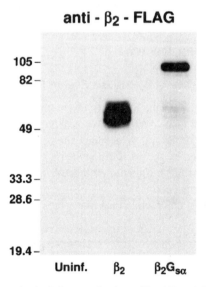

Figure 10.4. Immunological characterization of β_2-AR and β_2-AR–Gsα with anti-β_2-FLAG Ig. Sf9 membrane proteins (15 μg per lane) from uninfected cells (uninf.) and cells expressing β_2-AR (β_2, 9.0 pmol/mg) or β_2-ARGsα (β_2Gsα, 5.0 pmol/mg) were separated by SDS-polyacrylamide gel electrophoresis and probed with anti-β_2-Ig (M1 antibody, 1:1,000). Numbers on the left are molecular masses of marker proteins (kD). Shown is the autoluminogram of a gel containing 8% (w/v) acrylamide. (Modified from Seifert et al. [1998b], with permission of the publisher.)

competition curve (Fig. 10.6A). Even when the expression level of β_2-AR was decreased to 0.2 pmol/mg (i.e., when the ratio of available endogenous G proteins per GPCR was increased, no high-affinity agonist binding could be detected (data not shown). These results are indicative of poor coupling of the β_2-AR to the endogenous Gsα-like G proteins of insect cells.

Sf9 cells provide a sensitive system for reconstitution of GPCRs with their cognate G protein α subunits. Because the expression level of Gsα in β_2-AR–Gsα can be precisely determined by ³H-DHA saturation binding, immunoblots with anti-Gsα Ig and defined amounts of β_2-AR–Gsα can be used as a standard to estimate the expression level of nonfused Gsα. Using this approach, we estimated the expression level of Gsα in Sf9 membranes to be at least 100–150 pmol/mg (Seifert et al., 1998b). Other authors reported similar expression levels of mammalian G protein subunits in Sf9 cells (Butkerait et al., 1995).

We prepared membranes from Sf9 cells that expressed the β_2-AR at 1.4 pmol/mg and Gsα at ~100 pmol/mg (i.e., at a stoichiometry that is similar to the receptor G protein ratio in cell culture lines and organs) (Ransnäs and Insel, 1988; Kawai and Arinze, 1991). In Sf9 membranes expressing β_2-AR plus Gsα, the ISO-competition curve of ³H-DHA binding was biphasic (i.e., ~35% of the β_2-ARs formed ternary complexes) (Fig. 10.6B). The K_i values for low- and high-affinity agonist binding were 3.9 ± 1.5 nM and 198 ± 19 nM, respectively. GTPγS increased the high-affinity K_i value for ISO to 24.6 ± 5.6

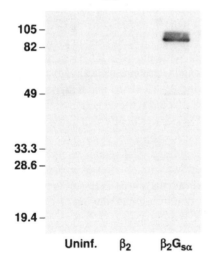

Figure 10.5. Immunological characterization of β_2-AR and β_2-AR–Gsα with anti-Gsα Ig. Sf9 membrane proteins (15 μg per lane) from uninfected cells (uninf.) and cells expressing β_2-AR (β_2, 9.0 pmol/mg) or β_2-AR–Gsα (β_2Gsα, 5.0 pmol/mg) were separated by SDS-polyacrylamide gel electrophoresis and probed with anti-Gsα Ig (1:1,000). Numbers on the left are molecular masses of marker proteins (kD). Shown is the autoluminogram of a gel containing 8% (w/v) acrylamide. (Modified from Seifert et al. [1998b], with permission of the publisher.)

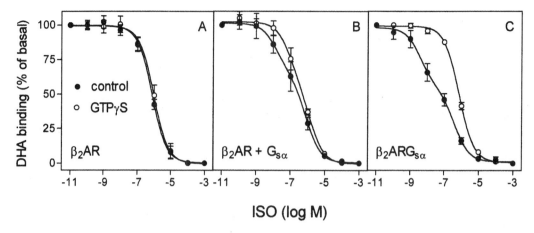

Figure 10.6. Competition by ISO of ^{3}H-DHA binding in Sf9 membranes: Comparison of membranes expressing β_2-AR without or with mammalian Gsα and membranes expressing β_2-AR–Gsα. ^{3}H-DHA competition binding in Sf9 membranes was perfomed as described in section 2.B. The ^{3}H-DHA concentration was 1 nM, and the GTPγS concentration was 10 μM. Data points shown are the means $\pm$SD of five to seven experiments performed in triplicate. Data were analyzed by nonlinear regression. For the data shown in **A**, the expression level of β_2-AR was 5.1–6.0 pmol/mg. For the data shown in **B**, the expression level of β_2-AR was 1.4 pmol/mg and the expression level of Gsα was $\sim$100 pmol/mg. For the data shown in **C**, the expression level of β_2-AR–Gsα was 3.0–7.5 pmol/mg. (Modified from Seifert et al. [1998b], with permission of the publisher.)

nM, but the agonist competition curve in the presence of GTPγS was still biphasic. These findings indicate that, following GTPγS binding to the G protein, Gsα uncouples from the β_2-AR only incompletely (i.e., even GTPγS-liganded Gsα can interact with the β_2-AR). Guanine nucleotide-insensitive interactions of receptors and G proteins have already been documented for several GPCRs, including the β_2-AR (Citri and Schramm, 1982; Childers et al., 1993; Gürdal et al., 1997). Notably, the detection of high-affinity agonist binding in the coexpression system critically depended on the β_2-AR/Gsα ratio. Specifically, when the expression level of β_2-AR was increased by $\sim$10-fold without changing Gsα expression, the agonist competition curve was virtually identical to that obtained with membranes expressing the β_2-AR alone (Fig. 10.6A). These results indicate that a vast excess of G protein compared with β_2-AR is required for detection of high-affinity agonist binding. Similar conclusions were obtained for the dopamine D2S receptor co-expressed with Gi proteins in Sf9 membranes (Grünewald et al., 1996).

Studying β_2-AR–Gsα fusion protein expressed in S49 cyc^{-} lymphoma cells, Bertin et al. (1994) found $\sim$50% of the β_2-ARs to display high agonist affinity. Ternary complex formation was completely guanine nucleotide sensitive (i.e., there was efficient receptor-G protein coupling and uncoupling in the fusion protein). We obtained similar results as Bertin et al. (1994) with β_2-AR–Gsα fusion protein expressed in Sf9 membranes. Specifically, $\sim$45% of the β_2-ARs displayed high agonist affinity, and the high-affinity agonist binding was abolished by GTPγS (Fig. 10.6C). The high- and low-affinity K_i values of ISO for

β_2-ARGsα expressed in Sf9 membranes were 1.0 ± 0.6 nM and 102 ± 38 nM, respectively, and corresponded well to the K_i values reported for the fusion protein expressed in S49 cyc$^-$ cells (Bertin et al., 1994) and for nonfused β_2-AR (Fig. 10.6B) (Samama et al., 1993). One might have expected that because of the 1:1 stoichiometry of GPCR to G protein in the fusion protein, virtually all β_2-ARs should display high agonist affinity. However, full conversion of β-ARs into a high-agonist-affinity state appears virtually impossible to achieve in almost any experimental system, including membrane systems with nonfused $\beta_{(2)}$-AR and Gsα (Figs. 10.6A,B) (Samama et al., 1993; Bertin et al., 1994), reconstituted systems with purified β_2-AR and Gsα (Cerione et al., 1984), fusion proteins (Fig. 10.6C) (Bertin et al., 1994), or cardiac tissue of transgenic mice overexpressing Gsα (Gaudin et al., 1995).

3.C. GTPase Studies

Sf9 cells do not express mammalian-type Gi proteins, and, compared with mammalian systems, the basal GTPase activity in Sf9 membranes is low (Koski and Klee, 1981; Okajima et al., 1985; Grandt et al., 1986; Richardson and Hosey, 1992; Kleymann et al., 1993; Quehenberger et al., 1992). Thus, Sf9 membranes provide a low background for studying β_2-AR regulation of Gs–GTPase.

In Sf9 membranes expressing β_2-AR, ISO had only a marginal stimulatory effect on GTP hydrolysis, and ICI was without inhibitory effect (Fig. 10.7A). These data corroborate the conclusion that the β_2-AR couples only poorly to Gsα-like G proteins of Sf9 cells.

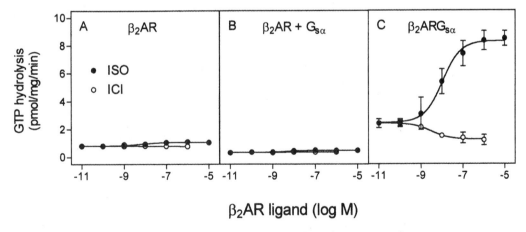

Figure 10.7. GTPase regulation by ISO and ICI: Comparison of membranes expressing β_2-AR without or with mammalian Gsα and membranes expressing β_2-AR–Gsα. GTPase activity was determined as described in section 2.C. Data points shown are the means $\pm$SD of two to three experiments performed in duplicate. Data were analyzed by nonlinear regression. For the data shown in **A**, the expression level of β_2-AR was 7.5 pmol/mg. For the data shown in **B**, the expression level of β_2-AR was 1.4 pmol/mg and the expression level of Gsα was ~100 pmol/mg. For the data shown in **C**, the expression level of β_2-ARGsα was 8.9 pmol/mg. (Modified from Seifert et al. [1998b], with permission of the publisher.)

Despite the high expression level of Gsα, the reconstitution of high-affinity agonist binding, and the low background for GTPase studies, we could not detect a significant agonist or inverse agonist regulation of GTP hydrolysis in membranes expressing β_2-AR and Gsα (Fig. 10.7B).

Compared with membranes expressing nonfused β_2-AR with or without Gsα, the basal GTPase activity in membranes expressing β_2-ARGsα was about fourfold higher (Fig. 10.7C). This finding, together with the fact that the inverse agonist ICI reduced basal GTP hydrolysis in membranes expressing β_2-AR–Gsα by about 50%, indicates that the basal GTPase activity in membranes expressing β_2-ARGsα is largely attributable to the fused Gsα. Moreover, we observed highly effective GTPase activation by ISO in membranes expressing β_2-AR–Gsα (up to 250% above basal). This is presumably the highest GTPase stimulation of Gs by a GPCR in a membrane system ever reported (Cassel and Selinger, 1976; Pike and Lefkowitz, 1980).

3.D. GTPγS Binding Studies

The GTPγS binding assay can be more sensitive than the GTPase assay at detecting GPCR regulation of G protein activity. Particularly in HL-60 leukemia membranes, which express Gi protein–linked chemoattractant receptors, and in SK-N-SH neuroblastoma cells, which express Gi/Go protein–linked μ- and δ-opioid receptors, agonist stimulates GTPγS binding up to 500%, whereas typical GTPase stimulations in these membranes amount only to about 30%–90% (Gierschik et al., 1989, 1991; Selley and Bidlack, 1992; Selley et al., 1997). GDP reduces basal GTPγS binding to a greater extent than agonist-stimulated GTPγS binding so that the relative stimulatory effect of agonist on GTPγS binding increases with increasing GDP concentration (Gierschik et al., 1991; Selley et al., 1997).

Manning and others (Barr et al., 1997; Barr and Manning, 1997) used agonist-stimulated GTPγS binding to G protein α subunits followed by quantitative immunoprecipitation of the GTPγS-liganded α subunits to monitor GPCR–G protein coupling in Sf9 cell membranes. However, because we did not have an unlimited source of anti-Gsα Ig, this assay was impractical. In addition, the method employed by Manning and coworkers did not appear to be suitable for processing large numbers of samples. We wished to establish a high-capacity GTPγS binding assay that would allow us to detect agonist stimulation without the need for immunoprecipitation of samples. Therefore, we used the GDP-quenching technique to unmask agonist stimulation of GTPγS binding.

In a concentration-dependent fashion, GDP reduced basal GTPγS binding in Sf9 membranes expressing β_2-AR with or without mammalian Gsα (Fig. 10.8A,B). ISO exhibited only marginal stimulatory effects on GTPγS binding in both systems, even with GDP at a concentration of 10 μM (i.e., a concentration that suppressed GTPγS binding by more than 90%). These data confirm the results obtained in GTPase studies showing that coupling of the β_2-AR to endogenous Gsα-like G proteins of Sf9 cells or nonfused mammalian Gsα is so weak that it is almost undetectable at the level of GDP/GTP exchange reactions.

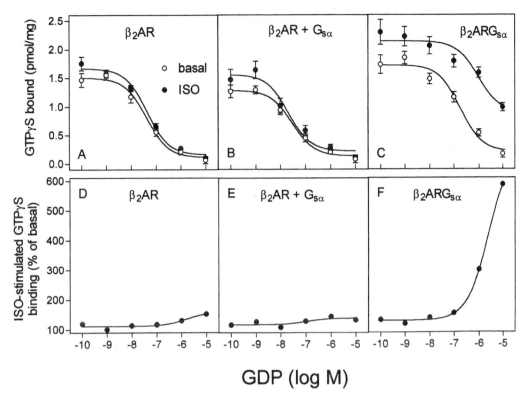

Figure 10.8. Effect of ISO on GTPγS binding in Sf9 membranes expressing β$_2$-AR without or with mammalian Gsα and in membranes expressing β$_2$-ARGsα. GTPγS binding in Sf9 membranes was determined as described in section 2.D. Reaction mixtures contained Sf9 membranes expressing various signal transduction components, 1 nM ^{35}S-GTPγS, and distilled water (basal) or ISO (10 μM). Reaction mixtures additionally contained GDP at the concentrations indicated on the abscissa. **A–C:** Absolute values of GTPγS binding. Data shown are the means ±SD of two to three independent experiments performed in duplicate. **D–F:** Relative stimulatory effects of ISO. For each GDP concentration, the stimulatory effect of ISO was expressed relative to basal GTPγS binding. β$_2$-AR, membranes expressing β$_2$-AR at 7.5 pmol/mg; β$_2$-AR + Gsα, membranes expressing β$_2$-AR at 1.4 pmol/mg plus Gsα at ~100 pmol/mg; β$_2$-ARGsα, membranes expressing β$_2$-ARGsα at 7.4 pmol/mg. (Modified from Seifert et al. [1998b], with permission of the publisher.)

Compared with the co-expression system of β$_2$-AR (1.4 pmol/mg) plus non-fused Gsα (~100 pmol/mg), basal GTPγS binding in membranes expressing β$_2$-ARGsα (7.4 pmol/mg) was about 40% higher (Fig. 10.8B,C). In the absence of GDP, ISO stimulated GTPγS binding in membranes expressing β$_2$-ARGsα by no more than 30%. However, as has been shown for Gi/Go-linked GPCRs (Gierschik et al., 1991; Selley et al., 1997), GDP quenched basal GTPγS binding in membranes expressing β$_2$-ARGsα more efficiently than ISO-stimulated GTPγS binding. The result of these differential effects of GDP was that, with GDP at 10 μM, ISO could increase GTPγS binding by up to 500% (Fig.

10.8C,F). These findings corroborate the view that receptor–G protein coupling in the β_2-ARGsα fusion protein is much more efficient than in a system with nonfused β_2-AR plus Gsα.

3.E. Adenylyl Cyclase Studies

AC activity is the classic readout for studying GPCR-Gs interaction (Gilman, 1987; Hein and Kobilka, 1995). The AC assay takes advantage of the signal amplification at the effector level. However, because the number of available AC molecules can be limiting, the relation between cAMP formation and efficiency of receptor–G protein coupling is not necessarily proportional (Alousi et al., 1991; MacEwan et al., 1996). In membranes expressing β_2-AR alone (Fig. 10.9A), β_2-AR plus Gsα (Fig. 10.9B), or β_2-AR–Gsα (Fig. 10.9C), qualitatively similar patterns of AC regulation were seen. Specifically, GTP per se increased basal AC activity by about 100%. The stimulatory effect of ISO on this GTP-dependent AC activity amounted to about 50%, and the inhibitory effect of ICI amounted to about 50% as well. The strong stimulatory effects of GTP on basal AC activity and the substantial inhibitory effect of ICI on GTP-dependent cAMP formation reflect the considerable constitutive activity of the β_2-AR (Chidiac et al., 1994; Gether et al., 1995; Gürdal et al., 1997).

In contrast to the similar relative effects of GTP, ISO, and ICI on AC regulation in the three systems studied, there were large differences in the absolute AC activities. Thus, in membranes expressing β_2-AR alone, AC activities were

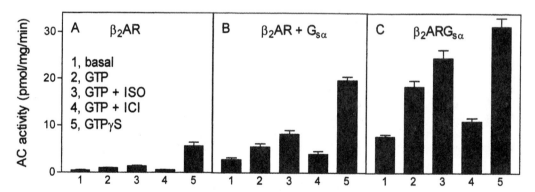

Figure 10.9. AC activities in Sf9 membranes expressing β_2-AR without or with mammalian Gsα and in membranes expressing β_2-AR–Gsα: Effects of GTP, ISO, ICI, and GTPγS. AC activity in Sf9 membranes was determined as described in section 2.E. Reaction mixtures contained Sf9 membranes (15–75 μg of protein per tube) and distilled water (basal) (1), GTP (1 μM) (2), GTP (1 μM) plus ISO (10 μM) (3), GTP (1 μM) plus ICI (1 μM) (4), or GTPγS (10 μM) (5). Data shown are the means ±SD of two to three independent experiments performed in duplicate. **A:** Membranes expressing β_2-AR at 7.5 pmol/mg. **B:** Membranes expressing β_2-AR at 1.4 pmol/mg plus Gsα at ~100 pmol/mg. **C:** Membranes expressing β_2-AR–Gsα at 7.4 pmol/mg. (Modified from Seifert et al. [1998b], with permission of the publisher.)

near the detection limit of the assay, while the absolute AC activities in membranes expressing β_2-AR plus Gsα were much higher. These AC activities, in turn, were surpassed by the AC activities observed with membranes expressing β_2-AR–Gsα. These data corroborate the data of Bertin et al. (1994) showing that AC activation by β_2-AR–Gsα is more efficient than by β_2-AR with non-fused Gsα.

3.F. The Importance of Stoichiometry and Relative Orientation With Respect to Each Other and of Signaling Components for Their Interaction

The differences in the effects of ISO in the agonist competition, GTPase, GTPγS binding, and AC assays in the various membranes studied provide important insights into the mechanisms of receptor–G protein coupling. There was uniformly poor coupling in all four assays when the β_2-AR was expressed alone (Figs. 10.6A, 10.7A, 10.8A,D, and 10.9A). However, when the β_2-AR was co-expressed with mammalian Gsα, coupling was efficient with respect to high-affinity agonist binding and AC activation (Figs. 10.6B and 10.9B) but inefficient with respect to GTPγS binding and GTPase stimulation (Figs. 10.7B and 10.8B). This discrepancy can be explained by a model according to which β_2-ARs interact only with a very small portion of the available Gsα molecules. However, these few Gsα molecules are sufficient to convert a major fraction of the β_2ARs into a state of high agonist affinity and to efficiently activate AC. Because only a small fraction of expressed Gsα molecules is actually engaged in the coupling process, there may be so few GDP/GTP exchange processes occurring that agonist effects, despite high assay sensitivity, are still near the detection limit.

In the case of β_2-AR–Gsα, the maximum stimulatory effects of ISO on GTPγS binding and GTPase amounted to about 500% and 250% above basal, respectively (Figs. 10.7C and 10.8C), whereas AC stimulation was less than 50% (Fig. 10.9C). These findings can be explained by a scenario in which a substantial portion of the expressed β_2-AR–Gsα fusion proteins is functionally uncoupled from AC. The functionally uncoupled fusion protein molecules can, nonetheless, undergo "private" GDP/GTP exchange processes, resulting in much more effective stimulation of GTPγS binding and GTP hydrolysis than of AC. One reason for the partial uncoupling of β_2-AR–Gsα could be that the number of available AC molecules is limited (Alousi et al., 1991; MacEwan et al., 1996). This view is supported by the fact that in membranes expressing β_2-AR plus Gsα, the maximum agonist-stimulated AC activity is much lower than the maximum GTPγS-stimulated AC activity (Fig. 10.9B), while in membranes expressing β_2-AR–Gsα the ISO-stimulated AC activity approaches the GTPγS-stimulated AC activity (Fig. 10.9C). Alternatively, or in addition, β_2-AR–Gsα fusion proteins and AC molecules may be localized in different membrane microcompartments, and their effective coupling is prevented by diffusion barriers. Differential compartmentalization of signal transduction components has been amply documented (Neubig, 1994; Neer, 1995).

3.G. Uses of Fusion Proteins

Compared with a conventional co-expression system of GPCR with G protein, the fusion protein technique offers the advantage of precisely defined stoichiometry of signaling partners and higher coupling efficiency (Bertin et al., 1994; Seifert et al., 1998b,c). Moreover, there is no evidence that fusion of a GPCR to a G protein α subunit grossly changes the functional properties of either of the fused proteins (Bertin et al., 1994; Wise et al., 1997a,b). Finally, the fusion protein technique can apparently be applied to a broad variety of GPCRs and G protein α subunits (Bertin et al., 1994; Seifert et al., 1998a,c; Wise et al., 1997b; Medici et al., 1997).

The fusion protein approach can be used to address numerous questions in signaling research. Fusion proteins have been employed to study the differential coupling of a given GPCR to closely related G protein α subunits (Seifert et al., 1998c), to precisely assess the intrinsic activities of partial agonists (Wise et al., 1997a) and to estimate the GTP turnover of a G protein α subunit in a membrane system (Wise et al., 1997b; Seifert et al., 1998b). Fusion proteins have also been used to explore the role of βγ subunits in GPCR–G protein coupling (Wise et al., 1997b; Seifert et al., 1998b), to study the role of covalent modifications in G protein function (Wise and Milligan, 1997), to identify functional domains of G protein α subunits (Medici et al., 1997), to test models of GPCR activation (Seifert et al., 1998a) and to study GPCR desensitization (Bertin et al., 1997a,b; Seifert et al., 1998b).

4. SUMMARY

The inefficient interaction of the β_2-AR with endogenous Gsα-like G proteins of Sf9 cells provides an excellent background for functional reconstitution studies of the β_2-AR with mammalian Gsα. Intriguingly, the efficiency of reconstitution of β_2-AR–Gsα coupling is not uniform, but critically depends on the specific parameter analyzed. Our data highlight the importance of the stoichiometry, and orientation with respect to each other, of signal transduction components for their effective interaction. In organs and cultured cell lines, Gsα is present in ~100-fold excess relative to the β_2-AR (Ransnäs and Insel, 1988; Kawai and Arinze, 1991). We reconstituted the β_2-AR and Gsα at a similar stoichiometry in Sf9 cells. This co-expression system allowed us to detect ternary complex formation and effective agonist activation of AC, but we failed to detect substantial GPCR–G protein interaction at the level of GDP/GTP exchange. Thus, only a minor portion of the expressed Gsα molecules may actually interact with the β_2-AR to form high-affinity ternary complexes. However, these few receptor-activated Gsα molecules are sufficient to transduce a stimulatory signal to AC, resulting in substantial cAMP formation.

The β_2-ARGsα fusion protein ensures a defined 1:1 stoichiometry of GPCR to G protein and can be expressed at rather high levels in Sf9 cells. The fusion protein is much more effective than the co-expression system consisting of nonfused β_2-AR and Gsα with regard to activation of GTPase, GTPγS binding, and AC despite the fact that the absolute concentration of Gsα in membranes expressing β_2-ARGsα was far lower than in membranes expressing β_2-AR and Gsα as separate proteins. Thus, fusion of the β_2-AR to Gsα induces an optimal positioning of the receptor relative to the G protein so that Gsα can efficiently shuttle between the β_2-AR and AC.

ACKNOWLEDGMENTS

The authors thank Drs. K. Wenzel-Seifert, E. Sanders-Bush, U. Gether, T.W. Lee, and V.T. Lam for their collaboration in the fusion protein project. R.S. was supported by a research fellowship of the Deutsche Forschungsgemeinschaft.

REFERENCES

Alousi AA, Jasper JR, Insel PA, Motulsky HJ (1991): Stoichiometry of receptor–Gs–adenylate cyclase interactions. FASEB J 5:2300–2303.

Alvarez R, Daniels DV (1990): A single column method for the assay of adenylate cyclase. Anal Biochem 187:98–103.

Barr A , Manning DR (1997): Agonist-independent activation of G_z by the 5-hydroxytryptamine$_{1A}$ receptor co-expressed in *Spodoptera frugiperda* cells. Distinguishing inverse agonists from neutral antagonists. J Biol Chem 272:32979–32987.

Barr AJ, Brass LF, Manning DR (1997): Reconstitution of receptors and GTP-binding regulatory proteins (G proteins) in Sf9 cells. A direct evaluation of selectivity in receptor–G protein coupling. J Biol Chem 272:2223–2229.

Bertin B, Freissmuth M, Jockers R, Strosberg AD, Marullo S (1994): Cellular signaling by an agonist-activated receptor/$G_s\alpha$ fusion protein. Proc Natl Acad Sci USA 91:8827–8831.

Bertin B, Jockers R, Strosberg AD, Marullo S (1997a): Activation of a β2-adrenergic receptor/Gsα fusion protein elicits a desensitization-resistant cAMP signal capable of inhibiting proliferation of two cancer cell lines. Recept Channels 5:41–51.

Bertin B, Strosberg AD, Marullo S (1997b): Human β2-adrenergic receptor/Gsα fusion protein, expressed in 2 *ras*-dependent murine carcinoma cell lines, prevents tumor growth in syngeneic mice. Int J Cancer 71:1029–1034.

Brandt DR, Ross EM (1986): Catecholamine-stimulated GTPase cycle. Multiple sites of regulation by β-adrenergic receptor and Mg^{2+} studied in reconstituted receptor–G_s vesicles. J Biol Chem 261:1656–1664.

Butkerait P, Zheng Y, Hallak H, Graham TE, Miller HA, Burris KD, Molinoff PB, Manning DR (1995): Expression of the human 5-hydroxytryptamine$_{1A}$ receptor in Sf9 cells. Reconstitution of a coupled phenotype by co-expression of mammalian G protein subunits. J Biol Chem 270:18691–18699.

Cassel D, Selinger Z (1976): Catecholamine-stimulated GTPase activity in turkey erythrocyte membranes. Biochim Biophys Acta 452:538–551.

Cassel Z, Selinger Z (1978): Mechanism of adenylate cyclase activation through the β-adrenergic receptor: Catecholamine-induced dsplacement of bound GDP by GTP. Proc Natl Acad Sci USA 75:4155–4159.

Cerione RA, Codina J, Benovic JL, Lefkowitz RJ, Birnbaumer L, Caron MG (1984): The mammalian β_2-adrenergic receptor: Reconstitution of functional interactions between pure receptor and pure stimulatory nucleotide binding protein of the adenylate cyclase system. Biochemistry 23:4519–4525.

Chidiac P, Hebert TE, Valiquette M, Dennis M, Bouvier M (1994): Inverse agonist activity of β-adrenergic antagonists. Mol Pharmacol 45:490–499.

Childers SR, Fleming LM, Selley DE, McNutt RW, Chang KJ (1993): BW373U86: A nonpeptidic δ-opioid agonist with novel receptor–G protein-mediated actions in rat brain membranes and neuroblastoma cells. Mol Pharmacol 44:827–834.

Citri Y, Schramm M (1982): Probing of the coupling site of the β-adrenergic receptor. Competition between different forms of the guanyl nucleotide binding protein for interaction with the receptor. J Biol Chem 257:13257–13262.

Clawges HM, Depree KM, Parker EM, Graber SG (1997): Human 5-HT$_1$ receptor subtypes exhibit distinct G protein coupling behaviors in membranes from Sf9 cells. Biochemistry 36:12930–12938.

De Lean A, Stadel JM, Lefkowitz RJ (1980): A ternary complex model explains the agonist-specific binding properties of the adenylate cyclase-coupled β-adrenergic receptor. J Biol Chem 255:7108–7117.

Gaudin C, Ishikawa Y, Wight DC, Mahdavi V, Nadal-Ginard B, Wagner TE, Vatner DE, Homcy CJ (1995): Overexpression of G$_{s\alpha}$ protein in the hearts of transgenic mice. J Clin Invest 95:1676–1683.

Gether U, Lin S, Ghanouni P, Ballesteros JA, Weinstein H, Kobilka BK (1997): Agonists induce conformational changes in transmembrane domains III and VI of the β$_2$ adrenoceptor. EMBO J 16:6737–6747.

Gether U, Lin S, Kobilka BK (1995): Fluorescent labeling of purified β$_2$ adrenergic receptor. Evidence for ligand-specific conformational changes. J Biol Chem 270:28268–28275.

Gierschik P, Moghtader R, Straub C, Dietrich K, Jakobs KH (1991): Signal amplification in HL-60 granulocytes. Evidence that the chemotactic peptide receptor catalytically activates guanine-nucleotide-binding regulatory proteins in native plasma membranes. Eur J Biochem 197:725–732.

Gierschik P, Sidiropoulos D, Steisslinger M, Jakobs KH (1989): Na$^+$ regulation of formyl peptide receptor-mediated signal transduction in HL 60 cells. Evidence that the cation prevents activation of the G protein by unoccupied receptors. Eur J Pharmacol 172:481–492.

Gilman AG (1987): G proteins: Transducers of receptor-generated signals. Annu Rev Biochem 56:615–649.

Grandt R, Greiner C, Zubin P, Jakobs KH (1986): Bradykinin stimulates GTP hydrolysis in NG108-15 membranes by a high-affinity, pertussis toxin–insensitive GTPase. FEBS Lett 196:279–283.

Graziano MP, Casey PJ, Gilman AG (1987): Expression of cDNAs for G proteins in *Escherichia coli*. Two forms of G$_{s\alpha}$ stimulate adenylate cyclase. J Biol Chem 262:11375–11381.

Grünewald S, Reiländer H, Michel H (1996): *In vivo* reconstitution of dopamine D$_{2S}$ receptor-mediated G protein activation in baculovirus-infected insect cells: Preferred coupling to G$_{i1}$ versus G$_{i2}$. Biochemistry 35:15162–15173.

Gürdal H, Bond RA, Johnson MD, Friedman E, Onaran HO (1997): An efficacy-dependent effect of cardiac overexpression of β$_2$-adrenoceptor on ligand affinity in transgenic mice. Mol Pharmacol 52:187–194.

Hein L, Kobilka BK (1995): Adrenergic receptor signal transduction and regulation. Neuropharmacology 34:357–366.

Kawai Y, Arinze IJ (1991): Ontogeny of guanine-nucleotide–binding regulatory proteins in rabbit liver. Biochem J 274:439–444.

Kleymann G, Boege F, Hahn M, Hampe W, Vasudevan S, Reiländer H (1993): Human β₂-adrenergic receptor produced in stably transformed insect cells is functionally coupled via endogenous GTP-binding protein to adenylyl cyclase. Eur J Biochem 213:797–804.

Koski G, Klee WA (1981): Opiates inhibit adenylate cyclase by stimulating GTP hydrolysis. Proc Natl Acad Sci USA 78:4185–4189.

MacEwan DJ, Kim G-D, Milligan G (1996): Agonist regulation of adenylate cyclase activity in neuroblastoma × glioma hybrid NG108-15 cells transfected to co-express adenylate cyclase type II and the β₂-adrenoceptor. Evidence that adenylate cyclase is the limiting component for receptor-mediated stimulation of adenylate cyclase activity. Biochem J 318:1033–1039.

Medici R, Bianchi E, Di Segni G, Tocchini-Valentini P (1997): Efficient signal transduction by a chimeric yeast–mammalian G protein α subunit Gpa1–Gsα covalently fused to the yeast receptor Ste2. EMBO J 16:7241–7249.

Mumby SM, Gilman AG (1991): Synthetic peptide antisera with determined specificity for G protein α or β subunits. Methods Enzymol 195:215–233.

Neer EJ (1995): Heterotrimeric G proteins: Organizers of transmembrane signals. Cell 80:249–257.

Neubig RR (1994): Membrane organization in G protein mechanisms. FASEB J 8: 939–946.

Okajima F, Katada T, Ui M (1985): Coupling of the guanine nucleotide regulatory protein to chemotactic peptide receptors in neutrophil membranes and its uncoupling by islet-activating protein, pertussis toxin. A possible role of the toxin substrate in Ca^{2+}-mobilizing receptor-mediated signal transduction. J Biol Chem 260:6761–6768.

Ostrowski J, Kjelsberg MA, Caron MG, Lefkowitz RJ (1992): Mutagenesis of the β₂-adrenergic receptor: How structure elucidates function. Annu Rev Pharmacol Toxicol 32:167183.

Pike LJ, Lefkowitz RJ (1980): Activation and desensitization of β-adrenergic receptor-coupled GTPase and adenylate cyclase of frog and turkey erythrocytes. J Biol Chem 255:6860–6867.

Quehenberger O, Prossnitz ER, Cochrane CG, Ye RD (1992): Absence of G_i proteins in the Sf9 insect cell. Characterization of the uncoupled recombinant N-formyl peptide receptor. J Biol Chem 267:19757–19760.

Ransnäs LA, Insel PA (1988): Quantitation of the guanine nucleotide binding regulatory protein G_s in S49 cell membranes using antipeptide antibodies to $α_s$. J Biol Chem 263:9482–9485.

Richardson RM, Hosey MM (1992): Agonist-induced phosphorylation and desensitization of human m2 muscarinic cholinergic receptors in Sf9 insect cells. J Biol Chem 267:22249–22255.

Samama P, Cotecchia S, Costa T, Lefkowitz RJ (1993): A mutation-induced activated state of the β₂-adrenergic receptor. Extending the ternary complex model. J Biol Chem 268:4625–4636.

Samama P, Pei G, Costa T, Cotecchia S, Lefkowitz RJ (1994): Inverse agonist activity of β-adrenergic antagonists. Mol Pharmacol 45:390–394.

Seifert R, Gether U, Wenzel-Seifert, Kobilka BK (1998a): Differential effects of partial agonists on G protein coupling of the β₂-adrenoceptor and a constitutively active mutant of the β₂-adrenoceptor. Naunyn-Schmiedebergs Arch Pharmacol 358 (suppl):R611.

Seifert R, Lee TW, Lam VT, Kobilka BK (1998b): Reconstitution of β_2-adrenoceptor–GTP-binding-protein interaction in Sf9 cells: High coupling efficiency in a β_2-adrenoceptor–$G_{s\alpha}$ fusion protein. Eur J Biochem 255:369–382.

Seifert R, Wenzel-Seifert K, Lee TW, Gether U, Sanders-Bush E, Kobilka BK (1998c): Different effects of $G_s\alpha$ splice variants on β_2-adrenoceptor–mediated signaling. The β_2-adrenoceptor coupled to the long splice variant of $G_{s\alpha}$ has properties of a constitutively active receptor. J Biol Chem 273:5109–5116.

Selley DE, Bidlack JM (1992): Effects of β-endorphin on μ and δ opioid receptor–coupled G-protein activity: Low-K_m GTPase studies. J Pharmacol Exp Ther 263:99–104.

Selley DE, Sim LJ, Xiao R, Liu Q, Childers SR (1997): μ-Opioid receptor–stimulated guanosine-5-O-(γ-thio)-triphosphate binding in rat thalamus and cultured cell lines: Signal transduction mechanisms underlying agonist efficacy. Mol Pharmacol 51: 87–96.

Strader CD, Ming Fong T, Tota MR, Underwood D, Dixon RAF (1994): Structure and function of G protein–coupled receptors. Annu Rev Biochem 63:101–132.

Wieland T, Jakobs KH (1994): Measurement of receptor-stimulated guanosine 5'-O-(γ-thio)triphosphate binding by G proteins. Methods Enzymol 237:3–13.

Wise A, Carr IC, Groarke A, Milligan G (1997a): Measurement of agonist efficacy using α_{2A}-adrenoceptor-$G_{i1\alpha}$ fusion protein. FEBS Lett 419:141–146.

Wise A, Carr IC, Milligan G (1997b): Measurement of agonist-induced guanine nucleotide turnover by the G-protein $G_{i1}\alpha$ when constrained with an α_{2A}-adrenoceptor–$G_{i1}\alpha$ fusion protein. Biochem J 325:17–21.

Wise A, Milligan G (1997): Rescue of functional interactions between the α_{2A}-adrenoreceptor and acylation-resistant forms of $G_{i1}\alpha$ by expressing the proteins from chimeric open reading frames. J Biol Chem 272:24673–24678.

CHAPTER 11

PEPTIDES AS TOOLS FOR THE STUDY OF RECEPTOR–G PROTEIN INTERACTIONS

RICHARD R. NEUBIG

1. INTRODUCTION	206
2. DESIGN AND SYNTHESIS OF PEPTIDES	207
3. CHARACTERIZATION OF PEPTIDES	209
A. Determination of Identity and Purity	209
B. Care and Storage of Peptides	210
4. INHIBITION OF RECEPTOR–G PROTEIN COUPLING	211
A. Protocol for Inhibition of Receptor–G Protein Coupling by Peptides	211
B. Inhibition of Receptor-Stimulated Nucleotide Binding and Hydrolysis	213
C. Protocol for Inhibition of Receptor–Stimulated Binding and Hydrolysis by Peptides	213
D. Intact Cell Studies	215
5. RECEPTOR-MIMICKING PEPTIDES ACTIVATE G PROTEIN	215
A. Protocol for Fluorescent Nucleotide Binding Kinetics	217
B. Peptides Mimicking Receptor Activation of G Proteins	217
6. STRUCTURAL STUDIES	220
7. INTERPRETATION OF DATA AND CAVEATS	223
9. POTENTIAL THERAPEUTIC UTILITY	224
10. FUTURE PROSPECTS	225

Structure–Function Analysis of G Protein-Coupled Receptors, Edited by Jürgen Wess.
ISBN 0-471-25228-X   Copyright © 1999 Wiley-Liss, Inc.

I. INTRODUCTION

Signaling through G protein–coupled receptors (GPCRs) has become very complex. There are over 250 known human GPCRs (Stadel et al., 1997) and over 1,000 possible combinations of G protein subunits (Simon et al., 1991; Ray et al., 1995). In addition, the number and complexity of effector systems regulated by G proteins continues to increase (Gudermann et al., 1997). Thus, an understanding of the determinants of specificity in receptor–G protein (RG) coupling, or perhaps more importantly receptor–G protein effector (RGE) coupling, is a key question in the field. A wide variety of approaches have been used to examine this question, including reconstitution, mutagenesis, antibody blockade, antisense, and knockout methods. The focus of this chapter is on the use of synthetic peptides from receptors, G proteins, or effectors to better understand the structural basis of specificity of RGE coupling.

There is a tremendous amount of primary structural (i.e., sequence) information available for receptors, G proteins, and effectors due to the cloning of many types and subtypes of these proteins. Great progress has been made in determination of the three-dimensional structure of G proteins and effectors (Noel et al., 1993; Tesmer et al., 1997; Sprang, 1997). Direct structural information about receptors is also becoming available (Unger et al., 1997; Yeagle et al., 1997a) but has lagged behind that of the soluble proteins in G protein pathways. The conserved structures of receptors and G proteins as well as the crosstalk seen between receptors and G proteins from different species and the ability of different receptor types to activate a single G protein suggest that the general mechanisms of G protein activation by receptors are largely similar. Certain regions of receptors have been consistently shown to play a role in G protein activation. Specifically, the third intracellular loop is most often implicated, but substantial data indicate roles for the second intracellular loop and the carboxy terminus as well (Neubig, 1998).

The contact sites between receptor and G protein and between G protein and effector would be predicted to have complementary structures both geometrically and electronically. Thus, a peptide fragment from the region of a receptor that contacts the G protein might contain sufficient structural information to interact with the binding pocket on the G protein. These peptide fragments could bind to the G protein and competitively block the interaction between the G protein and receptor. This was first established (1988) by Hamm and colleagues. Alternatively, receptor peptides could mimic the receptor and activate the G protein as has been previously shown for the wasp venom peptide mastoparan (Higashijima et al., 1988). These properties have made peptides useful both as biochemical probes of the structural basis of the RG interaction and as lead compounds for the design of drugs to activate or inhibit G protein function (for review and references, see Taylor et al., 1994a; Taylor and Neubig, 1994; Hargrave et al., 1993; Voss et al., 1993; Hamm, 1991; Munch et al., 1991).

A receptor peptide could be in a random conformation that is induced to form the correct conformation upon interaction with the G protein. In this case, the specificity for the G protein site would be due to the linear arrangement of residues in the peptide. Alternatively, the peptide could adopt a significant secondary structure on its own (Yeagle et al., 1997a), which would reduce the entropic requirements for the subsequent interaction with the G protein. Because the peptide structure is not likely to contain all of the receptor structure, the affinity of these peptides for the G protein is typically not as high as that of the receptor itself.

In addition to receptor peptides, peptides derived from sequences of many other G protein–coupled signal transduction proteins have been used to probe mechanisms of signaling. Peptides from G protein α subunits such as transducin (Hamm et al., 1988; Dratz et al., 1993), αs (Palm et al., 1990; Rasenick et al., 1994), and αq (Akhter et al., 1998) have been shown to uncouple a GPCR from its associated G protein. Generally, peptides from the extreme carboxy terminus and some internal regions of the Gα subunit are active in this regard. Similarly, peptides from the conformationally sensitive switch regions of G protein α subunits can disrupt Gα–effector interactions. This has been demonstrated for transducin coupling to cGMP phosphodiesterase (Rarick et al., 1992) and Gq coupling to PLC-β (Arkinstall et al., 1995). In the latter case, the specificity of these effects was shown by the synthesis of 188 overlapping 15 residue peptides, with peptides from only two regions being active. These two regions corresponded to effector-coupling regions of Gs and transducin. Peptides from putative G protein–coupling regions of effectors have also been identified that can block G protein–effector coupling via both the α and βγ subunits. A peptide from rod cGMP phosphodiesterase (Artemyev et al., 1992) blocks transducin (Gt)-α binding to the PDE-γ subunit. Interestingly, peptides from a number of Gβγ effectors are able to prevent productive interactions between Gβγ and effector. This has been shown for type II adenylyl cyclase (Chen et al., 1995), phospholipase Cβ$_2$ (Sankaran et al., 1998), phosducin (Blüml et al., 1997), and voltage-gated calcium channels (Herlitze et al., 1997).

This chapter focuses on the approaches used for and some of the conclusions that can be drawn from the use of synthetic peptides derived from the sequence of GPCRs. While the chapter is limited to this aspect of peptide studies of G protein systems, the information presented in the preceding paragraph indicates that a much broader range of questions about G protein mechanisms can be addressed by use of synthetic peptides.

2. DESIGN AND SYNTHESIS OF PEPTIDES

The size of the peptide is one of the main considerations when planning peptide studies. There are at least two applications for the use of synthetic peptides to study GPCR signaling. One is to identify regions of the receptor or G protein that are involved in coupling. The other is as pharmacological inhibitors of signaling processes to dissect mechanisms. In the former case, the smallest peptide with good potency will provide the finest mapping of the interaction sites. In general, because peptides much smaller than about 10 amino acids have little activity, chemically synthesized peptides with 12–30 amino acids are typically used. The upper limit to the size of the peptides is generally determined either by technical feasibility or by cost of synthesizing long peptides. Current technology can often reach lengths of 40 to even 60 amino acids, but such peptides are expensive and difficult to purify. It is possible to chemically ligate unblocked peptides to generate larger proteins (Muir et al., 1998). As the size exceeds 40 amino acids, however, it may be more feasible to prepare the peptide biosynthetically by expression in bacteria, either alone or as a fusion protein (Okamoto et al., 1995; Maina et al., 1988; Guan and Dixon, 1991; Martin et al., 1996; Kempe et al., 1985).

The specific structural features that are included in a peptide will depend on the amino acid sequence of the active region of the protein being studied. Certain aspects of the structure are likely to be critical in obtaining a functional peptide. Both charged residues and hydrophobic residues play important roles in protein–protein contacts, and their order in the sequence will help determine the biological activity of the peptide. For peptides designed to interact with a plasma membrane signal transduction system, one would predict that hydrophobic residues that would target the peptide to the membrane would enhance potency. This has been shown for both the α_{2a}-adrenergic receptor (α_{2a}-AR) (Taylor et al., 1994a; Taylor, 1995; Ikezu et al., 1992) and the β-adrenergic receptor (β-AR) (Wakamatsu et al., 1993; Shinagawa et al., 1994). Similarly, numerous investigators have proposed that amphiphilicity or the ability to display hydrophobic residues on one side of the helix or strand and charged or hydrophilic residues on another face plays an important role in the ability of peptides to activate G proteins (Hillaire-Buys et al., 1992; Higashijima et al., 1990; Naim et al., 1994; Wakamatsu et al., 1992). When designing a peptide, it is important to know its degree of amphiphilicity. This can be determined with a variety of computer programs for protein structural analysis such as the Wisconsin Package Genetics Computer Group (GCG, Madison, WI). The generality of the requirement for an amphiphilic structure, however, has been questioned (Voss et al., 1993; Wade et al., 1996). A final consideration in peptide design is how to handle the termini of the peptides. Most peptides that are synthesized represent internal sequences in the proteins from which they are derived. Thus, the amino and carboxy termini of the sequences would be engaged in amide bonds to the adjacent sequences and would not be charged at physiological pH. This can be mimicked in the peptide by preparing acetylated amino- and amidated carboxy termini. If the peptide is derived from the extreme amino- or carboxy-terminal sequence of the protein, it should be left unmodified unless the native protein structure is known to contain a modification such as myristoylation (Mumby et al., 1990; Jones et al., 1990) or prenylation (Kisselev et al., 1994).

Certain well-known problems related to peptide chemistry should also be avoided in the design of peptides as probes in signal transduction. Some problems arise after synthesis and purification and are dependent in part on storage conditions. Cysteines are likely to oxidize to disulfides upon storage in aqueous solutions. This can be used to advantage when it is done under controlled circumstances (Wade et al., 1994) but leads to confusing results if not recognized (see Fig. 11.1, below). Thus cysteines should be avoided in peptide synthesis or their oxidation carefully monitored. This can be done by mass spectroscopy or by quantitating free cysteines with Ellman's reagent (Ellman, 1959). Similarly, methionine residues can oxidize to methionine sulfoxide upon storage. One can sometimes substitute norleucine for methionine to avoid this problem. Also, dry peptides oxidize less rapidly than ones stored in solution, and storage in an inert atmosphere in a dessicator at freezer temperatures will help limit these problems. Amino-terminal glutamines will spontaneously cyclize to pyroglutamate if the amino terminus of the peptide is not blocked by acetylation. Thus, preparing a peptide with an amino-terminal glutamine with a free amino terminus should be avoided if possible. Other problems occur during synthesis to yield the incorrect product. Sequences containing the amino acid pairs Asp–Asp, Asp–Asn, Asp–Gly, and Asn–Gly can cyclize internally during synthesis at a fairly high efficiency. When they open, they can reform the original peptide bond or an isopeptide bond. Avoiding these sequences will make the job of the peptide chemist easier. Other sequences that cause difficulties during synthesis are strings of β-branched amino acids (Val, Ile, and Thr), β-sheet formers (Val, Ile, Phe, Tyr, and Thr), and sequences with a propensity to form an α helix. Long homopolymer stretches (e.g., eight Glu residues in a row) should also be avoided.

3. CHARACTERIZATION OF PEPTIDES

3.A. Determination of Identity and Purity

It is critical to confirm the purity and identity of any peptides that are synthesized. During peptide synthesis many steps can go wrong, including skipped residues, side-chain isomerization, incomplete removal of blocking groups from residues, and modification of labile side chains. Furthermore, most peptides contain some degree of impurity, which can be a serious problem if peptides are used with only partial purification. In general, purification to 95% purity or greater should be done before concluding that effects observed are due to the predicted peptide.

Figure 11.1 illustrates some of the subtle complications that can occur in the preparation of peptides. High-performance liquid chromatography (HPLC)

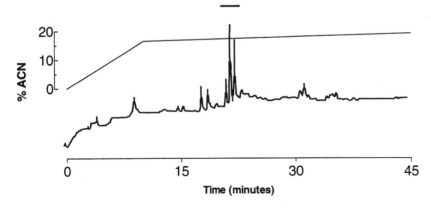

Figure 11.1. HPLC separation of a crude peptide synthesis mixture. The solid bar indicates a triplet of related peptides that were initially isolated as a single peak from a preparative HPLC. The triplet represents homo- and heterodimers of the expected peptide α_{2a}–i3c (Q) and the N-formyl tryptophan derivative of this peptide (Qf). The three peaks were shown to be Q–Q, Q–Qf, and Qf–Qf dimers (Wade et al., 1994). Only separation on a very shallow acetonitrile (ACN) gradient and mass spectroscopy were able to identify the formyl tryptophan, which significantly modified functional activity leading to heterogeneity in results from different batches of peptides.

separation of a crude synthetic peptide mixture of a fragment of the α_{2a}-AR (amino acids 361–373 + cys, RWRGRQNREKRFT-C) was performed on a very shallow acetonitrile gradient. The triplet of peaks marked with the thick horizontal bar was initially thought to be a single peptide because it eluted as a single peak on a standard preparative HPLC gradient. Only after a shallow gradient was done were the three peaks resolved. The three peptides were found to have masses of 2X, 2X + 28, and 2X + 56, where X = 1,890 was the expected mass. Further study indicated that all three were disulfide-linked dimers of the original cysteine-containing peptide. The two peaks with the higher masses had either one or two of the original peptides still containing a formyl-tryptophan (Wade et al., 1994). The formyl group is frequently used as a blocking group to prevent modification of tryptophan during synthesis, but it should be removed during the cleaving of the peptide from the resin used in synthesis. In this case removal of the blocking groups was incomplete. Interestingly, the diformylated peptide dimer had no activity (Wade et al., 1994), indicating that even such small modifications can be very important for function.

Table 11.1 lists a number of methods and criteria for determining the purity and identity of synthetic peptides. Reverse phase HPLC is the standard method; however, it is not sufficient to detect all possible impurities. At a minimum, mass spectroscopy of synthesized peptides should also be done to identify problems in synthesis such as simple errors in which amino acid was used for the synthesis, missing residues, or side chain modifications (oxidation, isomerization, or incomplete deblocking). An error in the order of amino acids in the sequence could be missed by both HPLC and mass spectroscopy and might only be detected by sequencing of the peptide either by Edman degradation or tandem mass spectroscopy (Angeletti et al., 1998).

3.B. Care and Storage of Peptides

Peptides should be stored as solids at −20°C or below in a dessicated container. It is optimal to prepare the peptide fresh from solid each day; however, when only small amounts are used, this can be wasteful. It may be necessary to store small aliquots of concentrated stock solutions frozen in water. These samples should be checked by HPLC and/or mass spectroscopy after the period of storage to ensure stability. For very hydrophobic peptides, storage in DMSO, DMF, or another organic solvent may be required. In general, stability is increased if the samples are stored at high concentration (e.g., 0.1–10 mM).

TABLE 11.1. Methods To Determine Identity and Purity of Peptides

Method	Problems Detected
Reverse-phase HPLC	A, B, C, E
Mass spectroscopy	A, B, C, E
Aminoacid analysis	C, B
Edman degradation, sequencing	B, C, D, E

A, incomplete deblocking; B, side-chain isomerization; C, incorrect sequence deletion; D, incorrect sequence scrambling; E, oxidation of side chains.

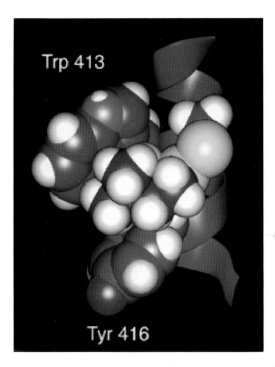

Figure 2.5. Molecular model showing the quaternary ammonium of MTSET after reaction with T412C in Van der Waals contact with the aromatic side chains of Trp[413] and Tyr[416].

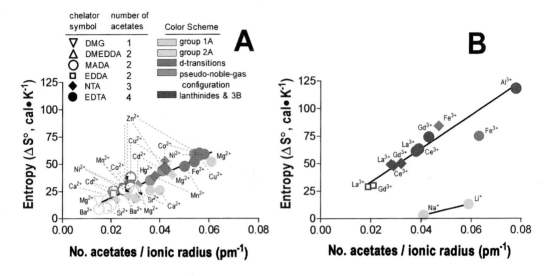

Figure 7.1. The stability of metal complexes is a function of complexation entropy. **A, B:** $\Delta S°$ means that the entropy measurements were taken at standard state, i.e., 1 atm and 298°K. The number of acetates refers to the number of negatively charged groups at the coordination site of the chelator, and the radius refers to the ionic radius of the metal cation. The arrowheads on some of the dashed lines refer to data points that are obscured by other data points (i.e., underneath). Studies with EDTA analogs have demonstrated that favorable entropy contributions can substantially increase the stability of metal complexes when the metals are small and the coordination site has multiple ligating groups. In addition, the magnitude of the favorable entropy change increases as the valence of the metal cation increases. DMEDDA, dimethylethylenediaminediacetate; DMG, dimethylglycinate; MADA, methylaminediacetate; EDDA, ethylenediiminodiacetate; NTA, nitrilotiracetate; EDTA, ethylenediaminetetraacetate. Metal complex entropy values are from Martell (1960) and Martell and Smith (1974). Ionic radius values are from Weast (1985).

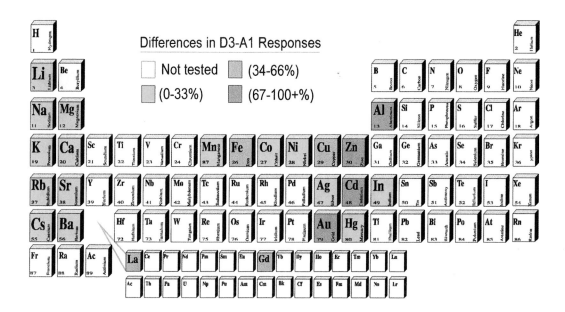

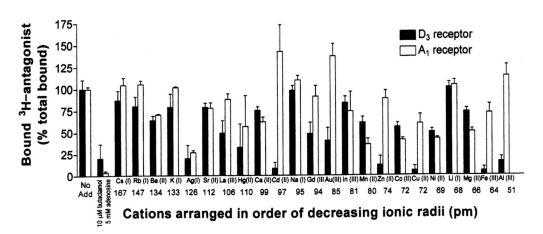

Figure 7.2. Screening for the effects of a single high concentration of metal cations on ^{3}H-antagonist binding to GPCRs. The effects of a single high concentration (0.05–5 mM) of metals on the binding of the dopamine D3 and adenosine A_1 receptors were screened using ^{3}H-methylspiperone (945 ± 350 pM) and ^{3}H-DPCPX (1,3-dipropyl-8-cyclopenylxanthine, 703 ± 260 pM), respectively. The pattern of metal sensitivity of ^{3}H-antagonist binding to dopamine D3 receptors is representative of the metal sensitivities of all cloned rat dopamine receptors (Schetz and Sibley, 1997, and unpublished results). The equilibrium binding data shown in the histogram were grouped in thirds according to the relative magnitude of metal sensitivities to D3 verses adenosine A_1 receptors. These relative differences in metal sensitivities are shown schematically as a function of their position in the periodic table.

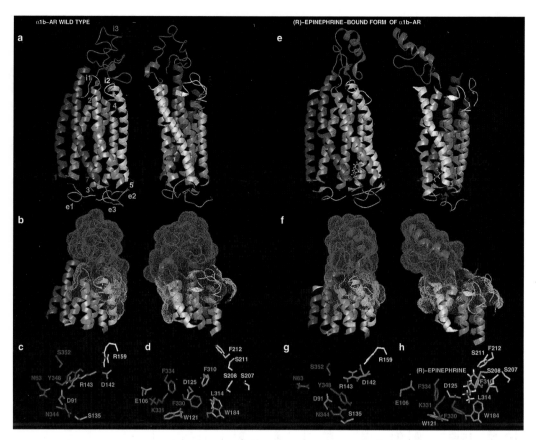

Figure 9.3. Ground state and agonist-bound α_{1b}-AR structures. **Left:** Wild-type α_{1b}-AR (ground state). **Right:** R(−)-epinephrine-bound α_{1b}-AR. All views are in a direction perpendicular to the helix main axis. **a, e:** Two views of the average minimized structure. The seven helices are colored in blue (1), orange (2), green (3), pink (4), yellow (5), sky blue (6), and violet (7), respectively. The i1, i2, and i3 loops are colored in green, white, and cyclamen, respectively; whereas e1, e2, and e3 are colored in carnation, respectively. **b, f:** Two views of the cytosolic half of the average minimized structure. The solvent-accessible surface areas computed on i2, i3, and the cytosolic extensions of helices 5 and 6 are displayed. **c, g:** Some amino acids (colored according to their location) in the environment of D91 (helix 2), D142, and R143 of the E/DRY. **d, h:** Some amino acids in the environment of D125 (helix 3) and S207 (helix 5).

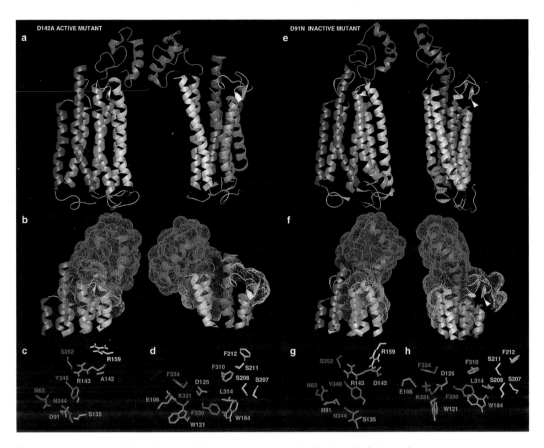

Figure 9.4. Active and inactive α_{1b}-AR mutant structures. **Left:** Constitutively active D142A mutant. **Right:** Inactive R143A mutant. Other details are as in Figure 9.3.

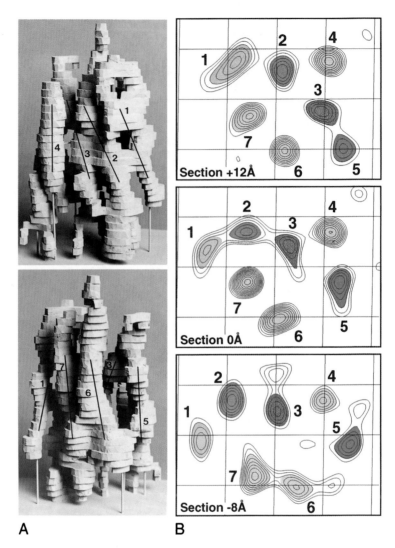

A   B

Figure 12.17. The seven helices in the rhodopsin structure. **A:** Structure of frog rhodopsin obtained by electron cryomicroscopy (Unger et al., 1997a). Two views of a solid model of the rhodopsin map are shown. **Top:** View from helix 2 toward helix 6. **Bottom:** View from helix 6 toward helix 3. The model was constructed from 33 contour sections 2 Å apart. The cytoplasmic side is at the top, and the intradiscal or extracellular side is at the bottom. The central sections of the seven transmembrane helices are marked with lines starting at section +12 Å at the top and ending at section −8 Å. The corresponding sections are shown in B. The peaks representing the seven helices are interpreted according to the sequence assignment (Baldwin, 1993) to the projection map of rhodopsin (Schertler et al., 1993). **B:** Three slices through the best part of the density map of rhodopsin (Unger et al., 1997a). In each of these sections, peaks can be seen for each of the seven transmembrane helices. The section closer to the cytoplasmic side is at z = +12 Å from the center and the last section at z = −8 Å from the center of the map. The least tilted helices (4, 6, and 7) are colored grey, and the most tilted ones are in four different colors. The grid spacing is 10 Å, with lines parallel to the **a** and **b** axes (+**b** is horizontal to the right, and +**a** points toward the bottom).

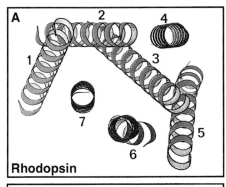

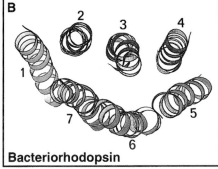

Figure 12.18. The arrangement of α helices in rhodopsin **(A)** and bacteriorhodopsin **(B).** A ribbon diagram of rhodopsin was drawn using the coordinates from a recently published α-carbon template for the transmembrane helices in the rhodopsin family of GPCRs (Baldwin et al., 1997). A similar diagram was generated for bacteriorhodopsin based on the coordinates from the bacteriorhodopsin structure (Grigorieff et al., 1996). The diagrams illustrate the different arrangement of helices in bacteriorhodopsin and rhodopsin (Schertler, 1998).

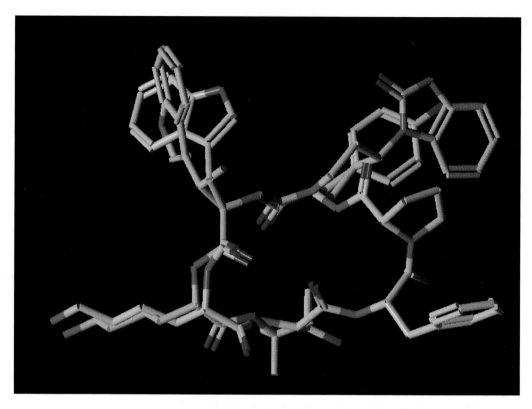

Figure 17.5. Comparison of the NMR-derived structure of **3** (carbon atoms in pink) with a low energy conformation of L-054,522 (carbon atoms in green). In both structures, nitrogens are blue and oxygens are red.

4. INHIBITION OF RECEPTOR–G PROTEIN COUPLING

Because the predicted mechanism of action of receptor-derived peptides is to bind to the G protein at the same site as the receptor, the peptides should result in an "uncoupling" of the receptor from its G protein. One commonly used method to examine RG protein coupling is the measurement of high affinity agonist binding in the "ternary complex" of agonist–receptor G protein (De Lean et al., 1980). This is often determined by the appearance of "flat" competition curves (i.e., Hill coefficient $n_H < 1$) of an unlabeled agonist for a radiolabeled antagonist. This is usually accompanied by a GTP-induced shift of the curve to higher K_i values for the agonist. The availability of radiolabeled agonists (e.g., ^{3}H-glucagon, ^{3}H-oxotremorine-M, and ^{125}I-iodo-clonidine) provides a direct and more sensitive measure of RG coupling. In this case one simply measures high affinity agonist binding. Manipulations that reduce RG coupling (GTP, pertussis toxin, sulfhydryl reagents, detergent solubilization, alkaline or urea treatment) reduce that binding (Hartman and Northrup, 1996; Kim and Neubig, 1985; Okajima et al., 1985; Limbird and Speck, 1983), while manipulations that increase RG coupling (reconstitution with exogenous G proteins) increase high-affinity agonist binding (Kim and Neubig, 1987). Parallel studies of antagonist binding are a useful control to show that the peptide effect is on RG coupling and not on receptor number. Thus, data may be presented as a ratio of agonist to antagonist binding at fixed ligand concentrations above their K_d values.

The use of this approach is illustrated in Figure 11.2 where the effect of synthetic peptides from the α_{2a}-AR are determined on agonist binding to the α_{2b}-adrenergic (α_{2b}-AR) and M_4 muscarinic receptors in NG10815 cell membranes (Neubig and Dalman, 1991). These peptides have been previously shown to uncouple the homologous α_{2a}-AR from G_i in platelet membranes. This experiment was designed to test the generality of the peptide effect. If the peptide interacts with the G protein, one would expect it to reduce agonist binding to any G_i coupled receptor. The binding of the agonists ^{125}I-p-iodoclonidine and ^{3}H-oxotremorine-M was dependent on G protein because addition of 10 μM GppNHp reduced binding by 60%–80% (horizontal lines). The i2 loop peptide from the α_{2a}-AR at 100 μM reduced binding of both agonists to nearly the same degree as GppNHp. The i3c peptide (derived from the carboxy-terminal segment of the third intracellular loop, i3) was somewhat less potent but also reduced the binding of both agonists. There was no significant effect of the peptides on binding of ^{3}H-yohimbine, indicating an agonist-specific effect at the α_{2b}-AR.

4.A. Protocol for Inhibition of Receptor–G Protein Coupling by Peptides

1. Thaw membrane samples in ice-cold binding buffer (TME; 50 mM Tris-C1, 10 mM $MgCl_2$, 1 mM EGTA, 1 mM DTT, pH 7.6).
2. Incubate membranes with 0.1–100 μM of peptide on ice for 30 minutes in half of the final reaction volume (50 or 250 μl).

3. Initiate the binding reaction by adding the agonist radioligand at twice the final concentration (2 nM *p*-iodoclonidine in 50 µl or 5 nM oxotremorine-M in 250 µl).

4. Include samples with excess unlabeled ligand to define nonspecific binding (e.g., 10 µM yohimbine or 10 µM atropine for the α-adrenergic and muscarinic receptors, respectively). Also, agonist binding in the presence of 10 µM GppNHp is tested to determine what fraction is dependent on G protein.

5. Incubate at room temperature (22°–24°C) for the appropriate time (60–90 minutes).

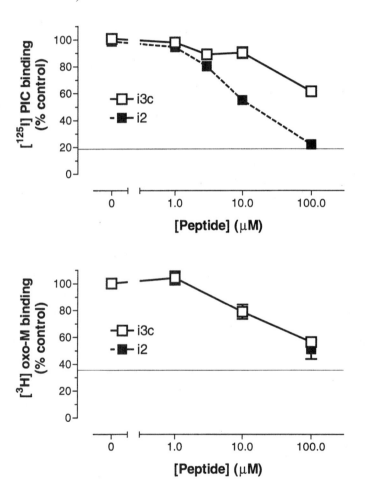

Figure 11.2. Peptide inhibition of agonist binding to α_{2b}-adrenergic and M_4 muscarinic receptors in NG-10815 cell membranes. Membranes from NG-10815 cells were preincubated with the indicated concentrations of a peptide from the i2 or i3c region of the α_{2a}-AR. Subsequently, binding of either 1 nM ^{125}I-*p*-iodoclonidine (PIC) **(top)** or 2.5 nM ^{3}H-oxotremorine-M **(bottom)** was measured. The horizontal lines indicate binding in the presence of 10 µM GppNHp. Nonspecific binding in the presence of 10 µM yohimbine or atropine, respectively has been subtracted.

6. Dilute the reaction mixtures with 4 ml of ice-cold filter buffer (50 mM Tris-Cl, 10 mM $MgCl_2$, pH 7.6), and then filter on glass-fiber filters (Whatman GF/B presoaked in 0.05% polyethylenimine to reduce nonspecific binding).
7. Wash the filters twice with 4 ml of cold filter buffer.
8. Count the filters in a scintillation counter.
9. Calculate specific binding and plot against peptide concentration.

As noted above, parallel studies to rule out peptide effects on antagonist binding should also be done. At high concentrations of peptides, nonspecific effects may be observed. Because one would not expect a reduction of antagonist binding from uncoupling receptor and G protein, reduced antagonist binding provides a warning about nonspecific effects. In general, it is a good idea to keep peptide concentrations at or below 100 μM. Effects observed at mM concentrations are more likely to be nonspecific.

4.B. Inhibition of Receptor-Stimulated Nucleotide Binding and Hydrolysis

In addition to peptide effects on ligand binding, it is also important to show that peptides can alter receptor function. This raises an important caveat about the use of synthetic peptides (i.e., access to their site of action). Peptides are expected to act at the RG interface on the cytoplasmic surface of the plasma membrane. Because most peptides are hydrophilic, they would not be expected to gain access to intact cells. This difficulty can be circumvented experimentally by use of GTPase or other enzymatic measurements of receptor function in broken cell preparations. Agonist-stimulated GTPase activity can be used to monitor activation of many GPCRs. Dalman and Neubig (1991) showed that a peptide from the carboxy-terminal end of the third intracellular loop of the α_{2a}-AR ($\alpha2$–i3c) was able to inhibit UK-14304–stimulated GTPase in human platelet membranes (Fig. 11.3).

4.C. Protocol for Inhibition of Receptor-Stimulated Nucleotide Binding and Hydrolysis by Peptides

1. Prepare 2 ml of GTPase cocktail and distribute 40 μl per point into test tubes and store on ice. To prepare GTPase cocktail, mix
   a. 1 ml incubation cocktail (50 μM propranolol, 0.5 mM isobutyl-methylxanthine [IBMX], 5 mM cAMP in H_2O)
   b. 100 μl regenerating system (100 mM phosphocreatine, 10 mg/ml creatine phosphokinase)
   c. 450 μl 2 mM ATP
   d. 250 μl 2 M NaCl
   e. 5 μl 1 mM GTP
   f. 1–5 μl ^{32}P-GTP (0.1–0.5 mCi/tube)
   g. 194 μl H_2O

2. Thaw cell membranes. Dilute to 0.5–1.5 mg/ml in TME (50 mM Tris-HCl, 10 mM MgCl$_2$, 1 mM EDTA), preparing sufficient material for 10 μl (5–15 μg protein) per point.

3. Dilute with 3 vol of 1 mM freshly prepared DTT in ice-cold water, and homogenize in a glass-teflon homogenizer and distribute 40 μl per tube.

4. Preincubate membranes for 10–30 minutes on ice with 0.1–100 μM peptide.

5. Add receptor agonists and/or antagonists in 20 μl of H$_2$O (e.g., 50 μM UK 14304 and/or yohimbine as α$_{2a}$-AR agonist and antagonist, respectively) to the tubes containing membranes.

6. In some tubes also include 50 μM unlabeled GTP with the agonists to determine low-affinity GTPase.

7. Prewarm the tube with membranes, peptide, receptor ligands, and/or unlabeled GTP and the tube with GTPase cocktail for 2 minutes.

8. Initiate reaction by addition of 40 μl GTPase cocktail to each membrane sample.

9. Incubate for 15 minutes at 30°C.

10. Quench the reaction by addition of 1 ml 25% activated charcoal in ice-cold phosphate buffer (10 mM, pH 2.3).

11. Spin for 15 minutes in a microcentrifuge in a cold room to pellet the charcoal. Let samples stop without brake to avoid disturbing pellet.

12. Carefully remove 500 μl of supernatant to count.

13. Subtract low-affinity GTPase values from total GTPase to calculate high-affinity GTPase.

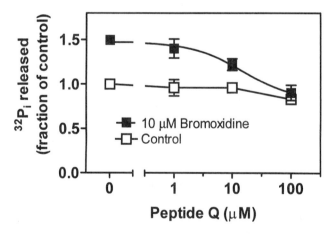

Figure 11.3. Peptide inhibition of α$_{2a}$-AR-stimulated GTPase activity in human platelet membranes. Basal and UK-14304–stimulated GTPase activities of human platelet membranes were measured in the presence of the i3c (Q) peptide from the α$_{2a}$-AR. The i3c peptide inhibited the receptor-stimulated GTPase but not basal GTPase, indicating that it disrupted receptor–G protein coupling. (Reproduced from Dalman and Neubig [1991], with permission of the publisher.)

14. Plot agonist-stimulated high-affinity GTPase as a function of peptide concentration.

These GTPase measurements were done under the same conditions as adenylyl cyclase assays to permit comparison of the two measurements. The propranolol was added to prevent any effect of agonist on β-ARs in platelet membranes and could be left out in nonadrenergic receptor systems. Similarly, the cAMP and IBMX are not required if comparison to adenylyl cyclase results is not desired. Details of the GTPase theory and methods have been described in detail by Gierschik et al. (1994).

4.D. Intact Cell Studies

The use of peptides in intact cell studies is somewhat more complicated because peptides generally will not cross the plasma membrane. Electrophysiological measurements, however, are well suited to peptide studies. The whole-cell patch clamp method provides access to the intracellular milieu, and the low molecular weight of peptides permits ready diffusion of the reagent into the cell (Chen et al., 1995; Herlitze et al., 1997; Kang et al., 1995; Zhu et al., 1997).

Peptides from Gi- and Gq-coupled receptors have also been introduced into cells genetically by transient expression of minigenes (Luttrell et al., 1993; Hawes et al., 1994; Thompson et al., 1998). These peptides inhibited receptor-mediated G protein function, and in some cases the inhibition extended to other receptors coupled to the same G protein as the receptor from which the minigene was derived. A minigene from the α_{1b} receptor blocked signals from the α_{1b} receptor as well as those from M_1 muscarinic (Hawes et al., 1994) and angiotensin AT1a receptors (Thompson et al., 1998). As expected, it had no effect on adenylyl cyclase coupling by the D1 dopamine, α_{2a}-adrenergic, or M_2 muscarinic receptors. This approach has been extended to an in vivo system in which a peptide from the carboxy terminus of αq was found to prevent cardiac hypertrophy presumably by blocking Gq interactions with receptors (Akhter et al., 1998).

5. RECEPTOR-MIMICKING PEPTIDES ACTIVATE G PROTEIN

The binding of nonhydrolyzable guanine nucleotide analogs to G proteins is commonly used as a measure of G protein activation. The release of GDP from the G protein is accelerated by receptors and receptor-mimetic peptides. This permits the binding of either radiolabeled or fluorescent GTP derivatives. GTPγS has a very high affinity for G proteins, and once bound it is not hydrolyzed, leading to a nearly irreversible bound state. The rate of GTPγS binding to G proteins is limited by GDP release so that the time course of the approach to maximal binding can be used to evaluate changes in GDP release rates. Methyl-anthraniloyl (mant) derivatives of guanine nucleotides have been used with ras family G proteins for some time (Eccleston et al., 1991). More recently, Remmers and colleagues (Remmers and Neubig, 1996; Remmers et al., 1994; Remmers, 1998) demonstrated their utility for the study of heterotrimeric G proteins. Wade et al. (1996) utilized this technique to assess the time course of mant-GTPγS binding and activation of Go heterotrimer in the presence of receptor-derived synthetic peptides. The mant-GTP fluorescence method is equally applicable to the study of purified heterotrimers in detergent solutions and lipid vesicles.

The large fluorescence signal obtained with the mant-guanine nucleotides is due to energy transfer from two or more conserved tryptophans in Go or Gi to the mant moiety (Remmers et al., 1994; Lan et al., 1997). Mant-GTP has minimal excitation at 280 nm but has an excitation peak at 362 nm, which overlaps well with the emission of tryptophan. Thus an excitation wavelength of 280 nm produces little excitation of free mant-GTP in solution but excites tryptophans in the Gα subunit resulting in resonance energy transfer and excitation of the bound mant-GTP. Because the emission of mant-GTP at 450 nm is sufficiently far away from the tryptophan emission peak at 350 nm, there is a minimal contribution from the G protein fluorescence itself. In more complex systems such as cell membranes, however, other fluorescent components may cause interference. Also, synthetic peptides may contain tryptophan, and, if used at very high concentrations, such peptides could lead to higher baseline fluorescence or absorbance artifacts. Because the parameter being evaluated is the *rate* of nucleotide binding, this is not as large a problem as it may seem. The rate constant for the time course of binding will be the same whether or not the absolute fluorescence level is reduced by absorbance of some of the exciting light by the peptide being tested (see inset to Fig. 11.4).

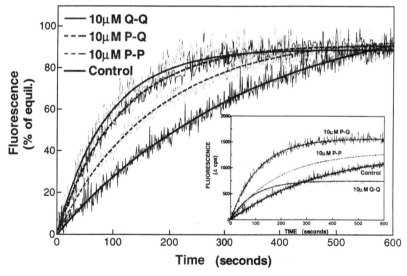

Figure 11.4. Fluorescence assay of peptide stimulation of mant-GTPγS binding to Go in lipid vesicles. The fluorescence signal from binding of 500 nM mant-GTPγS to 100 nM Go heterotrimer in asolectin vesicles was measured in the presence of different peptides from the α_{2a}-AR as described elsewhere (Wade et al., 1994). The P peptide is a 13-mer from the i3n region, and the Q peptide is a 13-mer from the i3c region. Disulfide-linked dimers were prepared and tested for their ability to stimulate GDP release. (Reproduced from Wade et al. [1994], with permission of the publisher.)

5.A. Protocol for Fluorescent Nucleotide Binding Kinetics

1. Add 200 μl of GTP binding buffer (50 mM HEPES, 1 mM EDTA, 1 mM DTT, 100 mM NaCl, 1.1 mM $MgCl_2$, and 0.0002% Lubrol [i.e., 20 ppm]) to a round quartz fluorescence cuvette (4 mm inner diameter) with stirring.

2. Equilibrate at 20°C for 3–5 minutes.

3. Dilute a concentrated stock solution of purified G protein heterotrimer (e.g., bovine brain Go) to a final concentration of 100 nM in the cuvette, and add the desired concentration of peptide activator.

4. Measure the baseline fluorescence for 0.5–1 minute with an excitation wavelength of 280 nm and an emission wavelength of 450 nm.

5. Add 300 nM mant-GTPγS, and follow the fluorescence with continuous data recording until it reaches a plateau (after approximately 10–20 minutes). Some fluorometers permit acquisition of time course data from multiple samples at once (e.g., in a four-sample turret).

6. Fit the curve with a nonlinear least-squares analysis program to an exponential function $F(t) = F_0 + \Delta F \cdot [1 - \exp(-k \cdot t)]$, where F_0 is the fluorescence immediately after the mant-GTPγS was added, ΔF is the change in fluorescence, and k is the rate constant for the increase in min^{-1}.

7. The effects of peptide can be expressed in a plot of rate of binding vs. peptide concentration.

The low concentration of free Mg^{2+} (0.1 mM) slows the spontaneous release of GDP from the G protein. This permits a greater increase above the basal release upon stimulation by receptor peptides. The assay can be carried out in a larger volume (400 μl) in 5 × 5-mm quartz fluorescence cuvettes, which are commercially available, instead of the custom-made 4-mm round cuvettes.

The use of these fluorescent nucleotide techniques with crude cell membrane preparations has not yet been established. The existence of multiple G proteins in a cell and the fluorescence from other membrane constituents may lead to a low signal to noise ratio. Thus, more work will be needed to establish the feasibility of this approach in such systems.

5.B. Peptides Mimicking Receptor Activation of G Proteins

The interaction of receptors with G proteins includes at least two components. There is the binding and recognition of G protein and a separate process of activation. Because intracellular peptides are derived from the regions thought to be involved in both of these processes, one might expect some peptides to block and others to mimic receptor activation. Munch et al. (1991) showed that peptides derived from the β-AR illustrate both types of phenomena. Peptides derived from the turkey erythrocyte β-AR i2, i3n (derived from the amino-terminal segment of the i3 loop), and carboxy-terminal regions interfere with receptor-stimulated adenylyl cyclase. In contrast, a peptide (284–295) derived from the β-AR i3c stimulates adenylyl cyclase by itself (Fig. 11.5). This

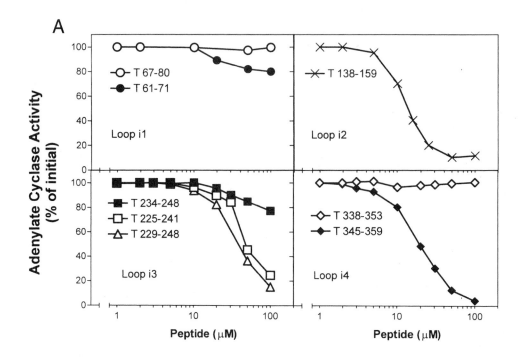

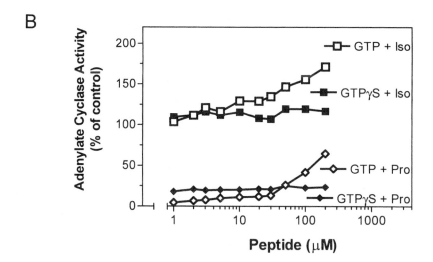

Figure 11.5. Adenylyl cyclase activation by β-receptor peptide. **A:** Peptides from the intracellular loops of the turkey erythrocyte β-adrenergic receptor (i1, i2, amino-terminal portion of i3, and i4 [e.g., carboxy tail]) inhibited β-adrenergic receptor-stimulated adenylyl cyclase (10 μM isoproterenol and 0.5 μM GTPγS). The residue numbers of the peptides are indicated. **B:** In contrast, the peptide from the carboxy-terminal portion of i3 (T 284–295) activated adenylyl cyclase. Assays were conducted in the presence of 10 μM isoproterenol (squares) or 10 μM propanolol (diamonds) with 10 μM GTP (open symbols) or 0.5 μM GTPγS (filled symbols). (Reproduced from Munch et al. [1991], with permission of the publisher.

suggests that for the β-AR, the i3c region is a G protein activator region, while i2, i3n, and the carboxy-terminal region appear to be more involved in G protein recognition and/or selectivity (Munch et al., 1991). A similar distribution of functions has been proposed for the α_{2a}-AR in which i3c is a G protein activator, while i2 and i3n were important for coupling and specificity (Wade et al., 1994). In contrast to the β-AR, there was no evidence for a role of the carboxy terminus of the α_{2a}-AR in G protein coupling.

An important caveat to this analysis should be mentioned. G protein activation by receptors is likely to involve multiple intracellular loops (and perhaps transmembrane domains as well; Abell and Segaloff, 1997). Thus, a single peptide from a G protein activator region may not be as effective at activating the G protein as is the intact receptor. Such a peptide would be expected to produce a weak activation of G protein but might block the stronger activation mediated by the receptor. Such a peptide could be considered as a "partial agonist" for G protein activation. Indeed the α_2-AR i3c peptide activates purified Go and Gi in lipid vesicles but blocks Gi activation by the α_2-AR in platelet membranes. The mechanism of this dual effect is illustrated by the interactions of the α_2-AR i3c peptide and the strong G protein activator mastoparan (Fig. 11.6). The α_{2a}-i3c peptide at 30 μM increases the basal GTPase activity about twofold, while the maximum effect of mastoparan is a fivefold increase. Combining the α_{2a}-i3c peptide with increasing concentrations of mastoparan results in the same maximum stimulation fivefold over the unstimulated basal level, but the EC_{50} for mastoparan is increased twofold (from 21 to 42 μM), consistent with a competitive interaction between the α_{2a}-i3c peptide and mastoparan.

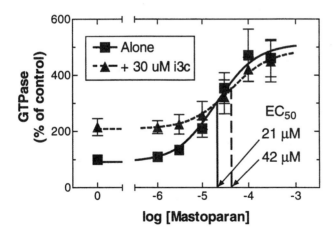

Figure 11.6. Mastoparan-Q peptide competition. The GTPase activity of Go heterotrimer in phospholipid vesicles was measured as described elsewhere (Wade et al., 1994). The effects of mastoparan with and without the α_{2a} i3c peptide (Q) are shown. The maximum GTPase activity in the presence of both peptides is the same. The EC_{50} for mastoparan in the presence of 30 μM Q peptide is twice that in its absence, consistent with a competitive interaction of the two peptides.

6. STRUCTURAL STUDIES

A final and quite powerful use of synthetic peptides is to define the contact sites on the target protein by chemical crosslinking of the peptide to its target. It is theoretically possible to crosslink the whole receptor with the G protein and identify the site on both proteins that becomes crosslinked. However, a small peptide derived from the receptor provides a much more tractable system on which to do this chemistry. A large number of methods have been developed to permit crosslinking of peptides to proteins. There are three general strategies. First, a nonspecific protein–protein crosslinker is added to the complex of peptide and its target. This is most effective when the affinity of peptide for target is very high so that the free peptide can be washed away and only bound peptide is present. This approach has been used often in the study of ligands at peptide receptors where the K_d is in the nM range. For synthetic peptides derived from internal sequences of larger proteins, the interactions are usually weaker—in the μM range—and this approach leads to much nonspecific crosslinking. The other two alternatives incorporate the crosslinking chemistry into the structure of the peptide. This can be achieved either by covalently attaching a crosslinker to the peptide during the synthesis or after synthesis is complete. The former is particularly simple and has been used to advantage recently.

Table 11.2 lists a number of crosslinking agents that have been used in peptide crosslinking studies. The probes that are chemically incorporated into the peptide after synthesis usually modify either a free amine (on a lysine or the amino terminus of the peptide) or a sulfhydryl group (on a cysteine). A variety of such probes are commercially available, with the Pierce chemical company providing a large selection and excellent information in their catalog (www .piercenet.com). Some of the newer photoaffinity probes in Table 11.2 have the benefit of higher photochemical yield or easier handling. To identify targets of peptide action, a photoaffinity peptide–crosslinker is incubated with the target and exposed to light. The linking of peptide to the target is usually monitored by radioactivity that has been incorporated into the peptide itself by iodination of tyrosine. Further identification of the site of incorporation is done by peptide mapping of the crosslinked complex. A number of the newer crosslinkers can be directly radioiodinated, which eliminates concerns about cleavage of the peptide between the crosslinker and the site of peptide labeling. Other methods to detect incorporation of a peptide into a protein target include a gel shift where a peptide of molecular weight 2–4 kD produces a signficant decrease in mobility of the target protein on SDS gels (Fig. 11.7A,B). Also, staining with antipeptide antibodies would provide a useful approach. One novel approach to identifying contact sites between peptides and their target proteins utilizes a label transfer technique (Liu et al., 1996). In this case, the photoaffinity label includes an easily iodinated phenyl ring that is attached to the peptide by a disulfide bond. After photoincorporation of the peptide into the target, the disulfide is reduced, releasing the peptide but leaving the radiolabeled photoprobe covalently attached to the target. This particularly simplifies identification of the site of incorporation by peptide mapping and sequencing of the protease-digested target protein.

TABLE 11.2. Crosslinking Reagents Used To Identify Site of Contact of Synthetic Peptides and Their Protein Target

Functional Class	Name	Site of Incorporation	Radio-iodinatable
Nonspecific protein crosslinkers	Formaldehyde	$-NH_2$	No
	Disuccinimidylsuberate	$-NH_2$	No
Chemically incorporated photoaffinity labels	p-Nitrophenyl 3-diazopyruvate (DAPpNP) (Harrison et al., 1989)	$-NH_2$	No
	N-bromoacetyl-N-(3-diazopyruvoyl)-m-phenylene-diamine (Br-DAP) (Taylor et al., 1994a,b; Mosier and Lawton, 1995)	$-SH$	No
	N-bromoacetyl-N-(4-hydroxybenzyl)-N-(3-diazopyruvoyl)-1,3-phenylenediamine (pHBDAP) (Taylor et al., 1996; Mosier and Lawton, 1995)	$-SH$	Yes
	1-(p-Azidosalicylamido)-4-(iodoacetamido)butane (Fernandez et al., 1996)	$-SH$	Yes
	N-hydroxysuccinimide ester of 4-azidobenzoylglycine (NHS-ABG) (Kundu et al., 1996)	$-NH_2$	Yes
Label transfer method	N-[3-iodo-4-azidophenylpropionamido-S-(2-thiopyridyl)]cysteine (ACTP) (Liu et al., 1996)	$-SH$	Yes
Synthetically incorporated photoaffinity labels	Benzophenone (Artemyev et al., 1993; Herblin et al., 1987)		No
	p-(4-Hydroxybenzoyl)phenylalanine (Wilson et al., 1997)		Yes
	p-Nitrotyrosine (Ji et al., 1997)		No
	N-beta-4-azidosalicyloyl-L-2,3-diaminopropionic acid (Ji et al., 1997; Anjuere et al., 1995)		Yes

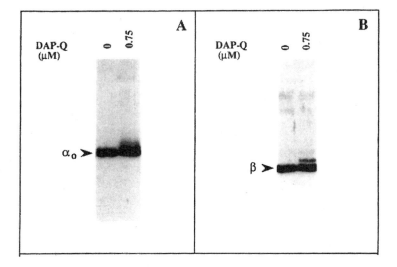

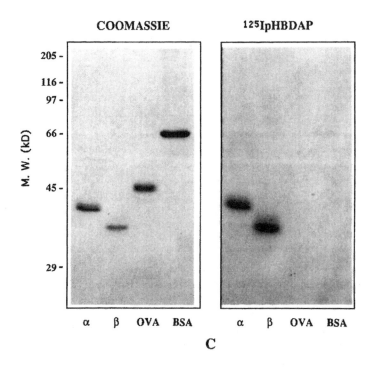

Figure 11.7. Crosslinking of DAP-Q to G protein. A diazopryuvolyl-conjugate of the i3c peptide from the α_{2a}-AR (DAP-Q) was photochemically crosslinked to G protein α or $\beta\gamma$ subunits as reported by Taylor et al., (1994a). The α- and β-subunit bands were shifted up by approximately 2 kD on the SDS gels **(A,B)** while the γ subunit was not affected (data not shown). The direct incorporation of a radioiodinated derivative of the i3c peptide (^{125}I-pHBDAP-Q) is shown in **C,** which also includes the negative controls bovine serum albumin (BSA) and ovalbumin (OVA), which did not incorporate the photoaffinity peptide.

7. INTERPRETATION OF DATA AND CAVEATS

The use of synthetic peptides to probe structural aspects of protein–protein interactions has enjoyed great success since its introduction by Hamm and colleagues (1988) 10 years ago. The concept is relatively simple, and the technical aspects of peptide synthesis have become routine. Thus, it provides a very easy way to begin to define the molecular nature of protein–protein contact sites, which has undoubtedly contributed to its popularity.

There are a number of aspects that should be considered when interpreting peptide inhibition data. First, it is important to distinguish between structural specificity and simple physicochemical properties such as charge or hydrophobicity as the determinants of peptide action. Including controls such as scrambled peptide sequences can help with this distinction. Testing the specificity of peptide effects on related but functionally different systems provides another means to help evaluate this issue. For example, a peptide preventing Gi coupling to receptors would be more likely to reveal structurally relevant information if it did not prevent coupling of Gs or Gq to their receptors. Another factor that is often overlooked in peptide studies is that at very high concentrations nonspecific effects are more likely to become a problem. It is desirable to identify peptides that exhibit their effects in the low micromolar range (1–30 μM) and to use extreme caution when interpreting results where peptide concentrations above 100 μM or in the mM range are used. Because the potential for nonspecific effects is always present, it is critical to use multiple approaches to identify functional regions of signal transducing proteins. The investigator should make full use of tools that have been developed for evaluating signaling mechanisms such as the ability of receptor function inhibitors (such as pertussis toxin) to modify actions of synthetic peptides. In addition, mutagenesis of the protein itself to confirm conclusions from peptide studies is important. Thus synthetic peptides provide one piece of the mosaic of information used to develop the full picture of complex signal transduction mechanisms.

Another factor that should be considered when evaluating peptide studies is the possible dependence on noncontiguous sequences. In the three-dimensional folding of a protein, many amino acids that are distant from one another in sequence are brought into contact. Thus protein interaction sites will include contributions from a number of distinct stretches of linear sequence. This is probably a major reason why peptides show relatively low affinities for their target. The synergistic effects of the rhodopsin intracellular loops on transducin (König et al., 1989) highlight this point. A similar observation has been made for the α_{2a}-AR in that dimeric peptides containing information from both the i3n and i3c regions are much more potent than the monomeric peptides in uncoupling this receptor from Gi or in activating purified Gi (Wade et al., 1994). This limitation to peptide methods is related to another point. The tight contacts between proteins are often dependent on the secondary structure of those proteins. In general, small peptides (e.g., 15 amino acids) exhibit little secondary structure. They may adopt a specific secondary structure upon interaction with a lipid bilayer (Wakamatsu et al., 1992) or with their target protein (Dratz et al., 1993; Sukumar et al., 1997; Kisselev et al., 1998), but this is not as engergetically favorable as in the case where they already have the appropriate secondary structure prior to binding. An exciting development in this area is the recent observation by Yeagle and colleagues (1995a,b, 1997b) that rhodopsin intracellular fragments do exhibit significant secondary structure in aqueous solution, and they even associate to generate a defined complex (Yeagle et al., 1997a). If this phenomenon is true for other signal transduction proteins, it should provide a powerful tool to enhance the utility of synthetic peptides.

8. POTENTIAL THERAPEUTIC UTILITY

The large family of GPCRs represents an important target for many drugs. Nearly all of those drugs are directed toward the ligand binding site either as agonists or antagonists. One of the exciting prospects for the use of synthetic peptides is as proof of principle and possible structural leads for targeting drugs to other steps in the signal transduction cascade. Specifically, the receptor–G protein interface, the G protein–effector interface, and possibly other regulatory molecules such as receptor kinases (Freedman and Lefkowitz, 1996) or the recently described members of the RGS family (regulators of G protein signaling) (Dohlman end Thorner, 1997). The fact that synthetic peptides can block some of these protein–protein interactions represents a small first step toward targeting these sites for therapeutic purposes. What is the rationale for such a therapeutic strategy, and what are its advantages and disadvantages? There are several reasons why one might want to target signaling at the receptor–G protein or G protein–effector interface. Many physiological processes are regulated by multiple converging receptor signals. One clear example is the regulation of vascular tone by the angiotensin AT1a, α_1-adrenergic, and endothelin receptors acting through a phospholipase C mediated (Gq/11 coupled) response. Physiologically, numerous mechanisms lead to increased blood pressure, and multiple receptors are activated. By blocking the response distal to the receptors (e.g., at the level of G protein) one could theoretically obtain a more complete inhibition of elevated blood pressure regardless of the specific physiological state that led to the increase. A second advantage of targeting the RG interface is that the response to constitutively activated receptors could be inhibited by this strategy. While the known pathophysiology of constitutively activated GPCRs is relatively limited (Lefkowitz, 1993; Shenker et al., 1993), it is likely that other examples will be identified, and drugs to block their signaling will be needed. The main disadvantage of targeting the RG interface is the potential of limited specificity for a Gq antagonist, for example. Because there are only 21 G protein α subunits, many of which are ubiquitously expressed, there could be substantial effects on signaling processes other than the intended target. This difficulty may be overcome by targeting specific G protein heterotrimers in which a particular β or γ subunit is expressed in a tissue-specific manner. Recent evidence indicates the importance of $\beta\gamma$ subunits to determine specificity in receptor–effector coupling (Gudermann et al., 1997). Because synthetic receptor peptides bind to both the α and β subunits (Taylor et al.,

1994a, 1996), the potential to enhance specificity by targeting unique heterotrimers should be examined.

In addition to their use to block receptor-mediated G protein activation, peptides have the potential to directly activate G proteins, thus becoming G protein agonists. Their use in this manner suffers from the same problems of specificity as when used as blockers of signal transduction. Furthermore, the structural features of peptides that most consistently convey the ability to activate G proteins are their positive charges (Higashijima et al., 1990; Wade et al., 1996). This points out one of the more difficult problems with the use of synthetic peptides as drugs. Peptides themselves are not going to be absorbed if given orally, and, in most cases, they do not enter cells when applied extracellularly. Thus, any significant utility of G proteins as therapeutic targets will require that small molecule drugs be developed against the structural targets initially defined by the synthetic peptide approach. The peptides themselves may serve as structural leads, but it is more likely that direct screening for small molecule compounds will be a more efficient strategy.

9. FUTURE PROSPECTS

Synthetic peptides will continue to represent an important tool in studies of protein–protein interactions, especially for complex systems such as receptors and G proteins. The challenges facing future work with synthetic peptides include (1) development of better strategies to identify specific versus nonspecific effects of peptides, (2) definition of the structure of peptides bound to their targets and development of structurally constrained peptides to enhance target affinity and specificity, and (3) development of peptidomimetic compounds that interact with peptide target sites yet have better properties for use as drugs, such as improved absorption and cell penetration.

The genetic delivery of peptides may provide an approach to address a number of these problems as larger protein fragments may be introduced into cells, improving affinity and specificity of the delivered reagent. In mechanistic studies, it is important to carefully evaluate potency and specificity of peptide effects and to check the hypotheses generated from such studies with other approaches such as mutagenesis of the proteins themselves.

ACKNOWLEDGEMENTS

The author acknowledges the outstanding efforts of all of the members of his laboratory, especially Hiroko Dalman, Joan Taylor, and Sue Wade, who contributed greatly to the development of the synthetic peptide studies. The enthusiasm and hard work of all of the members of the laboratory have made this effort fun. The unpublished work presented in Figures 11.1, 11.2, and 11.6 was done by Hiroko Dalman and Sue Wade. Drs. Phil Andrews and Ron Haaseth provided helpful advice about peptide synthesis and storage. The support of grant NIH HLGM 46417 is also gratefully acknowledged.

REFERENCES

Abell AN, Segaloff DL (1997): Evidence for the direct involvement of transmembrane region 6 of the lutropin/choriogonadotropin receptor in activating Gs. J Biol Chem 272:14586–14591.

Akhter SA, Luttrell LM, Rockman HA, Iaccarino G, Lefkowitz RJ, Koch WJ (1998): Targeting the receptor–Gq interface to inhibit *in vivo* pressure overload myocardial hypertrophy. Science 280:574–577.

Angeletti RH, Bonewald LF, Fields GB (1998): Six-year study of peptide synthesis. Methods Enzymol 289:697–717.

Anjuere F, Layer A, Cerottini JC, Servis C, Luescher IF (1995): Synthesis of a radioiodinated photoreactive MAGE-1 peptide derivative and photoaffinity labeling of cell-associated human leukocyte antigen-A1 molecules. Anal Biochem 229:61–67.

Arkinstall S, Chabert C, Maundrell K, Peitsch M (1995): Mapping regions of $G_{\alpha q}$ interacting with PLCβ1 using multiple overlapping synthetic peptides. FEBS Lett 364: 45–50.

Artemyev NO, Mills JS, Thornburg KR, Knapp DR, Schey KL, Hamm HE (1993): A site on transducin α-subunit of interaction with the polycationic region of cGMP phosphodiesterase inhibitory subunit. J Biol Chem 268:23611–23615.

Artemyev NO, Rarick HM, Mills JS, Skiba NP, Hamm HE (1992): Sites of interaction between rod G-protein α-subunit and cGMP-phosphodiesterase γ-subunit. Implications for the phosphodiesterase activation mechanism. J Biol Chem 267:25067–25072.

Blüml K, Schnepp W, Schroder S, Beyermann M, Macias M, Oschkinat H, Lohse MJ (1997): A small region in phosducin inhibits G-protein βγ-subunit function. EMBO J 16:4908–4915.

Chen J, DeVivo M, Dingus J, Harry A, Li J, Sui J, Carty DJ, Blank JL, Exton JH, Stoffel RH, Inglese J, Lefkowitz RJ, Logothetis DE, Hildebrandt JD, Iyengar R (1995): A region of adenylyl cyclase 2 critical for regulation by G protein βγ subunits. Science 268:1166–1169.

Dalman HM, Neubig RR (1991): Two peptides from the α_{2A}-adrenergic receptor alter receptor G protein coupling by distinct mechanisms. J Biol Chem 266:11025–11029.

De Lean A, Stadel JM, Lefkowitz RJ (1980): A ternary complex model explains the agonist-specific binding properties of the adenylate cyclase–coupled β-adrenergic receptor. J Biol Chem 255:7108–7117.

Dohlman HG, Thorner J (1997): RGS proteins and signaling by heterotrimeric G proteins. J Biol Chem 272:3871–3874.

Dratz EA, Furstenau JE, Lambert CG, Thireault DL, Rarick H, Schepers T, Pakhlevaniants S, Hamm HE (1993): NMR structure of a receptor-bound G-protein peptide. Nature 363:276–281.

Eccleston JF, Moore KJM, Brownbridge GG, Webb MR, Lowe PN (1991): Fluorescence approaches to the study of the p21ras GTPase mechanism. Biochem Soc Trans 19:432–436.

Ellman GL (1959): Tissue sulfhydryl groups. Arch Biochem Biophys 82:70–77.

Fernandez AM, Molina A, Encinar JA, Gavilanes F, Lopez-Barneo J, Gonzalez-Ros JM (1996): Synthesis of a photoaffinity labeling analogue of the inactivating peptide of the Shaker B potassium channel. FEBS Lett 398:81–86.

Freedman NJ, Lefkowitz RJ (1996): Desensitization of G protein–coupled receptors. Recent Prog Hormone Res 51:319–351.

Gierschik P, Bouillon T, Jakobs K-H (1994): Receptor-stimulated hydrolysis of guanosine 5′-triphosphate in membrane preparations. Methods Enzymol 237:13–26.

Guan KL, Dixon JE (1991): Eukaryotic proteins expressed in *Escherichia coli:* An improved thrombin cleavage and purification procedure of fusion proteins with glutathione S-transferase. Anal Biochem 192:262–267.

Gudermann T, Kalkbrenner F, Dippel E, Laugwitz KL, Schultz G (1997): Specificity and complexity of receptor–G-protein interaction. Adv Second Messenger Phosphoprot Res 31:253–262.

Hamm HE (1991): Molecular interactions between the photoreceptor G protein and rhodopsin. Cell Mol Neurobiol 11:563–578.

Hamm HE, Deretic D, Arendt A, Hargrave PA, Koenig B, Hofmann KP (1988): Site of G protein binding to rhodopsin mapped with synthetic peptides from the α subunit. Science 241:832–835.

Hargrave PA, Hamm HE, Hofmann KP (1993): Interaction of rhodopsin with the G-protein, transducin. Bioessays 15:43–50.

Harrison JK, Lawton RG, Gnegy ME (1989): Development of a novel photoreactive calmodulin derivative: Cross-linking of purified adenylate cyclase from bovine brain. Biochemistry 28:6023–6027.

Hartman IV JL, Northup JK (1996): Functional reconstitution *in situ* of 5-hydroxytryptamine2c (5HT2c) receptors with αq and inverse agonism of 5HT2c receptor antagonists. J Biol Chem 271:22591–22597.

Hawes BE, Luttrell LM, Exum ST, Lefkowitz RJ (1994): Inhibition of G protein–coupled receptor signalling by expression of cytoplasmic domains of the receptor. J Biol Chem 269:15776–15785.

Herblin WF, Kauer JC, Tam SW (1987): Photoinactivation of the μ opioid receptor using a novel synthetic morphiceptin analog. Eur J Pharmacol 139:273–279.

Herlitze S, Hockerman GH, Scheuer T, Catterall WA (1997): Molecular determinants of inactivation and G protein modulation in the intracellular loop connecting domains I and II of the calcium channel α1A subunit. Proc Natl Acad Sci USA 94: 1512–1516.

Higashijima T, Burnier J, Ross EM (1990): Regulation of Gi and Go by mastoparan, related amphiphilic peptides, and hydrophobic amines. Mechanism and structural determinants of activity. J Biol Chem 265:14176–14186.

Higashijima T, Uzu S, Nakajima T, Ross EM (1988): Mastoparan, a peptide toxin from wasp venom, mimics receptors by activating GTP-binding regulatory proteins (G proteins). J Biol Chem 263:6491–6494.

Hillaire-Buys D, Mousli M, Landry Y, Bockaert J, Fehrentsz JA, Carrette J, Rouot B (1992): Insulin releasing effects of mastoparan and amphiphilic substance P receptor antagonists on RINm5F insulinoma cells. Mol Cell Biochem 109:133–138.

Ikezu T, Okamoto T, Ogata E, Nishimoto I (1992): Amino acids 356–372 constitute a G_i-activator sequence of the α_2-adrenergic receptor and have a Phe substitute in the G protein–activator sequence motif. FEBS Lett 311:29–32.

Ji Z, Hadac EM, Henne RM, Patel SA, Lybrand TP, Miller LJ (1997): Direct identification of a distinct site of interaction between the carboxyl-terminal residue of cholecystokinin and the type A cholecystokinin receptor using photoaffinity labeling. J Biol Chem 272:24393–24401.

Jones TLZ, Simonds WF, Merendino JJ Jr, Brann MR, Spiegel AM (1990): Myristoylation of an inhibitory GTP-binding protein α subunit is essential for its membrane attatchment. Proc Natl Acad Sci USA 87:568–572.

Kang J, Richards EM, Posner P, Sumners C (1995): Modulation of the delayed rectifier K^+ current in neurons by an angiotensin II type 2 receptor fragment. Am J Physiol 268:C278–C282.

Kempe T, Kent SB, Chow F, Peterson SM, Sundquist WI, L'Italien JJ, Harbrecht D, Plunkett D, DeLorbe WJ (1985): Multiple-copy genes: Production and modification of monomeric peptides from large multimeric fusion proteins. Gene 39:239–245.

Kim MH, Neubig RR (1985): Parallel inactivation of α_2-adrenergic agonist binding and N_i by alkaline treatment. FEBS Lett 192:321–325.

Kim MH, Neubig RR (1987): Membrane reconstitution of high-affinity α2 adrenergic agonist binding with guanine nucleotide regulatory proteins. Biochemistry 26:3664–3672.

Kisselev OG, Ermolaeva MV, Gautam N (1994): A farnesylated domain in the G protein γ subunit is a specific determinant of receptor coupling. J Biol Chem 269:21399–21402.

Kisselev OG, Kao J, Ponder JW, Fann YC, Gautam N, Marshall GR (1998): Light-activated rhodopsin induces structural binding motif in G protein α subunit. Proc Natl Acad Sci USA 95:4270–4275.

König B, Arendt A, McDowel JH, Kahlert M, Hargrave PA, Hofmann KP (1989): Three cytoplasmic loops of rhodopsin interact with transducin. Proc Natl Acad Sci USA 86:6878–6882.

Kundu GC, Ji I, McCormick DJ, Ji TH (1996): Photoaffinity labeling of the lutropin receptor with synthetic peptide for carboxyl terminus of the human choriogonadotropin α subunit. J Biol Chem 271:11063–11066.

Lan K-L, Remmers AE, Neubig RR (1997): Role of Goα tryptophans in GTP hydrolysis, GDP release, and fluorescence signals. Biochemistry 37:837–843.

Lefkowitz RJ (1993): G protein–coupled receptors: Turned on to ill effect. Nature 365:603–604.

Limbird LE, Speck JL (1983): N-ethylmaleimide, elevated temperature, and digitonin solubilization eliminate guanine nucleotide but not sodium effects on human platelet α_2-adrenergic receptor-agonist interactions. J Cyclic Nucleotide Protein Phosphor Res 9:191–201.

Liu Y, Arshavsky VY, Ruoho AE (1996): Interaction sites of the COOH-terminal region of the γ subunit of cGMP phosphodiesterase with the GTP-bound α subunit of transducin. J Biol Chem 271:26900–26907.

Luttrell LM, Ostrowski J, Cotecchia S, Kendall H, Lefkowitz RJ (1993): Antagonism of catecholamine receptor signaling by expression of cytoplasmic domains of the receptors. Science 259:1453–1457.

Maina CV, Riggs PD, Grandea AG, Slatko BE, Moran LS, Tagliamonte JA, McReynolds LA, Guan CD (1988): An *Escherichia coli* vector to express and purify foreign proteins by fusion to and separation from maltose-binding protein. Gene 74:365–373.

Martin EL, Rens-Domiano S, Schatz PJ, Hamm HE (1996): Potent peptide analogues of a G protein receptor–binding region obtained with a combinatorial library. J Biol Chem 271:361–366.

Mosier GGJ, Lawton RG (1995): Development of a new family of biospecific photoactivatable cross-linking agents. J Org Chem 60:6953–6958.

Muir TW, Dawson PE, Kent SBH (1998): Protein synthesis by chemical ligation of unprotected peptides in aqueous solution. Methods Enzymol 289:266–298.

Mumby SM, Heukeroth RO, Gordon JI, Gilman AG (1990): G-protein α-subunit expression, myristoylation, and membrane association in COS cells. Proc Natl Acad Sci USA 87:728–732.

Munch G, Dees C, Hekman M, Palm D (1991): Multisite contacts involved in coupling of the β-adrenergic receptor with the stimulatory guanine-nucleotide binding regulatory protein. Structural and functional studies by β-receptor-site–specific peptides. Eur J Biochem 198:357–364.

Naim M, Seifert R, Nürnberg B, Grünbaum L, Schultz G (1994): Some taste substances are direct activators of G-proteins. Biochem J 297:451–454.

Neubig RR (1998): Specificity of receptor G protein coupling: Protein structure and cellular determinants. Semin Neurosci 9:189–197.

Neubig RR, Dalman HM (1991): Effect of α_{2a}-adrenergic receptor peptides on agonist binding to α_{2b}-adrenergic, muscarinic (M4) and opiate (δ) receptors in NG108-15 membranes (abstracted). FASEB J 5:A1594.

Noel JP, Hamm HE, Sigler PB (1993): The 2.2 Å crystal structure of transducin-α complexed with GTPγS. Nature 366:654–663.

Okajima F, Katada T, Ui M (1985): Coupling of the guanine nucleotide regulatory protein to chemotactic peptide receptors in neutrophil membranes and its uncoupling by islet-activating protein, pertussis toxin. A possible role of the toxin substrate in Ca^{2+}-mobilizing receptor-mediated signal transduction. J Biol Chem 260:6761–6768.

Okamoto H, Iwamoto H, Tsuzuki H, Teraoka H, Yoshida N (1995): An improved method for large-scale purification of recombinant human glucagon. J Prot Chem 14:521–526.

Palm D, Munch G, Malek D, Dees C, Hekman M (1990): Identification of a G_s-protein coupling domain to the β-adrenoceptor using site-specific synthetic peptides. Carboxyl terminus of $G_{s\alpha}$ is involved in coupling to β-adrenoceptors. FEBS Lett 261:294–298.

Rarick HM, Artemyev NO, Hamm HE (1992): A site on rod G protein α subunit that mediates effector activation. Science 256:1031–1033.

Rasenick MM, Watanabe M, Lazarevic MB, Hatta S, Hamm HE (1994): Synthetic peptides as probes for G protein function. Carboxyl-terminal $G_{\alpha s}$ peptides mimic G_s and evoke high affinity agonist binding to β-adrenergic receptors. J Biol Chem 269:21519–21525.

Ray K, Kunsch C, Bonner LM, Robishaw JD (1995): Isolation of cDNA clones encoding eight different human G protein γ subunits, including three novel forms designated the γ4, γ10, and γ11 subunits. J Biol Chem 270:21765–21771.

Remmers AE (1998): Detection and quantitation of heterotrimeric G proteins by fluorescence resonance energy transfer. Anal Biochem 257:89–94.

Remmers AE, Neubig RR (1996): Partial G protein activation by fluorescent guanine nucleotide analogs—Evidence for a triphosphate-bound but inactive state. J Biol Chem 271:4791–4797.

Remmers AE, Posner R, Neubig RR (1994): Fluorescent guanine nucleotide analogs and G protein activation. J Biol Chem 269:13771–13778.

Sankaran B, Osterhout J, Wu D, Smrcka AV (1998): Identification of a structural element in phospholipase C β2 that interacts with G protein βγ subunits. J Biol Chem 273:7148–7154.

Shenker A, Laue L, Kosugi S, Merendino JJ, Jr, Minegishi T, Cutler GB Jr (1993): A constitutively activating mutation of the luteinizing hormone receptor in familial male precocious puberty. Nature 365:652–654.

Shinagawa K, Ohya M, Higashijima T, Wakamatsu K (1994): Circular dichroism studies of the interaction between synthetic peptides corresponding to intracellular loops of β adrenergic receptors and phospholipid vesicles. J Biochem (Tokyo) 115:463–468.

Simon MI, Strathmann MP, Gautam N (1991): Diversity of G proteins in signal transduction. Science 252:802–808.

Sprang SR (1997): G protein mechanisms: Insights from structural analysis. Annu Rev Biochem 66:639–678.

Stadel JM, Wilson S, Bergsma DJ (1997): Orphan G protein–coupled receptors: A neglected opportunity for pioneer drug discovery. Trends Pharmacol Sci 18:430–437.

Sukumar M, Ross EM, Higashijima T (1997): A Gs-selective analog of the receptor-mimetic peptide mastoparan binds to $G_s\alpha$ in a kinked helical conformation. Biochemistry 36:3632–3639.

Taylor JM (1995): Molecular Interactions of Receptor-Derived Peptides With G Proteins. Ph.D. Thesis, University of Michigan.

Taylor JM, Jacob-Mosier GG, Lawton RG, Remmers AE, Neubig RR (1994a): Binding of an α_2 adrenergic receptor third intracellular loop peptide to Gβ and the amino terminus of Gα. J Biol Chem 269:27618–27624.

Taylor JM, Jacob-Mosier GG, Lawton RG, Neubig RR (1994b): Coupling an α_2-adrenergic receptor peptide to G-protein: A new photolabeling agent. Peptides 15:829–834.

Taylor JM, Jacob-Mosier GG, Lawton RG, VanDort M, Neubig RR (1996): Receptor and membrane interaction sites on Gβ—A receptor-derived peptide binds to the carboxyl terminus. J Biol Chem 271:3336–3339.

Taylor JM, Neubig RR (1994): Peptides as probes for G protein signal transduction. Cell Signal 6:841–849.

Tesmer JJG, Sunahara RK, Gilman AG, Sprang SR (1997): Crystal structure of the catalytic domains of adenylyl cyclase in a complex with Gsα·GTPγS. Science 278:1907–1916.

Thompson JB, Wade SM, Harrison JK, Salafranca MN, Neubig RR (1998): Co-transfection of second and third intracellular loop fragments inhibits angiotensin AT1a receptor activation of phospholipase C in HEK-293 cells. J Pharmacol Exp Ther 285:216–222.

Unger VM, Hargrave PA, Baldwin JM, Schertler GFX (1997): Arrangement of rhodopsin transmembrane α-helices. Nature 389:203–206.

Voss T, Wallner E, Czernilofsky AP, Freissmuth M (1993): Amphipathic α-helical structure does not predict the ability of receptor-derived synthetic peptides to interact with guanine nucleotide-binding regulatory proteins. J Biol Chem 268:4637–4642.

Wade SM, Dalman HM, Yang S-Z, Neubig RR (1994): Multisite interactions of receptors and G proteins. Enhanced potency of dimeric receptor peptides in modifying G protein function. Mol Pharmacol 45:1191–1197.

Wade SM, Scribner MK, Dalman HM, Taylor JM, Neubig RR (1996): Structural requirements for G_0 activation by receptor-derived peptides: activation and modulation domains of the α_2 adrenergic receptor i3c region. Mol Pharmacol 50:351–358.

Wakamatsu K, Okada A, Miyazawa T, Ohya M, Higashijima T (1992): Membrane-bound conformation of mastoparan-X, a G-protein–activating peptide. Biochemistry 31:5654–5660.

Wakamatsu K, Shinagawa K, Tanaka T, Oya M, Sukumar M, Higashijima T (1993): Interaction of cytoplasmic loop peptides of G protein–coupled receptors with G proteins and with phospholipid membranes (abstracted). J Cell Biochem 17C:286.

Wilson CJ, Husain SS, Stimson ER, Dangott LJ, Miller KW, Maggio JE (1997): p-(4-Hydroxybenzoyl)phenylalanine: A photoreactive amino acid analog amenable to radioiodination for elucidation of peptide–protein interaction. Application to substance P receptor. Biochemistry 36:4542–4551.

Yeagle PL, Alderfer JL, Albert AD (1997a): Three-dimensional structure of the cytoplasmic face of the G protein receptor rhodopsin. Biochemistry 36:9649–9654.

Yeagle PL, Alderfer JL, Albert AD (1995a): Structure of the carboxy-terminal domain of bovine rhodopsin. Nature Struct Biol 2:832–834.

Yeagle PL, Alderfer JL, Albert AD (1995b): Structure of the third cytoplasmic loop of bovine rhodopsin. Biochemistry 34:14621–14625.

Yeagle PL, Alderfer JL, Salloum AC, Ali L, Albert AD (1997b): The first and second cytoplasmic loops of the G-protein receptor, rhodopsin, independently form β-turns. Biochemistry 36:3864–3869.

Zhu M, Neubig RR, Wade SM, Posner P, Gelband CH, Sumners C (1997): Modulation of K^+ and $Ca2^+$ currents in cultured neurons by an angiotensin II type 1a receptor peptide. Am J Physiol 273:C1040–C1048.

CHAPTER 12

ELECTRON-CRYSTALLOGRAPHIC ANALYSIS OF TWO-DIMENSIONAL RHODOPSIN CRYSTALS

GEBHARD F.X. SCHERTLER

I.	INTRODUCTION	235
	A. Rhodopsin, a Representative G Protein–Coupled Receptor	235
	B. Comparison of Electron Crystallography With X-Ray Crystallography	236
	a. Single Particles	238
	b. Helical Assemblies of Proteins	238
	c. Two-Dimensional Crystals	238
	C. Image Processing	238
2.	PREPARATION OF TWO-DIMENSIONAL CRYSTALS	239
	A. Crystallization of Membrane Proteins in the Cell Membrane	239
	B. Purification of Rhodopsin	239
	C. Reconstitution Experiments	240
	D. Crystal Induction by Selective Extraction of Membranes	241
	E. Screening for Two-Dimensional Crystals	243
	a. Gels, Density Gradients, Spectra, and Assays for Function	243
	b. Electron Microscopy	243
	c. Taking Overview Pictures	243
	d. Imaging Single Crystals	245
	e. Optical Diffraction	245
3.	DATA COLLECTION WITH LOW-DOSE ELECTRON CRYOMICROSCOPY	245
	A. Vitrification of Specimens for Cryomicroscopy	246
	B. The Electron Microscope	246
	C. Low-Dose Imaging	247
	D. Spot Scan for Imaging Tilted Specimens	247
	E. Contrast Transfer Function	248
	F. Data Evaluation With Optical Diffraction	248

Structure–Function Analysis of G Protein-Coupled Receptors, Edited by Jürgen Wess.
ISBN 0-471-25228-X   Copyright © 1999 Wiley-Liss, Inc.

G. Preliminary Characterization of a Two-Dimensional Crystal   248
   a. Finding Elements of Crystal Symmetry: Unit Cells,
      Rotation Axes, and Screw Axes   248
   b. Finding Crystal Symmetry Elements in Unsymmetrized
      Maps of Negatively Stained and Unstained Specimens   250
   c. Determination of Resolution by Optical Diffraction   250
H. Properties of the Fourier Transform of a Two-Dimensional
   Crystal   250

4. IMAGE PROCESSING   254
  A. Extracting Amplitudes and Phases From Low-Dose Images   254
   a. Scanning Micrographs With a Precision Scanner   254
   b. Unbending Crystal Lattices With a Small Reference Area   256
   c. Unbending Procedure   257
   d. Evaluating Cross-Correlation Maps   257
   e. Correcting for Distortions in the Image   257
   f. Selection of a Coherent Crystalline Area and Extraction
     of Amplitudes and Phases   258
   g. Evaluation of Image Quality After Unbending   258
  B. Determination of Astigmatism and CTF   258
  C. Determination of the Plane Group   260
   a. Finding Crystal Symmetry Using Amplitudes   260
   b. Finding Crystal Symmetry Using Phase Relations   261
   c. Confirmation of Crystal Symmetry Using p1 Maps   261

5. MERGING AMPLITUDES AND PHASES FROM UNTITLED
  IMAGES FOR A PROJECTION MAP   261
  A. Merging Data From Different Images   261
  B. Refining the CTF Correction for Each Image
   Using a Merged Set of Images   263
  C. Scaling of Amplitudes Using Diffraction Data   263
  D. Evaluating the Resolution Limit of the Merged
   Projection Data   264
  E. Calculating a Projection Map From a List of Amplitudes
   and Phases   266
  F. Interpretation of Projection Structures   267

6. MERGING AMPLITUDES AND PHASES FROM TILTED
  IMAGES TO CALCULATE A THREE-DIMENSIONAL MAP   267
  A. Merging Data From Tilted Images   267
  B. Fitting Lattice Lines   269
  C. Evaluation of Resolution and Data Anisotropy   269
  D. Calculating a Three-Dimensional Map   269
  E. Representing the Map Using Contour Plots and
   Surface Representations   271

7. INTERPRETATION OF DENSITY MAPS   273
  A. The Arrangement of Rhodopsin α Helices in the Membrane   273
  B. Comparison of Rhodopsin With Bacteriorhodopsin   274
  C. The Retinal Binding Site in Rhodopsin   276
  D. The Intracellular G Protein Binding Region   277

8. ASSIGNING THE AMINO ACID SEQUENCE TO THE DENSITY
   FEATURES IN THE MAP   277
   A. Helical Wheel Plots   277
   B. Structural Constraints From Sequence Comparison   278
   C. Assignment of Amino Acid Sequence to Low-Resolution Maps   279
   D. Prediction of the Position of α Carbon Atoms   280
9. FUTURE PROSPECTS FOR ELECTRON CRYSTALLOGRAPHY
   OF RHODOPSIN   280
   A. A New Crystal Form Providing Higher Resolution Data 280
   B. Electron Diffraction   280
   C. Determination of the Structure of Other GPCRS
      and Membrane Proteins   283
10. CONCLUSIONS   283

I. INTRODUCTION

I.A. Rhodopsin, a Representative G Protein–Coupled Receptor

G protein–coupled receptors (GPCRs) enable the primary reactions by which cells sense alterations in their external environment and convey that information to the interior of the cell. To date more than 800 GPCRs have been cloned from a variety of species, ranging from fungi to human. They all share some degree of sequence similarity and are likely to have a similar arrangement of their seven transmembrane helices (Baldwin, 1993; Baldwin et al., 1997). In the last 20 years more than 100 new drugs have been registered that "activate" or "inhibit" GPCRs (Stadel et al., 1997).

Rhodopsin, the dim-light photoreceptor molecule, is one of the most intensively studied GPCRs (Hargrave and McDowell, 1992). Some unique properties make it a very good choice for structural investigations. The covalently bound 11-*cis*-retinal is an ideal reporter located in the center of the molecule that gives rhodopsin an absorption spectrum that is characteristic for its conformation. GPCRs are normally present only in small quantities on the cell surface. In contrast to this, rhodopsin is present at very high levels in the rod outer segment of the photoreceptor cell. The 11-*cis*-retinal is covalently bound to lysine 296 via a protonated Schiff base, and it acts like a covalently bound antagonist, keeping the photoreceptor in a nonsignaling conformation. In addition, rhodopsin has a rigid extracellular domain that might also help to reduce spontaneous activation of the receptor in the absence of light (Khorana, 1992; Unger et al., 1997a). Rhodopsin is one of the most stable and detergent-tolerant GPCRs known, and it can be isolated from retinas in large quantities, making it an ideal candidate for structural investigations.

In the following sections electron crystallography is compared with other methods that could potentially be used for the structure determination of rhodopsin. Why two-dimensional crystals are currently the best specimens for structure determination by electron crystallography is also discussed, and experimental steps that are required to calculate a three-dimensional map with the help of image processing are shown (Fig. 12.1). Detailed accounts of crystallization, data collection with the electron microscope, processing of images, merging data of untilted and tilted images, map calculation, and the presentation and interpretation of the resulting density maps are given.

I.B. Comparison of Electron Crystallography With X-Ray Crystallography

Nuclear magnetic resonance (NMR), x-ray crystallography, or electron microscopy can be used to determine the structure of proteins. However, so far no structure of any protein with a molecular weight of more than 35 kD has been solved by solution NMR, although new methods will probably extend the range of this method. Solid-state NMR may have some prospect of delineating structures in the future, but so far it has only been used for single distance measurements in membrane proteins.

In contrast, several structures of membrane proteins have been solved to atomic resolution with x-ray crystallography (photosynthetic reaction center [Deisenhofer et al., 1985]) porin [Weiss et al., 1991], cytochrome $bc1$ complex [Xia et al., 1997], and cytochrome-C oxidase [Ostermeier et al., 1995; Tsukihara et al., 1996]). Well-ordered three-dimensional crystals are a prerequisite, but are difficult to generate from membrane proteins. Because only the amplitudes are recorded in an x-ray diffraction experiment, the phases have to be determined indirectly by using heavy atom derivatives, molecular replacement, or possibly direct methods in the future.

Electron crystallography has allowed the structures of several important membrane proteins to be solved to a comparable resolution (bacteriorhodopsin [Henderson et al., 1990; Grigorieff et al., 1996], light harvesting complex II [Kuhlbrandt et al., 1994]). In addition, quite a number of low-resolution structures of membrane proteins have been obtained using electron microscopy (rhodopsin [Unger et al., 1997a], halorhodopsin [Havelka et al., 1995], aquaporin [Walz et al., 1997], gap junction membrane channel [Unger et al., 1997b], sarcoplasmic reticulum Ca^{2+} ATPase [Zhang et al., 1998], and plasma membrane H^+ ATPase [Auer et al., 1998]). Two-dimensional crystals are necessary, and these are often easier to obtain than three-dimensional ones. A major advantage of the method is that both amplitudes and phases can be obtained from images of the same crystal directly.

Electron microscopy and x-ray diffraction differ in that electrons interact more strongly with matter than do x-rays. Therefore, much smaller and thinner samples can be studied (e.g., large protein complexes or a helical assembly of proteins and two-dimensional crystals consisting of a monomolecular layer of protein molecules). Unfortunately, this strong interaction of electrons also causes damage to biological specimens, and therefore cryotechniques and low-dose strategies that reduce beam damage are essential for imaging proteins (Henderson, 1995). However, the low dose of electrons that can be used for proteins without damaging the specimen limits the amount of information obtained from a single molecule. If the protein is smaller than 200 kD, not enough information can be recorded to find the orientation of the particle, but this is essential to average it with other particle images. Therefore, a larger assembly, for

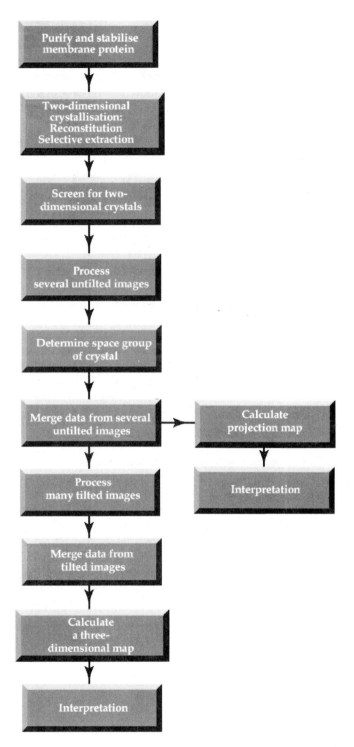

Figure 12.1. Flow chart for electron crystallography of two-dimensional crystals. All the different steps necessary to obtain a structure by a combination of electron microscopy, image processing, and electron crystallography are shown.

example, a two-dimensional crystal, is necessary for smaller proteins like rhodopsin. With the electron microscope we can record not only diffraction patterns and measure amplitudes as with x-rays but also images from which we can extract phases as well as amplitudes by Fourier analysis. Furthermore, computer processing of images allows us to remove some of the disorder present in the crystal lattice (Chiu, 1993; Henderson et al., 1986). In this way, we can obtain important structural information even from nonideal two-dimensional crystals by using electron microscopy and image processing.

I.B.a. Single Particles. Electron microscopy of single particles in combination with image processing is able to give structural information about protein complexes. After alignment, images from thousands of particles are averaged, and a three-dimensional reconstruction of the protein structure is calculated (Frank, 1996). However, images of smaller particles do not contain enough information to determine the orientation of the particle. Consequently, the molecular weight of rhodopsin and other GPCRs is too low for this approach. Therefore, we need a large, regular, and rigid assembly to obtain structural information for a protein of 30–200 kD with electron microscopy.

I.B.b. Helical Assemblies of Proteins. Helical arrangements have been used successfully to obtain the structure of the nicotinic acetylcholine receptor (Toyoshima and Unwin, 1990) and the Ca-ATPase of the sarcoplasmic reticulum (Toyoshima et al., 1993; Zhang et al., 1998). Narrow tubular rhodopsin crystals have been obtained, and therefore helical reconstruction could be used to obtain a map of rhodopsin. However, the observed helical diffraction did not extend to high resolution (Schertler, unpublished observation).

I.B.c. Two-Dimensional Crystals. Quite a few structures of membrane proteins were obtained from two-dimensional crystals in which the membrane protein is arranged in a regular array that can be described by two lattice vectors, **a** and **b** (Henderson et al., 1990; Kuhlbrandt et al., 1994; Unger et al., 1997a; Walz et al., 1997). These two-dimensional crystals are held together by protein–protein, lipid–lipid and lipid–protein interactions (Engel et al., 1992; Kühlbrandt, 1992). Two-dimensional crystals provide information about thousands of particles in one single image and therefore allow to extract structural information despite a noisy background. The structures of bovine, squid, and frog rhodopsin have been determined to low resolution by electron crystallography (Davies et al., 1996; Krebs et al., 1998; Schertler and Hargrave, 1995; Schertler et al., 1993; Unger et al., 1997a; Unger and Schertler, 1995).

I.C. Image Processing

The low dose of electrons used to take electron micrographs of biological samples means that an image taken from a single molecule is noisy. Only an average of many well-aligned images can give us structural detail. Therefore, the aim of image processing is to achieve the best possible alignment before averaging. In this way, image processing can improve the signal to noise ratio (Crowther et al., 1996; Frank, 1996; Henderson et al., 1986).

In an image of a nonideal two-dimensional crystal there are several sources of distortions. One is the disorder in the crystal that can be caused by the supporting carbon film and by the mechanical stress induced by shrinkage during freezing, or it reflects the disorder present in the specimens. The second source of distortion is the imaging system of the electron microscope itself. With image processing, all types of distortions can be corrected for. A small reference area of the best part of the imaged crystal is used to find distortion vectors for every part of the crystal, which are then used to correct the image. All unit cells are more accurately aligned, and an improved signal to noise ratio can be obtained. Therefore, higher resolution amplitudes and phases can be extracted by Fourier analysis from an image of nonideal two-dimensional crystals after unbending the lattice.

2. PREPARATION OF TWO-DIMENSIONAL CRYSTALS

Two-dimensional crystallization has been described in several reviews in detail (Engel et al., 1992; Jap et al., 1992; Kuhlbrandt, 1992; Rigaud et al., 1997; Walz and Grigorieff, 1998). In this chapter, a brief description of two-dimensional crystallization procedures is given, with the emphasis on extraction methods. Subsequently, how to search for a two-dimensional crystal with the electron microscope and how a first characterization of a crystal is carried out are described in detail.

2.A. Crystallization of Membrane Proteins in the Cell Membrane

In some rare cases, membrane proteins form crystals in the cell membrane. The purple membranes of *Halobacterium* are examples of naturally occurring two-dimensional crystals of bacteriorhodopsin, whereas halorhodopsin (Havelka et al., 1993; Heymann et al., 1993) and gap junction channels (Yeager, 1994) form crystalline arrays in the cell when they are overexpressed to a high level. Because the density of the membrane is increased by the high concentration of the protein in a two-dimensional crystal, density gradients can be used to obtain a purified crystalline sample. However, most membrane proteins do not form crystals in the membrane spontaneously. Therefore, the membrane is solubilized in a buffer, containing detergent, and the protein is purified by chromatography or density centrifugation in the presence of detergent.

2.B. Purification of Rhodopsin

Rhodopsin is a very abundant protein in the rod outer segment of the photoreceptor cell. After isolating the retinas in dim red light, the rod outer segments are separated from the rest of the retina by vigorous shaking or homogenization and floatation on a sucrose cushion. The rod outer segments are then solubilized in a buffer containing detergent and applied to a lectin affinity column (DeGrip, 1982). After elution from a concanavalin A matrix, the protein is 95% pure and can be used directly for reconstitution experiments, or it can be subjected to further purification by ion exchange chromatography.

2.C. Reconstitution Experiments

We have adapted several dialysis systems for two-dimensional crystallization. For large- scale (100–1,000 μl) crystallization of rhodopsin, we use Slide-A-Lyzers (Pierce). For small-scale crystallization (10–50 μl), we use the x-ray crystallographer's dialysis buttons with two additional holes to fill the pre-assembled button.

The aim of a reconstitution experiment is to form a large single membrane with as high a protein concentration as is feasible to induce crystallization (Engel et al., 1992; Jap et al., 1992; Kühlbrandt, 1992; Rigaud et al., 1995, 1997). Electron diffraction experiments need coherent crystalline areas larger than 1 μm in diameter, which are sometimes prepared by fusion of smaller crystals in an additional step (Baldwin and Henderson, 1984). Smaller membranes can only be studied by imaging.

Most membrane protein purification procedures are only partially delipidating. Affinity chromatography or density gradients in particular do not efficiently remove lipid. Therefore, membranes and two-dimensional crystals can sometimes be obtained by removing the detergent without addition of lipid. However, after several chromatography steps, often less than five lipids per molecule are associated with the protein and therefore additional lipid has to be added before the detergent is removed by dialysis (Engel et al., 1992) or by adsorption to Bio-Beads (Rigaud et al., 1997).

Reconstitution as well as crystallization are influenced by many parameters (Rigaud et al., 1995; Kühlbrandt, 1992). The nature of the remaining lipid is very important in most two-dimensional crystallizations, and the final mixture of endogenous lipid and lipid added prior to reconstitution is often difficult to control. This might explain some of the variability observed in reconstitution experiments. The nature, the stability, and the purity of the protein, as well as the stabilizing agents, such as cholesterol, glycerol, and sugars, are crucial. Furthermore, the nature and concentration of the detergent strongly influences the kinetics of the detergent removal process, which leads to the reconstitution of the protein into a lipid detergent phase. Temperature affects the state of the lipids as well as the stability and the mobility of the protein. Salt concentration and pH also affect the aggregation state and the stability of the proteins. Divalent cations such as Mg^{2+}, Mn^{2+}, and Ca^{2+} interact with lipid head groups, and high concentrations can induce oligomerisation of the protein and membrane fusion. Many of these parameters have to be optimized in an empirical way to find a suitable crystallization protocol.

The most important parameter to optimize is the protein to lipid ratio. In the purple membrane, the two-dimensional crystal of bacteriorhodopsin, there are 10 lipids per bacteriorhodopsin molecule. We therefore need to devise a method that allows the protein to be incorporated in a minimal amount of lipid. A series of conditions with decreasing lipid concentrations should be tested. Conditions that use the smallest amount of lipids and still give extended membrane sheets are most likely to succeed. The membranes can be observed either with optical phase contrast microscopy or with electron microscopy at low magnification (Figs. 12.2 and 12.3A,B) (Krebs et al., 1998; Schertler and Hargrave, 1995).

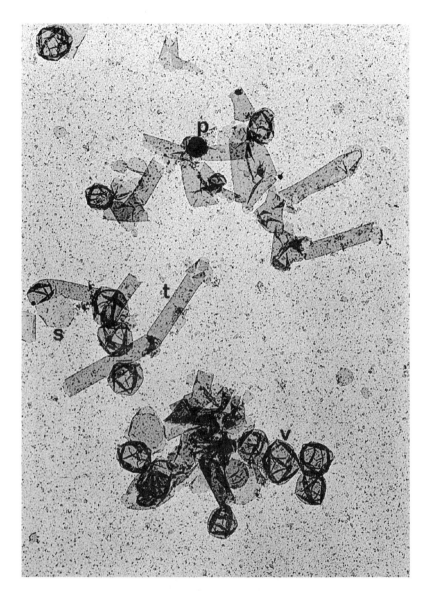

Figure 12.2. Electron micrograph overview. A picture of a sample obtained from a dialysis reconstitution experiment was taken at low magnification and with strong defocus (300 μm) to obtain high contrast. Single layer membranes (s), tubular membranes (t), vesicles (v), and protein aggregates (p) are visible. The single layers and the tubes are two-dimensional crystals of bovine rhodopsin (Krebs et al., 1998).

2.D. Crystal Induction by Selective Extraction of Membranes

From the rod outer segments in the retina we can isolate the disc membranes using Ficoll density floatation. These photoreceptor membranes are unique in that they contain a very high concentration of rhodopsin and about 70 lipids per protein (Litman and Mitchell, 1996). However, the protein is still highly mobile in

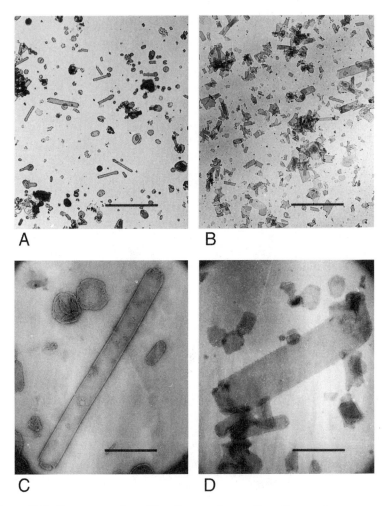

Figure 12.3. Tween extraction of frog disc membranes. Frog disc membranes were extracted with either Tween 80 **(A, C)** or a mixture of Tween 80 and Tween 20 **(B, D)** at a molar ratio of 1 to 1,000. The membrane suspensions were put on carbon-coated grids and stained with 1% uranyl acetate. In the Tween 80–extracted sample, tubular structures with variable widths (0.1–0.2 μm) can be seen at low magnification. Some of these structures are still attached to round remnants of disc vesicles at one end (A). In the sample extracted with a mixture of Tween 80 and Tween 20, less vesicular structures can be seen, and a number of wider double sheet structures are observed (B). The bars in A and B indicate 5 μm. A single tubular crystal obtained by extraction of disc membranes with Tween 80 that contains rhodopsin in the p2 crystal form is shown C. A $p22_12_1$ crystal obtained with a mixture of Tween 80 and Tween 20 is shown at higher magnification in D. The bars in C and D represent 0.5 μm.

this lipid bilayer, which contains many multiple unsaturated fatty acid side chains. Corless et al. (1982) discovered that Tween detergents induce the formation of two-dimensional crystals of frog rhodopsin in frog disc membranes. We were able to improve the reproducibility and crystal quality by optimizing the Tween to the protein ratio, the pH, and the buffer composition. A further improvement was made by combining Tween 80 and Tween 20 (Schertler and Hargrave, 1995). Frog rhodopsin crystals (p2 and p22$_1$2$_1$) obtained by this method are shown in Figure 12.3. The process of crystal formation is not completely understood. For example, if bovine disc membranes are incubated under the same conditions with Tween detergents, no stable two-dimensional crystals are observed (Dratz et al., 1985). However, it is likely that the Tween is extracting some lipids from the membrane and is inducing crystallization in this way. In most experiments we obtain crystals with a randomized orientation of the protein, which is probably due to membrane fusion (Schertler and Hargrave, 1995).

Tween extraction methods have recently been used to improve gap junction channel crystals (Unger et al., 1997b). When deoxycholate was used to extract purple membranes, crystals with a smaller unit cell and more densely packed protein molecules were produced (Glaeser et al., 1985; Tsygannik and Baldwin, 1987). Lipases have been used in a similar way to reduce the lipid content and induce crystallization or improve the two-dimensional order (Jap, 1988).

Extraction methods are especially attractive because they can be used as a second step and thus separate two of the processes involved in the generation of two-dimensional crystals. First, the membrane protein is incorporated into the bilayer, and, second, the proteins rearrange to form a crystalline patch. Each process has an independent set of optimal parameters that have to be matched in crystallization experiments, meaning that a successful solution to the problem is hard to find. Thus it is appealing to first optimize protein incorporation to obtain a membrane with a high concentration of protein in the native state and then try to induce crystallization in a second step using an extraction procedure.

2.E. Screening for Two-Dimensional Crystals

2.E.a. Gels, Density Gradients, Spectra, and Assays for Function.

Membrane fractions obtained from reconstitution experiments can be characterized in several ways. Protein degradation and sometimes aggregation can be evaluated by using gel electrophoresis. Functional assays such as binding assays for receptors or an absorption spectra for rhodopsin (Fig. 12.4B) can provide information concerning whether the protein is still in a native state in the obtained membranes. Density gradients can give information about whether a sample is homogeneous and can be used to separate crystalline from noncrystalline membranes (Havelka et al., 1993, 1995; Heymann et al., 1993). This has been used in the preparation and characterization of the p22$_1$2$_1$ crystal form of bovine rhodopsin (Fig. 12.4) (Krebs et al., 1998).

2.E.b. Electron Microscopy.

The most informative assay for the formation of a two-dimensional crystal is electron microscopy itself. We put the specimen on a carbon-coated electron microscopy grid that has been activated by glow discharging and thus strongly adsorbs the studied protein. The specimen is then stained with 1% uranyl acetate, without an additional wash step, to lose as little of the sample as possible.

2.E.c. Taking Overview Pictures.

Overview pictures are electron micrographs taken at low magnification with high contrast. They are a very effective

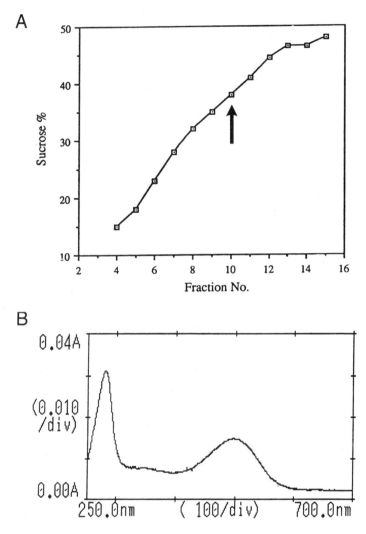

Figure 12.4. Purification of bovine rhodopsin p22₁2₁ crystals with a sucrose gradient. **A:** Tubular crystalline membranes were recovered in a sharp peak in fraction 10. **B:** The crystals were dissolved in detergent and an absorption spectrum was obtained, indicating that the retinal is in the 11-*cis* ground state conformation. Fraction 10 density, 1.166 g/ml.

way to screen different conditions and to document the results of reconstitution experiments, and they are a quick way to find the most promising condition in an experimental series.

Overview pictures are taken at low magnification (×1,000–3,000) and strong defocus (300 μm) to enhance image contrast. Alternatively, the best contrast can be achieved with a defocused diffraction mode. The high contrast is essential because a single molecular membrane is a low-contrast object and can be easily missed. About one-fourth of a grid square (400 mesh grid) is imaged (see examples in Figs. 12.2 and 12.3A,B), and the adsorption time of the sample is chosen so that a large number of particles and membranes are visible.

With some experience these overview pictures can be used to get a good impression of the bulk properties of the sample. Is the protein aggregated? Are there vesicles and precipitate present at the same time? Have larger membranes been formed? Which part of the sample is crystalline? Often a mixture of various objects and vesicles are observed (Figs. 12.2 and 12.3A,B) that should be classified from the overview picture and subsequently studied by imaging at higher magnification followed by optical diffraction.

2.E.d. Imaging Single Crystals. The same negatively stained samples are used to obtain high-magnification images of membranes for optical diffraction. To ensure that a crystal screen relying on optical diffraction is successful, the imaging process needs to be standardized. All pictures should be taken at the same magnification ($\times$20,000–40,000). The defocus level (5,000–10,000 Å underfocus) needs to be accurately set for every image to guarantee good contrast and to prevent the first zero of the contrast transfer function from coinciding with diffraction maxima. This can be done by focusing on the granularity of the carbon at very high magnification ($\times$170,000–300,000). This is best achieved in a low-dose procedure that also guarantees that the stain is not altered by a high electron dose. Focusing is done on a part of the grid close to the imaged area (3–5 μm). Reasonable coherence is achieved by using a moderately excited condenser lens and a small condenser aperture (50 μm). A 50-μm object aperture can be used to enhance contrast, but it will increase the need for astigmatism correction before every image is taken. Two examples of negatively stained crystals are shown in Figure 12.3C,D (Schertler and Hargrave, 1995).

2.E.e. Optical Diffraction. A magnified image of a two-dimensional crystal is a periodic modulation of the recorded optical density and can therefore act as a diffraction grating for light. The optical diffractometer is a laser setup that allows us to find diffracting parts in a negative recorded from a crystal (DeRosier and Klug, 1972; Spence, 1988). The diffraction pattern indicates whether there is a single molecular array or several. After calibrating the diffractometer with a grating of known spacing, lattice constants can be determined from optical diffraction patterns. First indications can be found for crystal symmetry elements and the plane group of the crystal lattice (discussed in detail later). In images taken from negatively stained samples, the resolution is normally limited by the stain to 15–20 Å. However, in images taken from vitrified specimens, we often need the diffractometer to find a crystal because of the low contrast. The outermost spots of a diffraction pattern define a resolution limit, and the intensity and sharpness of these spots can be compared with patterns from other images. Optical diffraction has proven to be essential in screening for two-dimensional rhodopsin crystals and in evaluating image quality before computer processing.

3. DATA COLLECTION WITH LOW-DOSE ELECTRON CRYOMICROSCOPY

We will describe next how we can embed our crystal in vitrified water. A very brief description of an electron cryomicroscope follows and a procedure for taking low-dose images with a spot scan method from tilted specimens is given.

The contrast transfer function that affects all images taken is introduced in the next section. Optical diffraction is used to evaluate the quality of the obtained images and can also be employed for a preliminary characterization of the two-dimensional crystal. In the next section a representation of the Fourier transform of a two-dimensional crystal in reciprocal space is discussed. This representation helps to explain how a three-dimensional structure can be obtained from a set of images taken from tilted crystals.

3.A. Vitrification of Specimens for Cryomicroscopy

Biological specimens such as two-dimensional crystals of proteins are highly beam sensitive (Henderson, 1995), and traditionally the only way to preserve the structure of protein assemblies in the microscope was to stain or metal shadow the proteins. The resulting replica either in a heavy metal salt or metal could then be observed and studied in the high vacuum of the electron microscope at high electron dose (>50 e$^-$ per Å^2). This approach was very successful in allowing a resolution of 15–20 Å and is simple and easy to use in extensive screening of two-dimensional crystals.

To preserve a beam-sensitive, hydrated specimen in the microscope to molecular detail, vitrification techniques and low-dose imaging methods have been developed. The sample is preserved by freezing it very rapidly in a thin film of water either in a hole of the carbon film or adsorbed to the carbon film. If the freezing rate is fast enough, the water does not form ice crystals, and the specimen is embedded in an amorphous, glass-like, water layer. The vitrification process is not very sensitive to the buffer components, and therefore the biological specimen can be applied in optimal buffer conditions.

The vitreous state of water is stable below a temperature of $-160°C$ in the microscope. A crystal batch can be frozen, and images can be obtained from the grids stored in liquid nitrogen over a long period of time (Dubochet et al., 1988).

3.B. The Electron Microscope

The most commonly used microscopes have an acceleration voltage of 100 kV and are equipped with a thermionic electron gun (heated tungsten filament or a heated lanthanum hexaboride crystal). More advanced microscopes used for electron crystallography should have an acceleration voltage of 200 kV or more and should be equipped with a field emission gun. Compared with a conventional electron source a field emission gun produces a beam with an improved spatial coherence due to the higher brightness and smaller effective size of the electron source and has a better temporal coherence due to the smaller energy spread. This leads to stronger image amplitudes at higher resolution (Spence, 1988; Fujiyoshi, 1998).

For low-temperature work with vitrified samples, an additional cold trap needs to be positioned near the specimen to prevent contamination build up on the cryospecimen. The frozen specimen is put under liquid nitrogen into a cold-stage in a loading station and is then transferred into the microscope at temperatures below $-160°C$.

3.C. Low-Dose Imaging

The vitrification keeps the specimen completely hydrated in the high vacuum of the electron microscope. The low temperature also decreases the beam sensitivity of the specimen, but, because of the strong interaction of electrons with matter, images must be obtained with a low electron dose (<20 e$^-$ per Å^2) to prevent beam damage (Dubochet et al., 1988; Henderson, 1995; Unwin and Henderson, 1975). To achieve this, three independent imaging modes on the microscope are needed: search mode, focus mode and exposure mode.

Interesting membranes can be found using a low-dose search mode and a magnification of less than $\times 3,000$. Either a strong defocus (300–500 μm) or a defocused diffraction mode can be used to identify membranes that have low contrast. The pointer in the microscope (pin) is aligned such that the same image area is correlated in exposure and search modes. In this way, a membrane can be selected for imaging without looking at it in high magnification, which would destroy the beam-sensitive sample.

To focus the sample without destroying the object, an area located a few microns (3–5 μm) away from the imaging area is selected. At high magnification ($\times 170,000$–300,000), the granularity of the supporting carbon film is used to focus and to correct the astigmatism. The defocus is set to a known underfocus value (5,000–10,000 Å). Finally, to obtain high-resolution information in the image, specimen drift must be minimal (<1 Å/sec).

After all corrections have been made and the drift has calmed down, the magnification is reduced ($\times 25,000$–60,000) and a low-dose image (<20 e$^-$ per Å^2) of the specimen is recorded.

3.D. Spot Scan for Imaging Tilted Specimens

Several reasons for the deterioration of the image quality at high resolution have been identified, including radiation damage, beam-induced specimen movement, charging, and attenuation of the envelope function of the phase contrast transfer function (described in the next section) due to imperfect coherence of the electron beam (Walz and Grigorieff, 1998; Spence, 1988). While the low-dose procedures can avoid radiation damage, spot scan procedures have been successfully used to reduce the effects of charging and specimen drift. The beam is reduced to a small diameter and scanned over the specimen to build up the image (Bullough and Henderson, 1987; Henderson and Glaeser, 1985). Spot scanning produces better images because at any given time during the build up of the image only a small area of specimen is illuminated, thereby reducing beam-induced movement and charging. The use of an anticontaminator or an object aperture was also reported to reduce charge build up, possibly due to neutralization by secondary electrons ejected from the anticontaminator (Dubochet et al., 1988; Walz and Grigorieff, 1998). For rhodopsin crystals we used a combination of a low-dose search procedure and a spot scan method. Spot scan was important for imaging tilted rhodopsin crystals preserved in vitrified water (Unger et al., 1997a; Unger and Schertler, 1995).

3.E. Contrast Transfer Function

The electron microscope does not transmit all spatial frequencies equally well. Particular spatial frequencies are lost completely, and the amplitudes of all others are modulated in a characteristic way. This phenomenon is analogous to a poor radio amplifier, which does not transmit all audio frequencies equally well and therefore can reproduce only a distorted version of a piece of music. The modulation function of the electron microscope is called the *contrast transfer function* (CTF). The main cause for the modulation of different amplitudes is the fact that we are recording a phase contrast image with an electromagnetic lens. The CTF is dependent on the defocus and astigmatism, which vary from image to image, and also on the spherical aberration and the high tension, which are constant for a series of images taken on a particular electron microscope. Therefore, the exact defocus and astigmatism have to be determined for every image to correct for CTF effects. A good estimate for the defocus and astigmatism can be obtained from the first zero of the CTF (Thon ring), which can be observed by optical diffraction (Fig. 12.5A,B) or in a calculated Fourier transform of the image (Thon, 1966; Wade, 1992).

3.F. Data Evaluation With Optical Diffraction

Optical diffraction is still the method of choice for preliminary data evaluation. No digitization is needed, and a large number of negatives can be dealt with promptly. However, fast computers with improved graphical output plus fast and user-friendly scanners will enable the development of programs that could replace the optical diffractometer. The program Ximdisp (Crowther et al., 1996) can already be used to do everything described below, but at present the optical diffractometer is much faster for screening negatives.

3.G. Preliminary Characterization of a Two-Dimensional Crystal

The negatives obtained from a low-dose session have low contrast, and often the imaged membrane is not visible. At this point, a well-aligned optical diffractometer is essential. The negatives are screened with an aperture that selects an area of 1–3 cm in diameter from the film for diffraction. First, the Thon rings caused by the CTF can be used to judge defocus level, astigmatism, and specimen drift. The strength and sharpness of a diffraction pattern obtained from a two-dimensional crystal is used to localize the crystal on the film. Two examples of optical diffraction patterns for a p2 and a $p22_12_1$ crystal of frog rhodopsin are shown in Figure 12.5 (Schertler and Hargrave, 1995). In both patterns the Thon rings caused by the CTF are clearly visible (see section on CTF, above).

3.G.a. Finding Elements of Crystal Symmetry: Unit Cells, Rotation Axes, and Screw Axes. When a diffraction pattern is found, we can often evaluate visually whether we have imaged a single crystal or a stack of crystals. The pattern from a single crystal is easy to index on the optical diffractometer. Super-

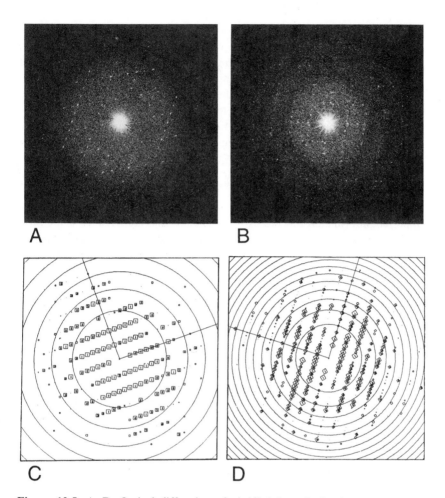

Figure 12.5. A, B: Optical diffraction of vitrified frog rhodopsin crystals. Frog rhodopsin crystals were rapidly frozen in liquid ethane. Electron micrographs of frozen hydrated crystals were taken at liquid nitrogen temperature using a Gatan 626 cold stage. The optical diffraction from an image of a crystalline tube embedded in vitrified water is shown. In A, the diffraction spots can be indexed on two independent p2 lattices with **a** = 32 Å, **b** = 83 Å, and γ = 91° for both layers. B shows an optical diffraction pattern of a frog $p22_12_1$ crystal. Two lattices can be indexed with **a** = 40 Å, **b** = 146 Å, and γ = 90°. Thon rings are clearly visible in both patterns. Only one Thon ring is visible in B because the image was taken closer to focus than in A. **C, D:** Calculated Fourier transforms of distortion corrected frog rhodopsin crystal layers. The best areas of each image were selected and digitized, 3,000 × 3,000 pixels with a 10 μm step size. Distortions in the images were corrected for in the crystal lattice using a 200 × 200 pixel reference area. For a second pass of correction, a 150 × 150 pixel area from the filtered image of the initially corrected area was used as reference. A third pass, with an extended list of Fourier components for the reference, improved the higher resolution components further. Finally, the area representing the extent of the crystal was boxed off before calculating the amplitudes and phases from each crystal. C shows the output for a p2 crystal, and D shows the output for a $p22_12_1$ crystal. The rings indicate where the CTF is zero.

imposed diffraction patterns from more than one crystal are better evaluated from a Fourier transform calculated from the digitized image. After calibration of the diffractometer with a diffraction grating, the basic lattice vectors **a** and **b** and the angle γ between them can be measured.

The symmetry of the diffraction pattern can indicate the presence of rotation axes in the crystal. Twofold, threefold, fourfold, and sixfold rotation axes can be found in two-dimensional crystals. Missing reflections can indicate centered plane groups, or, if they occur along the principal axes, they can indicate twofold screw axes in the **a–b** plane of the crystal. Together, the symmetry elements found can suggest one plane group out of the 17 possible groups listed in Table 12.1 (Amos et al., 1982).

It is important to determine the correct plane group because this will give important information concerning how the molecules are arranged in the crystal. The crystal symmetry can be used to average several molecules in the unit cell to give an improved asymmetric unit from which the whole unit cell or crystal can be reconstructed.

3.G.b. Finding Crystal Symmetry Elements in Unsymmetrized Maps of Negatively Stained and Unstained Specimens.

The presence of these symmetry elements should be confirmed by calculating unsymmetrized maps either from images obtained from negatively stained or vitrified specimens. For this purpose, a crystalline area is digitized with a very accurate image scanner (Amos et al., 1982), amplitudes and phases are extracted using Fourier analysis, and, after correction for lattice distortions, a low-resolution map can be calculated. Because the specimen might be affected by the stain or the adsorption to the carbon, all available information has to be carefully evaluated before a decision in favor of a particular plane group is made. This is often not a simple matter. This is discussed in more detail in the section on plane group determination.

3.G.c. Determination of Resolution by Optical Diffraction.

The intensity and sharpness of spots in the optical diffractometer can be used to select the most promising negatives. Often only a small percentage of all images is used for computer processing. With a small aperture on the optical diffractometer, a crystalline area can be scanned for its most ordered part, which then should be used as the center for the area to be scanned.

The outermost spots observed in the optical diffraction pattern can be used to compare images for the maximum of the crystal resolution, but it is rarely possible to observe optical diffraction beyond 5 Å resolution even if these spatial frequencies are present. It is therefore better to compare the intensity of specific reflections from different images. The real resolution limit can only be found after image processing.

3.H. Properties of the Fourier Transform of a Two-Dimensional Crystal

Representations of the Fourier transform in reciprocal space are important and helpful to understand data collection strategies in crystallography (Fig. 12.6). The reciprocal lattice constants **a*** and **b*** are proportional to 1/**a** and 1/**b**, and therefore

TABLE 12.1. Internal Phase Residuals for All Plane Groups Calculated for a Squid Rhodopsin Lattice

Plane Group	Phase Residue (No.) Other Spots (90 random)		Versus Theoretical (45 random)		Phase Origin Ox	Oy	Target Residual Based on Statistics
1 pl	33.8	186	25.3	186			
2 p2	50.9*	93	25.4	186	−39.0	−131.0	50.7
3b pl2_b	75.5	63	43.0	10	−183.4	−84.0	35.1
3a pl2_a	37.4!	65	19.5	14	114.0	139.0	35.6
4b pl21_b	30.4*	63	7.9	10	141.0	60.0	35.1
4a pl21−a	74.2	65	52.1	14	−180.0	−47.8	35.6
5b cl2_b	75.5	63	43.0	10	−183.4	−84.0	35.1
5a cl2_a	37.4!	65	19.5	14	114.0	−139.0	35.6
6 p222	67.5	221	41.4	186	6.3	41.0	40.9
7b p2221_b	69.9	221	25.8	186	140.7	−131.7	40.9
7a p2221_a	41.1*	221	25.4	186	−39.0	−131.0	40.9
8 p22121	64.6	221	38.6	186	50.7	101.5	40.9
9 c222	67.5	221	41.4	186	6.3	41.0	40.9
10 p4							
11 p422							
12 p4212							
13 p3	The plane group can not be square or hexagonal in this case,						
14 p312	because the unit cells are not equal and are not 90 or 120°.						
15 p321							
16 p62							
17 p622							

The asterisks mean that for the marked plane group the calculated phase residual is in "acceptable" agreement with the target phase residual, and the symmetry is very likely to be present in the specimen. The exclamation marks indicate that the marked plane group is a good possibility and should be considered.

The nomenclature for the two-sided plane groups used in the table is an extension of the nomenclature proposed by Holser (1958):

The cell type is indicated by a small letter p for primitive or c for centered.
The axis perpendicular to the plane is always chosen as the z-axis.
The first symbol following the cell type always describes the symmetry along the z-axis.
The second symbol indicates the symmetry parallel to one of the plane axes.
The small letter (a or b) at the end indicates along which axis the second symmetry element operates.

For example, p2221_a indicates a primitive cell, with a twofold axis perpendicular to the plane, a twofold rotation axis parallel to the **b**-axis, and a twofold screw axis parallel to the **a**-axis.

we call this the *reciprocal space.* The representations are used to predict which spatial frequencies (reflections) we expect to measure on a single film in an image or diffraction pattern. In addition, the representation of the Fourier transform of a two-dimensional crystal in reciprocal space helps us understand how data from a set of tilted and untilted images can be combined and how a three-dimensional structure of a two-dimensional crystal can be calculated (Amos et al., 1982).

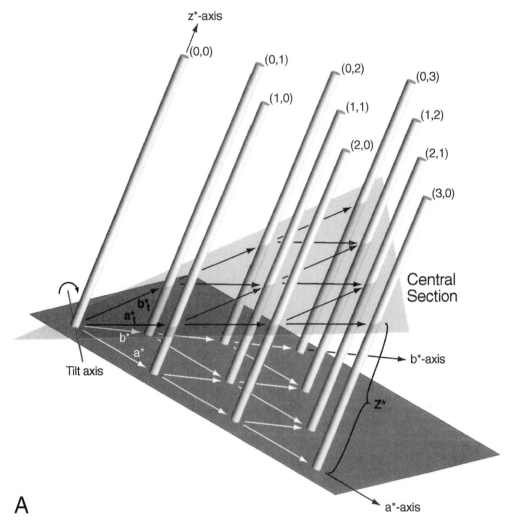

Figure 12.6. Fourier transform of a two-dimensional crystal. **A:** Fourier transform of a two-dimensional crystal. This transform is a set of lattice lines (represented by columns) arranged on a two-dimensional array defined by the reciprocal lattice vectors **a*** and **b*** and the angle γ*. Each lattice line is identified by two indices (h and k). The transform of a two-dimensional crystal is continuous along z* (normal to the crystal plane) because there is no periodicity in this direction. The orientation of the transform in reciprocal space is dependent on the orientation and tilt of the crystal in real space. The central section is the imaging plane of the electron microscope. **B:** The Fourier transform of a three-dimensional crystal is a set of lattice points arranged on a three-dimensional array defined by the reciprocal lattice vectors **a***, **b***, and **c*** and the angles α*, β*, and γ*. Each lattice point or structure factor is identified by three indices (h, k, and l) and has an amplitude and a phase. In B, a Fourier transform of a three-dimensional crystal (spheres in B) is superimposed on the Fourier transform of a two-dimensional crystal (columns in A and B). The lattice vector **c*** coincides with z*-axes of the lattice lines. This transform of a three-dimensional crystal samples the lattice lines of the two-dimensional crystal at regular intervals. The resulting list of structure factors can be used to calculate a three-dimensional map.

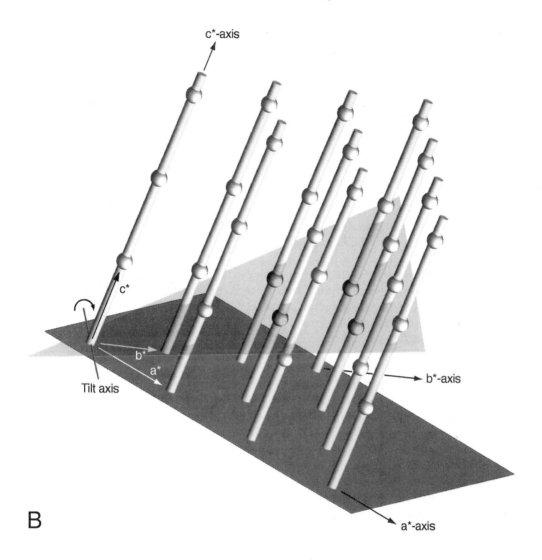

B

A representation of the Fourier transform of a two-dimensional crystal is shown in Figure 12.6A and compared to the three-dimensional transform of a three-dimensional crystal in Figure 12.6B. The Fourier transform of a two-dimensional crystal is a set of lattice lines arranged on a two-dimensional array defined by the reciprocal lattice constants **a*** and **b***. Each lattice line (symbolized by a column in Fig. 12.6A) is identified by two indices, h and k. The amplitudes and phases are changing continuously along the lattice line in the z*-direction (see Fig. 12.15).

The Fourier transform of a three-dimensional crystal is a set of lattice points (symbolized by the spheres in Fig. 12.6B) arranged on a three-dimensional array defined by the reciprocal lattice constants **a***, **b***, and **c***, which means it is discontinuous in all three directions of the reciprocal space. Each lattice point is identified by three indices, h, k, and l, and has one amplitude and one phase.

The position and the distance between lattice lines and lattice points, and the symmetry of the Fourier transform in reciprocal space, are dependent only on the crystal geometry and symmetry. The information about the structure of the molecule repeated in the crystal lattice is contained in the amplitudes and phases of the transform.

The central section (Fig. 12.6A) is the imaging plane of the microscope. It corresponds to the Ewald sphere in x-ray crystallography. The very small wavelength of the electron means that the sphere has a very large diameter in reciprocal space relative to the reciprocal lattice vectors, and therefore the surface of the sphere can be approximated by a plane. The central section intersects the lattice lines at positions that are determined by the crystal tilt and orientation. A single image or diffraction pattern contains only the spatial frequencies (amplitudes and phases) at the intersection points between the central section and the lattice lines. The Fourier synthesis of all these spatial frequencies is a projection image of the tilted crystal. By increasing the tilt, amplitudes and phases higher up on the lattice lines can be measured. The continuous Fourier transform of a two-dimensional crystal can be reconstructed from many tilted images.

The aim of a structure determination with a two-dimensional crystal is to measure the amplitudes and phases along the lattice lines as accurately as possible by collecting a set of tilted images.

4. IMAGE PROCESSING

The first part of this section describes in detail how every image of a tilted or untilted crystal is processed, and this procedure is outlined in Figure 12.7. The goal is to extract the amplitudes and phases of the periodic structure that was imaged as accurately as possible. The second part of this section discusses how we can use the amplitudes and phases from untilted images to verify the plane group of the two-dimensional crystal. Programs from the MRC Image processing software (Crowther et al., 1996) or the CCP4 x-ray crystallography program package (Bailey, 1994) used in the following procedures are indicated by words in capital letters.

4.A. Extracting Amplitudes and Phases From Low-Dose Images

This section describes how every image of a tilt series is processed as outlined in Figure 12.7. The micrograph is digitized, lattice distortions are removed, and then the amplitudes and phases are extracted. The defocus and astigmatism are determined and CTF correction is applied.

4.A.a. Scanning Micrographs With a Precision Scanner. Image manipulation and Fourier transformation require digitization of the image into numbers proportional to the optical density on the film, which should in turn be proportional to the number of electrons that were hitting the film. The step size used for digitization must be sufficiently small to sample the detail present in the image (Amos et al., 1982; DeRosier and Klug, 1968; DeRosier and Moore, 1970). The availability of inexpensive storage media and fast computers allows large

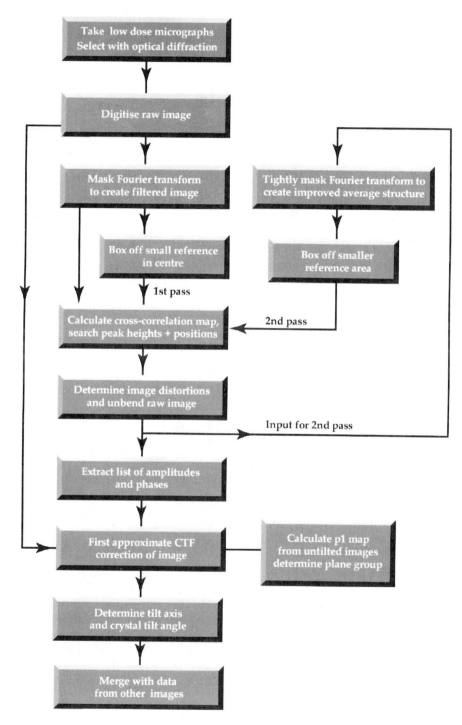

Figure 12.7. Image processing. The scheme show the steps needed to remove the lattice distortions in the image. Then the amplitudes and phases are extracted; image parameters, defocus, astigmatism, tilt, and crystal orientation are determined; and the effects of the CTF are corrected for. A projection structure can be calculated from the resulting data.

areas to be processed. For small, poorly ordered rhodopsin crystals, we used $1,000 \times 1,000$ pixel areas densitometered with a 10-μm step size (Schertler et al., 1993) from a negative that was taken at a magnification of $\times 36,000$. For the better-ordered bovine $p22_12_1$ rhodopsin crystals, images were recorded at a magnification of $\times 60,000$ and $6,000 \times 6,000$ pixel areas were scanned with a step size of 7-μm (Krebs et al., 1998).

4.A.b. Unbending Crystal Lattices With a Small Reference Area. The images obtained by low-dose electron cryomicroscopy are inherently noisy, and the amount of signal retrieved from a single molecule is very small. In an image of a nonideal, two-dimensional crystal, the lattice is distorted from the interaction of the collapsed vesicle with the carbon. The aim of image processing is to align all views obtained from a molecule as perfectly as possible and to calculate an average structure. By first averaging all images of the molecule in untilted crystalline patches, and then the images of various tilted views, we can improve the signal to noise ratio dramatically. In contrast to single-particle work, we do not have to pick single molecules for processing. The location of every molecule in the imaged lattice is determined by indexing the lattice and by fitting lattice parameters (XIMDISP). Because of the periodic nature of the two-dimensional crystal, we can use well-established Fourier filtering methods to separate the periodic signal from random noise. Correlation maps between the filtered image and a reference can be calculated by a Fourier transform method that can then be used to define the distortions of the image (Fig. 12.8).

A   B

Figure 12.8. Cross correlation map and lattice distortion vectors. **A:** Cross correlation map in which a more intense signal indicates a stronger correlation with the reference taken from the center of the image. Outside the crystal, the correlation falls to a background value. **B:** Distortion vectors found by the lattice unbending procedure. To make the distortions more visible, the vectors have been magnified. The modulation seen on the crystal in A and B is due to the spot scan procedure that was used to obtain the image.

4.A.c. Unbending Procedure. The best part of an image of a crystalline patch is scanned, for instance, 6,000 pixel squared, so that the image can be represented by a large array of densities (36 million pixels) by the computer. The image is Fourier transformed (FFTRANS), a plot of the Fourier transform is inspected, and the lattice is indexed and refined with the image display program XIMDISP to obtain accurate lattice constants. A list of all the diffraction spots visible above the level of the background is made. The list includes all visible spots but excludes weak spots near the minima in the CTF. The transform is then masked by allowing the parts of the transform within a certain radius from the center of each strong diffraction peak to go through unchanged and by setting all other parts of the transform, including the F(0,0) term, to zero. The filtered image is then back-transformed (FFTRANS). This step removes a large part of the random, nonperiodic noise from the image (Henderson et al., 1986).

4.A.d. Evaluating Cross-Correlation Maps. A small part (100–300 pixels squared) from the center of a filtered image representing the best part of the crystal is boxed off to provide an initial reference (BOXIMAGE) that is subsequently used to cross-correlate the entire scanned area to the reference area. The cross-correlation map is calculated by a Fourier transform method. This first involves the calculation of the Fourier transform of the reference area (FFTRANS), which is padded out to form precisely the same size area as the filtered image. Secondly this transform is multiplied with the complex conjugate of the filtered Fourier transform of the original image (TWOFILE). The product is then back-transformed to give a cross-correlation map (FFTRANS) (Berglaud, 1969). This procedure provides a map of the cross-correlation function that contains information about the alignment of crystalline parts further from the center. If the crystal is smaller than the scanned area, then the cross-correlation maps also show the extent of the crystalline area because the cross-correlation outside the crystal drops to the noise level (Fig. 12.8).

The peaks of the cross-correlation map are not exactly on the lattice points because of the displacements that we are trying to quantify and use. The exact position of the peaks is found using an algorithm in which the search begins at the origin peak and moves out along the x and y directions of the crystal by one unit cell at a time (QUADSEARCH). The amount of distortion is progressively learned. The position and height of the peaks are determined accurately by profile fitting. The profile is taken from the shape of the strong central autocorrelation peak (AUTOCORReL). The resulting set of vectors describe the distortions from the perfect lattice at each unit cell position, which can be used to remove the distortions from the original image (Henderson et al., 1986). An example of a cross-correlation map is shown in Figure 12.8A, and a corresponding set of distortion vectors is shown in Figure 12.8B.

4.A.e. Correcting for Distortions in the Image. The vectors describing the image distortions are used to correct the original densitometered image. A simple bilinear interpolation uses the four nearest spots in the cross-correlation map to interpolate at the scan points in the original image (CCUNBEND). In this way, the molecules in the crystal lattice are more perfectly aligned over the entire crystalline area. For all rhodopsin crystals one, two, or three additional passes of unbending were used to remove the distortions, whereby a reference with a better signal to noise ratio is taken from the previously unbent image (Fig. 12.7) (Davies et al., 1996; Havelka et al., 1995; Krebs et al., 1998; Schertler and Hargrave, 1995; Unger and Schertler, 1995).

4.A.f. Selection of a Coherent Crystalline Area and Extraction of Amplitudes and Phases. The actual crystals of rhodopsin were usually smaller than the square scanned area either because the crystalline patch had an elongated shape or because only part of the membrane turned out to be crystalline. In this case, a significant part of the image does not contain any information on the structure of the molecule and contributes to noise only (Fig. 12.8). Therefore, the signal to noise ratio can be improved further by setting the density to the average density of the entire image outside the crystal (BOXIMAGE). This selection of the best crystalline area has been especially important for the small and rather disordered crystals of rhodopsin that we first obtained (Schertler and Hargrave, 1995; Schertler et al., 1993).

After the image has been corrected for distortions and the crystal has been boxed, a final Fourier transform is calculated in which the diffraction peaks from the crystal are generally much sharper. The amplitudes and phases are extracted by fitting a sinc function (sin[x]/x), approximating the shape of the diffraction maxima, to the peaks in the Fourier transform. The phases are obtained by averaging the phases of the four center pixels of the peaks in the Fourier transform by the program MMBOX.

4.A.g. Evaluation of Image Quality After Unbending. A simple procedure can be used to find out to what resolution significant information about the periodic structure is present in an image after unbending. All the pixels in a box of a given size around every predicted reflection in a given resolution zone are summed. The resulting values are then scaled such that the average of all pixels in the perimeter of the box is adjusted to a preselected constant value (7 in Fig. 12.9) (MMBOX). The resulting peak in the center of the box is a criterion dependent only on the quality of the image itself and on the quality of the unbending procedure that was used to correct for distortions (Henderson et al., 1986; Krebs et al., 1998).

4.B. Determination of Astigmatism and CTF

As described earlier, the electron microscope does not transmit all spatial frequencies equally well. The very small granules of the carbon support film generate a nearly white spatial frequency spectrum, and, because not all frequencies are transmitted, areas of this transform are missing. A typical CTF of an image taken at underfocus with a reasonably astigmatism-corrected lens has a series of spatial frequencies that are zero and, therefore, not transmitted at all. The missing frequencies appear in the Fourier transform as a series of rings with an amplitude of zero or as a series of ellipses if astigmatism is significant (Fig. 12.5). The astigmatism can be described with two defocus values determined along the main axes of the ellipse and an angle between the x-axes of the scan and one of the principal axes of the ellipses. Much time and effort have to be spent determining

7	8	8	8	8	10	11	8	8	7
7	7	7	9	8	10	14	11	8	7
8	6	10	12	15	16	12	11	9	8
9	11	15	24	144	*204*	75	13	10	9
8	11	18	29	*251*	**481**	*249*	27	16	9
9	8	10	10	67	*194*	132	19	17	11
8	7	7	11	12	14	14	16	17	11
10	9	10	13	17	13	10	11	8	9
8	7	8	9	12	13	10	11	11	10
7	8	7	10	10	9	7	8	8	7

$$\infty - 9.0 \text{ Å}$$

6	6	6	7	7	6	7	7	8	7
7	8	7	7	9	9	8	7	7	7
8	8	8	9	19	26	16	8	8	7
8	8	7	9	*26*	**43**	29	11	8	7
7	7	8	8	13	*23*	18	8	9	9
6	7	7	8	7	8	8	8	9	8
7	8	7	8	9	7	7	7	8	7

$$9.0 - 5.0 \text{ Å}$$

7	6	6	7	7	7	7	8	8	7
7	6	7	8	8	7	7	8	7	7
6	6	6	7	8	*9*	8	7	7	7
7	7	7	7	7	**10**	*9*	7	8	7
7	7	7	8	8	*9*	8	7	8	7
8	7	7	8	8	8	7	7	7	7
7	7	7	7	8	7	7	7	7	7

$$5.0 - 3.5 \text{ Å}$$

Figure 12.9. Average intensities extracted from an image. The average intensities of peaks in the Fourier transform of an image after distortion corrections were determined by the program MMBOX in different resolution ranges. The small peak in the resolution range from 5.5 to 3.5 Å shows that there is periodic information from the crystal in the image beyond 5 Å resolution.

these three numbers for every image accurately, first by estimating them from the image as described above and then by refining them using a merged set of images as described below. Before we can calculate a meaningful map from the amplitudes and phases that we have extracted from our images, we need to correct for the modulation of amplitudes and for the phase shifts induced by the CTF. When this is done, we can calculate a projection structure for every processed image, which represents the average of all molecules in the crystalline patch.

4.C. Determination of the Plane Group

Crystal symmetry is very valuable when the data from several images are combined, and it is used to improve the quality of the map significantly. However, it is important that only the symmetries that are really present in the specimen are applied. Amplitudes and phases from untitled images can be used to verify the plane group of the crystal.

4.C.a. Finding Crystal Symmetry Using Amplitudes.
As mentioned earlier, there are only 17 combinations of crystal symmetry elements possible for two-dimensional protein crystals. They correspond to the plane groups listed in Table 12.1 (Amos et al., 1982). As mentioned above, we can obtain information on the symmetry of the two-dimensional crystal from optical diffraction of images from negatively stained or vitrified specimens. Three-, four- or sixfold rotational symmetry of the diffraction pattern or calculated Fourier transform indicates corresponding rotation axes normal to the crystal plane. The observation that the angle between the lattice vectors is very close to 90° can point toward an orthorhombic lattice. Missing reflections along the principal axes can indicate twofold screw axes along the **a**- or **b**-axis (Amos et al., 1982).

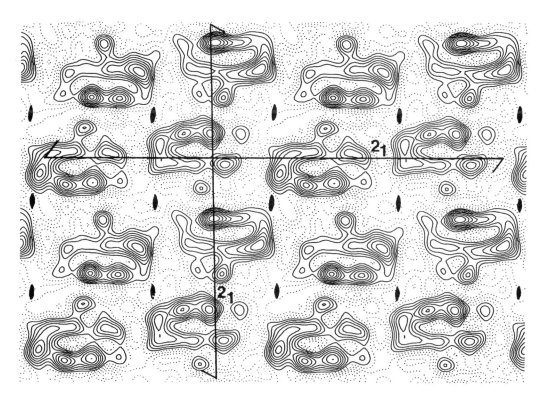

Figure 12.10. Unsymmetrized p1 map. An example of a p1 map from a p22₁2₁ bovine rhodopsin crystal. A set of twofold rotation axes normal to the crystal plane is indicated (▮). Sets of twofold screws along the **a**- and **b**-directions are indicated on the map (2₁).

4.C.b. Finding Crystal Symmetry Using Phase Relations. Rarely is a picture perfectly untilted, and often preservation can be variable. Therefore, some of the conclusions will be tentative and need to be checked carefully. This can be done by inspection of phase relations with the program ALLSPACE (Table 12.1) (Valpuesta et al., 1994) or by calculating a projection map without applying symmetry (p1 map) and inspecting it for symmetry elements.

ALLSPACE calculates alternative origins using an origin search of the image against itself, and the program is able to calculate target phase residuals to differentiate between the 17 possible plane groups (Table 12.1) (Schertler and Hargrave, 1995; Valpuesta et al., 1994). However, the results are affected by sample tilt, stain, or adsorption-induced distortions and have to be viewed critically in the context of other information such as absent reflections and the angle between cell parameters. In addition, the results have to be consistent with a visual inspection of an unsymmetrized map (p1 map, Fig. 12.10).

4.C.c. Confirmation of Crystal Symmetry Using p1 Maps. From a CTF-corrected list of amplitudes and phases we can calculate a map by Fourier synthesis (FFT CCP4 program). This map can then be analyzed for symmetry elements by visual inspection. An example of a p1 map for a $p22_12_1$ bovine rhodopsin crystal is shown in Figure 12.10, and some symmetry elements are indicated. Usually the first step is to determine the position and outline of single molecules. Then sets of twofold rotation axes normal to the crystal plane are found, and after this an estimate of the unit cell can be made. Sometimes twofold screws are found either on a line or in the middle between the rotation axes (Fig. 12.10 and Fig. 12.14).

5. MERGING AMPLITUDES AND PHASES FROM UNTITLED IMAGES FOR A PROJECTION MAP

This section describes in detail how the amplitudes and phases from several untilted images are combined to a projection structure. The method closely follows the outline of the general merging procedure given in Figure 12.11. The next sections cover amplitude scaling, map calculation and representation.

5.A. Merging Data From Different Images

To average several images of crystals, we first have to bring the images to the same scale and then shift them so that they are exactly on top of each other. In Fourier space this corresponds to scaling the amplitudes of all images to a reference image and finding a common phase origin. By shifting the phase origin of each image by a fraction of the unit cell (equivalent to adding a few degrees to the list of phases obtained from an image) and looking for maximal phase agreement (minimal phase residual) a common phase origin can be found. Thus phases of the Fourier components from each image can be compared (program ORIGTILT) against the averaged phases obtained from the measurements of all other images. The phase origin is indicated by maximal agreement (minimal

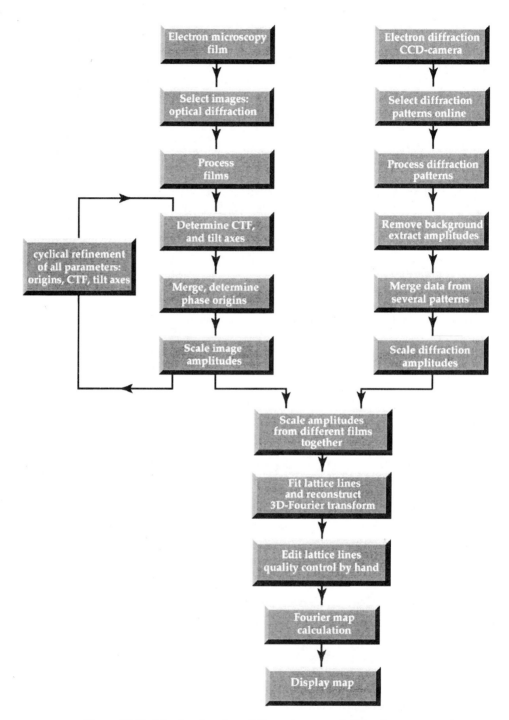

Figure 12.11. Merging data from different images and diffraction patterns. The steps needed to combine the data from different images are shown. The phase origins, defocus, astigmatism, and tilt parameters are refined in a cyclic procedure. The Fourier transform of the two-dimensional crystal is reconstructed by fitting lattice lines. After sampling the lattice lines at a regular interval, a three-dimensional map can be calculated.

phase residual) (Henderson et al., 1986; Amos et al., 1982). After the phase origin for every image has been determined, the average amplitudes and phases can be calculated with the program AVRGAMPHS.

5.B. Refining the CTF Correction for Each Image Using a Merged Set of Images

The first estimate for the CTF for every image is derived from defocus values obtained from Thon rings and missing amplitudes in the Fourier transform (see section on CTF, 3.E). The two defocus values and the angle between the scan axes, which define the CTF correction, and the phase origin can be refined using a merged set of amplitudes and phases containing the data from several images. The phase origin search for each image is repeated, whereby the two defocus values and the angle between the scan axes are varied until the phase residual against the data set is minimal. This can be done in an iterative loop until neither the defocus values nor the origin change any more. A correctly chosen resolution range and the inclusion of only reliable reflections are essential for a successful procedure (Henderson et al., 1986; Krebs et al., 1998).

5.C. Scaling of Amplitudes Using Diffraction Data

For fairly disordered crystalline arrays (e.g., all rhodopsin crystal forms) it is important to analyze the fall off of the amplitudes (Fig. 12.12). A strong amplitude fall off will prevent us from seeing any higher resolution features in the

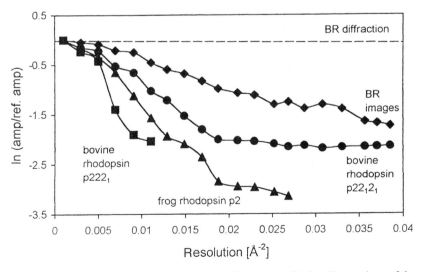

Figure 12.12. Resolution-dependent fadeout of image amplitudes. Comparison of the resolution-dependent fadeout of the image amplitudes for bacteriorhodopsin (BR) and frog and bovine rhodopsin projection data. Electron diffraction data from untilted bacteriorhodopsin crystals (dashed line) were used as a reference to determine the attenuation of image amplitudes with increasing resolution.

map even if the phases are clearly well defined (shown by small phase errors), because the higher resolution Fourier components will be weak (Schertler et al., 1993; Unger and Schertler, 1995). From a comparison of the fall off obtained for image amplitudes with diffraction amplitudes, either from the same sample or model data, an estimate can be made of the amount of rescaling needed.

Amplitudes are rescaled by comparing a ratio between diffraction amplitudes, either of the sample or a reference molecule (e.g., bacteriorhodopsin), with the measured amplitudes. The amplitudes obtained for a particular crystal form can be studied by sorting them according to resolution ranges (SCALI-MAMP3D). A plot of the amplitudes is very informative (Fig. 12.12) (Krebs et al., 1998; Unger and Schertler, 1995). The curve can be used to correct for the amplitude fall-off. This has been successfully used for rhodopsin and halorhodopsin crystals (Havelka et al., 1995; Schertler et al., 1993). In addition, a temperature factor used for isotropic scaling can be calculated from the fall off plot (Unger et al., 1997a; Unger and Schertler, 1995). The fall-off of the amplitudes indicates imperfections of the electron microscopic imaging system on the one hand and lack of order in the imaged crystal on the other. The program SCALIMAMP3D can rescale the amplitudes according to the measured fall off, and it will constrain the phases to 0 or 180 for projection data if required (Havelka et al., 1995).

5.D. Evaluating the Resolution Limit of the Merged Projection Data

It is important to evaluate the resolution and quality of a data set before calculating a projection map to prevent the inclusion of random amplitudes and phases.

As described earlier, averaged intensities can be used to find out to which resolution a single image contains periodic information from the crystal that was imaged (MMBOX). The resulting peak in the center of the box is a criterion dependent only on the quality of the image itself and on the quality of the unbending procedure (Fig. 12.11).

As soon as several images are merged together several measurements for all amplitudes, phases are listed, and errors can be estimated (ORIGTILT). A figure of merit is calculated from the combined phases ($m = \int P[\alpha] \cos d\alpha / \int P[\alpha] d\alpha$) for each measured phase where $P(\alpha)$ denotes the combined phase probability distribution (Henderson et al., 1986).

For single images, individual phase errors can be calculated by comparing them to a measured data set. The results not only are dependent on the unbending procedure but also reflect on the quality of the merge itself. Therefore, as we have seen, minimization of these phase errors is used extensively during origin and CTF refinement in the merging procedure.

An overall phase error can be used to estimate the reliability for the entire data set, and a plot of phase errors as a function of resolution (Fig. 12.13A) can be used to evaluate at which point the phases become random and to which resolution a map should be calculated (SCALIMAMP3D) (Krebs et al., 1998).

A   Overall average phase error

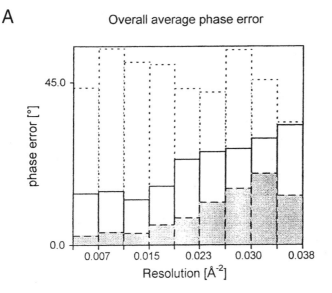

B   Comparison between two halves of data

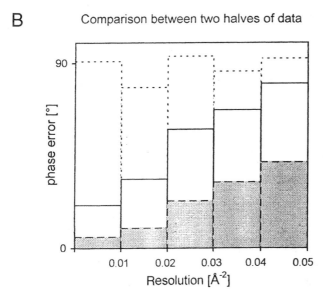

Figure 12.13. Statistical analysis of data set accuracy. **A:** Estimated phase error on the whole data as a function of resolution for the p22$_1$2$_1$ bovine rhodopsin crystal form. The phase error in given resolution bands represents the mean of all individual reflections within a resolution band compared with the theoretical value based on the constraints of the plane group. In plane group p22$_1$2$_1$ all phases are constrained to 0° or 180° in projection except for odd reflections (h,0 and 0,k), which are absent because of the twofold screw axes along **a** and **b**. The projection data (solid lines) are compared with a data set with random phases (dotted lines). Simulated phases with the distribution of error given by the signal to noise level were also analyzed (grey, dashed lines) (Krebs et al., 1998). **B:** Comparison of the final estimates of two independent halves of data with roughly equal information content.

A more sophisticated analysis of the accuracy of the final phases is to split a data set in two halves and compare phase differences between the two halves with the program HALFSTAT (Fig. 12.12B) (Baldwin et al., 1988; Henderson et al., 1986; Krebs et al., 1998).

5.E. Calculating a Projection Map From a List of Amplitudes and Phases

After a corrected list for all amplitudes and phases has been produced, a projection map can be calculated by Fourier synthesis with the program FFT (CCP4). This generates a map that contains one asymmetrical unit of the crystal that can be extended to represent the entire unit cell or a larger crystalline array using the program EXTEND (CCP4). The final map can then be presented as a contour plot with the program NPLT (CCP4). A projection structure calculated from 20 crystalline lattices with $p22_12_1$ symmetry is shown in Figure 12.14 (Krebs et al., 1998).

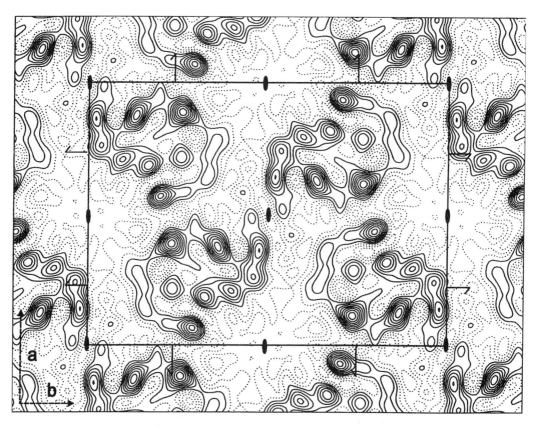

Figure 12.14. Projection map. A projection map of bovine rhodopsin in a $p22_12_1$ crystal was calculated to 5 Å resolution. The map represents the average of 20 images from virtually untilted crystals. One unit cell containing four rhodopsin molecules is indicated (rectangle), and the twofold rotational axes normal to the crystal plane and twofold screw axes along **a** and **b** are indicated.

5.F. Interpretation of Projection Structures

The projection map represents the sum of density along the z^*-axis normal to the crystal plane and therefore does not contain three-dimensional information, making projection data difficult to interpret. However, in simple arrangements, helices nearly perpendicular to the plane of the crystal (membrane plane) will give a more or less localized circular peak in the map. Tilted helices will produce much more diffuse, often overlapping density features (Krebs et al., 1998; Schertler and Hargrave, 1995; Schertler et al., 1993). A projection structure can very much restrain the possible arrangements of helices. In the case of rhodopsin, the projection structure was used in combination with extensive sequence comparisons to propose a three-dimensional arrangement of the helices (Baldwin, 1993).

6. MERGING AMPLITUDES AND PHASES FROM TILTED IMAGES TO CALCULATE A THREE-DIMENSIONAL MAP

A general merging procedure for tilted and untilted images is shown in Figure 12.11. A procedure to determine the tilt angle and the crystal orientation relative to the tilt axis is described in detail, and this is followed by an iterative refinement procedure for tilt parameters and CTF (astigmatism and defocus). In the next section the Fourier transform of the two-dimensional crystal (Fig. 12.6A) is reconstructed from the image data by fitting continuous curves to the lattice line data.

6.A. Merging Data From Tilted Images

The data for a projection map from several untilted images consist of a simple list of reflections, and several measurements are available for most of the amplitudes and phases. The data for a three-dimensional structure of a two-dimensional crystal are more complex. To understand how a three-dimensional structure of a two-dimensional crystal can be obtained from a set of tilted and untilted images, we need to study some properties of the Fourier transform of a two-dimensional crystal (Fig. 12.6A) (Amos et al., 1982). The task is to measure the continuously changing amplitudes and phases along the lattice lines as accurately as possible (Fig. 12.6A). This can be done by taking a large number of untilted as well as tilted images, which allows us to obtain measurements for amplitudes and phases along the lattice lines.

We have to record the images from tilted and untilted specimens with a low-dose procedure as described above, and the spot scan procedure improves the data for the tilted specimens especially. As for the untilted images, the distortions are corrected with an unbending algorithm that uses part of the crystalline array as a reference. The Fourier analysis of the corrected image gives a list of amplitudes and phases corresponding to the spatial frequencies contained in the image. For each tilted image we need to determine the tilt parameters. They allow us to calculate a z^*-value for each pair of amplitudes and phases that was determined. The tilt angle (TANGL) and the angle from the tilt axis to the $\mathbf{a}^*$

vector (TAXA) determine where the central section (Ewald sphere in x-ray crystallographic terms) is intersecting the lattice lines of the Fourier transform of the two-dimensional crystal. For each intersection point, an amplitude and a phase are recorded on the film (Fig. 12.6A).

The absolute tilt of the crystal in the microscope can be determined roughly from the Thon rings, which represent the zero values of the CTF (see sections on CTF 3.E and 4.B). The diametersof the Thon rings are dependent on the defocus levels. A first Thon ring with a larger diameter indicates that the specimen is closer to focus and therefore is physically higher in the microscope than a part of the specimen that shows a smaller Thon ring diameter (Fig. 12.5). If we divide our scanned image area into fourths (LABEL) and measure the diameter of the Thon rings (XIMDISP), then we can estimate the orientation of the tilt axes and calculate a tilt angle.

In the next step we have to determine the sign of the tilt angle. In a diagram, the four corners of the film are marked with "up" and "down" according to the defocus values determined from the Thon rings. This represents the absolute orientation of the crystal in the microscope. Subsequently, reciprocal lattice vectors a* and b* are drawn. From this diagram we establish the direction of the +z*-axis assuming a right-handed coordinate system. In the next step we determine whether the imaging plane (central section, Fig. 12.6A) is intersecting the z*-axes at the 1,0 lattice line with a positive or a negative z*-value. From the relation $z^* = \tan(TANGL)*(H/a)*\sin(TAXA)$ we can work out the absolute sign of the tilt angle, which will finally determine the hand of the molecule. If we choose TAXA to be positive ($<180°$), then a negative z* value for the intersection point of the 1,0 lattice line indicates a negative sign for the tilt angle.

It is important to realize that this is only true if the scanning method used does not affect the hand of the images. Otherwise, the absolute sign of the tilt angle has to be altered accordingly. Great care is necessary to ensure that the hand of the structure is correct.

In the next step, we reconstruct the Fourier transform of the two-dimensional crystal from the amplitudes and phases obtained from all images in the data set (see Fig. 12.6A and the section on Fourier transform of a two-dimensional crystal 3.H). In addition to the crystal orientation, a common phase origin for all these images must be found. This will align all the different projections in the right orientation for averaging (compare with section on merging data of untilted images 5). It has to be emphasized that every tilted image contains a subset of data along a line in the transform that can be compared with the projection data. To merge in the first tilted image, we find a common origin of this image with the already merged projection data. Unlike merging untilted images, the origin search is carried out with strong data (e.g. IQ 1–4) with a restriction of the resolution to e.g. 10 Å because these data are less affected by errors in the crystal orientation (TAXA), tilt (TANGL), and defocus, which are only roughly known at this point. All these parameters are refined as soon as a large enough three-dimensional data set has been merged together. The merging program ORIGTILT sorts spatial frequencies (amplitudes and phases) according to lattice lines (h,k) and it calculates z*-values from the crystal orientation (TAXA, TANGL). This creates a list in which each pair of amplitudes and phases is associated with a z*-value. The z*-value indicates how far up or down a lattice line the amplitudes and phases were measured (Fig. 12.6A). After this merging of data, phase origins, defocus and astigmatism, tilt angle, and beam tilt have to be refined for all images in an iterative procedure by maximizing the agreement of all phases with the program ORIGTILT (Henderson et al., 1986).

6.B. Fitting Lattice Lines

The measured amplitudes can be plotted for each lattice line and inspected visually. In the next step, the lattice lines are fitted either manually or with the program LATLINE (Agard, 1983). Examples of lattice lines are shown in Figure 12.15. Note that the higher resolution lattice lines extend further along the z*-axis for a given maximum tilt angle, as expected from the Fourier transform of a two-dimensional crystal (Fig. 12.6A). The data that cannot be measured are located on a double cone, often referred to as the *missing cone.* These missing data introduce anisotropy leading to a lower resolution normal to the crystal plane compared with the resolution within the plane (Fig. 12.16).

6.C. Evaluation of Resolution and Data Anisotropy

The data anisotropy can be represented directly by plotting all lattice lines and the figure of merit along the z*-axis (Unger et al., 1997a). A more sophisticated way to analyze the anisotropy of the resolution is to calculate a point spread function from the figures of merit (Henderson et al., 1990; Unger et al., 1997a; Unger and Schertler, 1995). This function describes how a single point is blurred by the quality and the incompleteness of the data. From these values, cut-off resolutions in different directions can be estimated (Fig. 12.16). Furthermore, the calculated maps often include features such as the distance between helices, or distances between bulky aromatic side chains within one helix, that can be used to verify the achieved resolution (Henderson et al., 1990; Kühlbrandt et al., 1994).

6.D. Calculating a Three-Dimensional Map

With the lattice lines, we have defined a continuous set of spatial frequencies. These define the Fourier transform of the two-dimensional crystal more or less correctly depending on the data available (Figs. 12.6A and 12.15). At this point, we can use some properties of Fourier transformations to simplify the calculation of a three-dimensional map.

The Fourier transform of a two-dimensional crystal, in contrast to that of a three-dimensional crystal, is a continuous function along z* (Fig. 12.6A,B). This transform is a result of the multiplication of a transform of a two-dimensional or three-dimensional lattice of points with the Fourier transform of a single molecule. The signal amplitude is proportional to the number of molecules present in the crystalline array. We can generate the Fourier transform of a hypothetical, three-dimensional crystal from the measured two-dimensional crystal by sampling all lattice lines at a regular interval along z* (at the center of the spheres in Fig. 12.6B). From the fitting error at each sampling point we can generate a figure of merit (Fig. 12.15). In this way we introduce a third lattice

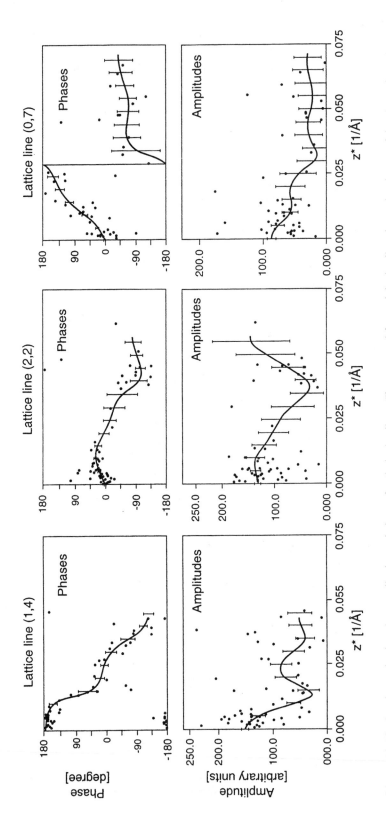

Figure 12.15. Continuously changing amplitudes and phases for six lattice lines. The top panel for each lattice line shows the change of the phase along the z*-axis. The bottom panel shows the changing amplitudes (in arbitrary units). The error bars show the fitting error at the sampling points and are spaced at 0.005 Å⁻¹ (1/200 Å) along the z*-axis normal to the crystal plane (Unger et al., 1997a).

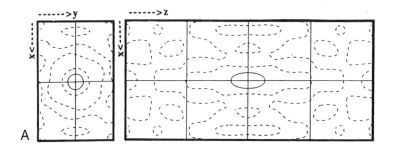

plane		half-width [Å]	effective resolution cut-off [Å]
x,y	x	5	7.4
	y	5.25	7.7
z,y	z	11.3	16.5
x,z	z	11.3	16.5

Figure 12.16. Point spread function. **A:** Sections through the point spread function of the experimental three-dimensional data set of frog rhodopsin. Solid contour lines indicate the contour at half height of the peak. The underlying grid lines have a spacing of 20 Å. **B:** Half-width data obtained from the point spread function is converted to effective resolution cut-off values (Unger et al., 1997a; Unger and Schertler, 1995).

constant c^* along the z^*-axis normal to the plane defined by a^*, b^* and can then generate a list that contains for each set of indices (h,k,l) an amplitude, a phase, and a figure of merit (see Fig. 12.6A). This list is equivalent to the data measured for a three-dimensional crystal. Therefore a Fourier synthesis program used for x-ray crystallography can be used to calculate the three-dimensional map (FFT from the CCP4 program package [Bailey, 1994]).

6.E. Representing the Map Using Contour Plots and Surface Representations

To analyze sections through the density map obtained from the Fourier synthesis, we can use a contouring program (NPLT CCP4) that allows us to plot densities with a standard printer (Fig. 12.17B). A useful three-dimensional representation of a three-dimensional density map can be obtained by printing sections on overhead transparencies and by arranging them on top of each other in a frame. The sections can also be used to build a solid model of the density. First, an informative contour level is chosen that emphasizes the important features (e.g., the helices); then contour plots are printed and pieces of balsa wood are cut and assembled accordingly. This is a very good way of looking at low-resolution density maps in detail (Fig. 12.17A). Finally, we can use chicken wire representations of the density in the program O (Jones et al., 1991), which is widely used to interpret density data in x-ray crystallography, or we can use

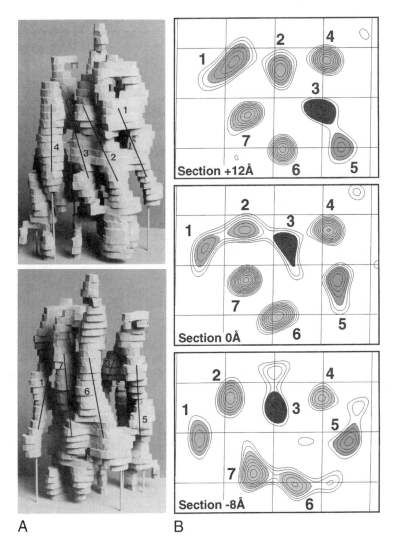

Figure 12.17. The seven helices in the rhodopsin structure. **A:** Structure of frog rhodopsin obtained by electron cryomicroscopy (Unger et al., 1997a). Two views of a solid model of the rhodopsin map are shown. **Top:** View from helix 2 toward helix 6. **Bottom:** View from helix 6 toward helix 3. The model was constructed from 33 contour sections 2 Å apart. The cytoplasmic side is at the top, and the intradiscal or extracellular side is at the bottom. The central sections of the seven transmembrane helices are marked with lines starting at section +12 Å at the top and ending at section −8 Å. The corresponding sections are shown in B. The peaks representing the seven helices are interpreted according to the sequence assignment (Baldwin, 1993) to the projection map of rhodopsin (Schertler et al., 1993). **B:** Three slices through the best part of the density map of rhodopsin (Unger et al., 1997a). In each of these sections, peaks can be seen for each of the seven transmembrane helices. The section closer to the cytoplasmic side is at z = +12 Å from the center and the last section at z = −8 Å from the center of the map. The least tilted helices (4, 6, and 7) are colored grey, and the most tilted ones are in four different colors. The grid spacing is 10 Å, with lines parallel to the **a** and **b** axes (+**b** is horizontal to the right, and +**a** points toward the bottom). (See color plates.)

a surface-rendering program like AVS (Sheehan et al., 1996) to give a surface representation at a certain contour level (Unger and Schertler, 1995).

7. INTERPRETATION OF DENSITY MAPS

The interpretation of density maps is exemplified in the next sections by looking at the p2 frog rhodopsin map in some detail. This is a good example of how a low-resolution map has contributed to our knowledge of the structure of GPCRs.

7.A. The Arrangement of Rhodopsin α Helices in the Membrane

Direct evidence for the arrangement of the seven α helices was first obtained from a 9 Å projection map of bovine rhodopsin (Schertler et al., 1993). Structural constraints obtained from a comparison of GPCR sequences were used to assign the seven hydrophobic stretches in the sequence to features in the projection map (Baldwin, 1993). A low-resolution three-dimensional structure of bovine rhodopsin (Unger and Schertler, 1995) and two projection structures of frog rhodopsin (Schertler and Hargrave, 1995) confirmed the positions of the three least tilted helices 4, 6, and 7. A more elongated peak of density for helix 5 indicated that it is tilted or bent (Schertler and Hargrave, 1995; Unger and Schertler, 1995), but helices 1, 2, and 3 were not resolved.

The extraction of frog rod cell membranes with Tween 80 resulted in the formation of two-dimensional crystals with p2 symmetry (Figs. 12.3A,C and 12.5A,C) (Schertler and Hargrave, 1995). The crystals were better ordered than two-dimensional crystals obtained previously by reconstitution of detergent-purified bovine rhodopsin in synthetic lipids (Schertler et al., 1993). The three-dimensional map that was calculated had an effective resolution of 7.5 Å in the membrane plane and 16.5 Å normal to it (Unger et al., 1997a). A single rhodopsin molecule in this map has planar dimensions of 28×39 Å. The molecule appears 64 Å high at the chosen contour level of Figure 12.17A, but this is only an approximation because of the low vertical resolution (Fig. 12.16). In contoured cross sections taken parallel to the membrane plane, the clearest features are close to the middle of the membrane. The central sections of the seven transmembrane helices are marked in Figure 12.17A with lines starting at section $+12$ Å and ending at section -8 Å. This part of the map is contained within the hydrophobic core of the bilayer.

The density peaks representing the seven helices have been interpreted according to the previous assignment of sequence segments (Baldwin, 1993) to the projection map of bovine rhodopsin (Schertler et al., 1993). The centers of the peaks on section $+12$ Å and -8 Å (Fig. 12.17B) were used to calculate tilt angles for the seven transmembrane helices. These tilt angles, which ignore possible helix curvature or kinks, give an indication of the tilt direction for each helix (Table 12.2). The density assigned to helix 3 has an inclination of 30°. This helix is the most tilted of all and is buried inside the molecule. Helix 3 closes the binding pocket of the retinal toward the cytoplasmic side and holds the vertical helices 4, 6, and 7 apart in the region of the structure that is closer to the intracellular surface. Currently, it is not possible to establish the precise

TABLE 12.2. Estimate of the Axes of the Seven Helices in Rhodopsin

Helix	Orientation		Position	
	Theta Degree	Phi Degree	x_0 (Å)	y_0 (Å)
1	28.4	141.0	-2.16	7.52
2	27.2	82.2	-6.24	15.08
3	29.6	50.7	-1.92	23.92
4	3.8	116.6	-7.04	30.08
5	22.7	-11.0	5.08	34.56
6	7.4	-90.0	10.40	23.16
7	13.4	-165.4	6.36	15.32

The axes are determined from the coordinates of the observed peaks on z-sections -8 Å and $+12$ Å of the rhodopsin map (Unger et al., 1997a). The crystallographic **b** axis is horizontal in Figure 12.16B, and the **c** axis is perpendicular to the plane of the membrane, with $+$**c** ($+$**Z**) pointing toward the intracellular side. Because the crystallographic axes are not orthogonal, coordinates are referred to orthogonal axes **X** and **Y,** where **Y** is parallel to **b** and **X** is perpendicular to **b** and **c.** Theta is the angle between the helix axis and the direction of the $+$**Z** axis; phi is the angle around the $+$**Z** axis measured from the direction of the $+$**X** axis; a positive phi angle indicates a right-handed rotation about $+$**Z** (i.e., positive phi toward $+$**Y,** negative phi toward $-$**Y**). x_0, y_0 are the coordinates in Å along the **X, Y** axes at which the helix axis intersects section $z = 0$.

starting and ending points of the helices because of the limited vertical resolution. Density observed towards the intradiscal end of helix 3 could be part of the loop connecting helices 4 and 5. This loop must reach between the extracellular ends of the helices to form the disulfide bond (C110–C187) (Karnik and Khorana, 1990) at the extracellular end of helix 3.

The density assigned to helix 4 is the least tilted feature and appears to be the shortest helix in the structure. It is separated from helices 6 and 7 by helices 2 and 3, producing a three-layered arrangement of the helices in the intracellular half of the molecule that interacts with transducin.

Helix 5 is less well resolved. It is seen as a tilted feature (23°) sloping from the bottom of helix 4. As helix 5 ascends toward the intracellular side, it appears to merge with helix 3 roughly 16 Å above the middle of the lipid bilayer.

Helix 6 is oriented nearly perpendicular to the membrane plane in the cytoplasmic half of the molecule. However, helix 6 appears bent toward helix 5 closer to the intradiscal side. This bend allows helix 6 to maintain contact with helix 5 and prevents the interior from becoming exposed to the lipid bilayer.

Helix 7 is assigned to a feature in the map oriented almost perpendicular to the plane of the membrane. It is close to helix 3 in the center of the molecule above the probable region where the retinal is likely to be attached. Near the intradiscal side close to the Schiff base, helix 7 appears to be distorted.

7.B. Comparison of Rhodopsin With Bacteriorhodopsin

Bacteriorhodopsin is a light-driven proton pump found in salt-loving archaebacteria. Like rhodopsin, it is a seven transmembrane helix protein. It has an all-*trans*-retinal bound via a protonated Schiff base to the ε-amino group of lysine (K261) in helix 7. There is no significant sequence similarity detectable between the two retinal proteins. The structure of bacteriorhodopsin has been determined by electron crystallography to near atomic resolution (Grigorieff et al., 1996; Henderson et al., 1990). Ribbon diagrams of rhodopsin and bacteriorhodopsin illustrate the different arrangements of helices in the two molecules (Fig. 12.18). Bacteriorhodopsin consists of three helices nearly perpendicular to the membrane plane (helices 2, 3, and 4) arranged parallel to a row of tilted helices (helices 1, 7, 6, and 5). In rhodopsin a band of tilted helices (helices 1, 2, 3, and 5) runs through the middle of the molecule with two nearly perpendicular helices on one side (helices 7, 6) and one helix on the other side (helix 4) (Baldwin et al., 1997; Unger et al., 1997b). Angles between pairs of helices have been calculated from the helix axes; these are compared in Table 12.3. The significantly different angles between pairs of helices 2^3, 3^4, 5^6, 7^1, 2^7, and 3^6 are indicative of the different packing of helices in rhodopsin and bacteriorhodopsin. They are independent of the overall orientation of either molecule in the membrane. Therefore, archaebacterial bacteriorhodopsin is a

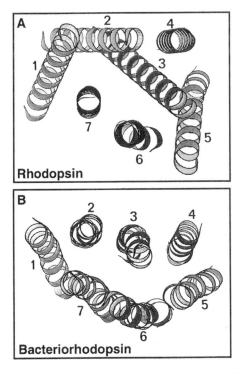

Figure 12.18. The arrangement of α helices in rhodopsin **(A)** and bacteriorhodopsin **(B)**. A ribbon diagram of rhodopsin was drawn using the coordinates from a recently published α-carbon template for the transmembrane helices in the rhodopsin family of GPCRs (Baldwin et al., 1997). A similar diagram was generated for bacteriorhodopsin based on the coordinates from the bacteriorhodopsin structure (Grigorieff et al., 1996). The diagrams illustrate the different arrangement of helices in bacteriorhodopsin and rhodopsin (Schertler, 1998). (See color plates.)

TABLE 12.3. Angles Between Pairs of Helices of Rhodopsin and Bacteriorhodopsin

Helix	Omega Degrees	
Pair	Rhodopsin	Bacteriorhodopsin
7^1	22.9	8.2
1^2	26.5	27.0
2^3	15.0	5.6
3^4	−28.2	−10.1
4^5	25.2	20.3
5^6	22.4	7.7
6^7	13.5	12.8
2^7	34.4	21.2
3^6	35.6	18.2

The angles between helix pairs were computed to compare the arrangement of helices in the seven helix retinal proteins rhodopsin and bacteriorhodopsin. Omega is the angle between the pair of helix axes specified. The omega angles are compared with those for the helix axes in bacteriorhodopsin, computed from the data repoted by Havelka et al. (1995). The significantly different angles for **7^1, 2^3, 3^4, 5^6, 2^7,** and **3^6** are indicative of the different packing of the helices in rhodopsin and bacteriorhodopsin. Positive angles indicate a left-handed relationship and negative numbers a right-handed one (Schertler, 1998).

much poorer model for GPCRs than mammalian rhodopsin, which is a member of that family.

7.C. The Retinal Binding Site in Rhodopsin

The 11-*cis*-retinal binds via a Schiff base linkage to the ε-amino group of lysine residue K296 in helix 7. The counterion for the protonated Schiff base is glutamic acid residue E113 in helix 3. This interaction might be mediated by a water molecule (Baldwin et al., 1997). Inspection of the rhodopsin structure shows that the helices are closely packed on the intracellular side of the molecule. The tilted helices 2 and 3 pack between the more perpendicular helices 4, 6, and 7, forming a three-layered structure (Fig. 12.17B, section +12 Å). The arrangement opens up toward the extracellular side, forming a cavity that serves as the binding pocket for retinal. It is formed by helices 3–7 (Fig. 12.17B, section −8 Å). This cavity is closed toward the intracellular side by the long and highly tilted helix 3 and must be closed toward the extracellular side by the loop linking helices 4 and 5, which is linked by a disulfide bridge (C110–C187) to the extracellular end of helix 3 (Karnik and Khorana, 1990). The retinal binding site is closer to the extracellular side of the molecule. The retinal chromophore lies at an angle of about 16° to the plane of the membrane (Liebman, 1962). The retinal is more likely to point toward the intracellular side from the Schiff base, placing the β-ionone ring of the retinal close to the conserved tryptophan W265 and the retinal polyene chain close to glycine G121 in helix 3 (Baldwin et al., 1997; Han et al., 1997; Han and Smith, 1995).

7.D. The Intracellular G Protein Binding Region

Less density is observed on the cytoplasmic side in comparison with the extracellular side, suggesting that the cytoplasmic loops are more loosely packed than the extracellular loops. The furthest extension of density on the cytoplasmic side is that corresponding to helix 6. This is supported by evidence from site-directed electron paramagnetic resonance measurements (Altenbach et al., 1996). Portions of the third cytoplasmic loop have been functionally implicated in interactions with transducin (Konig et al., 1989), arrestin (Krupnick et al., 1994), and rhodopsin kinase (Palczewski et al., 1991; Thurmond et al., 1997; Wilden et al., 1986). However, large segments of the second and third cytoplasmic loops can be deleted without affecting the ground state structure of the protein, as reflected by its ability to bind retinal (Franke et al., 1992).

The helix arrangement close to the cytoplasmic surface of rhodopsin is significantly more compact than near the extracellular surface (compare section $+12$ Å with section -8 Å in Fig. 12.17B). The formation of photoactivated rhodopsin (metarhodopsin II) was shown to be associated with the movement of helices (Farahbakhsh et al., 1995; Farrens et al., 1996; Sheikh et al., 1996). If this movement expands the cytoplasmic surface area, it would be consistent with the observation that formation of metarhodopsin II occurs with an increase in the overall volume of rhodopsin (Lamola et al., 1974) and provides a newly available binding site for transducin (König et al., 1989).

8. ASSIGNING THE AMINO ACID SEQUENCE TO THE DENSITY FEATURES IN THE MAP

During the last few years our laboratory has been involved in calculating projection structures and three-dimensional structures of rhodopsin from different species in a range of different crystal forms (Fig. 12.19) (Schertler et al., 1993; Davies et al., 1996; Schertler and Hargrave, 1995; Krebs et al., 1998; Unger and Schertler, 1995; Unger et al., 1997). These low-resolution structures of rhodopsin are affected by the resolution, contouring and rendering parameters, and errors in the data that introduce noise. In all these representations, the resolution is too low to assign the amino acid sequence to the density features in the maps. In the future we hope to reach a resolution in which density patterns on the helices will allow a direct assignment of sequence to at least part of the density map. Until then, we have to use information from the sequences of related GPCRs and other experiments as arguments to assign the density features to the protein sequence.

8.A. Helical Wheel Plots

Circular dichroism and Fourier transform infrared studies showed that a large part of the rhodopsin sequence is arranged in α helices that are more or less perpendicular to the membrane plane. Proteolysis data, topology studies with antibodies, site-directed cysteine spin labeling scans, and extensive sequence comparisons were used to examine where helices start and end, and several topology models have been suggested (summarized in Bladwin et al., 1997;

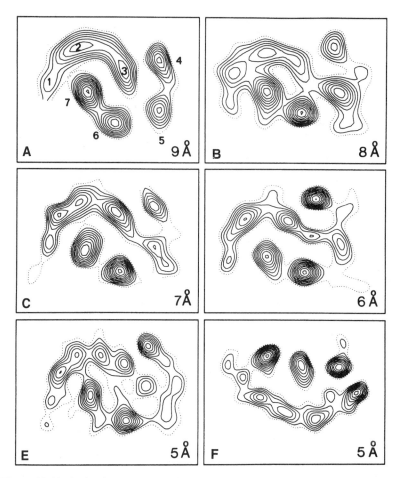

Figure 12.19. Projection structures of rhodopsin in five different crystal forms: bovine rhodopsin p222₁ **(A),** squid rhodopsin p222₁ **(B),** frog rhodopsin p22₁2₁ **(C),** frog rhodopsin p2 **(D),** and bovine rhodopsin p22₁2₁ **(E).** For comparison, a projection structure of bacteriorhodopsin to 5 Å resolution is shown in **(F).**

Hargrave and MacDowell, 1992). A helical wheel plot can be constructed that indicates the position of the C-α carbon atom on a cylinder. Helical wheel plots or polar plots are an excellent way to present stereochemical relations between amino acids in a helical stretch of sequence. For example, the amphipathic nature of a helix can be easily observed with polar residues clustering on one side of the helical wheel and the more hydrophobic residues clustering on the other side (Baldwin 1993, 1994; Baldwin et al., 1997).

8.B. Structural Constraints From Sequence Comparison

Rhodopsin is one of the best-studied examples of a GPCR, and it shares sequence homology with a large number of other receptors that activate heterotrimeric G proteins. More than 800 GPCR sequences are available, and they contain some interesting structural constraints that can be extracted. After comparing the length of loops and deciding where the helices start and end, it became clear that the helices must be arranged consecutively. The analysis indicated that helix 3 is the most buried helix in the structure, followed by helices 1, 4, and 5. Comparing closely related subtypes, the variability of side chains can be established and plotted in helical wheels. High variability should indicate exposure to the lipid rather than to the interior of the helix bundle. The same type of information is contained in sites where the variability is restricted and sites that accommodate polar residues. Conserved residues and functionally relevant residues should also be buried in the molecule rather than pointing toward the lipid (Baldwin, 1993, 1994; Baldwin et al., 1997).

8.C. Assignment of Amino Acid Sequence to Low-Resolution Maps

When the first projection map became available (Schertler et al., 1993), a comparison of GPCR sequences gave sufficient structural constraints to assign the seven hydrophobic stretches in the sequence of rhodopsin to features in the projection map (Baldwin, 1993). Helix 3 was found to be the most buried helix in the map and was therefore assigned to a feature in the center of the map. Because there were indications from the analysis of the length of hydrophilic loops that helices would be arranged in a consecutive way and crossovers could not occur, only two possibilities for the assignment were left. The sequence was arranged either clockwise or anticlockwise if the molecule was seen from the cytoplasmic side. However, only in the clockwise orientation was K296, which is known to form a protonated Schiff base with the 11-*cis*-retinal, in a position reasonably close to the known counterion, E113, of rhodopsin.

Helical wheel plots were then used to extend this prediction to a three-dimensional model of rhodopsin. The helical segments were split into three independent parts, and the helical wheels were then arranged in the best way to fit the projection structure, allowing several testable predictions. For example, it predicts that the cytoplasmic part of helix 1 is closer to helix 4 than its extracellular part, so it indicates the direction of tilt for the arc of helices 1–3. The first three-dimensional map of bovine rhodopsin did not have sufficient resolution to resolve the extracellular and intracellular part of helix 1 and therefore could not confirm the model. But it did show very clearly that helices 4, 6, and 7 were the least tilted helices and were arranged on a triangle (Unger and Schertler, 1995). However, a frog rhodopsin map, obtained at a resolution of 7.5 Å in the plane and at 16.5 Å normal to it, clearly showed that the intracellular density of helix 1 is closer to helix 4 than the extracellular part of the helix and therefore validated the prediction made by the model (Fig. 12.17) (Unger et al., 1997a). In these maps all seven helices are resolved, and the geometry of the helix bundle is defined. Assuming straight helices, tilt angles were also calculated (Table 12.2). This arrangement, while confirming the main features of the model, differs in that helices 1, 2, 3, and 5 are more tilted than originally predicted. Helix 3 is the most tilted helix, and its cytoplasmic part is closer to helix 5 than previously thought. In this structure, the helices are closely packed on the extracellular side forming a three-helix layer structure different from the intracellular part of bacteriorhodopsin. This part of the molecule is most likely rearranged during the activation of the receptor and is therefore directly involved in G protein activation. Toward the extracellular part the helices form a cavity that is likely to be the binding site for the retinal. This site is not in the middle of the molecule but is more toward the extracellular part of the structure.

8.D. Prediction of the Position of α Carbon Atoms

Further extensive sequence comparison, additional site-directed spin-labeling studies, and mutagenesis experiments have helped to refine the orientation and length of helices. There are also some indications of the relative stagger of the helices. Together with the frog rhodopsin low-resolution structure, this information was used to build a model template for GPCRs with a preliminary indication of the α carbon positions in the structure (Baldwin et al., 1997). A ribbon diagram is shown in Figure 12.18A. This model can be tested if we can obtain higher resolution maps that allow us to visualize the bulky aromatic amino acid side chains.

9. FUTURE PROSPECTS FOR ELECTRON CRYSTALLOGRAPHY OF RHODOPSIN

Electron cryomicroscopy is a fast developing area of structural biology. Expression systems, biochemical isolation methods, crystallization techniques, and image processing are being continuously improved, allowing a large number of membrane proteins to be studied with this approach in the future.

9.A. A New Crystal Form Providing Higher Resolution Data

A new crystal form of rhodopsin has been generated by using natural lipids in reconstitution experiments (Krebs et al., 1998). The crystals are better ordered than the ones we obtained earlier, and some images contain information to a resolution beyond 5 Å. In Figure 12.20A, a comparison of projection maps (calculated without applying the crystal symmetry or scaling the amplitudes) from a single image of a crystal taken with a conventional microscope (Fig. 12.20A) and with a field emission microscope (Fig. 12.20B) is shown. In the image taken with the more conventional microscope, we can mainly define the outline of the molecule, whereas in the image taken with the field emission microscope we can directly observe some features of the secondary structure of rhodopsin. It is quite remarkable that the three most perpendicular helices of rhodopsin are clearly resolved as peaks in this map obtained from a single image of a two-dimensional crystal from the bovine $p22_12_1$ crystal form. Therefore, we will be able to produce better three-dimensional structures of rhodopsin by using the new crystals and a field emission microscope.

9.B. Electron Diffraction

With the electron microscope we can either obtain images (by projecting the image plane of the objective lens on the film) or observe electron diffraction (by projecting the back focal plane of the objective lens on to the film or CCD detector).

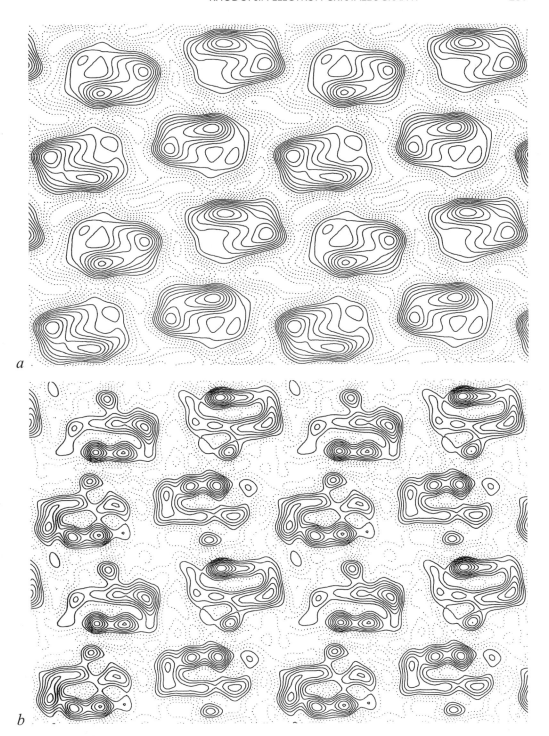

Figure 12.20. Advantage of a field emission microscope. A comparison of information obtained from a single image taken with a conventional microscope **(A)** and with a field emission microscope **(B)**. In A, only the outline of the rhodopsin molecule is visible. In B, individual peaks can be identified for the three least tilted helices of rhodopsin.

In diffraction mode, as in x-ray diffraction, we can only record the amplitudes of the Fourier components; the phases are lost when we use conventional detectors. However, measuring amplitudes with diffraction has some advantages. The defocus and spherical aberration do not affect the transmission of spatial frequencies, and therefore no correction for the CTF is necessary. Only a simple subtraction for the background is required. The diffraction pattern is insensitive to drift, which is a major concern in imaging. The limitation of diffraction is that a large, well-ordered crystalline membrane patch is needed to obtain a diffraction pattern with a sufficiently good signal to noise ratio. In contrast to imaging, we cannot correct for lattice distortions, and therefore the crystalline patch has to be well ordered and large enough to obtain a diffraction pattern that is not blurred. Recently, for the first time, we were able to obtain a promising diffraction pattern from rhodopsin crystals using a cooled CCD camera as the detector (Krebs et al., 1998; Faruqi et al., 1999; Faruqi et al., 1999). At present, only a small fraction of the membranes in our samples is large and well ordered enough to give diffraction. The highest resolution spots observed are at 3.5 Å (Fig. 12.21). We hope to obtain better structures of rhodopsin in the future by also including electron diffraction data.

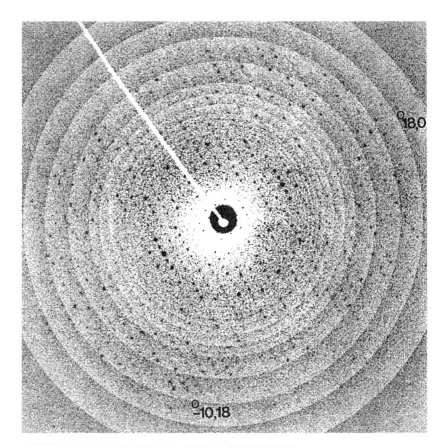

Figure 12.21. Electron diffraction analysis of bovine rhodopsin $p22_12_1$ (corrected). The best electron diffraction pattern obtained at present is shown. The circled reflections are at a resolution better than 3.5 Å (Krebs et al., 1998).

9.C. Determination of the Structure of Other GPCRs and Membrane Proteins

The biggest problem that needs to be overcome for most membrane proteins before direct structural methods can be applied is the overexpression of functional protein (Grisshammer and Tate, 1995; Schertler, 1992; Tate and Grisshammer, 1996). For a few GPCRs, procedures for the overexpression and purification have been described such that a few milligrams of protein can be purified on a regular basis. The next difficulty is to crystallize these proteins. Other GPCRs in contrast to rhodopsin are known to have only a few conditions and detergents in which they are stable for a longer period of time. Often functional assays are much more tedious than those used for rhodopsin. Therefore, it will take considerable effort to crystallize these important proteins. However, as our work has demonstrated, even rather disordered crystalline arrays can be used to extract useful structural information, and, as soon as the first crystals have been obtained, crystallization parameters can always be optimized. It is likely that several structures of interesting membrane proteins and GPCRs will be solved by electron cryomicroscopy in the future.

10. CONCLUSIONS

The arrangement of the α helices in rhodopsin has been determined by low-dose electron cryomicroscopy. The structure of rhodopsin is similar to that of bacteriorhodopsin around the retinal binding pocket, which is closer to the extracellular side of the molecule. However, near the intracellular side, which interacts with the G protein transducin, we observed three layers of helices arranged differently from those in bacteriorhodopsin. This arrangement changes after photoactivation and provides the G protein binding site. Movements of helices are thought to transmit changes caused by the isomerization of 11-*cis*-retinal in the retinal binding pocket to the intracellular surface of the rhodopsin molecule. This conformational change triggers the visual cascade. Recently, we obtained better ordered crystals of rhodopsin, and, with improved image processing methods, we hope to obtain higher resolution structures of rhodopsin and a structure of the photoactivated state, metarhodopsin II.

REFERENCES

Agard DA (1983): A least-squares method for determining structure factors in three-dimensional tilted-view reconstructions. J Mol Biol 167:849–852.

Altenbach C, Yang K, Farrens DL, Farahbakhsh ZT, Khorana HG, Hubbell WL (1996): Structural features and light-dependent changes in the cytoplasmic interhelical E–F loop region of rhodopsin—A site-directed spin-labeling study. Biochemistry 35:12470–12478.

Amos LA, Henderson R, Unwin PNT (1982): Three-dimensional structure determination by electron micrsocopy of two-dimensional crystals. Prog Biophys Mol Biol 39:183–231.

Auer M, Scarborough GA, Kuhlbrandt W (1998): Three-dimensional map of the plasma membrane H^+-ATPase in the open conformation. Nature 392:840–843.

Bailey S (1994): The CCCP4 suite—Programs for protein crystallography. Acta Crystallogr Sect D Biol Crystallogr 50:760–763.

Baldwin JM (1993): The probable arrangement of the helices in G-protein–coupled receptors. EMBO J 12:1693–1703.

Baldwin JM (1994): Structure and function of receptors coupled to G-proteins. Curr Opin Cell Biol 6:180–190.

Baldwin JM, Henderson R (1984): Measurement and evaluation of electron diffraction patterns from two dimensional crystals. Ultramicroscopy 14:319–336.

Baldwin JM, Henderson R, Beckman E, Zemlin F (1988): Images of purple membrane at 2.8Å resolution obtained by cryo-electron microscopy. J Mol Biol 202:585–591.

Baldwin JM, Schertler GFX, Unger VM (1997): An alpha-carbon template for the transmembrane helices in the rhodopsin family of G-protein–coupled receptors. J Mol Biol 272:144–164.

Berglaud GD (1969): Calculation of cross correlation maps. IEEE Spectrum 41.

Bullough P, Henderson R (1987): Use of spot-scan procedure for recording low-dose micrographes of beam-sensitive specimens. Ultramicroscopy 21:223–230.

Chiu W (1993): What does electron cryomicroscopy provide that x-ray crystallography and NMR spectroscopy cannot? Annu Rev Biophys Biomol Struct 22:233–255.

Corless JM, McCaslin DR, Scott BL (1982): 2-Dimensional rhodopsin crystals from disk membranes of frog retinal rod outer segments. Proc Natl Acad Sci USA 79:1116–1120.

Crowther RA, Henderson R, Smith JM (1996): MRC image-processing programs. J Struct Biol 116:9–16.

Davies A, Schertler GFX, Gowan BE, Saibil HR (1996): Projection structure of an invertebrate rhodopsin. J Struct Biol 117:36–44.

DeGrip WJ (1982): Purification of bovine rhodopsin over concanavalin a–sepharose. Methods Enzymol 81:197–207.

Deisenhofer J, Epp O, Miki K, Huber R, Michel H (1985): Structure of the protein subunits in the photosynthetic reaction center of rhodopseudomonas-viridis at 3Å resolution. Nature 318:618–624.

DeRosier D, Klug A (1968): Reconstruction of 3-dimensional structures from electron micrographs. Nature 217:130–134.

DeRosier DJ, Klug A (1972): Structure of the tubular variants of the head of bacteriophage T4 (polyheads). 1. Arrangement of subunits in some classes of polyheads. J Mol Biol 65:469–488.

DeRosier DJ, Moore PB (1970): Reconstruction of three dimensional images from electron micrographs of structures with helical symmetry. J Mol Biol 52:355–369.

Dratz EA, Vanbreemen JFL, Kamps KMP, Keegstra W, Vanbruggen EFJ (1985): Two-dimensional crystallization of bovine rhodopsin. Biochim Biophys Acta 832:337–342.

Dubochet J, Adrian M, Chang JJ, Homo JC, Lepault J, McDowell AW, Schultz P (1988): Cryo-electron microscopy of vitrified specimens. Q Rev Biophys 21:129–228.

Engel A, Hoenger A, Hefti A, Henn C, Ford RC, Kistler J, Zulauf M (1992): Assembly of 2-D membrane protein crystals: Dynamics, crystal order, and fidelity of structure analysis by electron microscopy. J Struct Biol 109:219–234.

Farahbakhsh ZT, Ridge KD, Khorana HG, Hubbell WL (1995): Mapping light-dependent structural-changes in the cytoplasmic loop connecting helix-c and helix-d in rhodopsin—A site-directed spin-labeling study. Biochemistry 34:8812–8819.

Farrens DL, Altenbach C, Yang K, Hubbell WL, Khorana HG (1996): Requirement of rigid-body motion of transmembrane helices for light activation of rhodopsin. Science 274:768–770.

Faruqi AR, Andrews HN (1997): Cooled CCD camera with tapered fiber optics for electron microscopy. Nucl Instr Methods Phys Res A 392:233–236.

Faruqi AR, Henderson R, Subrimanian S (1999): Cooled CCD detector with tappered fiber optics for recording electron diffraction patterns. Ultramicroscopy 75:235–250.

Frank J (1996): Three-Dimensional Electron Microscopy of Macromolecular Assemblies. San Diego: Academic Press.

Franke RR, Sakmar TP, Graham RM, Khorana HG (1992): Structure and function in rhodopsin—studies of the interaction between the rhodopsin cytoplasmic domain and transducin. J Biol Chem 267:14767–14774.

Fujiyoshi, Y (1998): The structural study of membrane proteins by electron crystallography. Adv Biophys 35:25–80.

Glaeser RM, Jubb JS, Henderson R (1985): Structural comparison of native and dexycholate-treated purple membrane. Biophys J 48:775–780.

Grigorieff N, Ceska TA, Downing KH, Baldwin JM, Henderson R (1996): Electron-crystallographic refinement of the structure of bacteriorhodopsin. J Mol Biol 259:393–421.

Grisshammer R, Tate CG (1995): Overexpression of integral membrane-proteins for structural studies. Q Rev Biophys 28:315–422.

Han M, Groesbeek M, Sakmar TP, Smith SO (1997): The C9 methyl group of retinal interacts with glycine-121 in rhodopsin. Proc Natl Acad Sci USA 94:13442–13447.

Han M, Smith SO (1995): High-resolution structural studies of the retinal-Glu113 interaction in rhodopsin. Biophys Chem 56:23–29.

Hargrave PA, McDowell JH (1992): Rhodopsin and phototransduction. Int Rev Cytol 137b:49–97.

Havelka WA, Henderson R, Heymann JAW, Oesterhelt D (1993): Projection structure of halorhodopsin from halobacterium-halobium at 6-Ångstrom resolution obtained by electron cryomicroscopy. J Mol Biol 234:837–846.

Havelka WA, Henderson R, Oesterhelt D (1995): 3-Dimensional structure of halorhodopsin at 7-Angstrom resolution. J Mol Biol 247:726–738.

Henderson R (1995): The potential and limitations of neutrons, electrons and x-rays for atomic-resolution microscopy of unstained biological molecules. Q Rev Biophys 28:171–193.

Henderson R, Baldwin JM, Ceska TA, Zemlin F, Beckmann E, Downing KH (1990): Model for the structure of bacteriorhodopsin based on high-resolution electron cryomicroscopy. J Mol Biol 213:899–929.

Henderson R, Baldwin JM, Downing KH, Lepault J, Zemlin F (1986): Structure of purple membrane from *Halobacterium halobium*—Recording, measurement and evaluation of electron-micrographs at 3.5 Å resolution. Ultramicroscopy 19:147–178.

Henderson R, Glaeser RM (1985): Quantitative analysis of image contrast in electron micrographs of beam-sensitive crystals. Ultramicroscopy 16:139–150.

Heymann JAW, Havelka WA, Oesterhelt D (1993): Homologous overexpression of a light-driven anion pump in an archaebacterium. Mol Microbiol 7:623–630.

Holser WT (1958): Point groups and plane groups in a two-sided plane and their subgroups. Z Kristallogr 110:266–281.

Jap BK (1988): High-resolution electron-diffraction of reconstituted phoe porin. J Mol Biol 199:229–231.

Jap BK, Zulauf M, Scheybani T, Hefti A, Baumeister W, Aebi U, Engel A (1992): 2D crystallization—From art to science. Ultramicroscopy 46:45–84.

Jones TA, Zou JY, Cowan SW, Kjeldgaard M (1991): Improved methods for building protein models in electron-density maps and the location of errors in these models. Acta Crystallogr Sect A Fundamentals of Crystallography 47:110–119.

Karnik SS, Khorana HG (1990): Assembly of functional rhodopsin requires a disulfide bond between cysteine residue-110 and residue-187. J Biol Chem 265:17520–17524.

Khorana HG (1992): Rhodopsin, photoreceptor of the rod cell—An emerging pattern for structure and function. J Biol Chem 267:1–4.

König B, Arendt A, McDowell JH, Kahlert M, Hargrave PA, Hofmann KP (1989): 3 cytoplasmic loops of rhodopsin interact with transducin. Proc Natl Acad Sci USA 86:6878–6882.

Krebs A, Villa C, Edwards PC, Schertler GFX (1998): Characterisation of an improved two-dimensional $p22_12_1$ crystal from bovine rhodopsin. J Mol Biol 282:991–1003.

Krupnick JG, Gurevich VV, Schepers T, Hamm HE, Benovic JL (1994): Arrestin–rhodopsin interaction—Multisite binding delineated by peptide inhibition. J Biol Chem 269:3226–3232.

Kühlbrandt W (1992): 2-Dimensional crystallization of membrane-proteins. Q Rev Biophys 25:1–49.

Kühlbrandt W, Wang DN, Fujiyoshi Y (1994): Atomic model of plant light-harvesting complex by electron crystallography. Nature 367:614–621.

Lamola AA, Yamane T, Zipp A (1974): Effects of detergents and high pressures upon the metharhodopsin I <=> metharhodopsin II equilibrium. Biochemistry 13:738–745.

Liebman PA (1962): In situ microspectrophotometric studies on the pigments of single retinal rods. Biophys J 2:162–178.

Litman BJ, Mitchell DC (1996): A role for phospholipid polyunsaturation in modulating membrane–protein function. Lipids 31:193–197.

Ostermeier C, Iwata S, Ludwig B, Michel H (1995): F-v fragment mediated crystallization of the membrane–protein bacterial cytochrome-c-oxidase. Nature Struct Biol 2:842–846.

Palczewski K, Buczylko J, Kaplan MW, Polans AS, Crabb JW (1991): Mechanism of rhodopsin kinase activation. J Biol Chem 266:12949–12955.

Rigaud JL, Mosser G, Lacapere JJ, Olofsson A, Levy D, Ranck JL (1997): Bio-beads: An efficient strategy for two-dimensional crystallization of membrane proteins. J Struct Biol 118:226–235.

Rigaud JL, Pitard B, Levy D (1995): Reconstitution of membrane–proteins into liposomes—Application to energy-transducing membrane–proteins. Biochim Biophys Acta Bioenerget 1231:223–246.

Schertler GFX (1992): Overproduction of membrane proteins. Curr Opin Struct Biol 2:534–544.

Schertler GFX (1998): Structure of rhodopsin. Eye 12:504–510.

Schertler GFX, Hargrave PA (1995): Projection structure of frog rhodopsin in 2 crystal forms. Proc Natl Acad Sci USA 92:11578–11582.

Schertler GFX, Villa C, Henderson R (1993): Projection structure of rhodopsin. Nature 362:770–772.

Sheehan B, Fuller SD, Pique ME, Yeager N (1996): AVS software for visualization in molecular microscopy. J Struct Biol 116:99–106.

Sheikh SP, Zvyaga TA, Lichtarge O, Sakmar TP, Bourne HR (1996): Rhodopsin activation blocked by metal-ion-binding sites linking transmembrane helice-C and helice-F. Nature 383:347–350.

Spence JCH (1988). Experimental High-Resolution Electron Microscopy, 2nd ed. Oxford: Oxford University Press.

Stadel JM, Wilson S, Bergsma DJ (1997): Orphan G protein–coupled receptors: A neglected opportunity for pioneer drug discovery. Trends Pharmacol Sci 18:430–437.

Tate CG, Grisshammer R (1996): Heterologous expression of G-protein–coupled receptors. Trends Biotechnol 14:426–430.

Thon F (1966): Zur Defokussierungsabhängigkeit des Phasenkontrastes bei der elektronenmikroskopischen Abbildung. Z Naturforsch 21a:476–478.

Thurmond RL, Creuzenet C, Reeves PJ, Khorana HG (1997): Structure and function in rhodopsin: Peptide sequences in the cytoplasmic loops of rhodopsin are intimately involved in interaction with rhodopsin kinase. Proc Natl Acad Sci USA 94:1715–1720.

Toyoshima C, Sasabe H, Stokes DL (1993): 3-Dimensional cryoelectron microscopy of the calcium-ion pump in the sarcoplasmic-reticulum membrane. Nature 362:469–471.

Toyoshima C, Unwin N (1990): 3-Dimensional structure of the acetylcholine-receptor by cryoelectron microscopy and helical image-reconstruction. J Cell Biol 111:2623–2635.

Tsukihara T, Aoyama H, Yamashita E, Tomizaki T, Yamaguchi H, Shinzawaitoh K, Nakashima R, Yaono R, Yoshikawa S (1996): The whole structure of the 13-subunit oxidized cytochrome-c-oxidase at 2.8 Ångstrom. Science 272:1136–1144.

Tsygannik IN, Baldwin JM (1987): 3-Dimensional structure of deoxycholate-treated purple membrane at 6 Å resolution and molecular averaging of 3 crystal forms of bacteriorhodopsin. Eur Biophys J 14:263–272.

Unger VM, Hargrave PA, Baldwin JM, Schertler GFX (1997a): Arrangement of rhodopsin transmembrane alpha-helices. Nature 389:203–206.

Unger VM, Kumar NM, Gilula NB, Yeager M (1997b): Projection structure of a gap junction membrane channel at 7 angstrom resolution. Nature Struct Biol 4:39–43.

Unger VM, Schertler GFX (1995): Low-resolution structure of bovine rhodopsin determined by electron cryomicroscopy. Biophys J 68:1776–1786.

Unwin PNT, Henderson R (1975): Molecular structure determination by electron microscopy of unstained crystalline specimens. J Mol Biol 94:425–440.

Valpuesta JM, Carrascosa JL, Henderson R (1994): Analysis of electron-microscope images and electron-diffraction patterns of thin-crystals of φ29-connectors in ice. J Mol Biol 240:281–287.

Wade RH (1992): A brief look at imaging and contrast transfer. Ultramicroscopy 46: 145–156.

Walz T, Grigorieff N (1998): Electron crystallography of two-dimensional crystals of membrane proteins. J Struct Biol 121:142–161.

Walz T, Hirai T, Murata K, Heymann JB, Mitsuoka K, Fujiyoshi Y, Smith BL, Agre P, Engel A (1997): The three-dimensional structure of aquaporin-1. Nature 387:624–627.

Weiss MS, Kreusch A, Schiltz E, Nestel U, Welte W, Weckesser J, Schulz GE (1991): The structure of porin from *Rhodobacter capsulatus* at 1.8 Å resolution. FEBS Lett 280:379–382.

Wilden U, Hall SW, Kuhn H (1986): Phosphodiesterase activation by photoexcited rhodopsin is quenched when rhodopsin is phosphorylated and binds the intrinsic 48-kda protein of rod outer segments. Proc Natl Acad Sci USA 83:1174–1178.

Xia D, Yu CA, Kim H, Xian JZ, Kachurin AM, Zhang L, Yu L, Deisenhofer J (1997): Crystal structure of the cytochrome $bc(1)$ complex from bovine heart mitochondria. Science 277:60–66.

Yeager M (1994): *In situ* two-dimensional crystallisation of a polytopic membrane protein: The cardiac gap junction channel. Acta Crystallogr D50:1–7.

Zhang PJ, Toyoshima C, Yonekura K, Green NM, Stokes DL (1998): Structure of the calcium pump from sarcoplasmic reticulum at 8-Ångstrom resolution. Nature 392:835–839.

CHAPTER 13

SITE-DIRECTED SPIN-LABELING (SDSL) STUDIES OF THE G PROTEIN–COUPLED RECEPTOR RHODOPSIN

DAVID L. FARRENS

1. INTRODUCTION   291
2. AN OVERVIEW OF SDSL   292
   A. What Can Be Learned From an SDSL Study   292
   B. Determining Protein Structures Using SDSL   292
      a. Nitroxide Scanning SDSL   292
      b. Interpreting Scanning SDSL Studies   293
      c. Distance Measurements Using SDSL   293
   C. Measuring Protein Conformational Changes Using SDSL   293
3. PRINCIPLES OF THE SDSL METHOD AND
   EPR SPECTROSCOPY   293
   A. Basic Principles of EPR Spectroscopy   293
      a. A Qualitative Description of EPR Spectroscopy   293
      b. The Basis of an EPR Spectrum   294
      c. Why Three Lines Are Observed in a Nitroxide EPR Spectrum   294
   B. How Information about the Spin Label Is Gained
      From EPR Studies   294
      a. Spin-Label Mobility   295
      b. Spin-Label Accessibility   296
      c. Measuring Distances Between Two Spin Labels   297
   C. Instrumentation for SDSL Studies   297
      a. EPR Instruments   297
      b. Loop-Gap Resonator   297
4. TYPICAL STEPS IN AN SDSL STUDY   298
   A. Constructing a Protein With No Reactive
      "Background" Cysteines   298
   B. Introducing Unique Reactive Cysteines Into the Region
      of Interest   298

Structure–Function Analysis of G Protein-Coupled Receptors, Edited by Jürgen Wess.
ISBN 0-471-25228-X   Copyright © 1999 Wiley-Liss, Inc.

C. Expressing, Purifying, and Spin Labeling the Cysteine
      Mutant Proteins   298
      a. Expression of Rhodopsin Cysteine Mutants   299
      b. Harvesting and Reconstitution of Rhodopsin Cysteine
         Mutants With 11-*cis*-Retinal   299
      c. Purification and Spin Labeling of Rhodopsin
         Cysteine Mutants   299
   D. Determining the Spin-Labeling Efficiency   300
      a. Double Integration of the EPR Spectrum and
         Comparison to a Known Standard   300
      b. Determining the Amount of Free Cysteines Remaining
         After Spin Labeling   300
      c. Comparison of Samples' Peak to Peak Height
         With a Standard Curve of Spin Labels   300
   E. Assessing the Consequence of the Cysteine Mutation
      and Spin Label on the Protein Structure   300
   F. Studying Each Spin-Labeled Mutant by EPR   301
      a. Spin-Label Mobility   301
      b. Spin-Label Accessibility   301
      c. Spin-Label Proximity to Another Spin-Label   302
   G. Constructing a Model Protein Structure Consistent
      With the SDSL Results   302
5. WHAT HAS BEEN LEARNED FROM SDSL STUDIES
   OF RHODOPSIN   302
   A. Background of the Rhodopsin System   303
      a. Structure of Rhodopsin   303
      b. Biochemistry of the Rhodopsin System   303
   B. Regions in the Cytoplasmic Domain Studied by SDSL   305
   C. Membrane–Aqueous Boundary   305
   D. Orientation of Individual Helices   305
   E. Three-Dimensional Packing of Helices   306
      a. Helix C Is Near Helix F   306
      b. Helix G is Near Helix A   306
   F. Structure of the Cytoplasmic Loops and Cytoplasmic Tail   306
      a. The Loop Connecting Helices C and D Forms
         a Nonperiodic Structure   306
      b. Helices E and F Form Extended Helices   306
      c. The Cytoplasmic Tail Is Near Helix F   307
   G. SDSL Studies of Conformational Changes at Rhodopsin's
      Native Cysteine Residues   307
      a. SDSL Studies of Wild-Type Rhodopsin   307
      b. The First SDSL Study of Cysteine Mutant Rhodopsins   308
   H. Conformational Changes in Rhodopsin Detected
      by Scanning SDSL Studies   308
   I. Conformational Changes in Rhodopsin Measured
      Using Spin-Labeled Double Cysteine Mutants   309
      a. Helix F Moves Away From C   309
      b. The Cytoplasmic Tail Moves   309

6. CONCLUSIONS AND FUTURE DIRECTIONS   310
   A. Summary   310
   B. Advances in the SDSL Technique   311
      a. Advances in Molecular Biology   311
      b. Advances in EPR Instrumentation   311
      c. Advances in EPR Data Analysis and Interpretation 311
   C. Future SDSL Rhodopsin Studies   311

I. INTRODUCTION

The goal of many laboratories (including our own) is to understand how inactive G protein coupled–receptors (GPCRs) are converted into active signaling proteins. To learn how these transmembrane receptors become "turned on," it is necessary to find out where changes occur in their structure upon activation. Unfortunately, no high-resolution GPCR crystal structure yet exists, requiring the use of other structural techniques to achieve this goal.

Many useful methods for studying the structure and function of GPCRs have been developed (see other chapters in this volume). One approach that has shown great promise for the study of membrane proteins such as GPCRs is site-directed spin labeling (SDSL) (for reviews of this technique, see Hubbell and Altenbach, 1994; Hubbell et al., 1996). SDSL involves the introduction of an electron paramagnetic resonance (EPR) spectroscopic reporter group into a protein, usually by engineering cysteines into the protein and then labeling them with the EPR probe. The attached EPR probes (called *spin labels*) are then studied by EPR spectroscopy. The structure of the most commonly used spin label is shown in Figure 13.1.

I discuss how SDSL and EPR spectroscopy can be used to study the structure and function of GPCRs. The examples given are limited to studies of rhodopsin, the only GPCR yet studied by this technique. However, SDSL should be readily applicable to the study of other GPCRs as well.

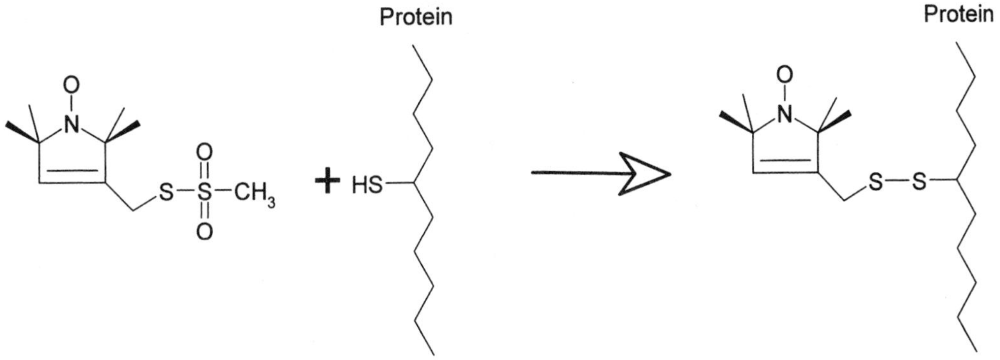

Figure 13.1. Reaction of the methanethiosulfonate spin label with a cysteine side-chain residue on a protein.

The chapter is organized into four parts: an overview of the SDSL technique, an explanation of the principles behind the SDSL approach, basic experimental procedures used in SDSL studies of rhodopsin, and a discussion of what has been learned about rhodopsin from SDSL studies.

2. AN OVERVIEW OF SDSL

2.A. What Can Be Learned From an SDSL Study

In an SDSL study, EPR measurements are carried out to learn the mobility of a spin label, its accessibility to the solvent, and its proximity to another spin-label (Hubbell and Altenbach, 1994). This information in turn allows conclusions to be drawn about the protein's structure in the region where the label is attached. Figure 13.2 summarizes the SDSL process.

2.B. Determining Protein Structures Using SDSL

2.B.a. Nitroxide Scanning SDSL. SDSL studies of a protein are most informative when a *nitroxide scan* is done. In such a study, a *series* of cysteine mutant proteins is first constructed, each of which contains a single reactive cysteine introduced sequentially throughout the region of the protein being studied (more information on introducing cysteines sequentially into proteins can be found in Chapter 2). Next, each cysteine mutant is reacted with a nitroxide accompanying spin label. The labeled mutants are then studied by EPR spectroscopy to determine the mobility and accessibility of each label. The nitroxide

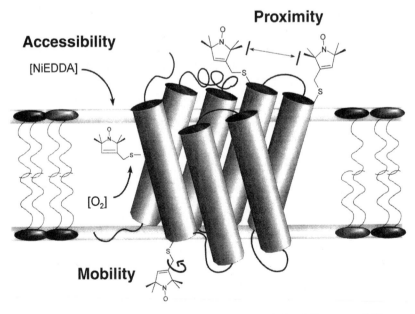

Figure 13.2. Representation of the types of data (spin-label mobility, accessibility, and proximity) that can be obtained from SDSL studies of a GPCR (see text for details).

scanning SDSL method thus produces an array of data about the region being studied, the analysis of which allows a probable structure to be modeled.

2.B.b. Interpreting Scanning SDSL Studies. The interpretations of SDSL results is based on firm ground: Spin labels on the outside surface of a protein are mobile and solvent exposed, whereas labels on the inside are always much less mobile and less accessible to the solvent (Altenbach et al., 1994; Mchaourab et al., 1996). Spin labels at tertiary contact sites lie somewhere in between. Thus, the structural class of a site (i.e., the fold of the peptide backbone) can be determined by comparing the mobility and solvent accessibility of a series of spin labels with their position in the amino acid sequence.

2.B.c. Distance Measurements Using SDSL. Measuring distances between sites in a protein helps to define the global packing of the protein structure. The scanning SDSL studies provide information on the types of secondary structure present (e.g., α helices, β sheets), and help to define possible tertiary interactions. However, to learn the arrangement of these elements in a three-dimensional fold requires further information. This information can be provided by using SDSL to measure distances between two nitroxide spin labels on the same protein (Rabenstein and Shin, 1995; Mchaourab et al., 1997b) and between a spin label and a paramagnetic metal bound to a protein (Voss et al., 1995). Finally, SDSL distance measurements can be used to study conformational changes in proteins, allowing the direction and size of domain movements in proteins to be determined (Farrens et al., 1996; Yang et al., 1996a; Thorgeirsson et al., 1997). This subject is explored further below.

2.C. Measuring Protein Conformational Changes Using SDSL

Perhaps one of the greatest strengths of SDSL for studying proteins is the fact that it provides real-time information about proteins in solution. Thus, SDSL can be used to monitor the *time course* of protein conformational changes in defined regions of a protein and to monitor folding processes in proteins (Farahbakhsh et al., 1993; Steinhoff et al., 1994; Hubbell et al., 1996).

3. PRINCIPLES OF THE SDSL METHOD AND EPR SPECTROSCOPY

A detailed description of spin labels and EPR spectroscopy is beyond the scope of this chapter. For a good introduction to these topics, many excellent articles are available (e.g., Brudvig, 1995; Millhauser et al., 1995). For the purposes of the discussion here, an overview of the EPR spectroscopy of spin labels is given below.

3.A. Basic Principles of EPR Spectroscopy

3.A.a. A Qualitative Description of EPR spectroscopy. An unpaired electron possesses a magnetic moment due to its quantum spin ($S = 1/2$). When placed in a static magnetic field, the dipole moments of an unpaired electron can orient either parallel or antiparallel with the static magnetic field, resulting in a splitting of the two allowed energy levels (spin states). Microwave radiation with an energy (frequency) that coincides with that required for spin reorientation will induce a resonance absorption that is registered as an EPR signal. This process can be described by the equation below:

$$h \cdot \nu = g \cdot \beta \cdot H$$

where h = the Planck constant, ν = the frequency of microwave radiation applied perpendicular to the magnetic field H, g is the free-electron g-factor, and β is the Bohr magneton.

3.A.b. The Basis of an EPR Spectrum.

An EPR spectrometer is similar to an absorption spectrometer in the sense that it measures the absorption of electromagnetic radiation. However, in EPR the resonance frequency depends on the magnitude of the magnetic field H. Furthermore, for practical reasons, rather than scanning the microwave frequency ν and measuring which wavelengths are absorbed by the sample, in an EPR spectrometer the *magnetic* field H is scanned while the microwave frequency ν is held constant. To improve the signal to noise ratio, while the magnetic field is being swept during a scan it is also modulated slightly. This process allows phase-sensitive detection to be used, increasing the overall sensitivity of the EPR measurements. The EPR spectra that are recorded look unique because they are plots of the *first derivative* of microwave absorption (y-axis) versus magnetic field strength (x-axis). First derivative spectra have the advantage in that they allow very small spectral changes to be observed.

3.A.c. Why Three Lines Are Observed in a Nitroxide EPR Spectrum.

While only one unpaired electron exists per nitroxide spin label, *three* absorption lines are observed in its EPR spectrum due to a phenomenon called *hyperfine splitting.* Hyperfine splitting is caused by an interaction (coupling) of the unpaired electron's spin with the spin of the nitrogen nucleus. Each nuclear spin coupling causes 2I + 1 "hyperfine" lines in the EPR spectrum (where I = the spin angular momentum quantum number of the nucleus). The nuclear spin (I) of [14]nitrogen is one. Thus, three nuclear hyperfine lines are observed in a nitroxide EPR spectrum. The position of these hyperfine lines varies depending on the molecular orientation of the nitroxide, which varies (in solution) depending on how the orientations are averaged due to molecular motions. This fact can be used to learn information about the spin label (described below).

3.B. How Information About the Spin Label Is Gained From EPR Studies

EPR studies of spin labels are usually carried out to learn at least one of the following three types of information: (1) the spin label's mobility, (2) the accessibility of the spin label, and (3) the proximity of one spin label to a second spin label. How these types of information are obtained is described below.

3.B.a. Spin-Label Mobility. The overall shape of the EPR spectrum is directly affected by the mobility of the nitroxide spin label. Often, the mobility of a spin label can be estimated simply by looking at the spectrum. An example is given in Figure 13.3. Figure 13.3 shows how the (simulated) EPR spectra of a spin label changes as a function of mobility (rotational correlation time, τ_R). One can see that at the greater spin label mobilities (shorter rotational correlation times), the three hyperfine lines are "narrow" and "sharp." At lower spin-label mobilities (longer rotational correlation times), the spectra become broader and the outer hyperfine lines become further separated. Such qualitative conclusions are often adequate to interpret the data from an SDSL study. However, more objective methods have been developed to assess mobility, two of which are described below (note that each method is usually valid only over a narrow range of cases):

1. The mobility of a spin label (essentially the rotational correlation time, τ_R) can be estimated by comparing the positions of the outer hyperfine

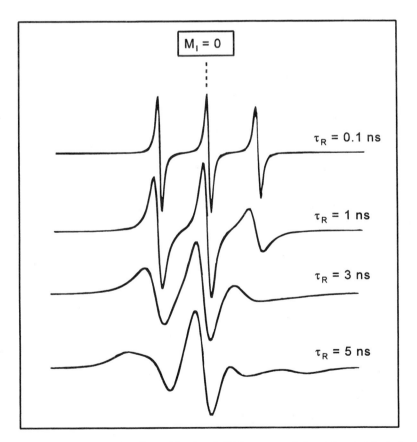

Figure 13.3. An example of how (simulated) EPR spectra of nitroxide spin labels depend on the rotational correlation time (τ_R) of the label. τ_R is directly related to the mobility of the spin label. Note the broadening of the central $M_I = 0$ hyperfine line as τ_R increases. (Adapted from Millhauser et al. [1995], with permission of the publisher.)

extrema peaks with those from spectral simulation of a spin label with different isotropic rotational rates (Freed, 1976; Altenbach et al., 1989).

2. The width of the central ($M_I = 0$) first derivative peaks can be used as a simple but reliable way to estimate a spin-label's mobility (Mchaourab et al., 1996). As the rotational rate of a spin label slows, this peak broadens (see Fig. 13.3). This value is often reported as a reciprocal (ΔH^{-1}) so that its value increases with increasing mobility (Altenbach et al., 1996; Mchaourab et al., 1996).

3.B.b. Spin-Label Accessibility. Whether a spin label is exposed or buried can be determined by measuring the rate of its collision with a paramagnetic agent such as O_2 or Ni. A greater rate of collision correlates with greater accessibility of the spin label, whereas a lower rate of collision suggests that the label is buried. Collision rates are measured on standard EPR instruments using a technique called *power saturation* (Subczynski and Hyde, 1981). Briefly, power saturation measurements are based on the fact that collisions of the paramagnetic agent with the nitroxide spin label results in a spin exchange between the two species. The spin exchange results in an increase in the spin-lattice relaxation rate of the nitroxide. The result of this is that the system under study is able to absorb more microwave energy before saturating (note that saturation is said to occur when the amplitude of the signal begins to deviate from the square-root dependency on the microwave power). Thus, the increased amount of microwave power required for power saturation is directly proportional to the collision rate of the nitroxide with the quenching agents and represents a measure of the accessibility of the spin label (Altenbach et al., 1989).

A power saturation measurement is usually carried out by measuring the peak-to-peak amplitude of the central $M_I = 0$ resonance (see Fig. 13.3) of the nitroxide spin label as a function of microwave power. Usually, a series of such spectra is recorded as a function of microwave power (usually in the range of 0.1–200 mW) in the presence and absence of the quenching agents (Ni or O_2).

The data from these measurements are then analyzed to find the $P_{1/2}$ value (the power at which the sample is said to be half-saturated). The $P_{1/2}$ value is determined for each case from a plot of the peak-to-peak amplitude of the central line ($M_I = 0$ spectrum) versus the square root of microwave power. Usually, the spin label's accessibility is measured by determining the *change* in $P_{1/2}$ values. This is accomplished by taking the difference of the $P_{1/2}$ values obtained in the presence of paramagnetic agents R (such as O_2 or Ni) and in their absence (by flushing with N_2):

$$\Delta P_{1/2} = P_{1/2}(R) - P_{1/2}(N_2)$$

To enable comparison of different spin-labeled sites that have differences in mobility, the $\Delta P_{1/2}$ values are usually divided by their peak-to-peak first derivative linewidths (ΔH).

Often, SDSL studies report the data as the "accessibility parameter," II, rather than reporting $\Delta P_{1/2}$ values. This is because $\Delta P_{1/2}$ values can be spectrometer dependent and vary depending on the nitroxide in question. These variables can be taken into account by normalizing the data, using a crystal of 2,2-diphenyl-1-

picrylhydrazyl (DPPH) as a "standard" and employing the following relationship to obtain the "accessibility parameter" II (Farahbakhsh et al., 1992; Altehbach et al., 1996):

$$II = \frac{\Delta P_{1/2}(R)}{P_{1/2}(DPPH)} \cdot \frac{\Delta H(DPPH)}{\Delta H(R)}$$

where $P_{1/2}$ (DPPH) is the $P_{1/2}$ for the DPPH crystal, $\Delta H(R)$ is the peak-to-peak central linewidth of the EPR spectrum from the spin-labeled mutant, and $\Delta H(DPPH)$ is the peak-to-peak central linewidth of the DPPH crystal standard. An advantage of this approach is that it results in dimensionless data.

3.B.c. Measuring Distances Between Two Spin Labels.

Distances of ~8–25 Å can readily be determined using EPR spectroscopy by measuring dipole–dipole interactions between a pair of spin labels. Experimentally, the magnetic interactions lead to a broadening of the EPR spectrum, which is observed as an apparent reduction in peak intensity. Unfortunately, interpreting the dipolar broadened spectra at room temperature is complicated. The amount of the spectral broadening induced by dipole–dipole coupling depends on the relative orientation of the interspin vector with respect to the magnetic field. Because the linewidth at room temperature is strongly dependent on relative mobility, distances are usually determined by measuring the spin-labeled samples in the frozen state where this motion is no longer a contributing factor to the lineshape.

Several methods to calculate distances between spin labels on frozen protein samples have been developed (Likhtenshtein, 1993; Farahbakhsh et al., 1995a; Rabenstein and Shin, 1995; Steinhoff et al., 1997; Hustedt et al., 1997). However, even at room temperature, important qualitative conclusions about the proximity of two spin labels can be reached (Mchaourab et al., 1997a; Cai et al., 1997). Furthermore, recent work has developed theory to enable measurements of distances between spin labels in smaller proteins at room temperature and showed that the calculated distances obtained agree well with the protein's known structure (Mchaourab et al., 1997b).

Another way to measure distances in proteins using SDSL involves measuring the distance between a nitroxide spin label and a bound paramagnetic metal. An example of this approach can be found in Voss et al., (1995).

3.C. Instrumentation for SDSL Studies

3.C.a. EPR Instruments.

EPR instruments are somewhat expensive, typically costing more than $100,000 for a new research grade instrument. New EPR instruments can be obtained from Bruker (Billerca, MA) or from JEOL USA (Peabody, MA). Varian no longer makes EPR instruments, but used models can sometimes be found in good working order and can often be interfaced to computers.

3.C.b. Loop-Gap Resonator.

An instrumental breakthrough that has enabled the use of the SDSL technique is the development of the loop-gap resonator (Froncisz and Hyde, 1982). This new form of microwave resonant structure increases the absolute sensitivity of the instrument by approximately 50-fold. The increased sensitivity allows for extremely small amounts of biological samples to be used, typically 20–40 μg of a spin-labeled sample in a working volume of ~3 to 5 μl, or ~0.5–1.0 nmol of spin label (Hubbell et al., 1987). Loop-gap resonators are manufactured by Jagmar (Krakow, Poland) and can be obtained through Medical Advances (Milwaukee, WI). Because of the small volume requirements of the loop-gap resonator, samples are usually placed in capillary tubes (~1 mm OD) usually made from Pyrex or from TPX, a gas-permeable plastic material (Popp and Hyde, 1981).

4. TYPICAL STEPS IN AN SDSL STUDY

The typical steps involved in an SDSL study are outlined below, in which rhodopsin is used as an example. In each step, the general principle is first described, followed by a more detailed description of the protocols.

4.A. Constructing a Protein With No Reactive "Background" Cysteines

Using site-directed mutagenesis, the reactive native cysteines present in the wild-type protein are mutated to neutral amino acids, usually serine or alanine. Removing all of the native cysteines in a protein is not always necessary (or possible), and the reactive cysteines may already be known from prior studies of the protein.

In rhodopsin, only 2 of the 10 native cysteines are highly reactive to cysteine reagents. Thus, it was found that by making cysteine to serine substitutions of these two native cysteines (C140S and C316S), and the two palmitoylation site cysteines (C322S and C323C), a background labeling with the methanethiosulfonate (MTSL) reagent of less than 0.3 per protein could be obtained (Resek et al., 1993).

4.B. Introducing Unique Reactive Cysteines Into the Region of Interest

The cysteines are systematically introduced into the "cys-less" background protein described above. Systematic cysteine scanning mutagenesis is made convenient in rhodopsin because of the availability of a synthetic rhodopsin gene (Oprian et al., 1987). For more details on mutagenesis of rhodopsin, see Chapters 5 and 6.

4.C. Expressing, Purifying, and Spin Labeling the Cysteine Mutant Proteins

One challenge in SDSL studies is to obtain sufficient amounts of purified, spin-labeled protein samples. Most of the SDSL studies of rhodopsin use mutant proteins that are first expressed in COS cells and then purified using an immunoaffinity method (Oprian et al., 1987). The spin labeling is carried out while the protein samples are bound to the antibody column (Resek et al.,

1993). This approach allows extensive washing of the samples to remove unbound labels before elution of the samples from the column.

4.C.a. Expression of Rhodopsin Cysteine Mutants.
The rhodopsin cysteine mutants are expressed in COS cells using a DEAE-dextran transient transfection protocol (Oprian et al., 1987; Karnik et al., 1993). Typically, 20 μg of plasmid mutant DNA is used per each 15-cm plate.

4.C.b. Harvesting and Reconstitution of Rhodopsin Cysteine Mutants With 11-cis-Retinal.
Two days after transfection, the cells are harvested and resuspended in PBSSC (137 mM NaCl/2.7 mM KCl/1.8 mM, KH$_2$PO$_4$/10 mM NaH$_2$PO$_4$) at pH 7.2. Two milliliters of PBSSC is used per one 15-cm plate. The resuspended samples are then treated with 11-cis-retinal (5 μM) for 2 hours at 4°C (usually five plates of transfected mutant are resuspended in 10 ml of PBSSC using a 15-ml falcon tube). The reconstituted samples are then spun down at 1,000g for 10 minutes and the supernatant removed and discarded. At this point, the cell pellets containing the samples are stored at either -78°C (for up to at least 3 months), or the protein purification and spin-labeling procedures described below are begun.

4.C.c. Purification and Spin Labeling of Rhodopsin Cysteine Mutants.
The purification and spin labeling of the rhodopsin cysteine mutants are usually carried out as previously described (Resek et al., 1993). The following buffers are used in these procedures: buffer A (PBSSC), buffer B (buffer A plus 1% [wt/vol] dodecyl maltoside detergent [DM]/0.5 mM phenylmethylsulfonylfluoride [PMSF]), buffer C (buffer B plus 1 M NaCl/2 mM ATP/2 mM MgCl$_2$), buffer D (buffer A plus 0.05% DM), buffer E (5 mM [2-(N-morpholino)ethanesulfonic acid (MES)], pH 6.0/0.02% DM), buffer F (buffer B plus 3mM DTT/50 mM MES, pH 6.0), and buffer G (buffer E plus 1 mM EDTA).

To begin purifying, the COS cells containing the regenerated rhodopsin cysteine mutants are first solubilized in buffer B for ~30 minutes at 4°C and then centrifuged for 10 minutes at 10,000g. To this supernatant is then added ~300 μl of a slurry of 1D-4-sepharose immunoaffinity beads (capacity 1 μg rhodopsin/1 μl beads). The mixture is incubated for 3 hours at 4°C with constant mixing, and then reagents are added to create buffer C. The antibody–bead bound rhodopsin samples are then washed by a series of centrifugation and resuspension cycles. These steps involve first centrifuging for 1 minute at ~2000g and then discarding the supernatant and adding back 10 ml of buffer to the beads and agitating them for 5–10 minutes. This washing process is carried out two times using buffer D and repeated five more times with 10 ml buffer E.

After this initial washing procedure, the MTSL spin label (Toronto Research Biochemicals) is added to the samples in 10 ml of buffer E. Usually, a spin-label concentration of 100 μM (from a 10 mM stock in acetonitrile) is used. During the labeling procedure, the resin is gently agitated at room temperature for 3 hours. The labeling reaction is stopped by washing the samples six times with buffer F using the centrifugation/resuspension cycles described above. The MTSL derivatized rhodopsin samples are then eluted from the beads three separate times in 500-μl aliquots of buffer F containing 100 μM of the competing nine-amino acid peptide oligomer.

Note that to purify and spin label double cysteine rhodopsin mutants (to measure spin-spin interactions), the procedure described above is used except the following modifications are employed (Farrens et al., 1996): All buffers are argon purged for 15–30 minutes before use (one must do this before adding the DM!). Buffer F is used to solubilize the cells, and buffer G is used in the washes after the spin labeling.

4.D. Determining the Spin-Labeling Efficiency

Determining the number of spin labels per protein requires an independent determination of the protein concentration and the amount of spin label present. This is often difficult because of the limited amounts of mutant samples available. Quantitating spin labels on rhodopsin in SDSL studies has been dealt with in the following ways.

4.D.a. Double Integration of the EPR Spectrum and Comparison to a Known Standard. This approach is difficult with rhodopsin studies because the low levels of protein available can result in poor signal to noise ratios, and thus some uncertainty is introduced into the results obtained from the double integration procedure.

4.D.b. Determining the Amount of Free Cysteines Remaining After Spin Labeling. The extent of spin labeling can be inferred by measuring the amount of free cysteine remaining on the samples after the spin labeling. This value can be determined using the cysteine reactive agent, 4,4′-dithiodipyridine (PDS). PDS reaction with cysteines can be quantitated by monitoring the absorption at 323 nm (Chen and Hubbell, 1978; Cai et al., 1997).

4.D.c. Comparison of Samples' Peak to Peak Height With a Standard Curve of Spin Labels. First, the samples are treated with a sulfhydryl-reducing agent to release the spin label off the protein. Releasing the spin labels off the protein results in a "sharpened" EPR spectrum that can then be compared with a standard curve generated using known concentrations of free spin label. Reducing agents such as triphenylphosphine (TPP, a membrane-soluble reducing reagent) and tris(2carboxyethyl)phosphine (TCEP, a water-soluble reducing agent) have been successfully employed, usually at a 100–1,000-fold excess. Under these conditions, the reducing agents do not appear to affect the nitroxide of the spin label itself, as no changes in the EPR signal intensity are observed for up to 1 hour after treatment (Farrens and Hubbell, unpublished data; Mchaourab et al., 1997a).

4.E. Assessing the Consequence of the Cysteine Mutation and Spin Label on the Protein Structure

Possible perturbation to the protein structure must be taken into consideration in SDSL studies. In rhodopsin studies, all of the cysteine mutants are carefully assessed for the impact of the cysteine mutation and spin labeling on protein function and characteristics by measuring both the rhodopsin absorption spectra for each mutant (Resek et al., 1993; Ridge et al., 1995; Yang et al., 1996b; Farrens et al., 1996; Cai et al., 1997) and the effect on the rate of MII decay/retinal release (Farrens and Khorana, 1995).

Recently, the effect of the introduced MTSL on the spin-labeled protein's stability and ability to function was quantitatively addressed in SDSL studies of T4 lysozyme (T4L) (Mchaourab et al., 1996). The study compared the effect of the introduced MTSL on T4L with previous studies of the effect caused by other amino acid substitutions at the same sites. Generally, it was found that (1) MTSLs introduced at loop regions and surface sites have no effect, (2) MTSLs at tertiary contact sites have only a small perturbing effect, and (3) MTSLs at buried sites of the protein have the greatest effect, although the effects are not much greater than those caused by the introduction of an alanine. One exception to these guidelines obviously occurs when a functionally important amino acid residue is changed.

4.F. Studying Each Spin-Labeled Cysteine Mutant by EPR

General outlines are given for measuring the spin-label's mobility, the accessibility of the spin label, and the proximity of one spin label to another (if applicable).

All of the rhodopsin EPR measurements described here were carried out in the Hubbell laboratory using a modified Varian E-109 spectrometer. Modifications to this instrument include an interface to a Nicolet 1280 computer, a low noise microwave preamplifier, and a loop-gap resonator (Hubbell et al., 1987). Before the EPR measurements, the rhodopsin samples are concentrated for use in the loop-gap resonator as follows: Typically, 500 μl of the spin-labeled rhodopsin sample ($\sim$20–100 μg/ml or a 0.5–2.5 μM concentration) is loaded into an Amicon Micro-concentrator (Microcon-10) and spun at $\sim$8,000 rpm using an Eppendorf 5415 microcentrifuge at 4°C in the dark. The samples should be checked every 10–20 minutes during this concentration procedure and stopped at a final volume of approximately 10–20 μl to avoid any "drying out" of the samples.

4.F.a. Spin-Label Mobility. The EPR measurements for SDSL studies of rhodopsin are carried out at X-band ($\sim$9.3 GHz microwave frequency), using 2 mW incident microwave power. The temperature is usually ambient (19°–21°C), and, to avoid spectral distortion, the modulation amplitude used in the rhodopsin studies is usually less than 2 Gauss. The spectra are typically obtained using a 100-Gauss sweep at a scan rate of 30 sec/scan. Typically, four to eight scans are thus obtained and averaged. The relative mobility of each spin label is determined from measuring ΔH, the peak to peak width of the central $M_I = 0$ line.

4.F.b. Spin-Label Accessibility. The accessibility of each spin label is determined from power saturation measurements using O_2 and NiEDDA as the paramagnetic collision agents (Farahbakhsh et al., 1995b; Altenbach et al., 1996). The power saturation measurements are carried out by measuring a series of scans of the central $M_I = 0$ resonance (usually over a 10–20 G range), as a function of microwave power (usually 0.1–36 mW). Three such measurements are made on each mutant. One measurement is made after equilibration with air (O_2), one after equilibration with a stream of N_2 gas (to remove all O_2), and one in the presence of 20 mM NiEDDA equilibrated with N_2. Note that these measurements are made using the gas-permeable TPX capillaries.

The data obtained are analyzed to find the $\Delta P_{1/2}$ and II values. These values are then used to determine the accessibility of each label. Most SDSL studies on rhodopsin are carried out in DM micelles; however, it is possible to compare them with data obtained for a series of spin-labeled phospholipids in DM micelles containing nitroxides at different locations on the phospholipid structure (Farahbakhsh et al., 1995b).

4.F.c. *Spin-Label Proximity to Another Spin Label.* To measure distances between nitroxide spin labels, double cysteine mutants are spin labeled and the EPR spectra measured with frozen samples. At low temperature, approximating the interspin distances using several different methods is possible (Farahbakhsh et al., 1995a; Rabenstein and Shin, 1995). Two rhodopsin SDSL studies have used this approach (Farrens et al., 1996; Yang et al., 1996a). Recently, another rhodopsin study has used the broadening caused by spin–spin interaction to detect qualitative changes in proximity between regions of rhodopsin at room temperature (Cai et al., 1997).

4.G. Constructing a Model Protein Structure Consistent With the SDSL Results

The results obtained from the SDSL studies are used to model the protein secondary and tertiary structures. The modeling is usually carried out by first inspecting the array of accessibility and mobility data generated from the SDSL studies. Periodicities of 3.6 in the data suggest an α-helical structure (Altenbach et al., 1990, 1996; Mchaourab et al., 1996). Periodicities of two suggest a β-sheet structure (Hubbell et al., 1996; Berengian et al., 1997; Mchaourab et al., 1997b; Klug et al., 1997). After thus assigning the secondary structural elements, the best possible tertiary packing scheme is assessed by comparing the SDSL data with the model. This process is greatly helped by any distance constraints obtained from distance measurements between pairs of spin labels. Using this approach, an evolving structural model of rhodopsin, based on SDSL studies, is being constructed by Dr. Christian Altenbach using the program Insight II (Molecular Simulations Inc., San Diego) on a Silicon Graphics Workstation (Altenbach et al., 1996; Farrens et al., 1996; Kim et al, 1997).

5. WHAT HAS BEEN LEARNED FROM SDSL STUDIES OF RHODOPSIN

An overview is given describing what has been learned about the cytoplasmic face of rhodopsin from SDSL studies.

5.A. Background of the Rhodopsin System

5.A.a. Structure of Rhodopsin. The visual photoreceptor rhodopsin represents perhaps the best characterized GPCR (Khorana, 1992). The sequence of bovine rhodopsin consists of a chain of 348 amino acids, of which approximately 50% are found in one of the seven membrane-spanning helices (Fig. 13.4). The light-absorbing part of rhodopsin (the 11-*cis*-retinal chromophore, with a maximal absorbance at 500 nm) resides in the middle of these helices attached to lysine-296. A model for the packing of the helices (Baldwin, 1993) derived from cryoelectron microscopy (Unger and Schertler, 1995) has been reported.

5.A.b. Biochemistry of the Rhodopsin System. Rhodopsin becomes activated when the 11-*cis*-retinal chromophore absorbs light, converting it to an all-*trans*-retinal form. This event induces rhodopsin to undergo a series of spectrally distinct photointermediates, finally forming a 380-nm absorbing species called Meta II (Wald, 1968). The cytoplasmic loops of the MII form of rhodopsin bind and activate the G protein transducin, which then goes on to stimulate and initiate the biochemical cascades responsible for vision. Rhodopsin is "turned off" after several of these activating cycles when rhodopsin kinase binds to the cytoplasmic loops and phosphorylates the carboxy-terminal tail of rhodopsin (Stryer, 1986, 1991).

Why does transducin only bind to the activated (MII) form of rhodopsin? Interestingly, the rhodopsin cytoplasmic loops alone can activate transducin

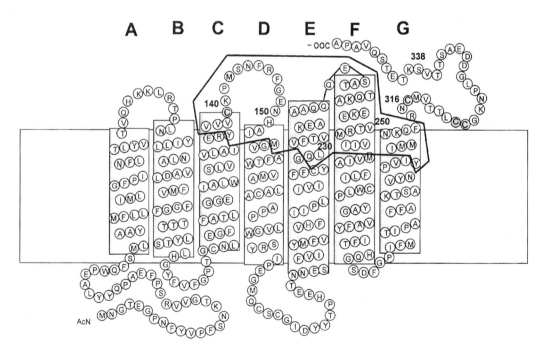

Figure 13.4. Secondary structure model of bovine rhodopsin. The cytoplasmic domain of the protein is at the top of the figure. The native cysteine residues in the cytoplasmic domain (C140, C316, C322 and C323) are shown by darker circles. The regions studied by SDSL are outlined.

(Konig et al., 1989), implying that dark state rhodopsin is unable to activate transducin because the cytoplasmic loops are inaccessible. Some conformational change must occur in rhodopsin to expose these loops upon rhodopsin activation. The nature of the changes that occur in rhodopsin upon photoactivation has recently been addressed by Khorana, Hubbell, and coworkers using the SDSL method. Using the methods described earlier in this chapter, a series of rhodopsin mutants were constructed, each containing a unique reactive cysteine residue in the cytoplasmic domain. The cysteine in each mutant was labeled with an MTSL, and the SDSL studies were carried out. From these measurements, the most probable protein structure of the cytoplasmic domain was modeled, and the conformational changes that occur upon MII formation were localized. These results are summarized below.

TABLE 13.1. Information About the Structure of the Cytoplasmic Face in Dark State Rhodopsin From SDSL Studies

Membrane/Aqueous Boundary	Residues at Boundary	Reference
C	137–140	Farahbakhsh et al. (1995b)
D	151–154	Farahbakhsh et al. (1995b)
E	227–231	Altenbach et al. (1996)
F	250–253	Altenbach et al. (1996)
G	309–311	Farrens et al. (unpublished data)

Orientation of Helices	Residues on Inside Face of Helix	Reference
C	136, 138, 139, 140	Farahbakhsh et al. (1995b)
D	153	Farahbakhsh et al. (1995b)
E	226, 230	Altenbach et al. (1996)
F	250, 251, 253	Altenbach et al. (1996)
G	306, 307, 309, 310	Farrens et al. (unpublished data)

Packing of Helices	Neighboring Residues	Reference
C–F	139–248, 39–250, 139–251	Farrens et al. (1996)
A–G	65–316	Yang et al. (1996a)

Cytoplasmic Loop Structure	Type of Structure	Reference
C–D	Unspecified structure, not simply α-helix or β-sheet	Farahbakhsh et al. (1995b)
E–F	E and F are extended helices, connected by a short loop at residues 238–239	Altenbach et al. (1996)

Cytoplasmic Tail	Proximity	Reference
Tail-F	Residues on helix F outer face (242, 245, 246, 249) are close to residue 338	Cai et al. (1997)

5.B. Regions in the Cytoplasmic Domain Studied by SDSL

Figure 13.4 shows a secondary structural model of bovine rhodopsin that indicates the regions studied by SDSL. Unless otherwise noted, the experimental conditions used in the studies reported here were carried out as described in the previous protocol section.

5.C. Membrane–Aqueous Boundary

The approximate membrane–aqueous interfaces determined from the SDSL studies are shown in Table 13.1 and Figure 13.4. These parameters were determined as described in previous sections. It must be kept in mind that the membrane–aqueous interface regions in the SDSL rhodopsin studies are approximate because the rhodopsin samples are in a detergent micelle rather than a true membrane.

5.D. Orientation of Individual Helices

The cytoplasmic ends of five of the seven rhodopsin helices can be arranged as shown in Figure 13.5 using the cryoelectron microscopy data of Unger and Schertler (1995) and oriented according to the accessibility and mobility data obtained from the SDSL studies (Farahbakhsh et al., 1995b; Altenbach et al.,

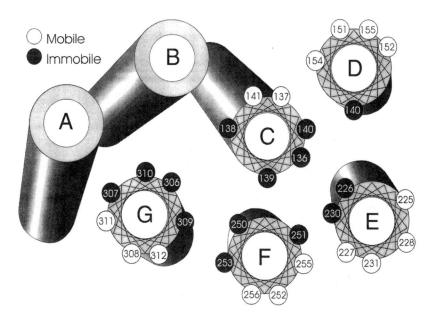

Figure 13.5. Model of helical packing on the cytoplasmic face of rhodopsin. The packing scheme is based on that of Unger and Schertler (1995) and the SDSL studies on those of Farahbakhsh et al. (1995b), Altenbach et al. (1996), and Farrens et al. (unpublished data). The packing shows the probable locations of the hydrophobic part of the rhodopsin sequence, including residues 136–141, 151–155, 226–231, 250–256, and 306–311. (Adapted from Altenbach et al. [1996], with permission of the publisher.)

1996; Farrens et al., unpublished data). The orientation is accomplished by simply making sure that positions in which the spin labels are immobilized and inaccessible are on the "inside" face of the helices in the packing scheme proposed from the cryoelectron microscopy data. Overall, the orientation of the helices is in excellent agreement with the residue packing model previously proposed by Baldwin (1993).

5.E. Three-Dimensional Packing of Helices

The probable packing of the seven transmembrane helices shown in Figure 13.5 is further supported by SDSL studies that measured spin–spin interactions between pairs of labels attached at different sites. These studies are discussed below.

5.E.a. Helix C Is Near Helix F.
A series of double cysteine mutants was constructed, each of which contained a single cysteine at residue 139, with the second cysteine placed in a series from 248 through 252 (Farrens et al., 1996). These mutants were then labeled and spin–spin interactions measured. It was found that in dark state rhodopsin, residues 248 and 251 are closest to 139 (~10–15 Å), in good agreement with their proposed location from the model shown in Figure 13.5. The proximity of these two helices was also suggested from the ability of several cysteine pairs to form a disulfide.

5.E.b. Helix G Is Near Helix A.
A double cysteine mutant containing one cysteine on helix A and the native cysteine on helix G (H65C + C316) was constructed and studied by SDSL (Yang et al., 1996a). It was found that the apparent distance between the two labels in dark state rhodopsin was quite close (~10 Å), suggesting that these two helices are in close proximity. This conclusion was supported by the ability of this cysteine pair to form a disulfide bond. Note that these data are again in good agreement with the model proposed by Baldwin (1993), and the model shown in Figure 13.5.

5.F. Structure of the Cytoplasmic Loops and Cytoplasmic Tail

5.F.a. The Loop Connecting Helices C and D Forms a Nonperiodic Structure.
The scanning SDSL study of the C–D loop region (from ~C140 to E150) showed no regular two-dimensional structure (Farahbakhsh et al., 1995b). However, from the accessibility studies, it was concluded that the C–D loop is close to the membrane interface.

5.F.b. Helices E and F Form Extended Helices.
The aqueous-exposed sequence in the E–F loop (residues 232–249) appears to consist of two α helices that extend up from the parent membrane helix and are connected by a short turn at residues 238 and 239 (Altenbach et al., 1996). This conclusion is based on SDSL mobility data and accessibility data, all of which showed a clear α-helical periodicity throughout the sequence of 225–256, interrupted only by a turn at residues 238/239.

5.F.c. The Cytoplasmic Tail Is Near Helix F. Recent SDSL studies have
shown that residue 338 in the cytoplasmic tail is in proximity to the outer face
of helix F of rhodopsin (Cai et al., 1997). Spin–spin interactions (and rates of
disulfide formation) were measured between a series of double cysteine mutants, each of which contained a cysteine residue at 338 with another in the sequence of 240–250. The SDSL studies suggested that the label at 338 was
closest to the labels at 245, 246, and 249, residues on the outer face of the extended helix F (Fig. 13.5). The proximity of these cysteine residues was also
suggested by their ability to form disulfide bonds.

5.G. SDSL Studies of Conformational Changes at Rhodopsin's Native Cysteine Residues

5.G.a. SDSL Studies of Wild-Type Rhodopsin. Wild-type bovine rhodopsin
has 10 native cysteine residues, only 2 of which are very reactive. A spin-
labeling study of wild-type rhodopsin was carried out by selectively labeling
the less reactive of these residues (C140 in helix C). This was done by first
blocking the more reactive residue (C316 in helix G) with PDS, then removing

**TABLE 13.2. Location of Conformational Changes in Rhodopsin Detected
by Scanning SDSL Studies**

Helix	Residue No.	Reference
C	Y136R1[a] $(+)$[b]	Farahbakhsh et al. (1995b)
	V139R1 $(++)$	Farahbakhsh et al. (1995b)
	C140R1 $(++)$	Farahbakhsh et al. (1995b)
D	R147R1 $(++)$	Farahbakhsh et al. (1995b)
	G149R1 $(++)$	Farahbakhsh et al. (1995b)
E	V227R1 $(---)$	Altenbach et al. (1996)
	K231R1 $(-)$	Altenbach et al. (1996)
	A235R1 $(-)$	Altenbach et al. (1996)
	Q238R1 $(-)$	Altenbach et al. (1996)
F	T242R1 $(-)$	Altenbach et al. (1996)
	Q244R1 $(+)$	Altenbach et al. (1996)
	K245R1 $(-)$	Altenbach et al. (1996)
	K248R1 $(+)$	Altenbach et al. (1996)
	V250R1 $(+++)$	Altenbach et al. (1996)
	T251R1 $(+++)$	Altenbach et al. (1996)
	I255R1 $(-)$	Altenbach et al. (1996)
	I256R1 $(+)$	Altenbach et al. (1996)

[a]R1 is defined as the cysteine residue labeled with the methanethiosulfonate spin label (see Fig.
13.1).
[b]$(+)$, Denotes an increase in spin-label mobility; thus $(++)$ denotes a greater increase in mobility.
$(-)$, Decrease in spin-label mobility, $(---)$ denotes a greater decrease in mobility.

the excess PDS and labeling C140 with the MTSL (Farahbakhsh et al., 1993). The EPR spectra of the spin label at C140 changed substantially upon MII formation, suggesting that a conformational change occurs in the protein at this location. Furthermore, the time course of this change was found to match the rate of light absorbance increase at 380 nm, suggesting that the conformational change at C140 occurs during formation of MII (Farahbakhsh et al., 1993).

5.G.b. *The First SDSL Study of Cysteine Mutant Rhodopsins.* The first SDSL study using cysteine mutant rhodopsin proteins (Resek et al., 1993) observed that changes in the EPR spectra of a spin label attached at C140 and at C316 only occurred upon MII formation. This study was also seminal in demonstrating that SDSL can be carried out on mutant rhodopsin proteins, and it established the conditions for labeling and purifying spin-labeled rhodopsin mutants.

5.H. Conformational Changes in Rhodopsin Detected by Scanning SDSL Studies

Of the regions in rhodopsin so far studied by the scanning SDSL technique (Fig. 13.4 and Table 13.1), only a minority of sites showed changes in spin-label mobility upon MII formation. Furthermore, the sites that showed changes were restricted primarily to helices C and F. Most of the changes showed the spin-label mobility increasing upon MII formation. However, an exception was found for the spin label at position 227 (helix E), which became much more immobilized after light activation. The locations of the conformational changes detected in scanning SDSL studies of rhodopsin reported to date are summarized in Table 13.2 and Figure 13.6. Based on these studies, it was speculated that helix F might move relative to helix C, a conclusion also reached from measuring distances between pairs of nitroxide labels attached to double cysteine mutants (described next).

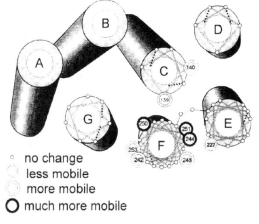

Figure 13.6. Location of changes in spin-label mobility upon MII formation, superimposed on the packing scheme shown in Figure 13.5. (Adapted from Altenbach et al. [1996], with permission of the publishers.)

5.I. Conformational Changes in Rhodopsin Measured Using Spin-Labeled Double Cysteine Mutants

5.I.a. Helix F Moves Away From C. Helix F movement away from helix C upon MII formation was suggested by a study measuring proximity between spin labels placed on helices C and F (Farrens et al., 1996). In this study, double cysteine mutants were constructed, each containing a cysteine residue at the cytoplasmic end of helix C (V139C) and one cysteine at various positions at the cytoplasmic end of helix F (ranging from K248C to R252C). The samples were spin labeled, and the spin–spin interaction between the two labels was measured in the dark and the MII states. The interspin distance between each pair of labels was estimated by simulation of the spectra. A range of likely distances was estimated, as shown in Table 13.3. From the changes in these distances upon MII formation, it was proposed that a rigid body movement occurs upon MII formation, which tilts helix F away from helix C (Fig. 13.7). The functional importance of this movement became evident when it was found that linking these helices together by a disulfide bond (Farrens et al., 1996) or a metal chelating agent (Sheikh et al., 1996) blocked the ability of rhodopsin to activate the G protein transducin. Note that movement in helix F (helix 6) has also been detected in the β_2-adrenergic receptor from site-directed fluorescence studies (see Chapter 12).

5.I.b. The Cytoplasmic Tail Moves. Two SDSL studies have suggested that the rhodopsin cytoplasmic tail moves upon MII formation. In the case of the double-mutant H65C/C316, the ~10 Å distance measured between nitroxides in the dark state was found to increase to ~15 Å after photobleaching the sample, suggesting that residue C316 (and perhaps the cytoplasmic tail) might move away from helix A upon MII formation (Yang et al., 1996a).

TABLE 13.3. Conformational Changes in Rhodopsin Measured Using Spin-Labeled Double Cysteine Mutants

Helices	Residue Numbers[a]	Distance (DS)[b]	Distance (MII)[b]	Δ Distance (DS→MII)[c]	Reference
A and G	H65R1 + C316R1	10 +/− 3	15 +/− 4	(+) 5	Yang et al., 1996a
C and F	V139R1 + K248R1	12–14	23–25	(+) 11	Farrens et al., 1996
	V139R1 + E249R1	15–20	15–20	0	Farrens et al., 1996
	V139R1 + V250R1	15–20	12–14	(−) 2–5	Farrens et al., 1996
	V139R1 + T251R1	12–14	23–25	(+) 11	Farrens et al., 1996
	V139R1 + R252R1	15–20	23–25	(+) 5–7	Farrens et al., 1996
F and tail	S338R1 + V245R1	Close	Farther	(+)	Cai et al., 1997
	S338R1 + E246R1	Close	Farther	(+)	Cai et al., 1997

[a]R1 is defined as the cysteine residue labeled with the methanethiosulfonate spin label.
[b]DS, Distances (in Å) measured between spin labels measured in dark state (DS) rhodopsin; MII, distances measured between spin labels in the activated, meta II form of rhodopsin.
[c](+), Increase in distance (in Å) between the pair of spin labels; (−), decrease in distance.

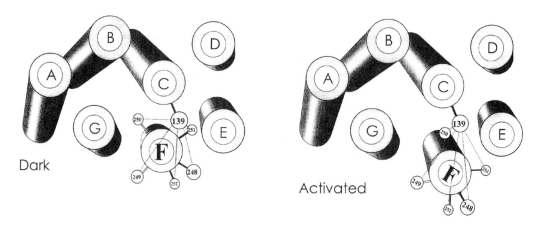

Figure 13.7. A model of a relative movement of helix F compared with helix C upon MII formation based on distance measurements between nitroxide spin labels at the indicated positions. The approximate locations of the nitroxide spin labels in the dark state rhodopsin are modeled using the packing scheme in Figure 13.5. The nitroxide labels are shown as circles sized depending on their distance (depth) from the viewer. For the activated (MII) model, the F helix was rotated and tilted to satisfy the new distance constraints from the spin–spin measurements. (Adapted from Farrens et al. [1996], with permission of the publisher.) Note that movement in other helices may also occur.

A subsequent study measured spin–spin interactions using a series of mutants containing a single spin label on helix F (ranging from position 240 to 250) and one on the cytoplasmic tail at S338C (Cai et al., 1997). The results from this study suggested that residues on the outer face of helix F (especially 245 and 246) are nearer residue 338 than are residues on the inner face of helix F. The high mobility of the spin label at S338C complicates the determination of absolute distances in these measurements, so only relative distances were presented. Upon MII formation, the distance between spin labels decreased, suggesting that the cytoplasmic tail moves away from helix F upon photoactivation. Similar conclusions were reached by measuring changes in the rate of disulfide bond formation between the double cysteine residues.

6. CONCLUSIONS AND FUTURE DIRECTIONS

6.A. Summary

This chapter has described how SDSL can be applied to a variety of problems, how it can be used in structure–function studies of GPCRs, and what has been learned about the GPCR rhodopsin from SDSL studies. Hopefully, this chapter has shown how SDSL can provide structural information about membrane proteins like GPCRs that are intractable to traditional methods such as x-ray crystallography and nuclear magnetic resonance spectroscopy. Below, some future directions of the SDSL technique are discussed, and areas of possible SDSL studies of rhodopsin are presented.

6.B. Advances in the SDSL Technique

As a technique, the SDSL approach continues to improve, aided by both advances in molecular biology techniques, and by advances in EPR spectroscopy.

6.B.a. Advances in Molecular Biology.
The construction of mutant proteins for SDSL studies continues to become a less expensive and easier undertaking. Oligonucleotides continue to drop in price, DNA sequencing is becoming automated, and the expression and purification of the mutant proteins can now be achieved using "kits." If this present trend continues, the construction and purification of spin-labeled protein samples for SDSL studies may no longer be the rate-limiting step that it is now.

6.B.b. Advances in EPR Instrumentation.
The introduction of the loop-gap resonator has been a huge advancement to the SDSL technique. Other technical improvements in EPR spectroscopy will likely continue to increase the power of the SDSL technique (though perhaps not as dramatically).

6.B.c. Advances in EPR Data Analysis and Interpretation.
The interpretation of SDSL data has been greatly advanced by the recent systematic studies of T4L in solution (Mchaourab et al., 1996, 1997b). Recently, interpretation of SDSL data has been advanced further by studies comparing the crystal structures of spin-labeled proteins with SDSL data (Steinhoff et al., 1997).

6.C. Future SDSL Rhodopsin Studies

The SDSL studies of rhodopsin are beginning to explain, on a molecular level, what happens to the cytoplasmic domain of rhodopsin when it forms MII. However, no information from SDSL studies has yet been reported about helices A and B or the cytoplasmic tail, although these regions are presently under study (Drs. H. Khorana and W. Hubbell, personal communication). Once these studies are completed, a more comprehensive picture will emerge of the structure and conformational changes that occur in this functionally important part of the GPCR rhodopsin.

Another exciting future direction of the SDSL technique is in its application to the study of abnormal mutant proteins. This approach will help determine whether the movements in rhodopsin detected by SDSL are impaired or altered in aberrant mutants, providing a partial explanation for their altered functional properties. A recent example of this type of study was reported in an SDSL study of constitutively active mutants of rhodopsin (Kim et al., 1997).

Our own laboratory is interested in determining what role conformational changes in GPCRs play in the binding and activation of G proteins. One way to approach this question will be to use SDSL to detect the precise location on the GPCR where the G protein binds by monitoring changes in spin label mobility and accessibility. This way of measuring protein–protein contact sites using SDSL has been demonstrated to be feasible by recent studies of the interaction of protein kinase C with calmodulin (Quin et al., 1996).

ACKNOWLEDGMENTS

The author thanks Dr. Christian Altenbach, Dr. Ralf Langen, and Dr. Hassane Mchaourab, as well as members of the Farrens laboratory (Tom Dunham and Steve Mansoor) for helpful comments and discussion of this manuscript. Preparation of this manuscript was supported in part by grants from the Medical Research Foundation of Oregon and the National Institutes of Health (R01EY12095-01).

REFERENCES

Altenbach C, Flitsch S, Khorana HG, Hubbell WL (1989): Structural studies on trans-membrane proteins. Biochemistry 28:7806–7812.

Altenbach C, Marti T, Khorana HG, Hubbell WL (1990): Transmembrane protein structure: Spin-labeling of bacteriorhodopsin mutants. Science 248:1088–1092.

Altenbach C, Steinhoff H-J, Greenhalgh DA, Khorana HG, Hubbell WL (1994): Factors that determine the EPR spectra of nitroxide sidechains in spin-labeled proteins and analysis by molecular dynamics simulation. Biophys J 66:A40.

Altenbach C, Yang K, Farrens DL, Farahbakhsh Z, Khorana HG, Hubbell WL (1996): Structural features and light-dependent changes in the E–F interhelical loop in rhodopsin: A site-directed spin-labeling study. Biochemistry 35:12470–12478.

Baldwin JM (1993): The probable arrangement of the helices in G protein–coupled receptors. EMBO J 12:1693–1703.

Berengian AR, Bova MP, Mchaourab HS (1997): Structure and function of the conserved domain in α-A-crystallin. Site-directed spin-labeling identifies a β-strand located near a subunit interface. Biochemistry 36:9951–9957.

Brudvig GW (1995): Electron paramagnetic resonance spectroscopy. 246:536–554.

Cai K, Langen R, Hubbell WL, Khorana, HG (1997): Structure and function in rhodopsin: topology of the C-terminal polypeptide chain in relation to the cytoplasmic loops. Proc Natl Acad Sci USA 94:14267–14272.

Chen YS, Hubbell WL (1978): Reactions of the sulfhydryl groups of membrane-bound bovine rhodopsin. Membrane Biochem 1:107–129.

Farahbakhsh ZT, Altenbach C, Hubbell WL (1992): Spin-labeled cysteines as sensors for protein–lipid interaction and conformation in rhodopsin. Photochem Photobiol 56:1019–1033.

Farahbakhsh ZT, Hideg K, Hubbell WL (1993): Photoactivated conformational changes in rhodopsin: A time-resolved spin-label study. Science 262:1416–1419.

Farahbakhsh ZT, Huang Q-L, Ding L-L, Altenbach C, Steinhoff H-J, Horwitz J, Hubbell WL (1995a): Interaction of alpha-crystallin with spin-labeled peptides. Biochemistry 34:509–516.

Farahbakhsh ZT, Ridge KD, Khorana HG, Hubbell WL (1995b): Mapping light-dependent structural changes in the cytoplasmic loop connecting helices C and D in rhodopsin: A site-directed spin-labeling study. Biochemistry 34:8812–8819.

Farrens DL, Altenbach C, Yang K, Hubbell WL, Khorana HG (1996): Requirement of rigid-body motion of transmembrane helices for light activation of rhodopsin. Science 274:768–770.

Farrens DL, Khorana HG (1995): Structure and function in rhodopsin. Measurement of the rate of metarhodopsin II decay by fluorescence spectroscopy. J Biol Chem 270:5073–5076.

Freed JH (1976): Theory of slow tumbling ESR spectra for nitroxides. In Berliner LJ (ed): Spin-Labeling Theory and Applications. New York: Academic Press, pp 53–132.

Froncisz W, Hyde J (1982): The loop-gap resonator: A new microwave lumped circuit ESR sample structure. J Magn Reson 47:515–521.

Hubbell WL, Altenbach C (1994): Investigation of structure and dynamics in membrane proteins using site-directed spin labeling. Curr Opin Struct Biol 4:566–573.

Hubbell WL, Froncisz W, Hyde JS (1987): Continuous and stopped flow EPR spectrometer based on a loop gap resonator. Rev Sci Instrum 58:1879–1886.

Hubbell WL, Mchaourab HS, Altenbach C, Lietzow MA (1996): Watching proteins move using site-directed spin labeling. Structure 4(7):779–783.

Hustedt EJ, Smirnov, AI, Laub C, Cobb CE, Beth A H (1997): Molecular distances from dipolar coupled spin-labels: The global analysis of multifrequency continuous wave electron paramagnetic resonance data. Biophys J 72:1861–1877.

Karnik SS, Ridge KD, Bhattacharya S, Khorana HG (1993): Palmitoylation of bovine opsin and its cysteine mutants in COS cells. Proc Natl Acad Sci USA 90:40–44.

Khorana HG (1992): Rhodopsin, photoreceptor of the rod cell. An emerging pattern for structure and function. J Biol Chem 267:1–4.

Kim J-M, Altenbach C, Thurmond RL, Khorana HG, Hubbell WL (1997): Structure and function in rhodopsin: Rhodopsin mutants with a neutral amino acid at E134 have a partially activated conformation in the dark state. Proc Natl Acad Sci USA 94: 14273–14278.

Klug CS, Su W, Feix JB (1997): Mapping of the residues involved in a proposed β-strand located in the ferric enterobactin receptor FepA using site-directed spin-labeling. Biochemistry 36:13027–13033.

Konig B, Arendt A, McDowell JH, Kahlert M, Hargrave PA, Hofmann KP (1989): Three cytoplasmic loops of rhodopsin interact with transducin. Proc Natl Acad Sci USA 86:68788–6882.

Likhtenshtein GI (1993): Biophysical Labeling Methods in Molecular Biology. New York: Cambridge University Press.

Mchaourab HS, Berengian AR, Koteiche HA (1997a): Site-directed spin-labeling study of the structure and subunit interactions along a conserved sequence in the α-crystallin domain of heat-shock protein 27. Evidence of a conserved subunit interface. Biochemistry 36: 14627–14634.

Mchaourab HS, Lietzow MA, Hideg K, Hubbell WL (1996): Motion of spin-labeled side chains in T4 lysozyme. Correlation with protein structure and dynamics. Biochemistry 35:7692–7704.

Mchaourab HS, Oh K-J, Fang CJ, Hubbell WL (1997b): Conformation of T4 lysozyme in solution. Hinge-bending motion and the substrate-induced conformational transition studied by site-directed spin-labeling. Biochemistry 36:307–316.

Millhauser GL, Fiori WR, Miick SM (1995): Electron spin labels. 246:589–610.

Oprian DD, Molday RS, Kaufman RJ, Khorana HG (1987): Expression of a synthetic bovine rhodopsin gene in monkey kidney cells. Proc Natl Acad Sci USA 84: 8874–8878.

Popp CA, Hyde J (1981): Effects of oxygen on EPR spectra of nitroxide spin label probes of model membranes. J Magn Reson 43:249–258.

Quin Z, Wertz SL, Jacob J, Savino Y, Cafiso DS (1996): Defining protein–protein interactions using site-directed spin-labeling: The binding of protein kinase C substrates to calmodulin. Biochemistry 35:13272–13276.

Rabenstein MR, Shin YK (1995): Determination of the distance between two spin-labels attached to a macromolecule. Proc Natl Acad Sci USA 92:8239–8243.

Resek JF, Farahbakhsh ZT, Hubbell WL, Khorana HG (1993): Formation of the meta II photointermediate is accompanied by conformational changes in the cytoplasmic surface of rhodopsin. Biochemistry 32:12025–12032.

Ridge KD, Zhang C, Khorana HG (1995): Mapping of the amino acids in the cytoplasmic loop connecting helices C and D in rhodopsin. Chemical reactivity in the dark state following single cysteine replacements. Biochemistry 34:8804–8811.

Sheikh SP, Zvyaga TA, Lichtarge O, Sakmar TP, Bourne HR (1996): Rhodopsin activation blocked by metal-ion-binding sites linking transmembrane helices C and F. Nature 383:347–350.

Steinhoff H-J, Mollaaghababa R, Altenbach C, Hideg K, Krebs M, Khorana HG, Hubbell WL (1994): Time-resolved detection of structural changes during the photocycle of spin-labeled bacteriorhodopsin. Science 266:105–107.

Steinhoff H-J, Radzwill N, Thevis W, Lenz V, Brandenburg D, Antson A, Dodson G, Wollmer A (1997): Determination of interspin distances between spin-labels attached to insulin: Comparison of electron paramagnetic resonance data with the x-ray structure. Biophys J 73:3287–3298.

Stryer L (1986): Cyclic GMP cascade of vision. Annu Rev Neurosci 9:87–119.

Stryer L (1991): Visual excitation and recovery. J Biol Chem 266(17):10711–10714.

Subczynski WK, Hyde J (1981): The diffusion-concentration product of oxygen in lipid biolayers using the spin-label T_1 method. Biochim Biophys Acta 643:283–291.

Thorgeirsson TE, Xiao W, Brown LS, Needleman R, Lanyi JK, Shin Y-K (1997): Transient channel-opening in bacteriorhodopsin: An EPR study. J Mol Biol 273:951–957.

Unger VM, Schertler GF (1995): Low resolution structure of bovine rhodopsin determined by electron cryo-microscopy. Biophys J 68:1776–1786.

Voss J, Salwinski L, Kaback HR, Hubbell WL (1995): A method for distance determination in proteins using a designed metal ion binding site and site-directed spin labeling: Evaluation with T4 lysozyme. Proc Natl Acad Sci USA 92(26):12295–12299.

Wald G (1968): The molecular basis of visual excitation. Nature 219:800–807.

Yang K, Farrens DL, Altenbach C, Farahbakhsh ZT, Hubbell WL, Khorana HG (1996a): Structure and function in rhodopsin. Cysteines 65 and 316 are in proximity in a rhodopsin mutant as indicated by disulfide formation and interactions between attached spin-labels. Biochemistry 35:14040–14046.

Yang K, Farrens DL, Hubbell WL, Khorana HG (1996b): Structure and function in rhodopsin. Single cysteine substitution mutants in the cytoplasmic interhelical E–F loop region show position-specific effects in transducin activation. Biochemistry 35:12464–12469.

CHAPTER 14

FLUORESCENCE SPECTROSCOPY ANALYSIS OF CONFORMATIONAL CHANGES IN THE β₂-ADRENERGIC RECEPTOR

ULRIK GETHER

1. INTRODUCTION	316
2. EXPRESSION AND PURIFICATION OF THE β₂-AR	317
A. Expression Vector and Transfection	318
B. Virus Amplification and Plaque Purification	318
C. Culturing of Sf9 Cells	318
D. Infections for Purification	318
E. Purification	319
F. Binding Assay	320
3. FLUORESCENT LABELING AND SPECTROSCOPIC ANALYSIS OF THE PURIFED β₂-AR RECEPTOR	320
A. Fluorescent Labeling of the Purified β₂-AR	323
B. Fluorescence Spectroscopy Analysis	324
4. SITE-SELECTIVE FLUORESCENT LABELING OF THE β₂-AR	325
A. Generation of Mutant Receptors	328
B. Expression in Sf9 Insect Cells	329
C. PCR Analysis of Virus Stocks	329
D. Purifications	329
E. Fluorescent Labeling and Fluorescence Spectroscopy	329
5. ANALYZING A CONSTITUTIVELY ACTIVE MUTANT RECEPTOR BY FLUORESCENCE SPECTROSCOPY	329
A. Expression, Purification, and Fluorescent Labeling of CAM	331
6. CONCLUDING REMARKS	331

Structure–Function Analysis of G Protein-Coupled Receptors, Edited by Jürgen Wess.
ISBN 0-471-25228-X   Copyright © 1999 Wiley-Liss, Inc.

315

I. INTRODUCTION

In spite of the impressive functional variability among G protein–coupled receptors (GPCRs), it is believed that the receptors share both a common overall tertiary structure and a common mechanism of activation. It is generally assumed that binding of the agonist to the receptor induces a set of finely orchestrated changes in the tertiary structure of the receptor that are recognized by the associated G protein. Many methodological approaches have been applied in an attempt to understand these structural changes, providing the critical link between agonist binding and G protein coupling (Gether and Kobilka, 1998). Until recently, models of how GPCRs are activated have been based on indirect evidence; hence, the conformation of the receptor has mostly been inferred from activation of messenger systems and/or from computational simulations (Samama et al., 1993; Luo et al., 1994; Ballesteros and Weinstein, 1995; Fanelli et al., 1995; Scheer et al., 1996). However, the possibility of establishing purification procedures and applying biophysical techniques in the study of this class of receptors has now allowed novel insight into the molecular mechanisms underlying activation of GPCRs (Altenbach et al., 1996; Farahbakhsh et al., 1995; Farrens et al., 1996; Gether et al., 1995, 1997a,b).

It is not surprising that a majority of these studies have been performed with rhodopsin. There are abundant natural sources of rhodopsin, and its inherent stability makes it possible to produce and purify relatively large quantities of recombinant protein. The elegant use of electron paramagnetic resonance (EPR) spectroscopy by Hubbell, Khorana, and coworkers has provided the most substantial insight into conformational changes associated with photoactivation of rhodopsin (Altenbach et al., 1996; Farahbakhsh et al., 1995; Farrens et al., 1996). Both the technique and the results of their work are described in Chapter 13 of this volume. Rhodopsin is also the only GPCR for which direct information about the tertiary structure is available. Projection maps at 8 Å resolution of bovine and frog rhodopsin based on electron microscopy of two-dimensional crystals have provided crucial information about the relative positioning of the transmembrane helices in the seven-helix bundle (Schertler et al., 1993; Unger et al., 1997) (see Chapter 12). Importantly, several models of other GPCRs have been developed based on this projection map (Baldwin, 1993; Ballesteros and Weinstein, 1995; Scheer et al., 1996).

Recently, we applied spectroscopic techniques to the β_2-adrenergic receptor (β_2-AR) (Gether et al., 1995, 1997a,b; Lin et al., 1996). As described in this chapter, we have taken advantage of the sensitivity of the emission from many fluorescent molecules to the polarity of their molecular environment. A sulfhydryl-reactive fluorescent probe was covalently incorporated into the purified β_2-AR and used as a molecular reporter for structural changes occurring following agonist binding to the receptor (Gether et al., 1995, 1997a,b). The background for applying spectroscopic approaches was a wish to develop methods that would allow direct time-resolved analysis of conformational changes accompanying ligand-induced activation of GPCRs. In contrast to rhodopsin, which is a highly specialized GPCR with its ligand covalently bound, the β_2-AR is a typical ligand-activated receptor. Developing spectroscopic techniques for the β_2-AR would thus enable analysis of differences between the conformational states of unbound receptors and receptors bound to different kinds of ligands, including full agonists, partial agonist, neutral antagonists, and inverse agonists.

Our data, together with data from other laboratories, strongly point to a critical role of helices 3 and 6 in receptor activation. Both the spin-labeling studies in rhodopsin and the fluorescence spectroscopy analyses of the β_2-AR suggest that a counterclockwise rotation of helix 6 is an essential part of the activation mechanism (Farrens et al., 1996; Gether et al., 1997b). Moreover, the spin labeling studies in rhodopsin suggest that helix 3 and 6 move away from each other during activation (Farrens et al, 1996). Helix 3 may also undergo a counterclockwise rotation (Gether et al, 1997b) (see Fig.14.4). It is noticeable that zinc binding to a bis-His zinc site constructed between the cytoplasmic end of transmembrane (TM) 3 and 6 in rhodopsin can prevent formation of the MII state of rhodopsin and transducin activation (Sheikh et al., 1996). A particularly important role of TM 6 in receptor activation has been further supported by evidence indicating a conformational rearrangement of TM 6 consistent with a counterclockise rotation in a constitutively activated β_2 receptor (Javitch et al., 1997). A cysteine in TM 6, which is not accessible to modification by charged hydrophilic sulfhydryl-specific methanethiosulfonate (MTS) reagent in the wild-type β_2 receptor, becomes accessible in the constitutively active mutant (Javitch et al., 1997) (see Chapter 2). In summary, many studies indicate that TM 3 and 6 are crucial for transition of the receptor to the active state supporting the idea that the activation mechanism is conserved among rhodopsin-like GPCRs.

The goal of this chapter is to describe the background and methodology involved in using fluorescence spectroscopic techniques for analyzing conformational changes in ligand-activated GPCRs. The currently available data are reviewed and discussed in the context of the experimental procedures.

2. EXPRESSION AND PURIFICATION OF THE β_2-AR

Spectroscopic analyses require a reliable expression system and preferably an easy purification procedure. Especially if the goal is to express and purify not only the wild type but also several mutant receptors, the workload involved in these procedures is a critical consideration. We have used the baculovirus/Sf9 cell system to express the β_2-AR. Sf9 cells are easy to grow, and Sf9 cell cultures can easily be scaled up. Sf9 cells do not require CO_2 and can grow in suspension cultures in a standard shaker using either glass or polyethylene Erlenmeyer/Fernbach flasks. The most prominent problem of the insect cell expression system may be the varying fraction of improperly folded and thus nonfunctional protein (Guan et al., 1992; Kobilka, 1995). In the case of the β_2-AR, approximately half of the synthesized receptor is nonfunctional (Kobilka, 1995). The fraction can vary dramatically from protein to protein and may hinder the use of the system. However, high levels of expression of several GPCRs in Sf9 insect cells have been reported (Grisshammer and Tate, 1995; Tate and Grisshammer, 1996).

We have expressed a modified form of the β_2-AR in the baculovirus/Sf9 cell system (Guan et al., 1992). A cleavable influenza-hemagglutinin signal-sequence followed by the M1 antibody "FLAG" epitope was inserted at the amino terminus (Guan et al., 1992; Kobilka, 1995). The signal-sequence resulted in an approximately twofold increase in expression (Guan et al., 1992). At the carboxy terminus the receptor was tagged with six histidines (SF-hβ_2-6H) (Kobilka, 1995). A three-step procedure was developed to purify the receptor from Sf9 cells, including an initial nickel chromatography step followed by anti-FLAG immunoaffinity and alprenolol affinity chromatography (Kobilka, 1995). Together these three steps ensure that only full-length, properly folded receptors are purified. It should be noted that for many purposes the immunoaffinity purification step can be omitted (Gether et al., 1997b). Although the purity of the protein is lower, the procedure still ensures that correctly folded protein is being purified.

2.A. Expression Vector and Transfection

Routinely we have expressed the β_2-AR under control of the polyhedrin promoter using the pVL1392 baculovirus expression vector (Pharmingen, San Diego, CA). In our hands, high and reproducible expression was consistently achieved with this vector. The cDNA encoding the human β_2-AR was epitope tagged at the amino terminus with the cleavable influenza-hemagglutinin signal-sequence followed by the FLAG epitope (Eastman Kodak, Rochester, NY) and tagged at the carboxy terminus with six histidines (SF-hβ_2-6H) as previously reported (Guan et al., 1992; Kobilka, 1995). The vector containing the cDNA encoding the modified β_2-AR is co-transfected with linearized Baculo-Gold DNA into Sf9 insect cells using the BaculoGold transfection kit according to the manufacturer's instructions (Pharmingen). This transfection kit is very reliable and usually gives rise to almost 100% recombinants.

2.B. Virus Amplification and Plaque Purification

The virus is harvested 4–5 days after transfection and amplified once before plaque purification. The plaque-purified viruses are usually amplified three times to obtain 500 ml of a high titer virus stock (about 1×10^9 pfu/ml). Each virus stock is tested in small-scale cultures to determine the optimal inoculum for large-scale infections.

2.C. Culturing of Sf9 Cells

Sf9 insect cells are maintained in SF900-II medium (Gibco Grand Island, NY) supplemented with 0.1 mg/ml gentamycin (Gibco) and 5% heat-inactivated fetal calf serum (Gibco). The optional addition of serum allows the cells to grow up to higher densities (7–8 million cells/ml). The cell stock is normally kept in 250-ml polypropylene Erlenmeyer flasks (Corning Costar, Acton, MA) at 27°C in a shaker set at 125 rpm. Each flask contains 70–100 ml medium, and the cells are kept at a density varying from 0.5 to 6 million cell/ml.

2.D. Infections for Purification

1. Cells are seeded in 2,800 ml triple-baffled Fernbach flasks (Bellco Glass Inc., Vineland, NJ) and grown until they reach a density of 5–7 million cells/ml in a total volume of 1,000–1,200 ml of medium.

2. The culture is removed from the incubator, and the flask is kept at room temperature for 1.5–2 hours to sediment the cells.

3. A majority of the medium is carefully aspirated, and the cells are resuspended in fresh media plus 1 μM of alprenolol to a cell density of 5 million cells/ml.

4. Cells are infected by adding virus stock (1:30 to 1:100 dilution). The optimal inoculum is determined for each virus stock by infecting small-scale suspension cultures (20 ml in 125 ml disposable Erlenmeyer flasks).

5. Cells are incubated for 48 hours at 27°C in a shaker set at 125 rpm and harvested by centrifugation for 10 minutes at 2,700g. The resulting cell pellets can be kept at −70°C until purification.

2.E. Purification

For spectroscopic studies, we have used the following purification procedure. This procedure has been described in detail previously (Kobilka, 1995).

1. One or two pellets of cells from infected 1,000-ml cultures are lysed in 100 ml/pellet of a 10 mM Tris-HCl buffer, pH 7.5, containing 1 mM EDTA, 10 μg/ml leupeptin (Boehringer, Mannheim, Germany), 10 μg/ml benzamidin (Sigma, St. Louis, MO), and 0.2 mM phenylmethylsulfonylfluoride (Sigma).

2. The lysed cells are centrifuged at 45,000g for 30 minutes, the supernatant is discarded, and the pellets are weighed. Pellets are resuspended in 20 mM Tris-HCl buffer, pH 7.5, containing 1.0% n-dodecyl-β-D-maltoside (DβM) (Anatrace Inc., Maumee, OH), 500 mM NaCl, 10 μg/ml leupeptin (Boehringer), 10 μg/ml benzamidin (Sigma), 0.2 mM phenylmethylsulfonylfluoride (Sigma), and 10^{-6} M alprenolol (Sigma). Ten milliliter of buffer is used per gram of lysed cells. The resuspended pellets are solubilized by douncing (20 strokes with tight pestle) followed by stirring at 4°C for 1.5–2 hours.

3. Nonsolubilized particulate is isolated from solubilized protein by centrifugation at 45,000g for 30 minutes. Imidazol is added to the supernatant to a final concentration of 50 mM from a 2.0 M stock solution (pH 8.0). Chelating Fast Flow Sepharose Resin (Pharmacia) (0.5 ml per gram of lysed cells) charged with nickel and equilibrated in *high salt buffer* (20 mM Tris-HCl, pH 7.5, with 500 mM NaCl and 0.08% DβM) is added followed by 2–3 hours of incubation at 4°C under gentle rotation.

4. Nickel resin is isolated by centrifugation for 5 minutes at 2,000g. The resin is washed once in badge in 4 vol of high salt buffer, loaded onto a column, and washed with 3 vol of high salt buffer and 2 vol of high salt buffer plus 25 mM imidazol. Elution is done in 1/8-column vol fractions with 200 mM imidazol in high salt buffer. Fractions are assayed for receptor binding activity, and peak fractions are pooled.

5. CaCl₂ is added to the pooled fractions to a final concentration of 2.5 mM. The pooled fractions are loaded onto an M1 antibody column (Eastman

Kodak) (0.2 ml per nmol of receptor) equilibrated with *low salt buffer* (20 mM Tris-HCl, pH 7.5, with 100 mM NaCl and 0.08% DβM) and recycled four times by gravity flow. The column is washed with 4 column vol of low salt buffer containing 2.5 mM CaCl$_2$ and eluted using low salt buffer with 1 mM EDTA in 1/4-column vol fractions. Fractions are analyzed for receptor binding activity and peak fractions pooled.

These two purification steps can produce almost pure protein (specific activity around 5 nmol/mg of protein). However, about half of the protein is nonfunctional (Kobilka, 1995). To separate the nonfunctional receptor from functional we use alprenolol affinity chromatography, which is a standard procedure for purification of the β$_2$-AR (Caron et al., 1979; Benovic et al., 1984; Parker et al., 1991). It is important to note that we have been able to exclude the M1 immunoaffinity chromatography in some applications. This results in a specific activity of the purified receptor around 5–10 nmol/mg compared with 10–15 nmol/mg for the three-step purification. Approximately 5 nmol of purified protein generally can be obtained from a 1,000-ml culture. Protein concentration is determined using the detergent-insensitive Bio-Rad DC protein assay kit (Bio-Rad, Hercules, CA). Purified receptors are analyzed by classical 10% SDS-polyacrylamide gel electrophoresis. Notably, samples should not be boiled before loading onto the gel to prevent receptor aggregation. The receptor is visualized by standard Coomassie staining.

2.F. Binding Assay

The amount of purified β$_2$ receptor was assessed in binding assays using ^{3}H-dihydroalprenolol (^{3}H-DHA) as radioligand (Amersham, Arlington Heights, IL). Purified β$_2$ receptor (10 μl appropriately diluted) is incubated with 10 nM ^{3}H-DHA (10 μl from a 1:100 dilution of ^{3}H-DHA) in a total volume of 100 μl low salt buffer (20 mM Tris-buffer, pH 7.5, containing 100 mM NaCl and 0.08 % DβM) for 1 hour. Nonspecific binding is determined in the presence of 10 μM alprenolol. The binding assay is stopped by separating free ^{3}H-DHA from bound by loading the binding mixture onto 2-ml Sephadex G50 (Pharmacia) columns (Poly Prep columns, BioRad). Columns are eluted with 1 ml of ice-cold low salt buffer directly into 20-ml scintillation vials. Scintillation fluid is added followed by counting in a scintillation counter.

3. FLUORESCENT LABELING AND SPECTROSCOPIC ANALYSIS OF THE PURIFIED β$_2$-AR

The emission from many fluorescent molecules is strongly dependent on the polarity of the environment in which they are located. Incorporation of fluorescent labels into proteins therefore can be used as sensitive indicators of conformational changes and protein–protein interactions that cause changes in polarity of the environment surrounding the probe (Cerione, 1994; Dunn and Raftery, 1993; Gettins et al., 1993; Phillips and Cerione, 1991). Nitrobenzdioxazol iodoacetamide (IANBD) is a highly fluorescent, cysteine-selective reagent (Dunn and Raftery, 1993; Gettins et al., 1993). The fluorescence from IANBD increases as the polarity of the solvent decreases and is more than 10-fold stronger in n-butanol and n-hexane than in aqueous buffer (Fig. 14.1A). There is a parallel blue shift in the emission maximum from 540 nm in aqueous buffer to 530 nm in n-butanol and 510 nm in n-hexane (Fig. 14.1A). Labeling of the β_2-AR purified from Sf9 insect cells with IANBD revealed a strong

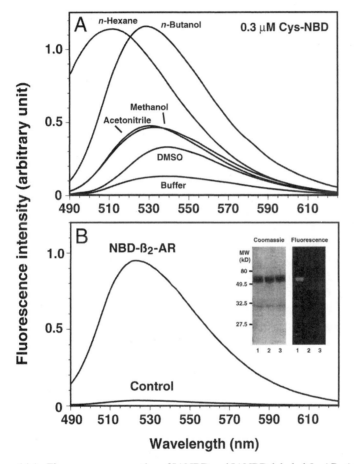

Figure 14.1. Fluorescence properties of IANBD and IANBD-labeled β_2-AR. **A:** Emission spectra of cysteine-reacted IANBD (0.3 μM) in solvents of different polarity. Excitation was set at 481 nm. **B:** Emission spectrum of IANBD-labeled β_2 receptor (0.15 μM receptor, 1.2 mol IANBD/mol receptor). Control is emission spectrum of 0.15 μM β_2 receptor "labeled" with IANBD prebound to free cysteine instead of free IANBD to assess possible nonspecific attachment of the probe to the receptor during labeling. **Insert:** 10% SDS-polyacrylamide gel electrophoresis of IANBD-labeled β_2 receptor. Lane 1, 150 pmol IANBD-labeled β_2 receptor; lanes 2 and 3, 150 pmol β_2 receptor preincubated before exposure to IANBD with iodoacetamide (lane 2) and N-ethylmaleimide (lane 3). **Insert left:** Commassie Blue staining of gel. **Insert right:** Gel photographed under UV light. The weak band with an apparent molecular weight of 32.5 kD is a degradation product of the receptor. (Reproduced from Gether et al. [1995], with permission of the publisher.)

fluorescence signal with an emission maximum at 523 nm (Fig. 14.1B). The blue shift in emission maximum, compared with cysteine-reacted IANBD in aqueous buffer, indicates that the modified cysteine(s) are located in an environment that, on the average, is of lower polarity than n-butanol but higher than n-hexane. This would likely involve labeling of one or more of the five cysteine residues that are located in the transmembrane hydrophobic core of the receptor (see Fig. 14.3). The covalent modification of the receptor was confirmed by SDS-polyacrylamide gel electrophoresis of the labeled receptor, and the specificity of labeling was verified by blocking the incorporation of IANBD with the cysteine-specific, nonfluorescent reagents, iodoacetamide and N-ethylmaleimide (Fig. 14.1B, insert). Importantly, the fluorescent labeling did not perturb the pharmacological properties of the receptor both in terms of agonist and antagonist binding (Gether et al., 1995).

To explore whether agonist binding to the receptor causes conformational changes leading to changes in the polarity of the environment surrounding the incorporated NBD fluorophore, we performed time-resolved spectroscopic analyses (Gether et al., 1995). As illustrated in Figure 14.2, binding of the full agonist isoproterenol to IANBD-labeled β_2 receptor caused a dose-dependent decrease in fluorescence, reaching a maximum amplitude below the extrapolated baseline after 10 minutes. The response to isoproterenol could be readily reversed by the active ($-$)-isomer of the antagonist propranolol but not by the less active ($+$)-isomer (Fig. 14.2). The response to isoproterenol was similarly reversed by several other antagonists, including alprenolol, ICI 118,551, pindolol, and dichloroisoproterenol (Gether et al., 1995). Moreover, the isoproterenol response was dose dependent and stereospecific (Gether et al., 1995). Prior to adding ligand, we normally observed a slight but constant decline in baseline fluorescence (Fig. 14.2). This loss of fluorescence over time is likely caused by bleaching of the fluorophore combined with some loss of protein possibly due to protein adhering to the inside of the cuvette. The decrease of fluorescence over time was unaffected by addition of 0.1% bovine serum albumin, 10% glycerol, or phospholipids to the cuvette (Gether et al., 1995). However, preincubation of the receptor in the cuvette for 15 minutes before performing the experiments minimized (but never eliminated) the constant decline in baseline fluorescence. It should be noted that the decline in fluorescence is unlikely due to denaturation of the protein because a similar loss of fluorescence was also observed with labeled receptor that was intentionally denatured in guanidinium chloride (Gether et al., 1995).

The observed agonist-induced decrease in fluorescence from the IANBD-labeled receptor is most likely due to movement of the fluorophore to a more polar environment upon agonist binding. Importantly, the magnitude of the fluorescence changes was found to correlate with the intrinsic biological efficacy of the ligand as demonstrated by comparing the effects of a series of partial and full agonists on adenylyl cyclase activity with their effect on the magnitude of the fluorescence changes (Gether et al., 1995). This suggests that the ligand-induced changes in fluorescence are relevant for the receptor activation mechanism.

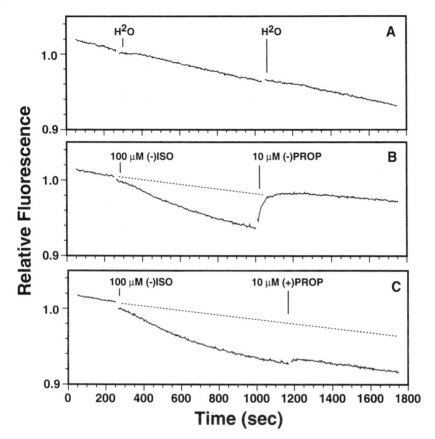

Figure 14.2. Isoproterenol induces a reversible decrease in fluorescence from the IANBD-labeled β$_2$-AR. **A:** Control addition of water (H$_2$O). **B,C:** Reversal of the response to isoproterenol (ISO) by the active (−)-isomer of the antagonist propranolol, (−)PROP (B), but not by the less active (+)-isomer, (+)PROP (C). Dotted lines indicate extrapolated baseline. Excitation was at 481 nm, and emission was measured at 523 nm. Fluorescence in all the individual traces shown was normalized to the fluorescence observed immediately after addition of ligand. All traces shown are representative of at least three identical experiments. (Reproduced from Gether et al. [1995], with permission of the publisher.)

3.A. Fluorescent Labeling of the Purified β$_2$-AR

We have used two different protocols for labeling of the purified β$_2$-AR (Gether et al., 1995, 1997b). Our results obtained with the two protocols are indistinguishable; however, we would recommend the second procedure because it generally would ensure best that all excessive dye is washed off the purified protein (Gether et al., 1997b, 1995).

1. Purified β$_2$ receptor (1–1.5 nmol) is incubated with 10–15-fold molar excess of IANBD (Molecular Probes, Eugene, OR) (150 μM) in a total volume of 100 μl buffer (20 mM Tris buffer, pH 7.5, containing 100 mM

NaCl and 0.08 % DβM). IANBD is added from a 10-mM stock solution in DMSO. The reaction is allowed to proceed for 1 hour at room temperature in the dark and is quenched by addition of 1 mM cysteine from a 100 mM stock, followed by a 5-minute incubation. Cysteine-reacted dye is removed using a Sephadex G50 gel filtration column (0.5 × 9 cm). The reaction mixture is applied directly to the column and eluted with 2.0 ml of buffer directly into a Centricon-30 filter device (Amicon, Beverly, MA). The eluate containing the labeled receptor is concentrated to approximately 50 μl in the Centricon-30 filter device by centrifugation for 45 minutes at 3,000g in a fixed angle rotor (Sorvall SS-34). The labeled receptor is either used directly for fluorescence spectroscopy analysis or stored at 4°C. Under these conditions the protein is stable for several days.

2. Purified receptor (usually a whole badge around 5 nmol) is bound to a 150-μl nickel column by recycling by gravity flow six times (Chelating Fast Flow Sepharose Resin from Pharmacia equilibrated in high salt buffer [20 mM Tris-HCl, pH 7.5, with 500 mM NaCl and 0.08% DβM]). IANBD labeling is achieved by recycling 1.0 ml of 0.5 mM IANBD in high salt buffer several times over the nickel column for 20 minutes. Excess dye is removed by extensive washing of the column with approximately 50 column vol of high salt buffer. Labeled receptor is eluted in 50-μl fractions with 200 mM imidazol in high salt buffer. Fractions are assayed for protein content, and peak fractions are pooled. The labeled receptor can be used directly for fluorescence spectroscopy analysis or stored on ice. Under these conditions the protein is stable for several days.

Both labeling procedures generally result in incorporation of 1.2–2 mol IANBD per mol receptor, as determined by measuring absorption at 481 nm and using an extinction coefficient of 21,000 $M^{-1} cm^{-1}$ for IANBD and a molecular weight of 50,000 D for the receptor. Protein concentration was determined using the BioRad *DC* protein assay kit (BioRad, Hercules, CA).

3.B. Fluorescence Spectroscopy Analysis

Fluorescence spectroscopy is performed at room temperature using a SPEX Fluoromax spectrofluorometer connected to a PC equipped with the Datamax software package. We use the photon counting mode and generally an excitation and emission bandpass of 4.2 nm (Gether et al., 1995, 1997a,b).

1. Emission scan experiments are done with 30–50 pmol IANBD-labeled receptor. Usually 10 μl of receptor is added to 390 μl buffer (20 mM Tris buffer, pH 7.4, containing 100 mM NaCl and 0.08 % DβM) in a 5 × 5 mm quartz cuvette and mixed by pipeting up and down. The excitation wavelength is set at 481 nm, and emission is measured from 490 to 625 nm, with an integration time at 0.3 sec/nm.

2. Like the emission scans, time-resolved fluorescence spectroscopy is performed using 30–50 pmol of labeled receptor. Usually 10 μl of receptor is added to 490 μl buffer (20 mM Tris buffer, pH 7.4, containing 100 mM NaCl and 0.08 % DβM) in a 5 × 5 mm quartz cuvette. To stabilize

the baseline, the mixture is preincubated for at least 10 minutes in the cu-vette before the experiment is started. Both during this period and during the time scan experiment the mixture is kept under constant stirring us-ing a 2×2 mm magnetic stir bar (Bel-Art Products, Pequannock, NJ). During time scan experiments the excitation wavelength is fixed at 481 nm and emission measured at a wavelength of 525 nm. The time scans are routinely performed over 30 minutes, and the first addition of ligand is usually done after 5 minutes. The volume of the added ligands is 1% of the total volume, and fluorescence is corrected for this dilution. The compounds tested in our fluorescence experiments have an absorbance of less than 0.01 at 481 nm and 525 nm in the concentrations used ex-cluding inner filter effects.

4. SITE-SELECTIVE FLUORESCENT LABELING OF THE β_2-AR

The β_2-AR contains 13 cysteines five of which are not expected to be available for chemical derivatization. In the extracellular loops, four cysteines (Cys[106], Cys[184], Cys[190], and Cys[191]) form two disulfide bridges (Fig. 14.3) (Dohlman et al., 1990; Fraser, 1989; Noda et al., 1994), and in the intracellular carboxy ter-minal tail, Cys[341] has been shown to be palmitoylated (Mouillac et al., 1992; O'Dowd et al., 1989). To identify the cysteine(s) responsible for the agonist-induced change in fluorescence and, thus, to establish a system that would al-low site-selective incorporation of the IANBD fluorophore, we mutated cysteines in the receptor and generated a series of mutant receptors with one, two, or three cysteines available for chemical derivatization (Gether et al., 1997b). All these mutants displayed minimal changes in pharmacological properties compared with the wild-type receptor with respect to both ligand binding and functional coupling to adenylate cyclase (Gether et al., 1997b). However, mutation of several cysteine residues led to a reduction in receptor expression (Gether et al., 1997b). Notably, a mutant receptor with all free cys-teines substituted was expressed so poorly that purification in quantities suffi-cient for fluorescence spectroscopy analysis was impossible (Gether et al., 1997b). Ideally, it should be possible to remove all endogenous cysteines and either reintroduce them one by one or introduce single cysteines in new posi-tions. Unfortunately this was not possible in the β_2-AR. Nevertheless, as illus-trated in Figure 14.3 and described below, it is possible to work with a system where not all cysteines can be substituted. It only requires the time-consuming inclusion of proper control mutants.

The mutant receptors containing one, two, or three of the naturally occurring cysteines were all purified and labeled with the IANBD fluorophore. As ex-pected, the IANBD-labeled mutants all gave emission maxima around 525 nm (Gether et al., 1997b). Time-resolved analysis of the mutants revealed that ag-onist-induced changes in fluorescence are observed only in receptors in which Cys[285] or Cys[125] is present (Fig. 14.3). A mutant lacking only these two cys-teines (Cys77,116,265,327,378,406) showed no response to agonist binding (Fig. 14.3).

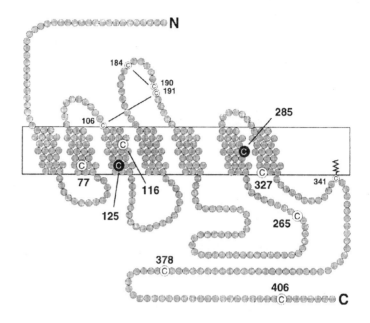

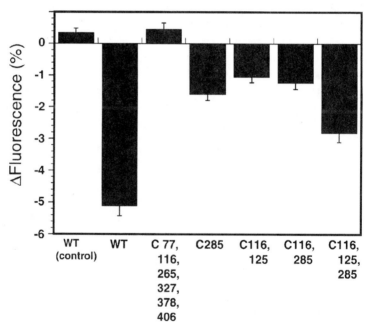

Figure 14.3. The effect of isoproterenol on fluorescence from IANBD-labeled wild type (WT) and mutant β_2-AR. **A:** Snake diagram of the β_2-AR. The receptor contains 13 Cys residues five of which (Cys^{77}, Cys^{116}, Cys^{125}, Cys^{285}, and Cys^{327}) are predicted to be in the transmembrane domain. Three Cys residues are predicted to be in the cytoplasmic regions (Cys^{265}, Cys^{378}, and Cys^{406}). Five Cys residues are not expected to be available for chemical derivatization (small white circles): four residues (Cys^{106}, Cys^{184}, Cys^{190}, and Cys^{191}) form two disulfide bridges (Dohlman et al., 1990; Fraser, 1989; Noda et al., 1994), and in the intracellular carboxy-terminal tail, Cys^{341} has been shown to be palmi-

The data suggest that agonist binding to the β_2-AR promotes a conformational change in the receptor that exposes NBD attached to Cys[125] in TM segment 3) and to Cys[285] (in TM 6) to a more polar environment. We have attempted to predict the actual structural changes using molecular modeling and computational simulations (Gether et al., 1997b). Most importantly, the simulations demonstrated a significant conformational restraint for NBD bound to Cys[285]-NBD and Cys[125]-NBD (Gether et al., 1997b). This suggests that the change in molecular environment around the bound NBD reflects movement of the transmembrane helix to which it is attached rather than movement of NBD relative to the transmembrane helix. As schematically illustrated in Figure 14.4, our data are consistent with a counterclockwise rotation of helices 3 and 6 in response to agonist binding. This is consistent with the suggested rigid body movements of the corresponding helices in rhodopsin (Farrens et al., 1996). It is interesting to note that Cys[285] in TM 6 is situated one α-helical turn below Pro[288], which is highly conserved among GPCRs and provides a flexible hinge in this helix (Gether et al., 1997b). It has therefore been speculated that the movement of Cys[285]-NBD to a more polar environment in the protein interior could be directly facilitated by this flexible hinge connecting the binding site with the putative G protein–coupling domain at the cytoplasmic end of the helix (Gether et al., 1997b).

A few issues should be emphasized when interpreting data from our fluorescence spectroscopy analysis. It is important to note that the amplitude of the fluorescent change is only a rough indicator of the magnitude of conformational changes. For example, we cannot assume that there is a linear correlation between change in fluorescence and magnitude of movements. Therefore, the movement of TM 3, for example, may not be of the same magnitude as that of TM 6. It should also be ensured that the fluorescent probe, when incorporated into the receptor, does not interfere with the binding of the ligands. In the β_2-AR this is highly unlikely. Labeling of the receptor with IANBD does not alter agonist or antagonist binding properties (Gether et al., 1995), as would be expected if the bound NBD were positioned within the ligand binding pocket. The results from mutagenesis studies have also provided substantial evidence that amino acids involved in forming the ligand binding pocket are on a different side of the transmembrane α helix and one to two α-helical turns closer to the membrane surface relative to Cys[125] and Cys[285].

---

toylated (Mouillac et al., 1992; O'Dowd et al., 1989). **B:** Bar diagram of changes in fluorescence in response to the full agonist isoproterenol (10^{-3} M) for the WT receptor and indicated mutants. The ligand concentration was chosen to ensure saturation of the receptors eliminating any influence from different agonist affinities. Excitation was set at 481 nm, and emission was measured at 525 nm. Data are given as percent change in fluorescence (mean $\pm$ SE, N = 3–6). The percent change was calculated as the change in fluorescence relative to the extrapolated baseline at t = 15 minutes after addition of ligand. The cysteine mutants are named according to the cysteines still present in the receptor and available for chemical derivatization. Thus, Cys[285] describes a construct where Cys[285] is present but where Cys[77], Cys[116], Cys[125], Cys[265], Cys[327], Cys[378], and Cys[406] have been mutated. Cys[106], Cys[184], Cys[190], and Cys[191] were excluded from the "name" because they are not available for chemical derivatization. (Reproduced from Gether et al. [1997b], with permission of the publisher.)

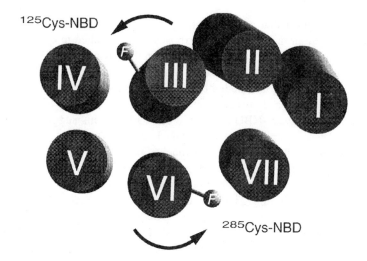

Figure 14.4. Simplified model of the β_2-AR indicating predicted movements in response to agonist binding. NBD bound to Cys[125] in TM 3 (III) and Cys[285] in TM 6 (VI) are predicted to lie at the protein–lipid interface, oriented predominantly towards the lipid (the NBD fluorophore is indicated by F) (Gether et al., 1997b). Cys[285]-NBD is predicted to be at the helix 6 (VI)–7 (VII) interface in a boundary zone between the lipid bilayer (or the hydrophobic tails of the detergent micelle) and the more polar interior of the protein (Gether et al., 1997b). An agonist-induced counterclockwise rotation seen from the extracellular side of helix 6 (VI) as indicated by the arrow would lead to movement of Cys[285]-NBD from the nonpolar environment of the lipid bilayer (or detergent micelle) to the more polar environment of the interior of the protein explaining the observed changes in fluorescence. This is consistent with spin-labeling studies in rhodopsin also suggesting a counterclockwise rotation of TM 6 (VI) upon photoactivation of rhodopsin (Farrens et al., 1996). In TM 3 (III), Cys[125]-NBD is predominantly exposed to the lipid bilayer (or detergent micelle) (Gether et al., 1997b). Our data indicate that the extent of lipid exposure is changed in response to agonist binding. This is consistent with an agonist-induced movement of TM 3 (III), causing the fluorophore to be exposed to a more polar face of TM 4 (IV) and/or the more polar interior of the receptor as indicated by the arrow.

4.A. Generation of Mutant Receptors

All molecular biology experiments were performed according to standard procedures. We used the cDNA encoding the epitope-tagged human β_2-AR described above as template for mutagenesis (Guan et al., 1992; Kobilka, 1995). To facilitate cloning procedures, an *Nhe*I site at bp 426 (as counted from the initiator codon) and a *Sac*I site at bp 1083 were introduced in SF-hβ2-6H by mutagenesis before constructing the cysteine mutants. The *Nhe*I and *Sac*I sites and cysteine mutations C77V, C116V, C125V, C265A, C285S, C327S, C378A, and C406A were all generated by polymerase chain reaction (PCR)-mediated mutagenesis using Pfu polymerase according to the manufacturer's instructions (Stratagene, La Jolla, CA). The generated PCR fragments were digested with the appropriate enzymes, purified by agarose gel electrophoresis, and cloned into the baculovirus expression vector pVL1392 containing SF-hβ2-6H (Guan et al., 1992). The constructs Cys[285], Cys[116,125,285], Cys[116,125], Cys[116,285], and Cys[77,116,265,327,378,406] were obtained by combining selected restriction enzyme fragments. All mutations were confirmed by restriction enzyme analysis and sequencing.

4.B. Expression in Sf9 Insect Cells

All mutants were expressed in Sf9 insect cells and virus generated as described in the previous sections for the wild-type receptor.

4.C. PCR Analysis of Virus Stocks

When handling virus stocks for several mutants simultaneously, it is crucial to ensure that no cross-contamination occurs during the many steps from transfection until a high titer virus stock is available in sufficient quantities. Therefore, virus stocks should be routinely checked for expression of the correct mutant by PCR analysis.

1. Isolate virus by centrifugating 5–10 ml virus stock for 20 minutes at 40,000g.
2. Discard supernatant and resuspend the pelleted virus in 400 μl TE buffer (10 mM Tris-HCl plus 0.1 mM EDTA).
3. Incubate for 30 minutes at 37°C with 10 μg/ml RNAse followed by incubation for 30 minutes at 37°C in the presence of 10 μg/ml proteinase K and 0.5% SDS.
4. Isolate DNA from the lysate by standard phenol–chloroform extraction and ethanol precipitation. Dissolve DNA in 50–100 μl TE-buffer.
5. Amplify by PCR using *Taq* polymerase and a set of primers that will amplify the entire coding region. Routinely, we use 2 μl template DNA in a standard reaction according to manufacturer's instruction (Perkin Elmer, Foster City, CA). The amplified fragments are analyzed by restriction enzyme analysis and agarose gel electrophoresis.

4.D. Purifications

The mutant β_2-ARs were purified by nickel chromatography followed by alprenolol affinity chromatography as described above.

4.E. Fluorescent Labeling and Fluorescence Spectroscopy

Fluorescent labeling and fluorescence spectroscopy were performed as described above for the wild-type receptor.

5. ANALYZING A CONSTITUTIVELY ACTIVE MUTANT RECEPTOR BY FLUORESCENCE SPECTROSCOPY

It has been shown that discrete mutations, in particular at the carboxy terminal end of the third intracellular loop can dramatically enhance agonist-independent receptor activity (Lefkowitz et al., 1993). However, the structural basis for this constitutive activation of the receptor is poorly understood. Our fluorescent assay represents the first direct assay to assess the conformational state of a ligand-activated GPCR. It was therefore attractive to apply the assay to a constitutively activated β_2-AR to explore the mechanism underlying constitutive receptor activation (Gether et al., 1997a). The constitutively activated mutant (CAM) of the β-AR, published by Samama et al. (1993), was expressed in Sf9 insect cells, purified, and fluorescently labeled as described in the previous sections. Subsequent analysis revealed two novel properties of the CAM: a marked structural instability of the receptor protein and an exaggerated conformational response to ligand binding (Gether et al., 1997a). The structural instability of CAM could be demonstrated by a fourfold increase in the rate of denaturation of purified receptor at 37°C as compared to the wild-type receptor (Gether et al., 1997a). As shown in Figure 14.5, spectroscopic analysis of purified CAM labeled with IANBD indicated that both agonist and antagonist elicit more profound structural changes in CAM than in the wild-type protein. The data directly support the idea that the receptor, in the absence of agonist, normally is constrained in an inactive conformation by stabilizing intramolecular interactions. Hence, mutations that confer constitutive activity to the receptor may remove some of these constraining interactions, allowing the receptor to more readily undergo transitions be-

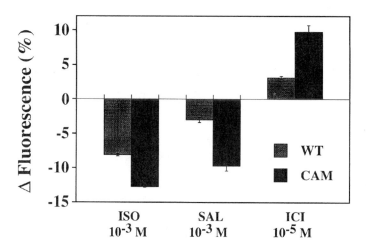

Figure 14.5. Ligand stimulation of IANBD-labeled CAM receptor causes greater changes in fluorescence than stimulation of IANBD-labeled WT receptor. Bar diagram of changes in fluorescence in response to the full agonist isoproterenol (ISO) (10^{-3} M), the partial agonist salbutamol (SAL) (10^{-3} M), and the inverse agonist ICI 118,551 (ICI) (10^{-5} M). Data are given as percent change in fluorescence (mean $\pm$ SE, N = 3). For ISO and SAL, the change was determined by the amplitude of the reversal by the antagonist alprenolol, which has a more than 100-fold higher affinity for the receptors than isoproterenol and salbutamol. For ICI, the percent change was calculated as the change in fluorescence relative to the extrapolated baseline at t = 15 minutes after addition of ligand. The ligand concentrations used were chosen to ensure saturation of the receptors eliminating any influence from different agonist affinities. Excitation was at 481 nm and emission was measured at 525 nm. (Reproduced from Gether et al. [1997a], with permission of the publisher.)

tween the inactive and active state and making the receptor more susceptible to denaturation (Gether et al., 1997a). These results underscore the advantage of biophysical assays and constitutively active mutants as important tools for delineating molecular mechanisms involved in activation of GPCRs.

5.A. Expression, Purification, and Fluorescent Labeling of CAM

CAM was expressed in Sf9 cells and purified essentially as described above for the wild-type receptor (Gether et al., 1997a). However, due to reduced expression and structural instability of the protein we had to make several changes in our protocols (Gether et al., 1997a). These changes may not only apply to CAM but may also provide hints for other receptors that may be more unstable than the wild-type β_2-AR and/or are expressed at low levels.

1. Amplifications of virus stocks were performed with 1 μM of the inverse agonist ICI 118,551 added to the media to lower constitutive cAMP production. High levels of cAMP production may select for inactive receptor mutants during virus amplification.

2. To obtain optimal expression, only 600 ml of Sf9 cell culture was grown and infected in a 2,800 ml Fernbach flask (Bellco) compared with the 1,000–1,200 ml used normally. This generally improved expression roughly 1.5-fold/ml of culture, most likely due to improved oxygenation of the cells.

3. In large-scale infections, 1 μM of the inverse agonist ICI 118,551 was added to the media to lower constitutive cAMP production and to stabilize the receptor protein.

4. The M1 anti-FLAG antibody column was omitted from the purification procedure. All incubations were shortened as much as possible to allow purification over 1 day instead of 2 days.

5. Immediately after elution from the alprenolol column, the receptor was bound to a 100–150 μl nickel column and IANBD-labeled on the column as described above. Spectroscopic analyses were performed within 24 hours after purification and labeling.

6. CONCLUDING REMARKS

Using a sulfhydryl-reactive and environmentally sensitive fluorescent probe, IANBD, as a molecular reporter we have been able to characterize and map ligand-induced conformational changes in the β_2-AR. The technique requires a convenient expression system that can be scaled up easily. The purification procedure should be reliable, relatively easy, and result in highly pure and active protein. The required yields from the purifications should be in the nanomolar range. Moreover, to achieve site-selective incorporation of the fluorophore it should be possible to mutate most endogenous cysteines without perturbing receptor function. Notably, the results described and discussed in the text have focused on the endogenous cysteines in the β_2-AR. Currently we are systematically introducing cysteines in new positions to further map the conformational changes accompanying agonist-induced activation of GPCRs. These new cysteines are introduced in a mutant β_2-AR still containing three endogenous cysteines to ensure a high level of expression. So far we have evidence from these new mutants for significant movement at the cytoplasmic site of TM 6 (A.D. Jensen and U. Gether, manuscript in preparation). These changes occur with the same kinetics as the changes discussed in this chapter, indicating that both responses reflect the same conformational change. The data underline that site-selective fluorescent labeling may be used as a general approach for mapping ligand-induced conformational changes in a purified receptor protein.

ACKNOWLEDGMENTS

The work described here was carried out in Dr. Brian Kobilka's laboratory at Howard Hughes Medical Institute, Stanford University, CA. Dr. Anne Dam Jensen is thanked for helpful comments on the manuscript.

REFERENCES

Altenbach C, Yang K, Farrens DL, Farahbakhsh ZT, Khorana HG, Hubbell WL (1996): Structural features and light-dependent changes in the cytoplasmic interhelical E–F loop region of rhodopsin: A site-directed spin-labeling study. Biochemistry 35: 12470–12478.

Baldwin JM (1993): The probable arrangement of the helices in G-protein–coupled receptors. EMBO J 12:1693–1703.

Ballesteros JA, Weinstein H (1995): Integrated methods for the construction of three-dimensional models and computational probing of structure–function relations in G protein coupled receptors. Methods Neurosci 25:366–428.

Benovic JL, Shorr RGL, Caron MC, Lefkowitz RJ (1984): The mammalian beta2-adrenergic receptor: Purification and characterization. Biochemistry 23:4510–4518.

Caron MG, Srinivasan Y, Pitha J, Kociolek K, Lefkowitz RJ (1979): Affinity chromatography of the beta-adrenergic receptor. J Biol Chem 254:2923–2927.

Cerione RA (1994): Fluorescence assays for G-protein interactions. Methods Enzymol 237:409–423.

Dohlman HG, Caron MG, DeBlasi A, Frielle T, Lefkowitz RJ (1990): Role of extracellular disulfide-bonded cysteines in the ligand binding function of the beta 2-adrenergic receptor. Biochemistry 29:2335–2342.

Dunn SMJ, Raftery MA (1993): Cholinergic binding sites on the pentameric binding site on the acetylcholine receptor of *Torpedo californica*. Biochemistry 32:8608–8615.

Fanelli F, Menziani MC, De Benedetti PG (1995): Molecular dynamics simulations of m3-muscarinic receptor activation and QSAR analysis. Bioorg Med Chem 3:1465–1477.

Farahbakhsh ZT, Ridge KD, Khorana HG, Hubbell WL (1995): Mapping light-dependent structural changes in the cytoplasmic loop connecting helices C and D in rhodopsin: A site-directed spin labeling study. Biochemistry 34:8812–8819.

Farrens DL, Altenbach C, Yang K, Hubbell WL, Khorana HG (1996): Requirement of rigid-body motion of transmembrane helices for light activation of rhodopsin. Science 274:768–770.

Fraser CM (1989): Site-directed mutagenesis of beta-adrenergic receptors. Identification of conserved cysteine residues that independently affect ligand binding and receptor activation. J Biol Chem 264:9266–9270.

Gether U, Ballesteros JA, Seifert R, Sanders-Bush E, Weinstein H, Kobilka BK (1997a): Structural instability of a constitutively active G protein-coupled receptor. Agonist-independent activation due to conformational flexibility. J Biol Chem 272:2587–2590.

Gether U, Kobilka BK (1998): G protein coupled receptors: II. Mechanism of agonist activation. J Biol Chem 273:17979–17982.

Gether U, Lin S, Ghanouni P, Ballesteros JA, Weinstein H, Kobilka BK (1997b): Agonists induce conformational changes in transmembrane domains III and VI of the beta2 adrenoceptor. EMBO J 16:6737–6747.

Gether U, Lin S, Kobilka BK (1995): Fluorescent labeling of purified beta2-adrenergic receptor: Evidence for ligand-specific conformational changes. J Biol Chem 270:28268–28275.

Gettins PGW, Fan B, Crews BC, Turko IV (1993): Transmission of conformational change from the heparin binding site to the reactive center of antithrombin. Biochemistry 32:8385–8389.

Grisshammer R, Tate CG (1995): Overexpression of integral membrane proteins for structural studies. Q Rev Biophys 28:315–422.

Guan XM, Kobilka TS, Kobilka BK (1992): Enhancement of membrane insertion and function in a type IIIb membrane protein following introduction of a cleaveable signal peptide. J Biol Chem 267:21995–21998.

Javitch JA, Fu D, Liapakis G, Chen J (1997): Constitutive activation of the beta2 adrenergic receptor alters the orientation of its sixth membrane-spanning segment. J Biol Chem 272:18546–18549.

Kobilka BK (1995): Amino and carboxyterminal modifications to facilitate the production and purification of a G protein coupled receptor. Anal Biochem 231:269–271.

Lefkowitz RJ, Cotecchia S, Samama P, Costa T (1993): Constitutive activity of receptors coupled to guanine nucleotide regulatory proteins. Trends Pharmacol Sci 14:303–307.

Lin S, Gether U, Kobilka BK (1996): Ligand stabilization of the beta 2 adrenergic receptor: Effect of DTT on receptor conformation monitored by circular dichroism and fluorescence spectroscopy. Biochemistry 35:14445–14451.

Luo X, Zhang D, Weinstein H (1994): Ligand-induced domain motion in the activation mechanism of a G-protein–coupled receptor. Protein Eng 7:1441–1448.

Mouillac B, Caron M, Bonin H, Dennis M, Bouvier M (1992): Agonist-modulated palmitoylation of beta 2-adrenergic receptor in Sf9 cells. J Biol Chem 267:21733-21737.

Noda K, Saad Y, Graham RM, Karnik SS (1994): The high affinity state of the beta 2-adrenergic receptor requires unique interaction between conserved and non-conserved extracellular loop cysteines. J Biol Chem 269:6743–6752.

O'Dowd BF, Hnatowich M, Caron MG, Lefkowitz RJ, Bouvier M (1989): Palmitoylation of the human beta 2-adrenergic receptor. Mutation of Cys341 in the carboxyl tail leads to an uncoupled nonpalmitoylated form of the receptor. J Biol Chem 264:7564–7569.

Parker EM, Kameyama K, Higashijima T, Ross EM (1991): Reconstitutively active G protein–coupled receptors purified from baculovirus-infected insect cells. J Biol Chem 266:519–527.

Phillips WJ, Cerione RA (1991): Labeling of the beta-gamma subunit complex of transducin with an environmentally sensitive cysteine reagent. Use of fluorescence spectroscopy to monitor transducin subunit interactions. J Biol Chem 266:11017–11024.

Samama P, Cotecchia S, Costa T, Lefkowitz RJ (1993): A mutation-induced activated state of the beta2-adrenergic receptor: Extending the ternary complex model. J Biol Chem 268:4625–4636.

Scheer A, Fanelli F, Costa T, De Benedetti PG, Cotecchia S (1996): Constitutively active mutants of the alpha 1B-adrenergic receptor: Role of highly conserved polar amino acids in receptor activation. EMBO J 15:3566–3578.

Schertler GF, Villa C, Henderson R (1993): Projection structure of rhodopsin. Nature 362:770–772.

Sheikh SP, Zvyaga TA, Lichtarge O, Sakmar TP, Bourne HR (1996): Rhodopsin activation blocked by metal-ion-binding sites linking transmembrane helices C and F. Nature 383:347–350.

Tate CG, Grisshammer R (1996): Heterologous expression of G-protein–coupled receptors. Trends Biotechnol 14:426–430.

Unger VM, Hargrave PA, Baldwin JM, Schertler GF (1997): Arrangement of rhodopsin transmembrane alpha-helices. Nature 389:203–206.

CHAPTER 15

BIOSYNTHETIC INCORPORATION OF UNNATURAL AMINO ACIDS INTO G PROTEIN–COUPLED RECEPTORS

ANDRÉ CHOLLET and GERARDO TURCATTI

1. INTRODUCTION   335
2. ENGINEERING AND CONSTRUCTION OF A MISACYLATED
   SUPPRESSOR tRNA   337
   A. Materials and Methods   338
   B. Mutagenesis and Preparation of cRNA   339
   C. Preparation of the Dinucleotide pdCpA   340
   D. Synthesis of pdCpA   340
   E. Synthesis of Protected Amino Acids   342
   F. Synthesis of the Dinucleotide–Amino Acid Conjugates   346
   G. Preparation of Chemically Misacylated Suppressor tRNAs   346
3. EXPRESSION OF RECEPTORS CONTAINING
   UNNATURAL AMINO ACIDS BY SUPPRESSION
   MUTAGENESIS: MICROINJECTION OF OOCYTES
   AND ELECTROPHYSIOLOGY   348
4. APPLICATIONS   349
5. INVESTIGATION OF MOLECULAR ARCHITECTURE OF
   LIGAND–RECEPTOR COMPLEXES BY FLUORESCENCE   350
6. FUTURE PROSPECTS   351

1. INTRODUCTION

Protein engineering is a key method for the investigation of the structure and function of proteins in general and of G protein–coupled receptors (GPCRs) in particular. Classic site-directed mutagenesis is limited to substitution of residues by natural amino acids only and therefore to a limited set of functional groups. Clearly, the ability to incorporate almost any chemical group at a

Structure–Function Analysis of G Protein-Coupled Receptors, Edited by Jürgen Wess.
ISBN 0-471-25228-X   Copyright © 1999 Wiley-Liss, Inc.

specific position in a GPCR would permit the detailed investigation of its structure and function. In particular, biophysical probes such as fluorescent or spin-labeled groups would provide valuable information on the molecular architecture of the receptor or ligand–receptor complexes and on the interacting proteins like G proteins or GPCR-specific kinases (GRK). Moreover, time-resolved methods can in principle be used to help understand the different conformations of GPCRs and how they interconvert on the receptor energy landscape. Biophysical probes can provide direct information about the orientation of helices, the water-exposed regions, and intra- or intermolecular distances in both steady state and real time.

Because GPCRs are large integral membrane proteins, usually of more than 350 amino acids, the introduction of reporter groups by classic chemical modification of reactive lysyl or cysteinyl side chains is not recommended. It requires tedious preliminary purification of the GPCR prior to labeling and then reconstitution in a membrane environment. Also, it usually results in multiple labeling in a nonregioselective manner.

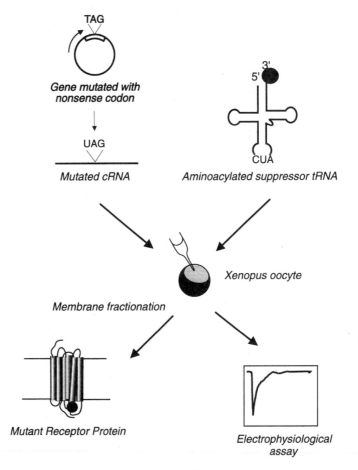

Figure 15.1. General scheme for the incorporation of unnatural amino acids into receptors expressed in *Xenopus* oocytes (see text for details).

A much more elegant approach involves unnatural suppression mutagenesis in which an engineered suppressor tRNA that recognizes the nonsense codon placed in the receptor gene at a selected site is used to incorporate an unnatural amino acid during biosynthesis. The translation machinery recognizes the exogenous aminoacylated suppressor tRNA and incorporates the unnatural functional group at sites preselected for by creation of a nonsense codon in the gene. This approach has been successfully developed *in vitro* using cell-free transcription-translation systems capable of sustaining protein synthesis (Heckler et al., 1984; Noren et al., 1989a; Bain et al., 1991). Although small soluble proteins can be expressed in cell-free systems, in our hands the production of biologically active integral membrane proteins such as GPCR failed under such conditions.

However, we have been able to achieve unnatural suppression mutagenesis in a whole-cell system as shown in Figure 15.1. This was accomplished by the following steps: (1) creation by mutagenesis of a nonsense stop codon (i.e., TAG) at the selected site in the cDNA encoding the GPCR; (2) transcription of the cDNA into cRNA; (3) construction by chemical and enzymatic steps of a UAG suppressor tRNA chemically misacylated with the unnatural amino acid; (4) co-injection of the cRNA and loaded suppressor tRNA into oocytes; and (5) assaying for receptor function by ligand binding and electrophysiology.

In this chapter, we describe the methodology of incorporation of unnatural amino acids into GPCR *in vivo* using nonsense codon suppression mutagenesis during functional expression in the native membrane environment of an intact cell system, the *Xenopus* oocyte. As a representative example, we demonstrate the incorporation of fluorescent amino acids into a prototypic GPCR, the neurokinin-2 (NK2) receptor, and describe a method allowing the measurement of intermolecular distances between sites on the NK2 receptor and a bound ligand by fluorescence resonance energy transfer.

2. ENGINEERING AND CONSTRUCTION OF A MISACYLATED SUPPRESSOR tRNA

The general scheme for the synthesis of the unnatural suppressor tRNA is shown in Figure 15.2. The functional, chemically acylated suppressor tRNA (Sup-tRNA^{Xaa}) was assembled from two RNA fragments, the dinucleotide 5'-phospho-2'-deoxycytidyl(3'-5')adenosine (pdCpA) and a 74-nucleotide transcript tRNA(-CA) containing the anticodon CUA, and the protected amino acid of interest using a combination of chemical and enzymatic steps. Following the procedures described below, this synthesis can be carried out by any competent chemist or biochemist and requires only standard laboratory equipment. It is particularly important to work in an RNAse-free environment and to follow pH and temperature indications carefully. The suppressor tRNA (Sup-tRNA) was derived from yeast tRNA^{Phe} by changing the anticodon to CUA to recognize UAG nonsense codon (Bruce et al., 1982) and by introducing G20U and A73G mutations (Sampson et al., 1989). Replacement of G by U at position 20 in the D stem-loop was introduced to reduce the rate of aminoacylation by Phe-tRNA synthetase, thus preventing the introduction of natural amino acid at the UAG site during protein synthesis. Mutation of the discriminator base at position 73 from A to G was made to further decrease the ability of *Xenopus* endogenous

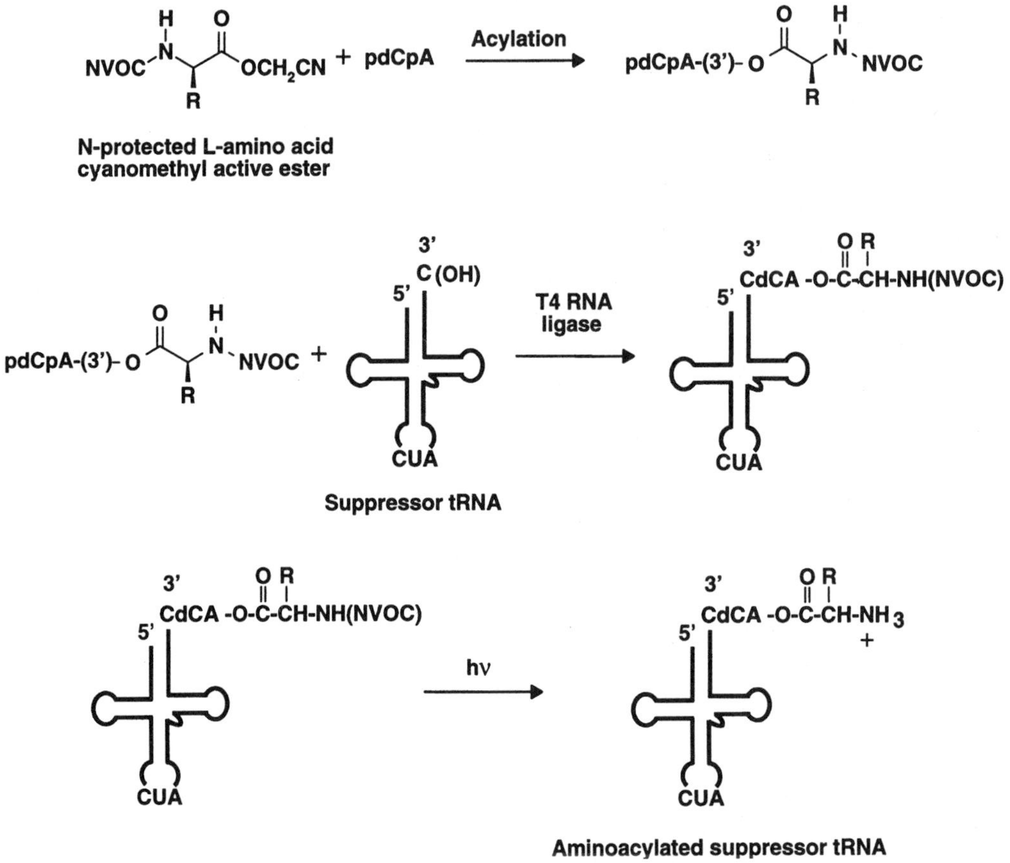

Figure 15.2. Scheme for the construction of aminoacylated suppressor tRNAs (aminoa-cyl-Sup-tRNA) (see text for details).

tRNA synthetases to reacylate Sup-tRNA. These changes were introduced by redesigning a vector described previously (Noren et al., 1989b) that allows for the synthesis of tRNA of defined length by run-off transcription. Both the 76-nucleotide full-length Sup-tRNA and a truncated 74-nucleotide form lacking the terminal pCpA in the 3' acceptor stem (Sup-tRNA [-CA]) were made by run-off *in vitro* transcription with T7 RNA polymerase of this DNA template, which was linearized at different restriction sites to control the 3' end of the transcript as described (Noren et al., 1989b). The 74-nucleotide Sup-tRNA (-CA) was used for the construction of the misacylated functional Sup-tRNAs, whereas the unloaded 76-nucleotide Sup-tRNA was used as a control in the un-natural suppression mutagenesis experiments. Recently, another suppressor tRNA functioning in oocytes has been described by Saks et al. (1996).

2.A. Materials and Methods

Anhydrous solvents were purchased from Aldrich; chemicals of the highest pu-rity grade were obtained from Fluka (Buchs, Switzerland), Aldrich (St. Louis,

MO), or Merck (Darmstadt, Germany). Acetonitrile and water used for the preparation of aqueous buffers in HPLC runs were HPLC grade from Baker (Gross-Gerau, Germany). Special care was taken for handling RNA due to the ubiquitous nature of RNA-destroying enzymes. All solutions were treated by the addition of 10% diethylpyrocarbonate (DEPC) (v/v in absolute ethanol) to a final concentration of 1% for 24 hours at room temperature (RT) and then sterilized by autoclaving at 125°C and 18 psi for 1 hour.

Thin-layer chromatography (TLC) was performed on 0.25-mm pre-coated silica gel plates (60F-254, E Merck, Germany) using the following solvent systems: *n*-butanol:acetic acid:water 4:1:1; and ethylacetate:acetic acid 98:2.

Two HPLC systems were used:

1. Waters (Millipore Corp.) 510 pumps, controlled with an automated gradient controller, Model 481 UV/Vis detector and equipped with a Macherey Nagel ET250/8/4 Nucleosil 5C18 (120 Å) column. The flow rate was 1 ml/minute. The mobile phase consisted of (A) 50 mM ammonium acetate (pH 4.5) and (B) acetonitrile. A gradient of 0% B for 1 minute, 0%–50% B over 24 minutes, 50%–85% B over 35 minutes, and 85% B for 10 minutes was used, and the detector was set at 260, 350, or 475 nm.

2. Gold System (Beckman) equipped with a programmable solvent module 126, a diode array detector module 168, and either a Rheodyne manual injector or an autosampler module 507. This system was equipped with a column Vydac C-4 (214TP54) 250 × 4.6 mm, flow rate 1 ml/minute. The mobile phases were (A) 100 mM TEAA (pH 7.0) in DEPC treated water and (B) acetonitrile. A gradient of 0% B for 5 minutes, 0%–15% B over 45 minutes, 15%–25% B over 10 minutes, 25% B for 5 minutes, and 25%–60% B over 10 minutes was used, and the detector was set at 260 and 280 nm.

2.B. Mutagenesis and Preparation of cRNA

The cDNA encoding the GPCR was transferred in a transcription vector (e.g., pGEM type) under the control of either SP6 polymerase or T7 polymerase. This vector was optimized for efficient transcription by engineering a short 5'-untranslated region with low secondary structure forming potential and a Kozak eukaryotic consensus sequence for efficient initiation of translation at AUG (Turcatti et al., 1996). Unique TAG nonsense (stop) codons were introduced at the sites selected for unnatural suppression mutagenesis. We used the method of Kunkel et al. (1991), but any other standard procedure for site-directed mutagenesis could have been employed. The termination codon for stopping protein synthesis should be either TAA or TGA. Capped cRNAs were produced by run-off transcription of linearized cDNA plasmids using commercially available kits (e.g., Promega, Madison, WI). The yield was approximately 10 μg RNA transcript (1.0–1.5 kb) from 1 μg cDNA. The cRNAs were checked on agarose gels and used directly for injection into oocytes. Detailed protocols can be found in Turcatti et al. (1996) or in manufacturers' instructions provided with the kits.

2.C. Preparation of the Dinucleotide pdCpA

The scheme for the construction of the fully deprotected pdCpA dinucleotide is shown in Figure 15.3. We found that automated synthesis in a commercial DNA synthesizer utilizing phosphoramidite chemistry was superior to the manual liquid phase method previously described (Robertson et al., 1991) in terms of speed of synthesis and purity of the final product. This also simplifies the whole strategy of misacylation of tRNA, rendering it more accessible to the nonchemist. Typically, one run would yield about 40 μmol, which is sufficient for 20 acylation reactions with appropriately protected and activated amino acids. The dinucleotide pdCpA was then extracted by n-butanol and subjected to a tetrabutylammonium ion exchange column to facilitate acylation reactions with N-protected cyanomethyl esters of amino acids. This process increases the solubility of the pdCpA in DMF and DMSO, the solvents used for the subsequent acylation reaction.

2.D. Synthesis of pdCpA

The synthesis of pdCpA was performed in an automated DNA/RNA synthesizer Model 394 (Applied Biosystems) using the phosphoramidite chemistry. The support used for the pdCpA synthesis was controlled-pore-glass (CPG) covalently derivatized with ribo-A. DMT-rA(Bz)-2′-tBuSi-CPG (1.0 g; 500 Å pore size; 51 μmol/g; Peninsula Laboratories, Inc.) was split and loaded in four 10-μmol scale columns and submitted to two standard coupling cycles following the classic conditions recommended by the manufacturer: the first cycle with the protected dC phosphoramidite and the second cycle with the cyanoethyl phosphoramidite, DMT-O-$(CH_2)_2SO_2$ $(CH_2)_2$-N,N-diisopropyl-β-cyanoethylphosphoramidite (5′ Phosphate-ON from Clontech) at 0.1 M in anhydrous acetonitrile. The dinucleotide pdCpA was cleaved from the solid support by treatment with ammonium hydroxide 25%(w/v) at RT and further deprotected in concentrated ammonia at 55°C for 5 hours.

n-Butanol (400 μl) was added to pdCpA (5 μmol) dissolved in 100 μl water. The mixture was vortexed for 1 minute and centrifuged for 3 minutes. The extraction was repeated three times. The pellet was taken up in water, and the amount of pdCpA was estimated from ϵ_{260} (pdCpA) = 23,000 cm^{-1} M^{-1}. pdCpA was lyophilized and stored at -80°C prior to the ion-exchange step.

To facilitate acylation reactions between N-protected cyanomethyl esters of amino acids and the pdCpA, the tetrabutylammonium salt of pdCpA was prepared. This process increases the solubility of the pdCpA in either DMF or DMSO, the solvents used for the acylation reaction by cyanomethyl active esters of unnatural amino acids. When 2.2 equiv. of tetrabutylammonium were added to 1 equiv. of pdCpA, the acylation reaction is catalyzed, proceeding to completion in a few hours at RT instead of a few days (Ellman et al., 1991). A strongly acidic Dowex cation exchange resin (Dowex-50W hydrogen form from Sigma) was loaded in a polypropylene column and washed sequentially with 70% ethanol in water, 0.1 M NaOH, 0.1 M HCl, and finally water until the pH was neutral. Then, 0.75 M tetrabutylammonium hydroxide (Aldrich, 10% w/v in water; premixed on a rotary shaker for 30 minutes) was loaded and the

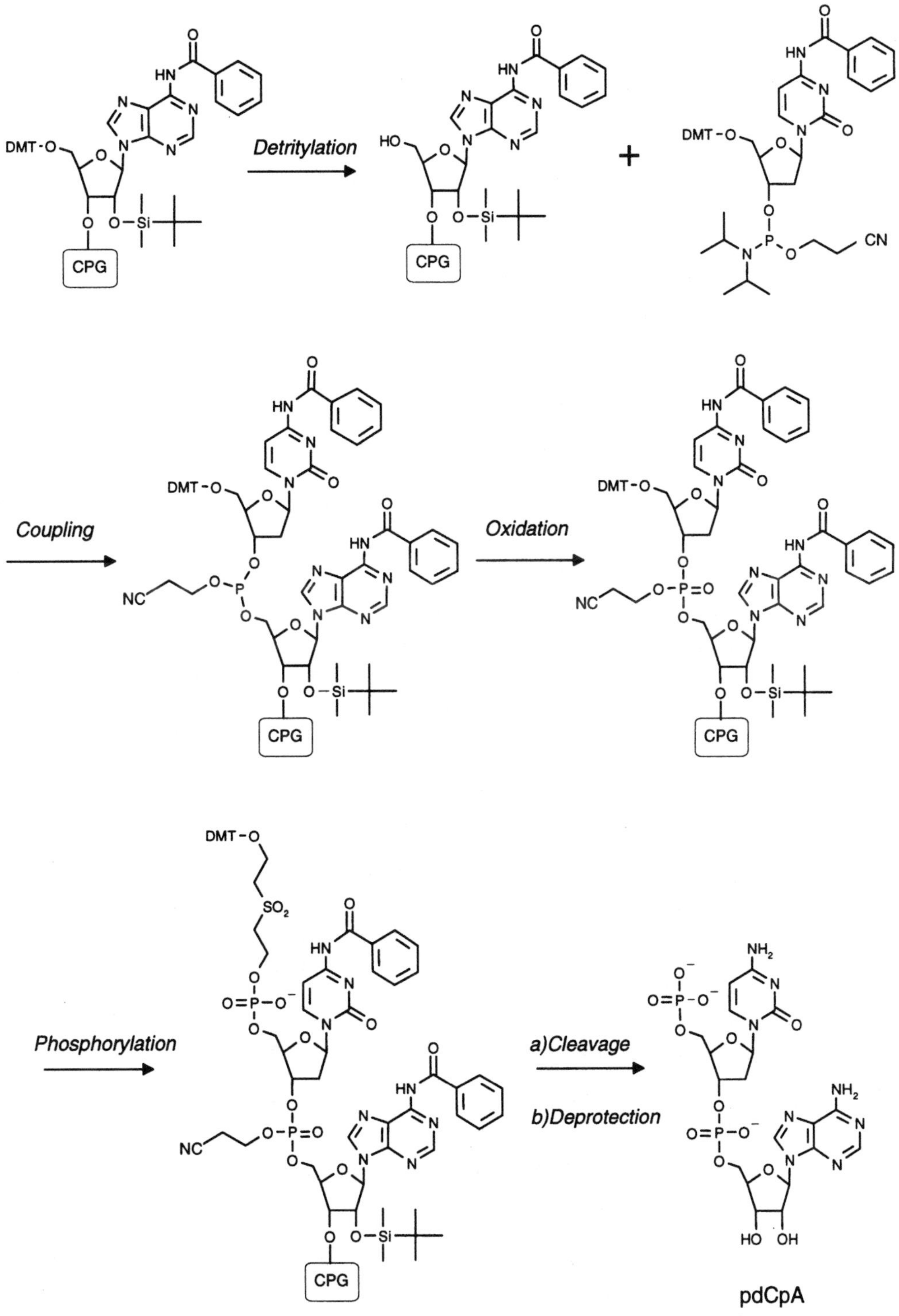

Figure 15.3. Scheme for the solid-phase synthesis of the dinucleotide pdCpA. CPG, controlled-pore-glass support; DMT, dimethoxytrityl. See text for details.

column further rinsed with water until the pH of the eluate was <8. Lyophilized pdCpA was then dissolved in water, applied to the resin, and eluted with water. An excess of tetrabutylammonium hydroxide (0.2 molar equiv. with respect to pdCpA) was added to the ion-exchanged product, followed by lyophilization. The tetrabutylammonium salt of pdCpA was stored in a dessicator at RT. For long-term storage it was kept at $-80°C$. The typical yield for the whole synthesis of pdCpA–tetrabutylammonium salt was around 80%. MS (ES) m/z (MH$^+$) was calculated to be 637.42 and found to be 637.36.

2.E. Synthesis of Protected Amino Acids

Cyanomethyl esters of N-α-protected amino acids were used for the selective monoacylation pdCpA at the 2′(3′)-O position. The use of these active esters obviates the need to protect pdCpA prior to aminoacylation. By limiting the reaction time (2.5 hours) and by using a fivefold molar excess of amino acid over pdCpA, the O,O-diacylation, the N-acylation, and racemization are avoided (Robertson et al., 1991). The interconversion between monoacylated isomers at either the 3′- or the 2′-position of adenosine is rapid in aqueous buffer, pH 7.3, 37°C with-a $t_{1/2} = 10^{-11}$ seconds (Schuber and Pinck, 1974).

Aminoacyl-pdCpA protected at the amino acid N-α-position has a longer half-life than free amino derivatives (Robertson et al., 1991). Ideally, α-amino protecting groups should permit efficient removal after the ligation reaction without modifying the tRNA structure. Standard amino protecting groups used in peptide chemistry cannot be used. For example, the acidic conditions required to remove the classic t-Boc group from N-Boc-aminoacyl-pCpA caused isomerization of the phosphodiester bond (Baldini et al., 1988) and depurination of tRNA. Also, base-labile protective groups (i.e., Fmoc) cannot be used because of the base sensitivity of the acyl linkage. The two chemical groups used as α-amino protecting groups were the Bpoc group, removable with mild acid treatment (Happ et al., 1987), and the photolabile nitroveratryloxycarbonyl (NVOC) protecting group (Pillai, 1980). NVOC can be photochemically cleaved with a 350-nm light source without affecting tRNA integrity and suppressor efficiency (Ellman et al., 1991). In our hands, we found that Bpoc amino acids were rather difficult to synthesize (Bodansky and Bodansky, 1994) and of poor stability due to slow autocatalyzed decomposition. Furthermore, this was exacerbated during HPLC purifications of Bpoc derivatives at pH 5.2. For these reasons, we recommend the use of the NVOC group. When photosensitive unnatural amino acids are introduced, photochemical deprotection of NVOC could be deleterious for the amino acid. We have developed NVOC photodeprotection conditions, using mild power light irradiation, that avoids photobleaching of photosensitive groups such as 7-nitrobenz-2-oxa-1,3-diazol-4-yl (NBD) or fluoresceinyl.

The reaction scheme for the synthesis of the protected representative fluorescent NBD-Dap amino acid is shown in Figure 15.4. Formation of cyanomethyl active esters of protected amino acids was achieved in typical yields of 75%–90%. Using the NVOC-amino protection and cyanomethyl ester strategy, the following amino acids were successfully used to form acylated suppressor tRNA: Phe, Ala, Tic, Dns-Dap, and NBD-Dap (Fig. 15.5).

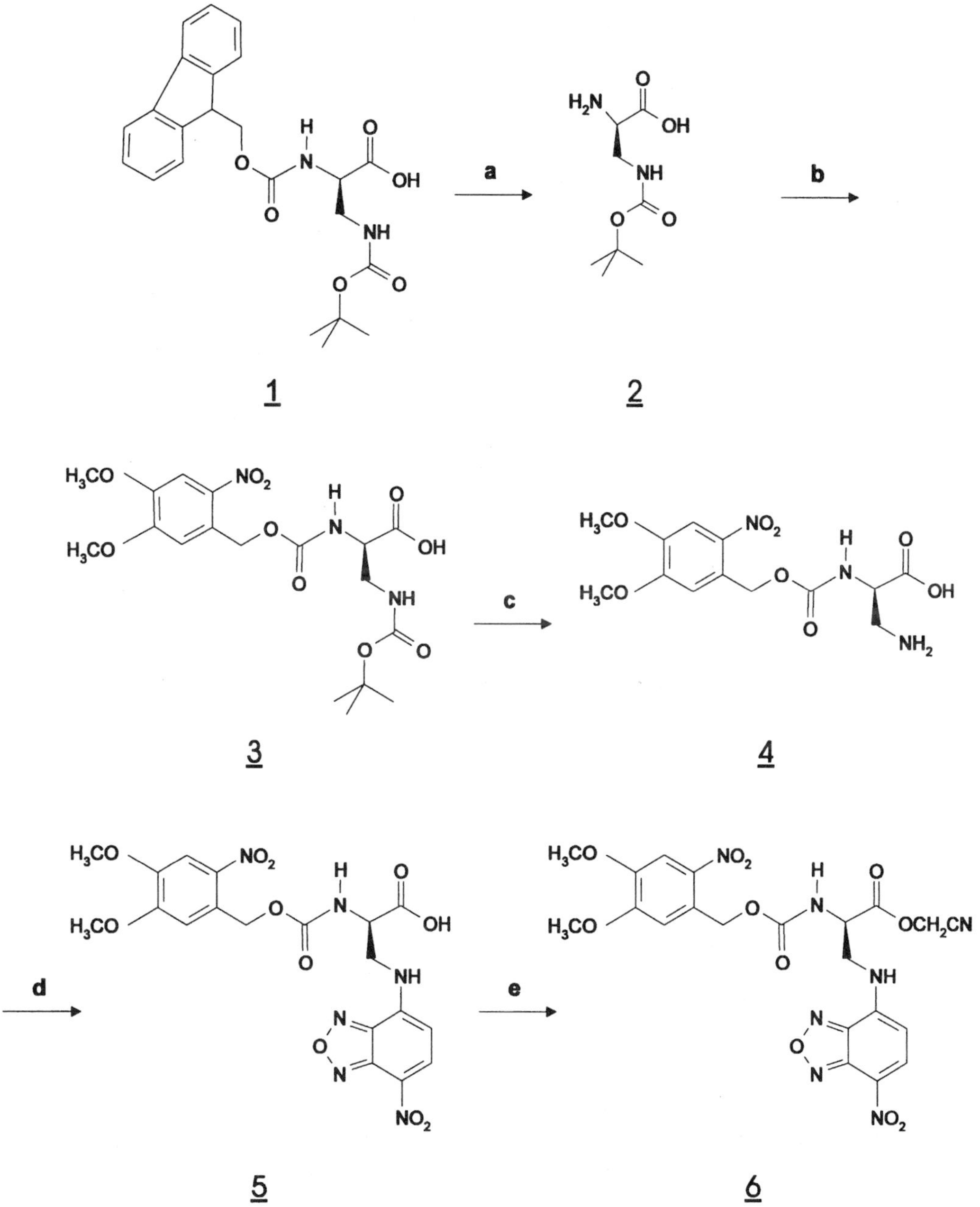

Figure 15.4. Representative chemical synthesis of a protected fluorescent amino acid (NBD-Dap). Reaction conditions: **a,** piperidine in DMF, 2 hours, RT; **b,** 1.0 equiv. NVOC-chloride in 0.15 M Na$_2$CO$_3$–DMF, 1 hour, RT; **c,** trifluoroacetic acid, 2 minutes, RT; **d,** 1.1 equiv. NBD-fluoride in 100 mM Na-borate, pH 9:CH$_3$CN:DMF 2:1:1, 1 hour, RT; and **e,** 3 equiv. ClCH$_2$CN, 2 equiv. triethylamine in CH$_3$CN, 24 hours, RT. See text for details.

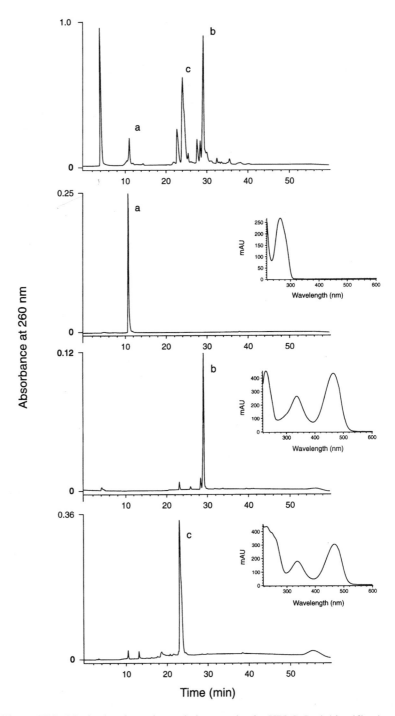

Figure 15.5. Monitoring the aminoacylation reaction by HPLC. Peak identification: **a,** pdCpA; **b, 6; c,** pdCpA–amino acid conjugate. The chromatogram at the top represents the reaction mixture after 3 hours (see text for details). The other chromatograms represent HPLC-purified compounds. Inserts are absorbance spectra recorded on-line at the apex of each peak using a diode-array detector.

Representative synthesis of an *N*-protected and activated amino acid is as follows: For synthesis of the fluorescent amino acid 2-[*N*-(6-nitroveratryl-oxy)]-3-[*N*-(7-nitrobenz-2-oxa-1,3-diazol-4-yl)]-2,3-(L)-diaminopropionic acid cyanomethyl ester (NVOC-NBD-Dap-OCH₂CN) (**6**), 2-[*N*-(9-fluorenyl-methyloxy)carbonyl]-3-[*N*-(terbutyloxycarbonyl)]-L-diaminopropionic acid 2-(*N*-Fmoc)-3-(*N*-Boc)-L-Dap-OH) (**1**) (Neosystems, Strasbourg) (435 mg, 1 mmol) was dissolved in 10 ml of piperidine (10% v/v) in DMF. The reaction mixture was stirred at RT for 2 hours, and then evaporated *in vacuo*. The solid residue was then triturated with 15 ml of diethyl ether:hexane 1:4 (v/v) and filtered to yield 190 mg (91%) of 3-[*N*-(terbutyloxycarbonyl)]-L-diaminopropionic acid 3-(*N*-Boc)-L-Dap-OH (**2**). TLC R_f (system 1) was 0.48; the R_f (system 2) was 0.27.

2 (150 mg, 0.705 mmol) and sodium carbonate (61.8 mg, 0.705 mmol, 1.0 equiv.) were dissolved in 2.33 ml of H₂O. NVOC chloride (194.3 mg, 0.705 mmol, 1.0 equiv.) in 2.33 ml of dioxane was added slowly with stirring to the aqueous solution. After 1 hour, the reaction mixture was diluted with 17 ml CH₂Cl₂ and acidified with 11.6 ml of 1 M NaHSO₄. The organic phase was collected, and the aqueous phase was washed with CH₂Cl₂. The organic extracts were combined and dried over sodium sulfate followed by concentration *in vacuo* to give a yellow solid. Recrystallization from ethylacetate:hexane (1:1 v/v) provided 310 mg (97%) of 2-[*N*-(6-nitroveratryl-oxy)]-3-(*N*-(terbutyloxy-carbonyl)]-L-diaminopropionic acid (2-(*N*-NVOC)-3-(*N*-Boc)-L-Dap-OH) (**3**) as a white solid. TLC R_f (system 2) was 0.35; HPLC (system 1) Retention time was 23.75 minutes.

3 (45 mg, 0.1 mmol) was dissolved in 0.2 ml trifluoroacetic acid and allowed to react for 2 minutes. The reaction product was then precipitated by adding the acidic solution to 20 ml ethyl ether to give 2-[*N*-(6-nitroveratryl-oxy)]-L-2,3-diaminopropionic acid (2-(*N*-NVOC)-L-Dap-OH) (**4**). The yield was determined by UV-VIS spectroscopy based on NVOC absorption at 350 nm by using an extinction molar coefficient; ϵ_{350nm} [NVOC] = 6336 M^{-1} cm^{-1}. Yield was 99%. HPLC (system 1) Retention time was 18.30 minutes.

4 (8.7 mg, 25 μmol) was dissolved in 6 ml of 100 mM sodium borate pH9:CH₃CN:DMF 2:1:1 (v/v). 4-Fluoro-7-nitrobenz-2-oxa-1,3-diazole (NBD-fluoride) (Fluka) (5 mg, 27.3 μmol) in 400 μl DMF was added to the solution, and the reaction mixture was stirred for 1 hour. The solution was concentrated and directly subjected to HPLC purification: the column was a Macherey Nagel ET250/8/4 Nucleosil 5 C18 (120 mm); flow rate was 1 ml/min. Buffer A was 50 mM NaOAc (pH 5.2) and (B) was acetonitrile. Gradient was 0% B over 1 minute, 0%–50% B over 24 minutes, 50%–85% over 35 minutes, and 85% B over 10 minutes. Detection was monitored at 475 nm. Fractions containing pure 2[*N*-(6-nitroveratryl-oxy)]-3-[*N*-4-(7-nitrobenz-2-oxa-1,3-diazol-4-yl)]-L-diaminopropionic acid (2-(*N*-NVOC)-3-(*N*-NBD)-L-Dap-OH) (**5**) were pooled and lyophilized. The compound was desalted using C18 "Sep-Pak" (Millipore) cartridges and eluted with acetonitrile:water 1:1 (v/v). The yield was estimated by UV spectroscopy in methanol. The molar extinction coefficients of **5** at 350 and 466 nm were calculated by addition of individual contributions of the NVOC and NBD chromophores: ϵ_{350nm} = 11,741 M^{-1} cm^{-1}; ϵ_{466nm} = 24,000 M^{-1} cm^{-1}. Yield was 87%. HPLC (system 1) retention time was 23.1 minutes.

To **5** (16 μmol) in 150 μl acetonitrile were added triethylamine (3.24 mg, 32 μmol) and chloracetonitrile (3.61 mg, 48 μmol). The solution was stirred for 24 hours at RT followed by dilution with 5 ml CH_2Cl_2 and extraction with 1M $NaHSO_4$. The organic phase was dried over sodium sulfate and evaporated to an oily residue. The product was dissolved in 1 ml DMF and purified by RP-HPLC using the HPLC system 1, followed by a desalting step using a Sep-Pak C18 cartridge. Elution of pure 2-[N-(6-nitroveratryl-oxy)]-3-[N-(7-nitrobenz-2-oxa-1,3-diazol-4-yl]-L-diaminopropionic acid cyanomethylester (2-(N-NVOC)-3-(N-NBD)-L-Dap-OCH₂CN) **(6)** was performed with acetonitrile/water 1:1 (v/v). The amount of final product was quantified by absorbance at 466 nm using ϵ_{466nm} = 24,000 M^{-1} cm^{-1}. Yield was 89%. HPLC (system 1) Retention time was 35.2 minutes. MS (ES) m/z (MH⁺) was calculated to be 546.78 and found to be 546.10.

2.F. Synthesis of the Dinucleotide–Amino Acid Conjugates

This reaction was performed on a 1–2 μmol scale. The tetrabutylammonium salt of pdCpA was dissolved in dry DMF at 50 mM. To this solution was added the unnatural amino acid cyanomethylester in an equal volume of dry DMF (threefold molar excess over pdCpA) and the mixture was stirred at RT. After 2 hours, aliquots of 5–10 nmol were taken, diluted with 100 μl 50 mM NH_4OAc, pH 5.1:CH₃CN 1:1 and analyzed by HPLC (system 1) to monitor the progress of the reaction (Fig. 15.6). When there was less than 50% acylation after 2 hours, aliquots of tetrabutylammonium acetate in dry DMF (0.2 equivalents per pdCpA) were added to the mixture. When the reaction was complete (usually after 6–18 hours) the mixture was diluted with 1 ml 50 mM NH_4OAc, pH 4.5:CH₃CN 1:1 and purified by HPLC (system 1). The yield of the aminoacylation reaction was calculated from purified material collected from the HPLC column by using extinction coefficients calculated by the addition of the individual contributions: ϵ_{260} (NVOC) = 2,140 M^{-1} cm^{-1}; ϵ_{350} (NVOC) = 6,336 M^{-1} cm^{-1}; ϵ_{260} (pdCpA) = 23,000 M^{-1} cm^{-1}. Typical yields were in the 30%–60% range. The NVOC-protected NBD amino acid conjugated to pdCpA gave a clear ESI-MS spectrum with a strong signal at 1,125.44, in good agreement with the calculated M+H⁺ of 1,125.40.

2.G. Preparation of Chemically Misacylated Suppressor tRNAs

Functional suppressor tRNAs loaded with unnatural amino acids were assembled by T4 RNA ligase enzymatic ligation of a 74-nucleotide tRNA transcript, Sup-tRNA(-CA) lacking the 3′ dinucleotide CA stem, and the amino acid–pdCpA conjugate described above. The 74-nucleotide 3′-truncated Sup-tRNA$_{CUA}$ was designed and prepared in milligram quantities by *in vitro* run-off transcription as described by Turcatti et al. (1996) and references cited therein. Crude tRNAs were dissolved in 100 mM triethylammonium acetate. pH 7.0, and purified by HPLC using system 2 elution gradient, which was optimized to obtain baseline resolution between full-length (76 nucleotides) and truncated (74 nucleotides) suppressor tRNAs. The typical yield of tRNA, after HPLC purification, was 500 μg from 100 μg DNA. The HPLC-purified tRNA was then

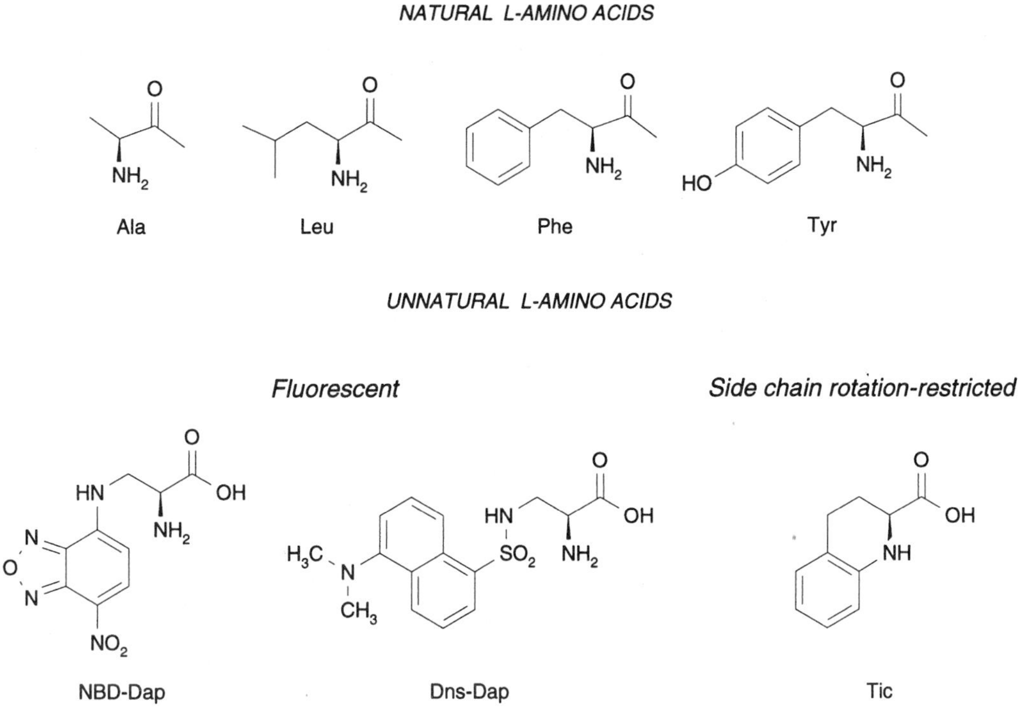

NATURAL L-AMINO ACIDS

Ala   Leu   Phe   Tyr

UNNATURAL L-AMINO ACIDS

Fluorescent   Side chain rotation-restricted

NBD-Dap   Dns-Dap   Tic

Figure 15.6. Chemical structure of selected amino acids that have been incorporated into GPCR by the nonsense suppression methodology. NBD-Dap, 3-[N-(7-nitrobenz-2-oxa-1,3-diazol-4-yl)]-2,3-diaminopropionic acid; Dns-Dap, 3-[N(5-N′,N″-dimethy-lamino-naphthalensulfon-2 yl)]-2,3diaminopropionic acid; Tic, 2-(L)-(1,2,3,4-tetrahydro) isoquinolic acid.

lyophilized and desalted using a Sephadex prepacked size exclusion column (NAP-5). Up to 250 μg of HPLC-pure tRNA was dissolved in 200 μl of DEPC-treated water, loaded, and eluted with DEPC-water. Quantitation of the eluate was carried out by UV absorption at 260 nm (1 OD at 260 nm corresponded to 40 μg/ml of RNA). The tRNA was divided in 10-μg aliquots and stored lyophilized at −80°C.

Although the ligation reaction has been previously described by other groups (Bain et al., 1991; Robertson et al., 1991; Ellman et al., 1991), we report here an improved protocol derived from a systematic study of all the parameters. The buffer composition is of paramount importance. Ligation buffer is prepared in advance, and aliquots are stored at −80°C. Buffer composition (subsequently diluted 4× in the ligation reaction) was 220 mM HEPES Na$^+$ (pM 7.5), 80 mg/ml BSA, 1 mM ATP, and 60 mM MgCl$_2$ in DEPC-treated water.

To 25 nmol of lyophilized amino acid–pdCpA conjugate was added 4 μl of DMSO, 21 μl of DEPC-treated water, and 10 μl of ligation buffer. This mixture was added to 10 μg (0.38 nmol) of desalted and lyophilized Sup-tR-NA$_{CUA}$(-CA), followed by the addition of 5 μl (100 units) of T4 RNA ligase (New England Biolabs). It is very important to add the reagents rapidly because acylated pdCpA and tRNA hydrolyze at an appreciable rate at basic pH. The reaction mixture was incubated for 10 minutes at 37°C, quenched by the addition of 100 μl 0.42 M NaOAc (pH 4.5) to prevent basic hydrolysis of aminoacyl tRNA, sequentially extracted with 140 μl of phenol/isoamyl alcohol/chloroform (25:1:24 v/v/v) and 140 μl isoamyl alcohol/chloroform (1:24 v/v), and then precipitated with 3 vol of o ethanol at −70°C. The resulting tRNA pellet was rinsed with 70% ethanol to remove the unreacted amino acid–dinucleotide and buffer salts and dried *in vacuo* for 5 minutes. The acylated tRNA was then desalted using prepacked size exclusion column (Sephadex-NAP-5) eluted with water, lyophilized, and stored at −80°C (up to several months) before use. The extent of the ligation reaction was monitored using HPLC (system 2) with an optimized gradient allowing a good resolution between the 74- and 76-nucleotide tRNAs. The yield of the ligation was estimated from peak area integration and found to be about 80%.

N-protected aminoacyl-tRNAs are much more stable than their free amino counterparts. Therefore, the removal of the *N*-protecting group was performed only immediately before the suppression experiments. For mild photodeprotection of the NVOC group the optimal conditions are 10 μg aminoacyl-tRNA in 20 μl of 1 mM KOAc in DEPC-treated water (pH 4.5) in a pyrex glass tube (the pyrex glass acts as a 300-nm filter) kept on ice and irradiated for 10 minutes with a TFP-35L lamp (Vilber-Lourmat, Marne-la-Vallée, France) (six tubes × 15W, power 180 W). Under these conditions, photobleaching of the NBD fluorophore of the amino acid NBD-Dap was estimated to be only 7%. The sample was then diluted with 0.3 M sodium acetate (80 μl) and the acylated tRNA was precipitated with ethanol at −70°C. The resulting deprotected acyl-tRNA pellet was rinsed with 70% ice-cold ethanol (in DEPC water) and dried *in vacuo* for 5 minutes. This material was stored at −80°C until required for injection in oocytes.

The Bpoc deprotection reaction has been optimized by using Bpoc-Ala-tRNA as a model system. Mild acid deprotection of Bpoc was performed by dissolving the tRNA pellet (10 μg) with 22 mM sodium citrate, 25 mM sodium phosphate (pH 2.65) (20 μl) at 37°C for 10 minutes. The reaction was then diluted with 0.3 M sodium acetate (80 μl), and the acylated tRNA precipitated with 300 μl ethanol at −70°C. The deprotected acyl-tRNA was then rinsed with 70% ice-cold ethanol, dried *in vacuo* for 5 minutes, and kept at −80°C until required.

3. EXPRESSION OF RECEPTORS CONTAINING UNNATURAL AMINO ACIDS BY SUPPRESSION MUTAGENESIS: MICROINJECTION OF OOCYTES AND ELECTROPHYSIOLOGY

Heterologous expression of GPCR was performed in *Xenopus laevis* oocytes. Oocytes are single large cells of about 1 mm diameter suitable for the expression of cloned genes. Expression can be achieved by microinjection of either DNA into the nucleus or mRNA into the cytoplasm and monitored within a few hours after injection. In the unnatural suppression mutagenesis experiments we have used intracytoplasmic co-injection of the cRNA and misacylated suppressor tRNA. It is possible to inject a volume of up to 50 nl into the cytoplasm. A skilled person can routinely inject 100–200 oocytes per hour. For the neurokinin receptors, the peak of expression was between 24 and 36 hours postinjection of RNA. Typically, receptor binding and functional assays were performed after 24 hours. The misacylated suppressor tRNA was used in large excess to compensate for hydrolysis and also to increase the incorporation rate. Detailed protocols for the maintenance of *X. laevis* and the preparation of oocytes can be found elsewhere (Goldin, 1992; Stühmer, 1992; Bertrand et al., 1991). Biological activity can be recorded electrophysiologically for the large number of GPCRs that are capable of mobilizing Ca^{2+} from intracellular stores after stimulation. This is achieved by stimulation of phospholipase C and production of inositol triphosphate, through Gq (Shapira et al., 1994), Gs (de la Pena et al., 1995), Go, or Gi (Kasahara and Sugiyama, 1994), with transduction of the signal by G$\beta\gamma$ subunits (Stehno-Bittel et al., 1995). Elevation of the intracellular Ca^{2+} concentration activates Ca^{2+}-dependent chloride channels. The inward chloride currents are easily detectable by standard voltage-clamp techniques (Turcatti et al., 1996).

Oocytes were prepared and injected into the cytoplasm with 25 ng cRNA and 100 ng aminoacyl-Sup-tRNA in a total of 50 nl of DEPC-treated water and incubated for 24 hours at 18°C as reported by Nemeth and Chollet (1995). It is important to centrifuge the tube containing the tRNA and cRNA before injection to avoid clogging the injection needle with particulate matter. A detailed protocol for the preparation of oocyte membrane fractions and ligand binding assays can be found in Nemeth and Chollet (1995).

Many factors control the efficiency of unnatural biosynthetic incorporation. We have reported here optimum experimental conditions. The effects of RNA stoichiometry, UAG codon context, and RNA structure on the efficiency of translation termination and suppression have been discussed elsewhere (Turcatti et al., 1997). Unfortunately, these factors are often unpredictable and thus limit the flexibility of the methodology. Nonetheless, we and others (Nowak et al., 1995) have shown that a variety of unnatural amino acid side chain structures can be successfully incorporated with efficiencies in the 5%–30% range.

4. APPLICATIONS

The methodology described here for the incorporation of unnatural amino acids endowed with special functions at known sites into GPCR will enable us to study the structure, function, and dynamics of GPCR either at steady state or in real time. Direct applications include the detailed structural analysis of ligand–receptor recognition and the relative contributions of contact points, the identification of solvent-exposed residues, the determination of the orientation of helices, the detection of time-dependent conformational changes, and the measurement of intra- or intermolecular distances with ligands, G proteins, or other regulator proteins.

5. INVESTIGATION OF MOLECULAR ARCHITECTURE OF LIGAND–RECEPTOR COMPLEXES BY FLUORESCENCE

Intermolecular distances were determined by measuring the fluorescence resonance energy transfer between unique fluorescent unnatural amino acid NBD-Dap residues incorporated at specific sites into the NK2 receptor and a bound NK2 peptide antagonist labeled with the fluorescent group tetramethylrhodamine in native oocyte membranes (Turcatti et al., 1996, 1997).

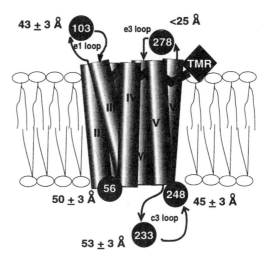

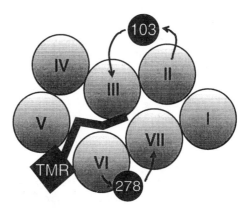

Figure 15.7. Structural model of NK2 receptor–ligand complex determined by FRET experiments with fluorescent NK2 receptor mutants prepared by unnatural suppression mutagenesis. FRET distances between sites on the NK2 receptor labeled with NBD fluorophore (black circles with position number in white) and bound heptapeptide antagonist (black rod) labeled with tetramethylrhodamine (TMR) (gray square) are shown. The transmembrane helices are numbered I through VII. Connecting loops are symbolized by arrows. The receptor is viewed either from the side **(top)** or from the top **(bottom)** (see text for more details).

These distance measurements confirmed the heptahelical structure of GPCR and suggested the structural model for NK2 ligand–receptor interactions shown in Figure 15.7. There is only one possible orientation for the peptide ligand that is inserted between the fifth and sixth transmembrane domains, suggesting that antagonist binding may prevent proper helix packing required for receptor function.

6. FUTURE PROSPECTS

The work described here demonstrates the feasibility of unnatural suppression mutagenesis of GPCR in an intact cell, the *Xenopus* oocyte, thus broadening the scope of protein engineering. Moving beyond this proof of concept, one future aim will be to render this technology more versatile by extending its applicability to mammalian cell lines that are currently used in transfection and expression studies such as CHO, COS, or HEK293 cells. Preliminary experiments in our laboratory indicate that the main challenge will be to develop an efficient system for the introduction of suppressor tRNA and cRNA into these cells.

Ultimately, the challenge will be to create a new tRNA–aminoacyl-tRNA synthetase pair, specific for a new unnatural amino acid. Recently, some encouraging preliminary work toward finding a functional synthetase–tRNA pair has been reported (Schimmel and Söll, 1997; Liu et al., 1997).

REFERENCES

Bain JD, Diala ES, Glabe CG, Wacker DA, Lyttle MH, Dix TA, Chamberlin AR (1991): Site-specific incorporation of nonnatural residues during *in vitro* protein biosynthesis with semisynthetic aminoacyl-tRNAs. Biochemistry 30:5411–5421.

Baldini G, Martoglio B, Schachenmann A, Zugliani C, Brunner J (1988): Mischarging *Escherichia coli* tRNA(Phe) with L-4′-(3-(trifluoromethyl)-3H-diazirin-3-yl)phenylalanine, a photoactivatable analogue of phenylalanine. Biochemistry 27:7951–7959.

Bertrand D, Cooper E, Valera S, Rungger D, Ballivet M (1991): Electrophysiology of neuronal nicotinic acetylcholine receptors expressed in *Xenopus* oocytes following nuclear injection of genes or cDNAs. In Conn PM (ed): Electrophysiology and Microinjection, Methods in Neurosciences, vol 4. New York: Academic Press, pp 174–193.

Bodansky M, Bodansky A (1994): The Practice of Peptide Synthesis, 2nd ed. Berlin: Springer Verlag.

Bruce AG, Atkins JF, Wills N, Uhlenbeck O, Gesteland RF (1982): Replacement of anticodon loop nucleotides to produce functional tRNAs: Amber suppressors derived from yeast tRNA[Phe]. Proc Natl Acad Sci USA 79:7127–7131.

de la Pena P, del Camino D, Pardo LA, Dominguez P, Barros F (1995): Gs couples thyrotropin-releasing hormone receptors expressed in *Xenopus* oocytes to phospholipase C. J Biol Chem 270:3554–3559.

Ellman J, Mendel D, Anthony-Cahill S, Noren CJ, Schultz PG (1991): Biosynthetic method for introducing unnatural amino acids site-specifically into proteins. Methods Enzymol 202:301–336.

Goldin AL (1992): Maintenance of *Xenopus laevis* and oocyte injection. Methods Enzymol 207:266–279.

Happ E, Scalfi-Happ C, Chladek S (1987): New approach to the synthesis of 2'(3')-O-aminoacyl oligoribonucleotides. J Org Chem 52:5387–5391.

Heckler TG, Chang L-H, Zama Y, Naka T, Chorghade MS, Hecht SM (1984): T4 RNA ligase mediated preparation of novel "chemically misacylated" tRNA^Phe s. Biochemistry 23:1468–1473.

Kasahara J, Sugiyama H (1994): Inositol phospholipid metabolism in *Xenopus* oocytes mediated by endogenous Go and Gi proteins. FEBS Lett 355:41–44.

Kunkel TA, Bebenek K, McClary J (1991): Efficient site-directed mutagenesis using uracil-containing DNA. Methods Enzymol 204:125–139.

Liu DR, Magliery TJ, Pastrnak M, Schultz PG (1997): Engineering tRNA and aminoacyl-tRNA synthetase for the site-specific incorporation of unnatural amino acids into proteins *in vivo*. Proc Natl Acad Sci USA 94:10092–10097.

Nemeth K, Chollet A (1995): A single mutation of the NK2 receptor prevents agonist-induced desensitization. Divergent conformational requirements for NK2 receptor signaling and agonist-induced desensitization in *Xenopus* oocytes. J Biol Chem 270:27601–27605.

Noren CJ, Anthony-Cahill SJ, Griffith MC, Schultz PG (1989a): A general method for site-specific incorporation of unnatural amino acids into proteins. Science 244:182–188.

Noren CJ, Anthony-Cahill SJ, Suich JD, Noren KA, Griffith MC, Schultz PG (1989b): *In vitro* suppression of an amber mutation by a chemically aminoacylated transfer RNA prepared by runoff transcription. Nucleic Acids Res 18:83–88.

Nowak MW, Kearney PC, Sampson JR, Saks ME, Labarca CG, Silverman SK, Zhong W, Thorson J, Abelson JN, Davidson N, Schultz PG, Dougherty DA, Lester HA (1995): Nicotinic receptor binding site probed with unnatural amino acid incorporation in intact cells. Science 268:439–442.

Pillai VNR (1980): Photoremovable protecting groups in organic synthesis. Synthesis 1:1–26.

Robertson SA, Ellman, JA, Schultz PG (1991): A general and efficient route for chemical aminoacylation of transfer RNAs. J Am Chem Soc 113: 2722–2729.

Saks ME, Sampson, JR, Nowak, MW, Kearney PC, Du F, Abelson, JN, Lester HA, Dougherty DA (1996): An engineered *Tetrahymena* tRNA^Gln for *in vivo* incorporation of unnatural amino acids into proteins by nonsense suppression. J Biol Chem 271:23169–23175.

Sampson JR, Di Renzo A, Behlen LS, Uhlenbeck OC (1989): Nucleotides in yeast tRNA^Phe required for the specific recognition by its cognate synthetase. Science 243:1363–1366.

Schimmel P, Söll D (1997): When protein engineering confronts the tRNA world. Proc Natl Acad Sci USA 94:10007–10009.

Schuber F, Pinck M (1974): On the chemical reactivity of aminoacyl-tRNA ester bond. I—Influence of pH and nature of the acyl group on the rate of hydrolysis. Biochimie 56:383–390.

Shapira H, Way J, Lipinsky D, Oron Y, Battey JF (1994): Neuromedin B receptor, expressed in *Xenopus laevis* oocytes, selectively couples to $G_{\alpha q}$ and not $G_{\alpha 11}$. FEBS Lett 348:89–92.

Stehno-Bittel L, Krapivinsky G, Krapivinsky L, Perz-Terzic C, Clapham D (1995): The G protein $\beta\gamma$ subunit transduces the muscarinic receptor signal for Ca^{2+} release in *Xenopus* oocytes. J Biol Chem 270:30068–30074.

Stühmer W (1992): Electrophysiological recording from *Xenopus* oocytes. Methods Enzymol 207:319–339.

Turcatti G, Nemeth K, Edgerton MD, Knowles J, Vogel H, Chollet A (1997): Fluorescent labeling of NK2 receptor at specific sites *in vivo* and fluorescence energy transfer analysis of NK2 ligand–receptor complexes. Receptors Channels 5:201–207.

Turcatti G, Nemeth K, Edgerton MD, Meseth U, Talabot F, Peitsch M, Knowles J, Vogel H, Chollet A (1996): Probing the structure and function of the tachykinin neurokinin-2 receptor through biosynthetic incorporation of fluorescent amino acids at specific sites. J Biol Chem 271:19991–19998.

USE OF NUCLEAR MAGNETIC RESONANCE TECHNIQUES TO STUDY G PROTEIN–COUPLED RECEPTOR STRUCTURE

PHILIP L. YEAGLE

1. INTRODUCTION 356

2. AVAILABLE STRUCTURAL INFORMATION 356

3. OUTLINE OF AN ALTERNATIVE APPROACH 358

4. EXPERIMENTAL PROCEDURES AND DIFFICULTIES 359

    A. NMR Spectroscopy 359

    B. Structure Refinement 360

    C. Difficulty in Obtaining Structures From Small Peptides 360

    D. Handling of Peptides From GPCRs 360

    E. Suitable Template for the Docking Experiments Described Below 361

5. STRUCTURAL RESULTS 361

    A. The Carboxy-Terminal β Sheet of Rhodopsin Is Preserved in Peptides of Different Sizes 361

    B. Carboxy-Terminal Domain of Rhodopsin 362

    C. The Cytoplasmic Loops of Rhodopsin 364

    D. Docking of Cytoplasmic Domains to the Transmembrane Domain of Rhodopsin 365

    E. Structure of a Complex of the Cytoplasmic Domains of Rhodopsin 366

6. VERIFICATION OF RESULTS 367

7. DOMAINS FROM THE INTRADISCAL FACE OF RHODOPSIN 370

8. FUTURE OUTLOOK 370

Structure–Function Analysis of G Protein-Coupled Receptors, Edited by Jürgen Wess.
ISBN 0-471-25228-X   Copyright © 1999 Wiley-Liss, Inc.

I. INTRODUCTION

In 1994, the Nobel prize in Physiology or Medicine was awarded to A.G. Gilman and M. Rodbell "for their discovery of G-proteins and the role of these proteins in signal transduction in cells" (Nobel Committee). This award culminated a long series of studies on the GTP binding proteins that carry signals from G protein–coupled receptors (GPCRs) to target enzymes within cells. The family of GPCRs has grown large as these studies have progressed. The members of this family now number in the hundreds.

Much has been revealed about the biochemistry and cell biology of these receptors. Upon binding ligand, the receptor can enter an activated state in which a heterotrimeric G protein can bind to the cytoplasmic face of the receptor. The mechanisms involved in receptor–G protein coupling and G protein activation have been described in previous chapters.

Regulation of receptor function has also been revealed by extensive studies. The receptor in the activated state is itself a substrate for receptor kinases. Phosphorylation of the receptor in the carboxy-terminal domain and other cytoplasmic regions leads to its inactivation. Phosphorylation also allows the binding of other regulatory proteins that accelerate the inactivation of the receptor. These reactions take place at the cytoplasmic face of the receptor and are part of the desensitization process that the receptor undergoes when strongly stimulated.

Because of the many roles that GPCRs play in cell biology, knowledge of the structure of these receptors is especially crucial to advancing our knowledge of signal transduction in cells. In particular the cytoplasmic face of these receptors provides the binding and activation site for the G protein. Thus information about the structure of the cytoplasmic face of GPCRs is critical to understanding the molecular mechanism of G protein activation and receptor desensitization. In spite of intense interest and critical need, the three-dimensional structure of no GPCR is known.

2. AVAILABLE STRUCTURAL INFORMATION

While the complete high-resolution structure is not known for any GPCR, some structural information has been obtained by less direct means. Much of this information is described in detail elsewhere in this volume.

Although very few structures, with atomic resolution, are available for transmembrane proteins because of their poor crystallization properties, some information is available for the GPCR rhodopsin. Circular dichroism (CD) measurements (Albert and Litman, 1978), primary sequence determination (Hargrave et al., 1983), and recently reported projection structures (Schertler et al., 1993; Unger et al., 1997) have placed rhodopsin in the class of membrane proteins that contain a bundle of seven transmembrane helices. Analysis of the primary sequence of rhodopsin also suggests the presence of three cytoplasmic loops connecting the putative transmembrane helices. The carboxy terminus is attached to helix 7 (Fig. 16.1). A low-resolution structure of the transmembrane helices of rhodopsin has been published (Schertler et al., 1993; Unger et al.,

cytoplasmic face

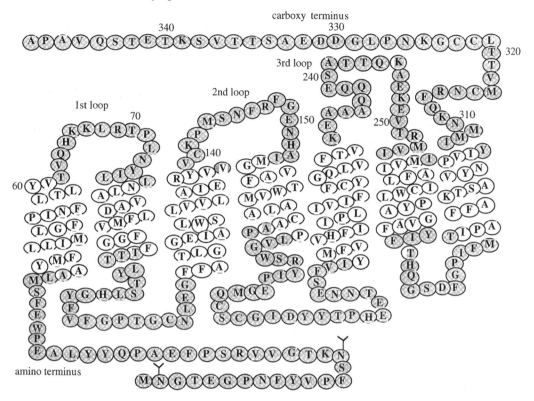

Figure 16.1. Primary sequence of bovince rhodopsin (Hargrave et al., 1983).

1997), but for technical reasons this approach cannot provide high-resolution structures for the extramembraneous portions of the receptor (i.e., those regions that interact with G protein and receptor kinase).

Fourier transform infrared (FTIR) spectroscopy data for intact rhodopsin revealed that the carboxy-terminal domain contained β structure and the third cytoplasmic loop contained α helix, respectively (Pistorius and deGrip, 1994). Spin labeling data showed that the sequence 232–245 of the third cytoplasmic loop in intact rhodopsin was likely in the form of an α helix (Altenbach et al., 1996). Previous work had shown that the Cys^{322} and Cys^{323} in the carboxy-terminal domain are palmitoylated, suggesting the presence of a fourth cytoplasmic loop (Morrison et al., 1991). A series of through-space distance measurements were made on intact rhodopsin using the dipolar interactions between two spin labels in defined sites in the protein. These measurements were made on rhodopsin in the dark state and in the active **R*** state (Farrens et al., 1996; Yang et al., 1996). Recently, a spin label was attached specifically to Cys^{140} and, in addition, a nuclear spin label in the form of a phosphorous was introduced into the carboxy-terminal domain by controlled phosphorylation by rhodopsin kinase. By analyzing dipolar interactions between the spin label and the phosphorus, the through-space distance could be determined between these two sites (Albert et al., 1997). Metal binding sites have been introduced by mutagenesis to localize adjacent or nearby helices (Sheikh et al., 1996). Cysteine crosslinking has produced information about nearest neighbors in rhodopsin helices (Yu et al, 1995). Solid-state nuclear magnetic resonance (NMR) studies on rhodopsin have revealed details about the retinal binding site (Smith et al., 1987, 1992). The cross-sectional area of the cytoplasmic face of rhodopsin was modeled with a cylinder of diameter approximately 36 Å (Sardet et al., 1976; Osborne et al., 1978). Previous work on boundary lipids using spin-labeled lipids measured the number of lipids required to make a single layer around the protein. This study predicts a protein circumference consistent with the size of the cytoplasmic face described here (Watts et al., 1979). This listing is only meant to be representative of the kinds of work being reported and, in many cases, reviewed in other chapters in this volume.

Finally, some significant modeling studies have been reported. Baldwin and co-workers have built a model of the transmembrane helices of rhodopsin, most recently based on the density maps of a structure of the transmembrane helix bundle from two-dimensional crystals of rhodopsin (Baldwin, 1993; Baldwin et al., 1997; Unger et al., 1997). In addition, some models of the transmembrane helices of rhodopsin have been obtained from simulated annealing using available information as constraints.

This listing emphasizes both the considerable effort that has been expended to obtain structural information on GPCRs and the lack of detailed structural information. To understand the molecular mechanism of G protein–mediated signal transduction, it is essential to find some means of obtaining high-resolution structural information on the receptor face that interacts with the G proteins. This chapter describes a new, alternative path toward obtaining such structural information that should work for any GPCR.

3. OUTLINE OF AN ALTERNATIVE APPROACH

Membrane proteins generally do not crystallize in a form suitable for x-ray crystallography. They are generally also not soluble due to their hydrophobic transmembrane domains. Therefore, multidimensional high-resolution NMR techniques are not useful for determination of intact integral membrane protein structure. Alternative approaches to elucidating membrane protein structure are required.

We have explored the effectiveness of an alternative approach to study the structure of the GPCR bovine rhodopsin. Specifically, we reported the structure of soluble peptides that represented two cytoplasmic domains of the receptor: an abbreviated carboxy-terminal domain (Yeagle et al., 1995a) and the third cytoplasmic loop (Yeagle et al., 1995b). Both these cytoplasmic domains exhibited biological activity and thus likely maintained crucial elements of native structure in the soluble peptides. In each case, compact globular structures were obtained using multidimensional ^{1}H-NMR. These data suggested that the cytoplasmic loops and the carboxy terminus of rhodopsin formed domains, with their individual structures determined by their amino acid sequence.

Based on these observations, we devised a comprehensive approach to determine the structure of the entire cytoplasmic face of this GPCR. We reasoned that if each of the cytoplasmic domains of rhodopsin formed compact structures in solution, it would be possible to determine the individual structures of these domains and then bring them together like building blocks to "construct" the cytoplasmic face of the receptor. First, the structures of all cytoplasmic domains of rhodopsin (the three cytoplasmic loops and the carboxy terminus) were determined by multidimensional NMR. Second, these four cytoplasmic domains were docked to the low-resolution structure of the transmembrane core of rhodopsin, which provided a template with which to associate the building blocks into an appropriate complex. The result was a structure for the entire cytoplasmic face of rhodopsin.

Third, a different path was needed to verify the original results. CD measurements and biological activity data suggested that the four cytoplasmic domains formed a complex in solution. The structure of this complex was determined by NMR. It was found to be similar to the structure determined using the docking procedure.

Fourth, the relationship between the structure of the cytoplasmic face of rhodopsin and the structure of the intact protein was tested. Several point to point distance measurements on intact rhodopsin were recently published. The structure obtained for the cytoplasmic face of the receptor was in good agreement with the measurements from the intact protein and with other data briefly referenced above. Details of these procedures and recent results derived from them are described below.

4. EXPERIMENTAL PROCEDURES AND DIFFICULTIES

The solution structures of the different receptor peptides were determined using standard NMR methods.

4.A. NMR Spectroscopy

Two-dimensional homonuclear ^{1}H-NMR spectra for the cytoplasmic loops (synthesized by solid-phase peptide synthesis with no stable isotope labeling) were recorded on a Bruker AMX-600 spectrometer at 10°C in 1 mM phosphate buffer, pH 5, and including 0.1 mM β-mercaptoethanol for peptides containing cysteine. Standard pulse sequences and phase cycling were employed to record: in H_2O (10% D_2O), double quantum filtered (DQF) COSY, HOHAHA (Braunschweiler and Ernst, 1983; Bax and Davis, 1985), and NOESY (400 msec mixing time) (Kumar et al., 1980). For larger domains that can be produced in an expression system, nOes were obtained from multidimensional NMR experiments, including (1) 3D ^{15}N and ^{13}C-resolved NOESYs and, if required, 4D ^{15}N, ^{13}C resolved NOESY; (2) coupling constants from E. COSY–like heteronuclear experiments to give torsion angle restraints (phi, psi, and chi-1); and (3) chemical shift information to give crude torsion angle restraints when necessary. All spectra were accumulated in a phase-sensitive manner using time-proportional phase incrementation for quadrature detection in F1. Chemical shifts were referenced to internal methanol.

4.B. Structure Refinement

The sequence-specific assignment of ^{1}H-NMR spectra was carried out using standard methods. Assigned nOe cross peaks were segmented using a statistical segmentation function and characterized as strong, medium, and weak corresponding to upper bounds distance range constraints of 2.7, 3.5, and 5.0 Å, respectively. Lower bounds between nonbonded atoms were set to the sum of their van der Waals radii (approximately 1.8 Å). Pseudoatom corrections were added to interproton distance restraints where necessary (Wüthrich et al., 1983). Distance geometry calculations were carried out using the program DIANA (Guntert et al., 1991) within the SYBYL 6.4 package (Tripos Software Inc., St. Louis). First-generation DIANA structures, 150 in total, were optimized with the inclusion of three REDAC cycles. Energy refinement calculations (restrained minimizations/dynamics) were carried out on the best distance geometry structures using the SYBYL program implementing the Kollman all-atom force field. Statistics on structures were obtained from Explore. These calculations were performed on a Silicon Graphics R10000 workstation. Other distance geometry approaches are also used. Assignment and structure generation protocols can be found elsewhere (Roberts, 1993; Evans, 1995).

4.C. Difficulty in Obtaining Structures From Small Peptides

In some cases, peptides corresponding to GPCR loops do not readily form a single conformation. As mentioned below, lowering the temperature can help in stabilizing a single conformation. Another way is to link the two ends of the turn. This was done for the third cytoplasmic loop of the PTH receptor (Mierke et al., 1996). Interestingly, the structure obtained within the cyclized peptide representing the third cytoplasmic loop from the PTH receptor is nearly identical to the structure of the third cytoplasmic loop of rhodopsin determined without cyclization (Yeagle et al., 1997a).

4.D. Handling of Peptides From GPCRs

In the laboratory, the peptides corresponding to the extramembraneous domains of GPCRs, including rhodopsin and the β-adrenergic receptor, have posed some challenges. Solubility of the peptides is often limited. Millimolar concentrations are required (in about 0.6 ml vol) in hydrogen-free buffer (usually phosphate buffer in 10% D_2O, 90% H_2O) at relatively low pH (pH 5–6 to reduce the exchange rate of the amide hydrogens of the peptide bonds). In addition, low temperatures are used to increase the stability of the peptide in a single conformation. These conditions and the sequence of the peptide can result in a peptide that is not stably soluble for the length of time required for the NMR experiments (48 hours, for example, may be typical for a COSY and a

NOESY). Furthermore, these peptides are known to form clear macroscopic gels, which can sometimes produce unacceptable NMR spectra.

Sometimes the addition of small amounts of DMSO, or even complete substitution of DMSO for water, becomes necessary to stabilize the peptides in solution. The use of DMSO raises obvious concerns, but some of these receptor domains are not very exposed to water when in the native structure, which may explain their limited solubility.

Aggregation of peptides can interfere with the structure determination of NMR. Therefore conditions need to be established under which no aggregation occurs. The extent of this problem can be assessed through several techniques. (1) The ^{1}H chemical shifts of the peptides can be examined as a function of concentration. Changes with increased concentration may mean aggregation is occurring. Temperature can also be explored as a variable, though lower temperatures are favored to stabilize the structure of the peptides in solution. (2) CD spectra can be obtained as a function of peptide concentration. Again, changes with increased concentration may be indicative of aggregation. (3) Finally, analytical ultracentrifugation can be used to determine the state of aggregation of the peptides in solution.

4.E. Suitable Template for the Docking Experiments Described Below

Below, docking of the cytoplasmic domains of rhodopsin to the transmembrane bundle of helices is described in which a model of the transmembrane core of rhodopsin is used to provide the template for the docking procedure. Two precautions should be noted. One is that even though high-resolution structures of the transmembrane domain of bacteriorhodopsin are available (Grigorieff et al., 1996; Pebay-Peyroula et al., 1997), they cannot be used as models for GPCRs. The reason for this is that the arrangement of helices in bacteriorhodopsin is not the same as in the dark-adapted state of rhodopsin (Schertler et al., 1993). Second, the activated state of rhodopsin has a different arrangement of the transmembrane helices than the dark-adapted state (Farrens et al., 1996). Thus one must be cautious in using the published rhodopsin structure if one wants to examine the activated state of the receptor.

5. STRUCTURAL RESULTS

5.A. The Carboxy-Terminal β Sheet of Rhodopsin Is Preserved in Peptides of Different Sizes

The approach to elucidate membrane protein structure described here is based on the domain structure of membrane proteins. Data described below support this approach because they demonstrate that short-range interactions are more important than long-range interactions in the sequences that are characteristic of these domains of membrane proteins. Studies on the carboxy terminus of rhodopsin were carried out to test the dominance of short-range interactions in this domain of rhodopsin.

We have examined the stability of β structure as a function of peptide size. In the carboxy-terminal domain of rhodopsin, the most carboxy-terminal portion forms a short, two-strand antiparallel β sheet. We have solved the structure of peptides with the sequence D330–A348 (19mer), G324–A348 (25mer), C316–A348 (33mer), and Y306–A348 (43mer). In each case the β sheet is preserved. As described in more detail below, short-range interactions dominate the formation of secondary structures. Our data support this conclusion. Therefore it is reasonable to expect that, if the cytoplasmic loops of rhodopsin have secondary structure (such as β turns, which they do; see below), they should be able to form the appropriate structure without the remainder of the amino acid sequence of the protein. This concept, along with the knowledge that the peptides derived from the cytoplasmic surface of rhodopsin exhibit biological activity, encouraged pursuit of the "building block" approach described here.

5.B. Carboxy-Terminal Domain of Rhodopsin

The carboxy-terminal domain of bovine rhodopsin was synthesized, containing the last 43 amino acids of the protein (rhoIVe; residues 306–348). This sequence included the entire putative fourth cytoplasmic loop as well as a significant portion of helix 7. The solution structure of rhoIVe was determined by multidimensional ^{1}H-NMR using standard methods.

The high-resolution NMR structure obtained was compared with other structural analyses of the peptide. We obtained CD data on the peptide in solution and analyzed this spectrum for the percent secondary structure. We analyzed the Cα–H chemical shifts according to the method of Wishart et al. (1992) and obtained a pattern of secondary structure for the peptide. Figure 16.2 compares these analyses for the peptide with the sequence C316–A348. As can be seen, all the analyses agree well with each other on the presence and quantity of secondary structure in this peptide. In addition, Figure 16.2 shows the output of a structure prediction program that also agrees with the experimental data (although this output was not used in the structural analysis). As described above, the β structure was also found in the carboxy terminus of the intact rhodopsin by FTIR. Therefore there can be no doubt that the structure determined by NMR is a true representation of the actual structure of the peptide in solution. In particular, the peptide is not a random structure in solution, whether assessed by NMR, CD, or biological activity.

There was no evidence of alternative conformations based on the NMR data for the carboxy-terminal domain of rhodopsin. We have now solved the solution structures of all six loops and the amino-terminal (references above and manuscript in preparation), and carboxy-terminal domains of rhodopsin. In only one case (loop 3 on the intradiscal face) was there evidence of a second conformation, which was likely due to *cis* and *trans* proline.

These experiments were performed at low temperature (5° or 10°C) to enhance the formation of a single stable structure in solution. We cannot rule out the presence of an unfolded form that is in rapid exchange with the one folded structure that is observed. However, it should be noted that shifts from the random coil value are observed in the Cα–H resonances for many of the residues

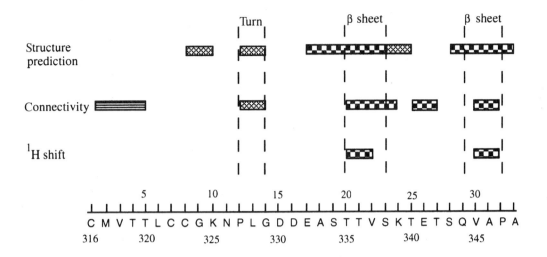

Figure 16.2. Analysis of secondary structure in aqueous buffer for the peptide containing the sequence of the carboxy terminul of rhodopsin (Yeagle et al., 1995a). Connectivity is from the sequential and long-range nOes in the NOESY map, and the ^{1}H shift is an analysis of the Cα–H ^{1}H-NMR chemical shifts according to Wishart, et al. (1992).

that are reasonable in magnitude. The observed shifts would be a weighted average of the shift in the random coil and the shift in the folded form. Therefore either the population of random coil conformations is modest or, if significant unfolded forms exist, the shifts in the Cα–H of the folded form are very large.

The structure of the carboxy-terminal domain of rhodopsin contained a portion of α helix corresponding to the cytoplasmic extension of transmembrane helix 7. Helix 7 is thus longer than suggested by hydropathy analysis. The structure also revealed the fourth cytoplasmic loop. The palmitoylation sites of rhodopsin are located at the bottom of this loop near the deduced membrane surface (Fig. 16.3). However, palmitoylation is not required for formation of this loop. A useful, animated color version of this structure can be seen at the web site <http://www.emory.edu/molvis/v2/yeagle>.

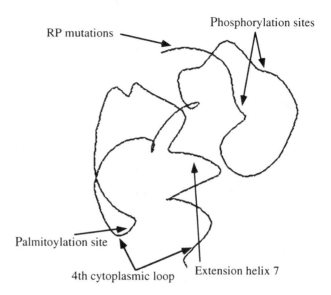

RP mutations

Phosphorylation sites

Palmitoylation site

4th cytoplasmic loop

Extension helix 7

Figure 16.3. Representation of the solution structure of the 43mer (residues 306–348 of rhodopsin) peptide of the carboxy trminus of bovine rhodopsin. The amino terminus is at the bottom, and the carboxy terminus is at the top. The trace of the peptide backbone is a smooth ribbon through the α-carbon positions of the stucture. RP, retinitis pigmentosa.

5.C. The Cytoplasmic Loops of Rhodopsin

The structure of peptides containing the amino acid sequence of the first and second cytoplasmic loops of rhodopsin were also determined. Peptides containing the sequences corresponding to these loops were synthesized: for the first cytoplasmic loop, a 17mer (residues 60–76), for the second cytoplasmic loop, a 16mer (residues 139–154). CD spectra showed secondary structure in solution, indicating that the peptides were ordered. The structures were determined by two-dimensional NMR techniques. Both loops show ordered structures in solution (Fig. 16.4). In both cases, the ends of the transmembrane helices unwind and form a β turn. The conformations of the two loops are remarkably similar, even though their sequences are not. These data suggest a structural motif for short loops in transmembrane proteins. The well-ordered structures of these loops, in the absence of the transmembrane helices, indicate that the primary sequences of these loops stabilize the β turn. These data further suggest that the loops may contribute to the folding of GPCRs during their synthesis and insertion into membranes.

To determine the structure of the third cytoplasmic loop, a 26mer (residues 231–256) was synthesized. The structure determined for this peptide is shown in Figure 16.4. It contains an α helix, which is an extension of transmembrane helix 5, in good agreement with results of a spin-labeling study (Farrens et al., 1996). This structure is in remarkable agreement with the structure of the third cytoplasmic loop of the PTH/PTHrP receptor, another GPCR (Mierke et al., 1996).

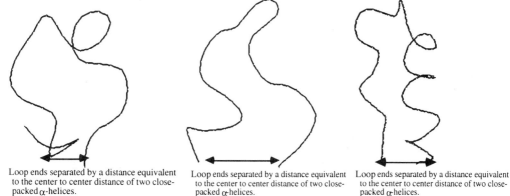

Loop ends separated by a distance equivalent to the center to center distance of two close-packed α-helices.

1st cytoplasmic loop

Loop ends separated by a distance equivalent to the center to center distance of two close-packed α-helices.

2nd cytoplasmic loop

Loop ends separated by a distance equivalent to the center to center distance of two close-packed α-helices.

3rd cytoplasmic loop

Figure 16.4. Structures of the peptides representing the three cytoplasmic loops of bovine rhodopsin. The first cytoplasmic loop peptide contains residues 60–76, the second cytoplasmic loop peptide residues 139–154, and the third cytoplasmic loop peptide residues 231–256. In each panel, the amino terminus is at the right, and the carboxy terminus is at t he left. The trace of the peptide backbone is a smooth ribbon through the a-carbon positions of the structure.

As described above, these observations open an alternate approach to study the structure of the cytoplasmic face of rhodopsin. Two methods were used that produced similar structures. (1) The experimentally determined structures of the four cytoplasmic domains of rhodopsin were docked to the structure of the transmembrane core to build the cytoplasmic face of this receptor. (2) Three of these domains (second and third cytoplasmic loops and carboxy-terminal domain) have biological activity in solution and are synergistic in their ability to inhibit G protein activation by light-activated rhodopsin (Konig et al., 1989). The latter data suggest that these domains may form a complex in solution. CD data confirm this suggestion (see below). The structure of this complex (with the addition of the first cytoplasmic loop) was determined from multidimensional NMR data. The obtained structure is similar to that achieved with the docking procedure (see below).

5.D. Docking of Cytoplasmic Domains to the Transmembrane Domain of Rhodopsin

Low-resolution information on the structure of the transmembrane domain of rhodopsin has been reported (Schertler et al., 1993). A model of this domain was constructed using the helix assignment and rotational orientation proposed by Baldwin (1993), which has found considerable recent experimental support (Zhou et al., 1994; Elling et al., 1995; Yu et al., 1995; Elling and Schwartz, 1996; Farrens et al., 1996; Mizobe et al., 1996; Sheikh et al., 1996; Yang et al., 1996). A structure of the cytoplasmic face of rhodopsin is obtained by docking the experimentally determined structures of the four cytoplasmic domains to the transmembrane domain. In several cases, such as the carboxy-terminal domain (Yeagle et al., 1996) and the third cytoplasmic loop (Yeagle et al., 1997a), docking is achieved by overlap of the helix in the structure of the cytoplasmic domain with the appropriate helix of the transmembrane core (a useful animation showing this docking can be seen at <http://www.emory.edu/molvis/v2/yeagle>). In the case of the second cytoplasmic loop, while the helices of the loop are more open than α helices, they nevertheless represent a logical unwinding of the transmembrane helices and can be readily docked with the appropriate transmembrane helices. The ends of the first cytoplasmic loop are not well ordered in the original structure determination, but it is possible to position the loop over the ends of helices 1 and 2 and thus locate it correctly relative to the other cytoplasmic domains. This docking procedure organizes the cytoplasmic domains relative to each other. The energy of the structure, without the transmembrane helices, was then minimized in the force field in Sculpt (Interactive Simulations, San Diego, CA). The resulting structure is shown in Figure 16.5.

5.E. Structure of a Complex of the Cytoplasmic Domains of Rhodopsin

Because activity data suggested that the three biologically active peptides (second and third cytoplasmic domains and carboxy terminus) might form a complex that would be useful in structure determinations, CD spectroscopy was used to explore whether such a complex actually formed. CD spectra were obtained separately for each biologically active peptide in solution and of a solution containing a mixture of the three peptides. The CD spectrum for the buffer was subtracted from all the other spectra. The difference between the spectrum of the sum of the CD of each of the peptides by themselves and the spectrum of the peptide mixture suggests that the three peptides interact in solution. The agreement between the two spectra in the range from 210 to 260 nm suggests

Figure 16.5. Representation of the three-dimensional structure of the cytoplasmic face of bovine rhodopsin, obtained as described in the text. The trace of the peptide backbone is a smooth ribbon through the α-carbon positions of the structure.

that the secondary structure (α helix, β turn, and β sheet) does not change significantly when the three peptides interact. The difference between the two spectra at about 200 nm suggests that separately the three peptides exhibit some disordered structure that is ordered upon interacting with each other (Venyaminov and Yang, 1996). These data indicate that a complex is formed, in agreement with the NMR data (see below).

The structure of this complex, with the addition of the first cytoplasmic loop, was determined by multidimensional NMR. All NMR spectra were accumulated in 10 mM phosphate buffer at pH 5.9 on a Bruker AMX-600 spectrometer at 10°C. Standard pulse sequences and phase cycling were employed to record in H_2O (10% D_2O), double quantum filtered (DQF) COSY, and NOESY (400 msec mixing time) (Kumar et al., 1980). All spectra were accumulated in a phase-sensitive manner using time-proportional phase incrementation for quadrature detection in F1. The sequence-specific assignment of the ^1H-NMR spectrum of the complex was carried out as described previously (Yeagle et al., 1995a,b, 1996). Differences in the Cα–H chemical shifts from the individual peptides and from the peptides in the complex were minor (most in the range of ± 0.03 ppm). Assigned nOe cross peaks were segmented using a statistical segmentation function and characterized as strong, medium, and weak corresponding to upper bounds distance range constraints of 2.7, 3.7, and 5.0 Å, respectively. Lower bounds between nonbonded atoms were set to the sum of their van der Waals radii (approximately 1.8 Å). Pseudoatom corrections were added to interproton distance restraints where necessary (Wüthrich et al., 1983). A total of 1,119 constraints were used: 653 intraresidue, 285 sequential, 174 long range, and 7 intermolecular. Four molecules, representing the four cytoplasmic domains solved previously, were loaded into SYBYL. Simulated annealing was performed, and 16 structures were generated using the nOe-derived constraints. Energy refinement calculations (restrained minimizations/dynamics) were carried out on each of the structures using the SYBYL program implementing the Kollman all-atom force field. The only major difference between this structure and the previously described "docked" structure is the greater exposure of the β sheet of the carboxy-terminal domain in the latter, which contains the rhodopsin kinase sites (Yeagle et al., 1997b).

6. VERIFICATION OF RESULTS

Does the structure obtained for the cytoplasmic face of rhodopsin represent that present in the intact protein? If so, then this approach should become generally useful. In the following, the validity of the obtained structure is tested against the available structural data for intact rhodopsin.

1. FTIR data for intact rhodopsin revealed that the carboxy-terminal domain contained β structure and the third cytoplasmic loop contained α helix, respectively (Pistorius and deGrip, 1994). Our structure shows β structure in the form of a β sheet and a β turn in the carboxy-terminal domain, as well as α helix in the third cytoplasmic loop.

2. Spin-labeling data showed that the sequence 232–245 of the third cytoplasmic loop in intact rhodopsin was likely in the form of an α helix (Altenbach et al., 1996). Likewise, our structure shows an α helix between residues 232–243.

3. Previous work had shown that Cys[322] and Cys[323] in the carboxy-terminal domain are palmitoylated, suggesting the presence of a fourth cytoplasmic loop. Our structure shows that this fourth cytoplasmic loop is in fact formed, even in the absence of palmitoylation. Recent spin labeling studies with intact rhodopsin showed that ordered structure occurs in the fourth cytoplasmic loop in the absence of palmitoylation, suggesting that palmitoylation was not necessary for formation of the fourth cytoplasmic loop (Resek et al., 1993).

4. Accessibility experiments showed that Cys[140] and Cys[316] are exposed in intact rhodopsin (Farahbakhsh et al., 1993). These residues are also exposed in our structure, pointing to the outside of the structure. In another study, Cys[140] was shown to be less exposed than Cys[316] (Resek et al., 1993). In our structure, Cys[140] is close to the bundle of seven transmembrane helices and the lipid bilayer and thus less exposed than Cys[316], which is on "top" of the fourth cytoplasmic loop.

5. Resek et al. (1993) showed that His[65] is also exposed. Similarly, in our structure His[65] points to the outside.

6. Spin-labeling data showed that residues 141, 151, and 152 were not in contact with the rest of the protein (Farahbakhsh et al., 1995). Our structure indicates that these residues point away from the protein.

7. A series of through-space distance measurements were made on intact rhodopsin analyzing the dipolar interactions between two spin labels in defined sites in the protein. These measurements were made on rhodopsin in the dark and in the **R*** state. In Table 16.1, these distances are compared with measurements made on our structure. Good agreement is seen between our structure and the structure of the **R*** state of intact rhodopsin.

8. Recently similar experiments with intact rhodopsin have demonstrated that residues 335–340 are exposed, residues 245–246 and 249 are close to residue 338, while residues 248 and 250 are not close to this residue (Cai et al., 1997). In our structure, residues 335–340 are on the "top" of the carboxy-terminal domain, well exposed to the aqueous medium. Furthermore, distance measurements on our structure show that residues 245, 246, and 249 are much closer to 338 than are 248 and 250.

9. Recently reported spin-labeling studies also suggested that residues 335–339 of the carboxy terminus of rhodopsin are in a β conformation (Langen et al., 1998). This is in excellent agreement with data from the structure of the carboxy-terminal domain in which residues 335–338 form one half of an antiparallel β sheet (Yeagle et al., 1995a).

10. As a further test of the validity of our structure, we introduced a spin label specifically on Cys[140]. In addition, a nuclear spin label in the form of

TABLE 16.1. Distances Between Side Chains of Specific Amino Acid Residues in Rhodopsin, measured on our structure of the cytoplasmic face of rhodopsin[a] and the same distances reported recently on intact rhodopsin[b]

	This Structure	R*	R
Val[139] → Lys[248]	25 Å	23–25 Å	12–14 Å
Val[139] → Glu[249]	20 Å	15–20 Å	15–20 Å
Val[139] → Val[250]	13 Å	12–14 Å	15–20 Å
Val[139] → Thr[251]	20 Å	23–25 Å	12–14 Å
Val[139] → Arg[252]	11 Å	23–25 Å	15–20 Å
His[65] → Met[316]	14 Å	12–15 Å	7–10 Å

[a]The distances were calculated between atoms on the termini of the individual residues to approximate the distances between the unpaired electrons on the spin labels covalently bonded to cysteines introduced at these same positions. These measurements were made on the average structure of the ensemble of structures reported previously. **R***, activated form of rhodopsin; **R,** dark-adapted state of rhodopsin.
[b]Farrens et al. (1996) and Yang et al. (1996).

a phosphorus was introduced into the carboxy-terminal domain by controlled phosphorylation by rhodopsin kinase. By analyzing dipolar interactions between the spin label and the phosphorous, the through-space distance could be determined between these two sites, which was in good agreement with our structure of the cytoplasmic face of rhodopsin (Albert et al., 1997).

11. Baldwin et al. (1997) have suggested, based on the low-resolution structure of the bundle of transmembrane helices and modeling, that helix 3 is relatively long and extends more beyond the hydrophobic region than some of the other helices. We have recently solved the structure of the loop that connects to helix 3 on the intradiscal face of rhodopsin and found an extension of helix 3 into the intradiscal space. We also found evidence for a modest extension of helix 3 on the cytoplasmic face. Baldwin et al. (1997) also suggest an extension of helix 5 into the cytoplasm. Our data, as well as those for the PTH receptor (see 12, below), are consistent with this concept (see also 2 above).

12. The structure of the third cytoplasmic loop of the PTH receptor has been solved (Mierke et al., 1996), and the reported structure is structurally homologous to the one reported here. This result encourages extension of the approach described for rhodopsin to other GPCRs.

Finally, the structure of the cytoplasmic face of rhodopsin exhibits several additional interesting features that are in agreement with suggestions from previously published data. The cross-sectional area of the cytoplasmic face of rhodopsin determined in our work has dimensions of about 33 by 36 Å, in good agreement with previous low-resolution measurements that modeled rhodopsin with a cylinder of diameter approximately 36 Å (Sardet et al., 1976; Osborne et al., 1978). Previous work on boundary lipids (see above) predicts a protein circumference consistent with the structure described here (Watts et al., 1979). The palmitoylation sites on rhodopsin (Cys^{322} and Cys^{323}) are found exposed at the putative membrane surface in this structure, where they can be readily acylated (Morrison et al., 1991). In the structure reported here, the fourth cytoplasmic loop forms without being acylated, providing a possible explanation why palmitoylation was found to be nonessential to the activation of rhodopsin (Karnik et al., 1993).

In conclusion, the above data represent structural information from all four cytoplasmic domains of intact rhodopsin. In all cases, our structure agrees well with data obtained with the intact protein.

7. DOMAINS FROM THE INTRADISCAL FACE OF RHODOPSIN

The intradiscal face of rhodopsin consists of the amino-terminal domain and three loops. Recently, the structures of these domains were also solved by two-dimensional NMR methods (details to be reported elsewhere). All four domains formed compact structures. Each of the peptides from the regions that connect the transmembrane helices formed loops, showing that (as in the case of the cytoplasmic loops) the structure of these receptor domains is determined by short-range interactions. The sequence of these regions seems to be sufficient to determine the structure without the influence of other portions of the protein.

These observations also have ramifications for understanding processes of GPCR folding during biosynthesis. It was suggested a number of years ago that polytopic membrane proteins might fold by forming hairpin turns, two helices at a time, followed by their insertion into membrane lipid bilayers (Engelman and Steitz, 1981). Although helix–helix interactions have long been thought to play a role in this folding process, the structural information on the loops suggests that the loops may also play an important role. The folding of membrane proteins such as GPCRs may occur initially through formation of secondary structure (the loops, as β turns, and the transmembrane α helices), followed by the organization of these elements of secondary structure in the tertiary structure of the receptor.

8. FUTURE OUTLOOK

In the future, molecular modeling techniques should be used to bring together all available experimental constraints in an attempt to define the entire structure of a GPCR. This has been done, for example, in the case of the transmembrane helices of rhodopsin (Herzyk and Hubbard, 1995). It will be important to determine how many experimental constraints are necessary to achieve this goal. Work on the extramembraneous domains as described here should provide important information required in this effort.

REFERENCES

Albert AD, Litman BJ (1978): Independent structural domains in the membrane protein bovine rhodopsin. Biochemistry 17:3893–3900.

Albert AD, Watts A, Spooner P, Groebner G, Young J, Yeagle PL (1997): A distance measurement between specific sites on the cytoplasmic surface of bovine rhodopsin in rod outer segment disk membranes. Biochim Biophys Acta 1328:74–82.

Altenbach C, Yang K, Farrens DL, Farahbakhsh ZT, Khorana HG, Hubbell WL (1996): Structural features and light dependent changes in the cytoplasmic interhelical E–F loop region of rhodopsin: A site-directed spin-labeling study. Biochemistry 35: 12470–12478.

Baldwin JM (1993): The probable arrangement of the helices in G protein–coupled receptors. EMBO J 12:1693–1703.

Baldwin JM, Schertler GFX, Unger VM (1997): An alpha-carbon template for the transmembrane helices in the rhodopsin family of G-protein–coupled receptors. J Mol Biol 272:144–164.

Bax A, Davis DG (1985): MLEV-17–based two dimensional homonuclear magnetic transfer spectroscopy. J Magn Reson 65:355–360.

Braunschweiler L, Ernst RR (1983): Coherence transfer by isotopic mixing: Application to proton correlation spectroscopy. J Magn Reson 53:521–528.

Cai K, Langen R, Hubbell WL, Khorana HG (1997): Structure and function in rhodopsin: Topology of the C-terminal polypeptide chain in relation to the cytoplasmic loops. Proc Natl Acad Sci USA 94:14267–14272.

Elling CE, Nielsen SM, Schwartz TW (1995): Conversion of antagonist-binding site to metal-ion site in the tachykinin NK-1 receptor. Nature 374:74–77.

Elling CE, Schwartz TW (1996): Connectivity and orientation of the seven helical bundle in the tachykinin NK-1 receptor probed by zinc site engineering. EMBO J 15:6213–6219.

Engelman DM, Steitz TA (1981): The spontaneous insertion of proteins into and across membranes: The helical hairpin hypothesis. Cell 23:411–422.

Evans JNS (1995): Biomolecular NMR Spectroscopy. Oxford: Oxford University Press.

Farahbakhsh ZT, Hideg K, Hubbell WL (1993): Photoactivated conformation changes in rhodopsin: A time-resolved spin label study. Science 262:1416–1419.

Farahbakhsh ZT, Ridge KD, Khorana HG, Hubbell WL (1995): Mapping light-dependent structural changes in the cytoplasmic loop connecting helices C and D in rhodopsin: A site-directed spin labeling study. Biochemistry 34:8812–8819.

Farrens DL, Altenbach C, Yang K, Hubbell WL, Khorana HG (1996): Requirement of rigid-body motion of transmembrane helices for light activation of rhodopsin. Science 274:768–770.

Grigorieff N, Ceska TA, Downing KH, Baldwin JM, Henderson R (1996): Electron-crystallographic refinement of the structure of bacteriorhodopsin. J Mol Biol 259:393–421.

Guntert P, Braunk W, Wüthrich K (1991): Efficient computation of three-dimensional protein structures in solution from NMR data using the program DIANA and the supporting programs CALIBA, HABAS, and GLOMSA. J Mol Biol 217: 517–530.

Hargrave PA, McDowell JH, Curtis DR, Wang JK, Juszczak E, Fong SL, Rao JKM, Argos P (1983): The structure of bovine rhodopsin. Biophys Struct Mech 9:235–244.

Herzyk P, Hubbard RE (1995): Automated method for modeling seven-helix transmembrane receptors from experimental data. Biophys J 69:2419–2442.

Karnik SS, Ridge KD, Bhattacharya S, Khorana HG (1993): Palmitoylation of bovine opsin and its cysteine mutants in COS cells. Proc Natl Acad Sci USA 90:40–44.

Konig B, Arendt A, McDowell JH, Kahlert M, Hargrave PA, Hofmann KP (1989): Three cytoplasmic loops of rhodopsin interact with transducin. Proc Natl Acad Sci USA 86:6878–6882.

Kumar A, Ernst RR, Wüthrich K (1980): A two-dimensional nuclear Overhauser enhancement (2D NOE) experiment for the elucidation of complete proton–proton cross-relaxation networks in biological macromolecules. Biochem Biophys Res Commun 95:1–6.

Langen R, Kai K, Khorana HG, Hubbell WL (1998): Structure and dynamics of the C-terminal domain in rhodopsin probed by site-directed spin labeling and disulfide cross-linking. Biophys J 74:A290.

Mierke DF, Royo M, Pelligrini M, Sun H, Chorev M (1996): Third cytoplasmic loop of the PTH/PTHrP receptor. J Am Chem Soc 118:8998–9004.

Mizobe T, Maze M, Lam V, Suryanarayana S, Kobilka BK (1996): Arrangement of transmembrane domains in adrenergic receptors. Similarly to bacteriorhodopsin. J Biol Chem 271:2387–2389.

Morrison DF, O'Brien PJ, Pepperberg DR (1991): Depalmitoylation with hydroxylamine alters the functional properties of rhodopsin. J Biol Chem 266:20118–20123.

Osborne HB, Sardet C, Michel-Villaz M, Charbre M (1978): Structural study of rhodopsin in detergent micelles by small-angle neutron scattering. J Mol Biol 123:177–206.

Pebay-Peyroula E, Rummel G, Rosenbusch JP, Landau EM (1997): X-ray structure of bacteriorhodopsin at 2.5 angstroms from microcrystals grown in lipidic cubic phases. Science 277:1676–1681.

Pistorius AM, deGrip WJ (1994): Rhodopsin's secondary structure revisited: Assignment of structural elements. Biochem Biophys Res Commun 198:1040–1045.

Resek JF, Farahbakhsh ZT, Hubbell WL, Khorana HG (1993): Formation of the meta II photointermediate is accompanied by conformational changes in the cytoplasmic surface of rhodopsin. Biochemistry 32:12025–12032.

Roberts GCK (1993): NMR of Macromolecules: A Practical Approach. Oxford: IRL Press.

Sardet C, Tardieu A, Luzzati V (1976): Shape and size of bovine rhodopsin: A small-angle x-ray scattering study of a rhodopsin-detergent complex. J Mol Biol 105:383–407.

Schertler GFX, Villa C, Henderson R (1993): Projection structure of rhodopsin. Nature 362:770–772.

Sheikh SP, Zvyaga TA, Lichtarge O, Sakmar TP, Bourne HR (1996): Rhodopsin activation blocked by metal-ion binding sites linking transmembrane helices C and F. Nature 383:347–350.

Smith SO, Groot HD, Gebhard R, Lugtenburg J (1992): Magic angle spinning NMR studies on the metarhodopsin II intermediate of bovine rhodopsin: Evidence for an unprotonated schiff base. Photochem Photobiol 56:1035–1039.

Smith SO, Palings I, Copié V, Raleigh DP, Courtin J, Pardoen JA, Lugtenburg J, Mathies RA, Griffin RG (1987): Low temperature solid state C-13 NMR studies of the retinal chromophore in rhodopsin. Biochemistry 26:1606–1611.

Unger VM, Hargrave PA, Baldwin JM, Schertler GFX (1997): Arrangement of rhodopsin transmembrane α-helices. Nature 389:203–206.

Venyaminov SY, Yang JT (1996): Determination of Protein Secondary Structure: Circular Dichroism and the Conformational Analysis of Biomolecules. New York: Plenum Press, pp 69–108.

Watts A, Volovski ID, Marsh D (1979): Rhodopsin–lipid associations in bovine rod outer segment membranes. Identification of immobilized lipid by spin-labels. Biochemistry 18:5006–5013.

Wishart DS, Sykes BD, Richards FM (1992): The chemical shift index: A fast and simple method for the assignment of protein secondary structure through NMR spectroscopy. Biochemistry 31:1647–1651.

Wüthrich K, Billeter M, Braun WJ (1983): Pseudo-structures for the 20 common amino acids for use in studies of protein conformations by measurements of intramolecular proton proton distance constraints with nuclear magnetic resonance. J Mol Biol 169:949–961.

Yang K, Farrens DL, Altenbach C, Farahbakhsh ZT, Hubbell WL, Khorana HG (1996): Structure and function in rhodopsin. Cysteines 65 and 316 are in proximity in a rhodopsin mutant as indicated by disulfide formation and interactions between attached spin labels. Biochemistry 35:14040–14046.

Yeagle PL, Alderfer JL, Albert AD (1995a): Structure of the carboxyl terminal domain of bovine rhodopsin. Nature Struct Biol 2:832–834.

Yeagle PL, Alderfer JL, Albert AD (1995b): Structure of the third cytoplasmic loop of bovine rhodopsin. Biochemistry 34:14621–14625.

Yeagle PL, Alderfer JL, Albert AD (1996): Structure determination of the fourth cytoplasmic loop and carboxyl terminal domain of bovine rhodopsin. Molecular Vision 2: http://www.emory.edu/molvis/v2/yeagle.

Yeagle PL, Alderfer JL, Albert AD (1997a): The first and second cytoplasmic loops of the G-protein receptor, rhodopsin, independently form β-turns. Biochemistry 36:3864–3869.

Yeagle PL, Alderfer JL, Albert AD (1997b): Three dimensional structure of the cytoplasmic face of the G protein receptor rhodopsin. Biochemistry 36:9649–9654.

Yu H, Kono M, McKee TD, Oprian DD (1995): A general method for mapping tertiary contacts between amino acid residues in membrane-embedded proteins. Biochemistry 34:14963–14969.

Zhou W, Flanagan C, Ballesteros JA, Knovicka K, Davidson JS, Weinstein H, Millar RP, Sealfon SC (1994): A reciprocal mutation supports helix 2 and helix 7 proximity in the gonadotropin-releasing hormone receptor. Mol Pharmacol 45:165–170.

CHAPTER 17

LEAD DISCOVERY AND DEVELOPMENT FOR G PROTEIN–COUPLED RECEPTORS

DENNIS J. UNDERWOOD and MARGARET A. CASCIERI

1. INTRODUCTION   375
2. OVERVIEW OF DRUG DISCOVERY AND DEVELOPMENT
   METHODS   376
3. GPCR STRUCTURE AND MODELS   380
4. DEVELOPMENT OF LIGAND MODELS AND RECEPTOR–
   LIGAND MODELS   383
   A. Endogenous Ligand Structure   383
   B. Privileged Structures   384
5. EXAMPLES   388
   A. Angiotensin II Antagonists   388
   B. Somatostatin Agonists   391
   C. Neurokinin Antagonists   394
6. CONCLUSIONS AND FUTURE DIRECTIONS   395

1. INTRODUCTION

The large and growing family of G protein–coupled receptors (GPCRs) are attractive pharmacological targets for effective therapeutic intervention for several reasons. Drugs of proven therapeutic utility including nonselective α- and β-adrenergic agonists and antagonists, histamine antagonists, angiotensin antagonists, and serotonin antagonists have been developed that interact with these receptor proteins. In addition, potent, orally bioavailable compounds have been identified as GPCR agonists and antagonists for receptors with a wide variety of endogenous ligands, including biogenic amines (i.e., epinephrine and serotonin), small lipophilic inflammatory mediators (i.e., leukotrienes), small neuropeptides (i.e., Substance P, angiotensin), and large peptides (i.e., neuropeptide Y). Thus, from a pharmaceutical perspective, development of novel

Structure–Function Analysis of G Protein-Coupled Receptors, Edited by Jürgen Wess.
ISBN 0-471-25228-X   Copyright © 1999 Wiley-Liss, Inc.

GPCR agonists and antagonists has an enviable record of success that other potential classes of therapeutic target choices have as yet not matched.

In addition, the molecular cloning of new GPCR family members provides increased opportunity for the development of more selective therapies. These molecular techniques have revealed the existence of many more receptor subtypes than has the characterization of receptor subtypes by classic pharmacological methods and have made the unambiguous determination of the tissue distribution of these proteins possible. In many instances it is now possible to mimic agonism or tailor antagonism of a physiological mediator to a specific tissue or organ system based on the selective tissue distribution of a particular receptor subtype. This knowledge, combined with the development of techniques for the deletion of specific genes by homologous recombination, has served to rapidly improve our ability to predict the physiological consequences of a given therapeutic intervention.

The large effort to sequence the human genome has also added to the list of potential therapeutic targets. It is now possible to "discover" novel GPCR sequences by scanning the ever-growing genomic databases to identify likely GPCR family members. In many instances it is possible to predict the potential endogenous ligand for these orphan receptors by the degree of sequence similarity to known receptors. A recent example of this is the identification of a third subtype of galanin receptor by Wang et al. (1997). In other cases it has been necessary to purify novel ligands for these orphan receptors by classic biochemical techniques, as exemplified by the identification of the endogenous ligands nociceptin (Meunier et al., 1995) and the orexins (Sakurai et al., 1998). The development of strategies to identify novel and important therapeutic targets from genomic information will likely increase the number of GPCR-targeted drugs dramatically.

The discovery of selective GPCR agonists and antagonists is principally an empirical process. The ability to rapidly evaluate large numbers of compounds by high throughput screening technology in order to identify appropriate lead compounds is key to the success of these efforts. Both radioactive ligand binding and functional assays have been successfully utilized to this end.

Previous chapters have discussed the information derived from mutagenesis and modeling studies of the interactions of ligands with receptors. The data suggest that the binding site for nonpeptide agents is within a common binding pocket within the transmembrane domain regions of GPCRs (Cascieri et al., 1995; Strader et al., 1994). This chapter outlines efforts and methodologies designed to utilize this information in the process of lead compound discovery and optimization. The ultimate goal is to be able to predict active pharmacophores based on the primary sequence of a given GPCR. While it will be necessary to have more accurate structural data on these proteins to achieve this goal, it is currently possible to utilize the information available to help the discovery and design process.

2. OVERVIEW OF DRUG DISCOVERY AND DEVELOPMENT METHODS

The discovery of compounds eliciting a selective response in a biochemical assay or a whole-organism biological model is a major challenge within the pharmaceutical industry. Recent developments in computational selection methods for prioritizing compounds sent to screens have dramatically improved the rate of discovery of active compounds and thus lead compounds for drug development programs.* The fundamental assumption in these selection methods is that chemical agents elicit biochemical responses in a selective, mechanism-based manner. A fundamental tenet of lead discovery is that compounds that are structurally and chemically alike exhibit similar biological responses. Experience confirms this supposition. Consequently, the concepts of structural and chemical similarity of ligands and receptor or active site homology are central to the strategies and methods of database mining.

Traditional approaches to new lead discovery have been to screen mixtures, broths, and ferments from natural sources in primary biological assays. These remain important approaches to the discovery of structurally unique biological agents. Random screening of large sample collections and combinatorial libraries with high throughput assays is a relatively recent development yielding active compounds that are usually structurally simpler than those found in nature. Directed screening of sample collections, or *database mining,* has surpassed natural products screening and random screening as a method of identifying active compounds in GPCR assays. The various kinds of information used in database mining strategies and methods are described below.

Active compounds in the assay of interest or related assays can be used as probes for selecting structurally similar compounds that might have similar biological responses. In addition, the development of structure–activity relationships (SAR) from compound-biological activity data provides the basis for elaborating pharmacophore hypotheses that abstract and define the manner in which chemical features are presented to the receptor. Finally, information determined from the structure of the biological target directs the selection of compounds to those that fit the shape and the chemical character of these binding sites. In the absence of a high-resolution GPCR structure, receptor homology models provide a coarse representation of the shape and character of the binding site.

Numerous database mining strategies have been developed (Borman, 1992; Carhart et al., 1985; DesJarlais et al., 1988; Kearsley and Smith, 1992; Kearsley et al., 1994; Kuntz, 1992; Martin, 1992; Martin and Willett, 1990; Miller et al., 1994a,b, 1995; Mosley et al., 1995; Sheridan et al., 1994, 1995; Sheridan and Venkataraghavan, 1987). The simpler methods use two-dimensional similarity filters based solely on the chemical graph and the character of each atom. The probe, generally, is composed of one or more compounds that have activity in the target assay. Each structure is represented by a collection of descriptors involving each atom and its connectivity to all other atoms. An extension of this approach has been to enhance and extend the atom descriptors to include atom properties such as charge, hydrophobicity, and polarity in addition to the atom type (Kearsley et al., 1996). Searches using these simple methods can quite effectively "capture" the essential features of the probe molecule and find similar compounds. Depending on the details of the target and the assay particulars, our

---

*In the context of this chapter, an active compound is a compound that elicits the desired biological effect, within set activity criteria, in an assay model. A lead compound is an active compound that has acceptable properties and is appropriate for chemical analoging and optimization.

experience has shown that filtering of compound collections using topological similarity increases the rate of finding active compounds from around 1% (random selection) to greater than 5%. Emphasizing the important chemical functionality required for biological activity, and considering the manner in which this functionality is displayed in three-dimensional space, refines the search and increases the rate of finding active compounds. For example three-dimensional similarity probes typically increase the biological hit rate to around 15%, but this depends on the details of the target and the assay circumstances.

Importantly, topological similarity measures can be used to select either similar or diverse compounds: In the former situation, compounds that have a high similarity index are grouped together, whereas to select for diversity, compounds are selected such that they are maximally dissimilar. It is the introduction of structural diversity that makes topological similarity methods particularly valuable for new lead discovery. Because three-dimensional methods are not confined to follow the topology of the probe, structurally diverse leads are more often found. Lead compounds that are structurally dissimilar will likely have different bioprofiles. This will provide alternate development routes when meeting the challenges of later phase drug development such as toxicity, rapid metabolism, and poor absorption of drug candidates.

More discriminating topological methods have been developed that weight the descriptors by the biological activity of the compound in which they are found (Sheridan and Venkataraghavan, 1987). The resultant vector in multidimensional descriptor space (TrendVector) points from chemical features not contributing to biological activity to those that can be associated with active compounds. This vector can be used to search sample collections for compounds containing descriptors that contribute to activity. New, more accurate methods using partial least-squares methods (Sheridan et al., 1994) permit the evaluation of compounds across panels of TrendVectors and across TrendVectors that have been vectorially added and subtracted. The addition of TrendVectors for two assays would search for compounds having activity in both assays. Vectorial subtraction searches for compounds that are active in one assay and not the other.

As mentioned previously, incorporation of three-dimensional information on the ligand structure enables the use of more sophisticated database mining methods and can significantly improve the rate of finding active compounds. The most relevant information on the ligand structure is the receptor bound conformation of the ligand: It is this conformation that provides information on the way the chemical functionality is presented to the receptor. The receptor bound conformation, in conjunction with the SAR, provides information on the key interactions between the ligand and the receptor; i.e., the pharmacophore. However, the receptor bound conformation is difficult to determine experimentally and to predict theoretically. Structural information can come from a variety of sources including x-ray crystallography, nuclear magnetic resonance (NMR) spectroscopy, among other spectroscopic methods, and molecular modeling. The latter can provide accurate ligand structures through energy minimization and conformational searching using methods such as systematic torsion search, distance geometry, stochastic methods such as Monte Carlo, and deterministic methods such as molecular dynamics (Saunders et al., 1990; Van Gunsteran et al., 1994). In addition, there are computational methods that enable one to extract the biologically relevant conformation of a ligand through a process of three-dimensional pharmacophore generation. One method of pharmacophore generation, the so-called active analog approach, involves an iterative process of conformation generation and structural comparison of active compounds until a consistent model is reached (Prendergast et al., 1994). The quality of this model is very much determined by the structural similarity of the chemical types used to define the pharmacophore and by the quality of the biological data. Ensembles of structures containing one or more constraints are a very effective means of reducing the conformational complexity. The scale of this conformational and alignment problem can also be dramatically reduced by incorporating structural information derived from transfer nOe experiments (Campbell and Sykes, 1993; Ni and Scheraga, 1994), REDOR experiments (Creuzet et al., 1991), or spin label experiments (Steinhoff et al., 1994). This information by itself does not necessarily define the receptor bound conformation of the ligand but taken together can be very effective in limiting possibilities to those of low energy (Underwood, 1995).

There are a number of methods that use the three-dimensional structure of the ligand to search sample collections (Christie et al., 1990; Guner et al., 1992; Moock et al., 1994). We have developed three-dimensional similarity scoring methods such as SEAL (Kearsley and Smith, 1992) and SQ (Miller et al., 1995), that use the pharmacophore information to search databases containing multiple conformations of each compound ("FlexiBases" [Kearsley et al., 1994]). These methods find compounds that present similar chemical functionality to the pharmacophore in approximately the same three-dimensional way. SQ uses atom type and partial charge information and assigns hydrophobic, aromatic, polar, anionic, cationic, and hydrogen-bond donor or acceptor character to each atom (Bush and Sheridan, 1993; Miller et al., 1995). An alignment rule is generated that orients each compound in the database onto the probe compound and scores the overlap. The match in chemistry and the three-dimensional alignment determine the score.

The three-dimensional structure of the receptor, determined either from experiment or from homology models, provides more information on ligand binding opportunities. Recent reviews highlight a flurry of activity in the general area of docking algorithms (Bohm, 1996; Jones and Willett, 1995; Lengauer and Rarey, 1996; Rosenfeld et al., 1995; Willett, 1995). Docking methods score compounds on how well they fit the shape and character of the binding site. Kuntz and coworkers described the first methods of receptor-based database mining (see DesJarlais et al., 1988), which have been further developed and applied to systems of pharmaceutical importance (Rutenber et al., 1993; Schoichet et al., 1993). We have extended the docking technology to address a number of the limitations of the method (Miller et al., 1994a,b). These methods, however, might find only limited use in GPCR lead discovery because ligand binding is not well described by a static lock and key model. Rather, binding will be better described as induced-fit: an adaptive phenomenon in which the structure of both the receptor and the ligand change, in a complimentary way, accommodating to each other's shape and chemical character. The strategy of calculating or modeling a "global minimum" for the receptor bound ligand conformation is fraught with difficulties: Thermodynamics demands equilibria between ensembles of ligand and receptor conformations, and it is difficult *a priori* to decide which structures best represent the biologically active species. Furthermore, because there are probably significant structural differences between energy minima determined for ligands free in solution and for ligands bound to a receptor, a better approach is to calculate three-dimensional pharmacophores; these models arise from comparing conformations for each active compound and deducing common, overlapping space. We have dealt with the inherent conformational flexibility of ligands by generating and storing multiple, conformationally diverse models for each ligand (FlexiBases) (Kearsley et al., 1994). The challenge of representing the conformational flexibility of the receptor in such searches has yet to be met.

The significance of these methodologies lies in their potential to discover diverse structural types rather than in finding close homologs. Thus, it is clear from comparing the structural types selected by each of these methods that although many compounds selected have obvious chemical similarities, each method emphasizes a different aspect of the probe's structure, allowing different structural types to be found. This, along with the ability to select diverse structures, is the real utility of these methods in new lead discovery.

3. GPCR STRUCTURE AND MODELS

Thus far there is no reported high-resolution structure of a GPCR, although rhodopsin has been well studied using electron diffraction (Unger et al., 1997). Recently a high-resolution electron diffraction study (Kimura et al., 1997) and an x-ray structure (Pebay-Peyroula et al., 1997) of bacteriorhodopsin have been published. Atomic and electron force microscopy have been used to study the extramembranous loop regions and soluble regions of integral membrane proteins, including bacteriorhodopsin: This technique allows examination of conformational changes that might occur in these regions, but the experimental method itself may alter the loop structure (Heymann et al., 1997). All of the modeling to date has been based on structures derived from electron diffraction of either bacteriorhodopsin or, more recently, bovine and frog rhodopsin. For the most part, models of GPCRs have been derived using the structural biological equivalent of a game of 20 questions (Underwood, 1995). In this process, information and data from a variety of experimental techniques are scaled and combined to generate a coherent three-dimensional model to be tested and further refined. A novel approach using a simulated annealing, Monte Carlo search method with geometric and topological restraints finds ways of packing rigid transmembrane helices (Herzyk and Hubbard, 1995). This method uses structural information derived from electron microscopy, neutron diffraction, disulfide bond detection, site-directed spin labeling, crosslinking experiments, FTIR and fluorescence quenching and polarization, ^{13}C-NMR, site-directed mutagenesis, and natural variation and calculations of periodicity, hydropathicity, sequence variation, and substitution patterns. These data are transformed into a set of geometric restraints on the helices: Each helix is treated as a rigid segment, the membrane structure provides helical orientation, the amphipathic character and residue variability of the helices provide rotational and packing restraints, and the length of the loops connecting the helices and disulfide bonds provides topological restraints. Application of this method successfully generated the bacteriorhodopsin footprint to within a root-mean-squared deviation of 1.9 Å from the electron microscopy structure. Models of bovine rhodopsin were also generated that are consistent with the assignment and arrangement of helices from the recent electron diffraction work of bovine (Unger and Schertler, 1995) and frog (Unger et al., 1997) rhodopsin. The major difference between the derived model and the structure based on the electron density lies in the tilt of helix 5 and the observed difference between the packing of the helices on both the cytoplasmic and extracellular sides of the membrane: The area enclosed by the seven helices on the cytoplasmic side is around 25% smaller than the area enclosed by the helices on the extracellular side. Furthermore, the cytoplasmic footprint shows a topology in which helix 3 packs against and is enclosed by helices 2, 4, 5, and 6, whereas the extracellular footprint shows a more open, bacteriorhodopsin-like topology. The open structure on the extracellular side has implications for the binding and isomerization of retinal to rhodopsin and the processes of signal transduction. This approach, combined with the three-dimensional pharmacophore models of the ligands, may find utility in defining aspects of the ligand bound receptor complex.

Original models of GPCRs used the low-resolution, electron diffraction data of bacteriorhodopsin (Henderson et al., 1990) as a template for the organization of the helical segments (Hibert et al., 1991). Electron diffraction studies of rhodopsin (Unger et al., 1997; Unger and Schertler, 1995) suggest that the organization and packing of the helical segments differs from bacteriorhodopsin, at least at the resolution of the electron density in these experiments. As mentioned above, frog rhodopsin structures suggest that because of helix tilt through the membrane, the helical packing on the extracellular side is different from the cytoplasmic side. Later models of GPCRs, based on the helical footprint of the rhodopsins (Baldwin 1993, 1994; Baldwin et al., 1997), have brought into question earlier bacteriorhodopsin-based models (Hibert et al., 1991; Hoflack et al., 1994). In many ways, the controversy over which structure best represents the organization and packing of GPCR helices disregards a fundamental and distinctive property of membranes: their fluidity. It is likely that the differences in the microenvironments of receptors can influence the packing of the helices.

Membranes provide a substantial, selectively permeable barrier, regulating cellular traffic through active and passive filtering mechanisms and modulating and attenuating the response of the cell to hormones whose site of action is intracellular. This compartmentalization ensures that cells maintain a large degree of autonomy within very complex regulatory surroundings. Membranes also provide a specialized medium for specific biological processes critical for cellular function such as the interaction between activated GPCRs and the heterotrimeric G proteins and specific kinases and arrestins. The structural details of integral membrane proteins and their complexation are determined by the composition and structure of this complex environment. The nature of this environment determines the thermodynamics of the events that occur during ligand and G protein binding and the details of signal transduction, and it is expected that these events are dominated by forces that partition between an aqueous and a hydrophobic medium. Recent evidence suggests that cell membranes have domains of structure, implying that the fluid mosaic model of lipid bilayers is inaccurate: In this model, the membrane is considered as a two-dimensional solution in which the components are in rapid, dynamic diffusion forming a homogeneous medium without structure or organization. Recent evidence, however, indicates that cell surface membranes are laterally heterogeneous with microdomains of lipid, protein, and other membrane components ranging from tens of angstroms to a few micrometers in size (Edidin, 1997; Mouritsen and Jorgensen, 1997; Saxton and Jacobson, 1997; Simons and Ikonen, 1997). The implications of microdomain and "raft" formation on various biochemical and biophysical processes such as signal transduction, receptor dimerization (Hebert et al., 1996), and receptor desensitization (Freedman and Lefkowitz, 1996) are not well understood.

Membrane composition and fluidity are critical to biological processes such as phosphorylation, signal transduction, the passage of ions and molecules into and out of the cell, as well as membrane protein function. Regions of cellular membranes are inherently fluid and can undergo phase transitions typical of liquid crystals when temperature and pressure are varied; such transitions depend on the lipid and protein composition and reflect a constant, dynamic structural reorganization (Collings, 1990). These changes are expected to significantly alter the interaction and diffusion of membrane components. Processes such as signal transduction across membranes, ion diffusion into and out of the cell, and active and passive transport are inextricably linked to the nature of this medium through the kinetics and thermodynamics of these biological processes. Membrane fluidity and receptor flexibility are expected to be important to the binding and activation of GPCRs by ligands through induced fit. Nonspecific hydrophobic interactions are expected to be less important in membranes than in water; the nonpolar environment potentiates the coulombic interactions between charged, polar, and aromatic moieties critical for molecular recognition by receptors and important to protein–protein interactions in general. Regions of membranes are effectively two-dimensional fluids, and reactions involving the association of biological components are augmented by this reduction in dimensionality. Reaction enhancement by localization in membranes through conjugation with fatty acids is common: e.g., palmitoylation of GPCRs (Strader et al., 1994), palmitoylation and myristoylation of G proteins (Hamm and Gilchrist, 1996), or farnesylation of CAAX boxes (Moores, 1991; Reiss et al., 1991).

Receptor ligands can be considered as allosteric modulators of the G protein GTP/GDP exchange reaction. As such, it is useful to distinguish between ligand components that lead to affinity and those components that activate the receptor. Mutational analysis of GPCRs has demonstrated that distinct epitopes are involved in ligand binding, coupling of receptors to G proteins, and mediation of ligand-induced receptor activation (Chicchi et al., 1997; Strader et al., 1994; Underwood and Prendergast, 1997). These data suggest that GPCRs can attain multiple conformational states within membranes, that one or more of these conformations have a higher affinity for G proteins, and that the equilibrium among these various conformations is receptor and environment dependent. Interaction of receptor with ligand modulates this equilibrium toward conformations with high affinity for G proteins (i.e., agonists) or with very low affinity for G proteins (i.e., inverse agonists). Partial agonists produce an intermediate state with an inherently lower affinity for the G protein. As such, increasing the ratio of G protein to receptor can shift partial agonists to full agonists (Newman-Tancredi et al., 1997).

In addition, the observed efficacy of a given ligand can vary dramatically with levels of receptor expression. This is frequently observed in experimentally derived systems in which receptors are heterologously expressed in cells and tissues. Both concentration dependence (i.e., EC_{50}) and maximal receptor activation of an experimental agent is dependent on level of receptor expression (George et al., 1988; Whaley et al., 1994) so that it is possible for a compound to behave as an antagonist under some conditions and as a partial agonist at higher levels of receptor expression (Groblewski et al., 1997). Thus, when using these expression systems for screening purposes, it is always necessary to be cognizant of the degree to which they will predict efficacy in native target tissues.

4. DEVELOPMENT OF LIGAND MODELS AND RECEPTOR–LIGAND MODELS

4.A. Endogenous Ligand Structure

Early NMR studies of peptide hormones free in solution often indicate that there are few, if any, reliable structural features. Typically these experiments were performed in water, DMSO, or methanol, and under these conditions the peptides are often found to be conformationally flexible with either an extended or a random coil ensemble of structures in rapid equilibrium. Recent NMR studies using SDS micelles or trifluoroethanol (TFE), conditions known to induce ligand structure, have shown that many peptide hormones can adopt a defined structure. For example, early NMR work on C5a (Williamson and Madison, 1990) indicates that the carboxy-terminal tail is disordered and thus presumed flexible. Recent NMR studies suggest a helical turn (Ni et al., 1996) or a helix (Zhang et al., 1997) for this region. Importantly, the carboxy-terminal 10 residues alone can behave as a full agonist and activate the receptor albeit at reduced potency ($\times 100$). In the parlance of "message and address" (Portoghese, 1989; Schwyzer, 1986), the carboxy-terminal tail is the message and the 60 amino-terminal residues provide the address binding to receptor exosites (Siciliano et al., 1994).

The question remains whether the structure generated in these solvent systems is close to the bioactive conformation. Nonetheless, it is clear that many small peptide hormones, while having a propensity for structure that may be induced by the receptor, are conformationally flexible. Induction of structure or induced fit has been observed for antigen–antibody interactions (Dyson et al., 1988; Dyson and Wright, 1995; Rini et al., 1992; Wright et al., 1988, 1990). Paradoxically, peptide hormones interact in a highly selective way with their receptors and in so doing initiate the signal transduction process resulting in specific cellular responses. The precise nature of this interaction is reflected in the observation that for many peptide hormones small chemical and conformational changes can lead to crossover from agonism to antagonism. For example, NMR studies of Substance P, the endogenous ligand for the NK_1 receptor, conclude that in DMSO there is a mixture of extended conformations, whereas in TFE/water or SDS micelles the peptide shows helical structure with a flexible or extended carboxy terminus (Young et al., 1994; Zhang and Wong, 1993). Unlike Substance P, peptides derived from NKA, the endogenous ligand for the NK_2 receptor, can adopt a β turn in the carboxy-terminal region ($FxGLM_N$) (Ward et al., 1990; Wong et al., 1993), maintaining agonist activity at their receptor but becoming antagonists at the NK_1 receptor. This sensitivity to change implies a subtle recognition mechanism that is dependent on the complementarity between the agonist and each specific receptor.

There are other examples of peptide hormones that show a propensity for structure, including endothelin, angiotensin II, somatostatin, neuropeptide Y, neurotensin, and bradykinin. SAR and structural studies expose structural and molecular features of the hormone that determine its recognition by its receptor. The angiotensin II receptor data described below exemplify the interrelationship between molecular recognition of the agonist functionality and the modulation of the pharmacological response that occurs when this pharmacophore changes.

4.B. Privileged Structures

Privileged structure refers to a compound that shows activity in a range of biological assays (Evans et al., 1988; Patchett et al., 1995). Such nonselectivity is thought to result from binding of specific functionality in the compound to accessory binding sites common to a class or family of receptors (Ariens et al., 1979). It is believed that privileged structures interact with common binding motifs or epitopes of GPCRs. Derivatization and modification of the privileged structure introduces selectivity and increases potency and efficacy. This interpretation is supported by extensive mutagenesis studies of GPCRs showing that many small molecule ligands interact with residues in the transmembrane domain (Cascieri et al., 1995; Strader et al., 1994). Nonpeptidyl antagonists of the neurokinin, cholecystokinin, angiotensin, somatostatin, GHS, C5a, GnRH, and opiate receptor families appear to interact with residues predicted to be in the top 30%–50% of the transmembrane domain and within transmembrane helices 3 through 7. Thus, there appears to be a ligand binding pocket that is common within the GPCR family.

Typically, privileged structures contain structural templates, cores, or scaffolds that position key functional groups in an appropriate manner to interact with the accessory binding sites common to families of receptors. An example of a structural core is diacyl-piperazine, discussed below, and the β-D-glucosides described by Hirshmann et al. (1996). Simple derivatization of core structures, principally with aromatic groups, creates compounds that bind to more than one receptor within a family. Further derivatization of privileged structures provides opportunities for these compounds to interact with additional binding sites leading to selectivity (see Table 17.1). In this discussion, the core provides the framework for presenting chemical functionality in a specific manner and thus may not be directly involved in interactions with the receptor. Clearly this is an oversimplification; structural studies of enzyme binding and inhibition indicate that hydrogen-bonding interactions, for example, are critical for positioning the ligand in the active site appropriately. Nonetheless, ligand binding to GPCRs is likely to be dominated by side-chain interactions because the backbone is a tightly hydrogen bonded, helical structure. It is presumed that the side chains displayed from the helical segment provide the accessory binding sites for privileged structures. Early discovery and lead development work in GPCRs indicates that benzodiazepines and 1,1-diphenylmethanes are structural elements that are recurring features among drugs showing CNS activity. A benzodiazepine core is central to the natural product asperlicin, which is a CCK receptor antagonist. Functionalization of this core structure led to the development of potent, orally active nonpeptide CCK antagonists (Evans et al., 1986). Similarly, Hirshmann's β-D-glucose analog (Fig. 17.1) provides a structural core for mimicking a β-turn that can be functionalized for the neurokinin, somatostatin, and β_2-adrenergic receptors and thus can be considered as a privileged structure (Hirschmann et al., 1996). Remarkably, selectivity for the NK_1 receptor improves with simple acetylation of the sugar as shown in Figure 17.1, which diminishes binding to the somatostatin and β_2-adrenergic receptors.

A simple and powerful demonstration of the derivitization of a privileged structure is given in Table 17.1. The diacypiperazines L-159,705, L-159,588,

TABLE 17.1. Alteration of the Affinity and Selectivity of a Diacylpiperazine Scaffold

Compound	R	IC_{50} (nM) NK$_1$	IC_{50} (nM) AT2
L-159,705	H	700	1600
L-159,588	CO_2	265	1.5
L-161,664	$CO-NH-(CH_2)_3-N(C_2H_5)_2$	11	520

Data are from Mills et al. (1993) and Wu et al. (1993). NK$_1$ receptor assays were performed using [125]I-(Tyr[8])–Substance P (100 pM) and human NK$_1$ receptors expressed in CHO cells. Angiotensin II (AT2) assays were performed using ^{3}H-angiotensin II (50 nM) and rat midbrain membranes.

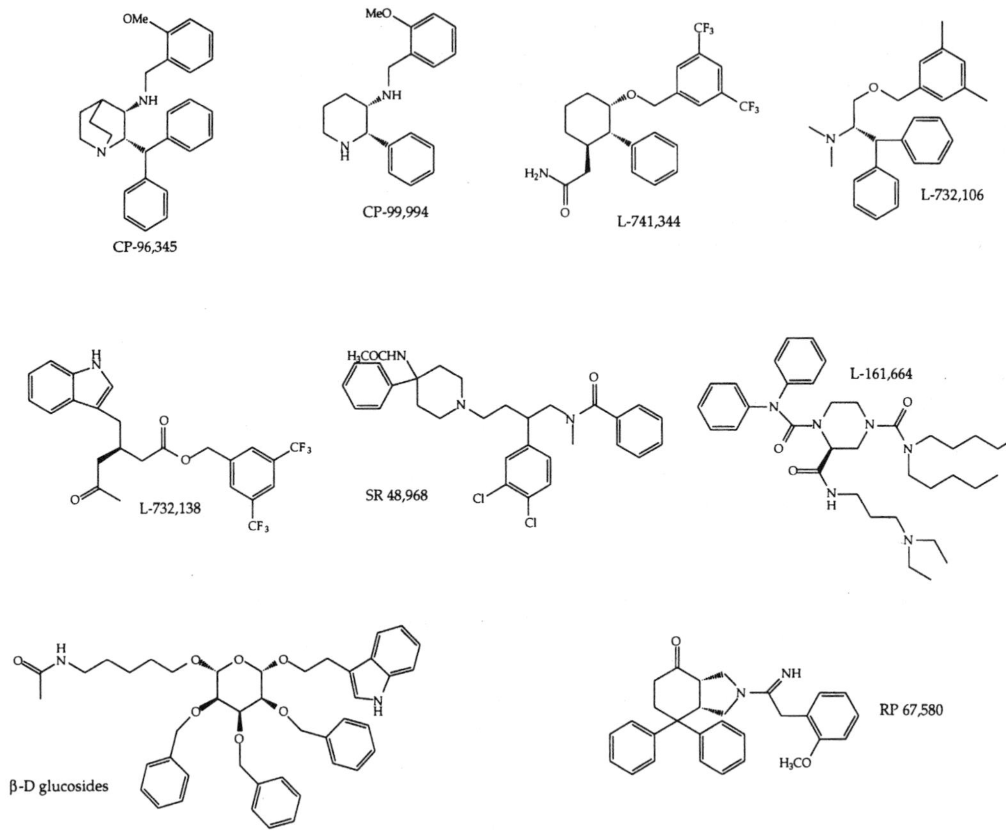

Figure 17.1. Cores and structural scaffolds for the neurokinins.

and L-161,664 share a common structural scaffold. The unsubstituted compound, L-159,705, shows submicromolar activity at the NK_1 receptor and low activity at the angiotensin II receptor and as such is a privileged structure. Derivitization of the 2 position of the piperazine ring with an acid, a pharmacophoric element for angiotensin II compounds, creates a potent and selective antagonist for this receptor. Replacing the acid functionality for a tertiary amine, as shown in L-161,664, reverses the selectivity of the scaffold, producing a potent NK_1 antagonist. Thus, high-affinity selective NK_1 antagonists or angiotensin antagonists can be derived from this privileged structure core. Other elements of molecular recognition such as acidic and basic functionality, hydrophobic groups, and hydrogen bond acceptors and donors may be components of privileged structures, but increasing the complexity of the structure by introducing further points of interaction with the receptor will likely introduce specificity and diminish the privileged nature of the compound as shown in this example.

Application of the privileged structure concept led to the design of potent, nonpeptide growth hormone (GH) secretagogue agonists that culminated in the development of MK-0677 (L-163,191), **1,** a clinical candidate for treatment of GH deficiency disorders (Patchett et al., 1995). The spiropiperidine moiety is a privileged structure and is observed in oxytocin (e.g., L-368,112, **2**) (Evans et al., 1992) and sigma (Chambers et al., 1992) receptor ligands, among others.

1   **2**

Comparison of privileged structures indicates that aromatic functionality is a common chemical feature that provides varied and substantial interactions with GPCRs. Comparison of optimized lead structures often shows structural diversity within the chemical framework and remarkable uniformity in the pharmacophoric features. For example, Figure 17.1 shows a collection of neurokinin antagonists, resulting from medicinal chemistry programs at a number of different sources, all of which maintain the important aromatic interactions using different structural cores. It is intriguing to observe that the importance of aromatics to privileged structures may reflect the tight conservation of aromatics across most receptor families (Underwood and Prendergast, 1997; Underwood et al., 1994). Furthermore, the interaction of aromatic compounds with receptors has a thermodynamic advantage. Aromatics, typical of other hydrophobic groups, are poorly solvated, and thus the free energy penalty paid for their desolvation on binding to receptors is small. However, aromatics favor interactions with certain protein side chains showing specific geometric interactions unlike nonaromatic hydrophobic functionality. Aromatics prefer stacking interactions with other aromatics forming specific structures reflecting overlap of electron density; e.g., "T" stacking and off-center, face-to-face stacking. Such stacking is observed with tryptophan, phenylalanine and tyrosine side chains. Basic groups such as lysine and arginine side chains interact with the center of aromatic rings where the electron density is highest. Hydrogen-bond donors such as the amide side chains of glutamine and asparagine also orient toward the center of the electron density. The imidazole side chain of histidine is both aromatic and able to donate a hydrogen bond to acceptors including aromatics. Even though the interaction energies are estimated to be small (2–3 kcal/mol), the overall free energy contributions to binding can be significant due to favorable desolvation. Aromatic functionality, therefore, provides an important component of privileged structures and selective GPCR agonists and antagonists.

5. EXAMPLES

Three examples are described. The nonpeptide angiotensin II antagonist example illustrates the use of database mining technology for the discovery of lead compounds for medicinal chemistry efforts. Typically this approach is used in all GPCR drug discovery programs. The second example concerning novel somatostatin agonists describes how NMR-derived structures for known ligands can accelerate and facilitate lead discovery. The final example, neurokinin antagonists, illustrates how the three-dimensional structure of the receptor in combination with mutagenesis experiments can facilitate the optimization of GPCR leads.

5.A. Angiotensin II Antagonists

As mentioned above, the goal of database mining is to prioritize the selection of compounds sent for assaying. The initial Takeda angiotensin II antagonist lead is shown in Figure 17.2 along with the potent and selective end products of lead optimization, Losartan, L-158,809 and SK-108566. Known angiotensin II receptor antagonists were identified from a TopoSim search across the Standard Drug File (SDF) (Derwent, 1991) database using the low-affinity ($K_i > 1$ μM) Takeda angiotensin II antagonist lead as the probe (Fig. 17.3). In fact, 10 of the top 50 hits from this search were recorded in the SDF as angiotensin II antagonists. Thus, searches using these simple methods can quite effectively "capture" the essential features of the probe molecule and score similar compounds. It is apparent from Figure 17.3 that topological similarity provides structural diversity, making topological similarity methods important for new

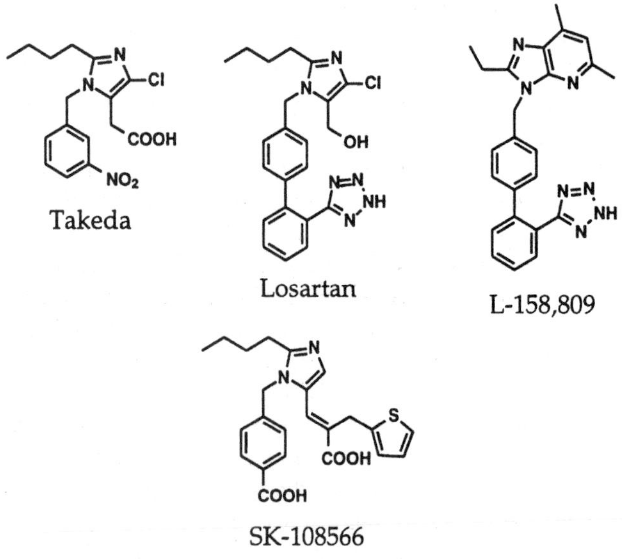

Figure 17.2. Typical angiotensin II antagonists.

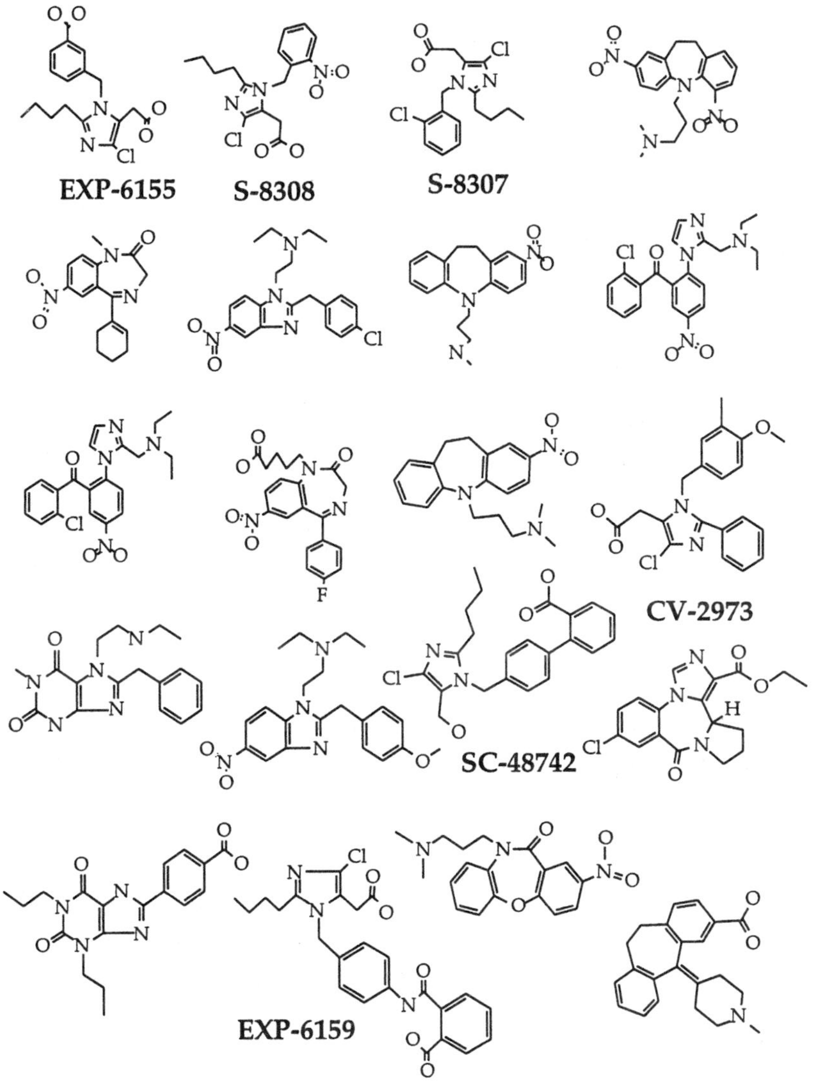

Figure 17.3. Top 20 compounds retrieved from a TopoSim search using the Takeda compound as a probe. Known angiotensin II antagonists are labeled.

lead discovery. A retrospective, computational analysis comparing the performances of TopoSim and SQ over random selection of compounds from the SDF database for angiotensin II antagonists shows that database mining methods increase the rate of discovery of angiotensin II antagonists over random selection but with a rate less than ideal (Fig. 17.4).

A number of ligands, prepared for the "balanced" angiotensin II antagonist program in which the goal was a compound with both angiotensins I and II receptor activity, show a surprising crossover in pharmacological behavior: The addition of branched alkyl or aryl groups on the terminal aromatic of the aryl-

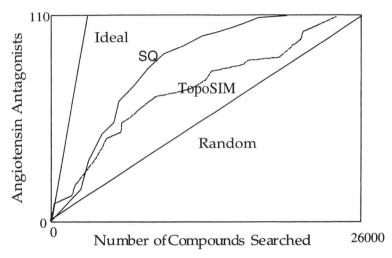

Figure 17.4. Plot percentage of angiotensin II antagonists discovered from the SDF database for ideal, random, TopoSim, and SQ selection methods. The "ideal line" represents the result if the known angiotensin II antagonists were assayed first.

or alkyl-sulfonamide class of compounds changed antagonists into partial agonists (Kivlighn et al., 1995). This subtle change in biological response upon modest structural changes has also been observed for CCK ligands (Dezube et al., 1995). Pharmocophore overlays of nonpeptides with peptides and SAR mapping indicate that the alkyl or aryl functionality is equivalent to the carboxy-terminal Phe in the endogenous ligand angiotensin II (Prendergast et al., 1994; Underwood et al., 1994). The importance of this position in determining the agonist response of the peptides is apparent from the sensitivity to change both in orientation and character of the chemical functionality (see Table 17.2) (Samanen and Regoli, 1994). Removing the aromatic character of this residue as shown in $[I^8]$-angiotensin II converts the compound to a potent antagonist (Table 17.2). Switching the chirality of the terminal residue from l to d ($[Sar^1,dF^8]$-angiotensin II) converts the peptide to an antagonist. The three -dimensional placement of the side chain aromatic at the 8 position, relative to other chemical features of the compound, is critical to agonist activity, e.g., fixing the aromatic side chain orthogonal to the backbone chain, as in $[Ind^8]$-angiotensin II, does not present the correct three-dimensional pharmacophore for agonism. Comparing $[Thiq^8]$-angiotensin II and $[dThiq^8]$-angiotensin II indicates that the pharmacophore requires the aromatic in the plane of the main chain as in $[Thiq^8]$-angiotensin II and changing the chirality at this position switches the agonist to an antagonist (Table 17.2). This parallels the SAR for the $[Sar^1,dF^8]$-angiotensin II peptide mentioned previously. It must be emphasized, however, that receptors that are precoupled to heterotrimeric G proteins are more readily activated by ligands and are consequently more tolerant of small changes in agonist structure (Underwood and Prendergast, 1997). Somatostatin is an example of a precoupled receptor in which most ligands are agonists.

TABLE 17.2. Structure–Activity Relationships (SAR) of the Carboxy-Terminal Residue of Angiotensin II (AII) Peptides. The functional response and the antagonist pA2 were measured in the rabbit aorta strip assay or the rat blood pressure assay as indicated by Samanen and Regoli (1994).

Peptide	Structure	Functional Response	pA2
AII	$_{NH_2}$DRVYIHPF$_{COOH}$	100%	
[I^8]-AII	$_{NH_2}$DRVYIHP**I**$_{COOH}$	Antagonist	8.32
[Sar^1,dF^8]-AII	$_{NH_2}$SarRVYIHP**dF**$_{COOH}$	Antagonist	9
[Thiq^8]-AII	$_{NH_2}$DRVYIHP	100%	
[dThiq^8]-AII	$_{NH_2}$DRVYIHP	Antagonist	7
[Ind^8]-AII	$_{NH_2}$DRVYIHP	Antagonist	
[Dip^8]-AII	$_{NH_2}$DRVYIHP	284% Super agonist	
[lDip^8]-AII	$_{NH_2}$DRVYIHP	48%	

5.B. Somatostatin Agonists

The three-dimensional structure of a somatostatin peptide agonist was used to search the available compound collections for small, pharmaceutically interesting compounds that would act in a similar way (Yang et al., 1998). The strategy utilized the three-dimensional structure of a known ligand, in this case a cyclic peptide, and the known SAR to develop a pharmacophore model. This model was used as a probe to search for similar compounds in compound collections. This approach identified a number of compounds with the desired features that showed good activity in the somatostatin assays and that provided leads for medicinal chemistry efforts.

The structure of the cyclic hexapeptide **3,** generated from NMR data (Veber, 1991; Yang et al., 1998) and distance geometry methods, contains the following features: residues Tyr^7-$dTrp^8$-Lys^9-Thr^{10} (numbering based on the somatostatin-14 peptide) are in a well-defined $\beta II'$ turn (nOe data); the side chains of $dTrp^8$ and Lys^9 are close (upfield shift of γ-CH_2-Lys^9); the side chains of Tyr^7 and $dTrp^8$ are close (temperature-dependent upfield shift of aromatic protons, see Mierke et al., 1990); and Pro^6 has a cis amide bond (nOe and SAR considerations).

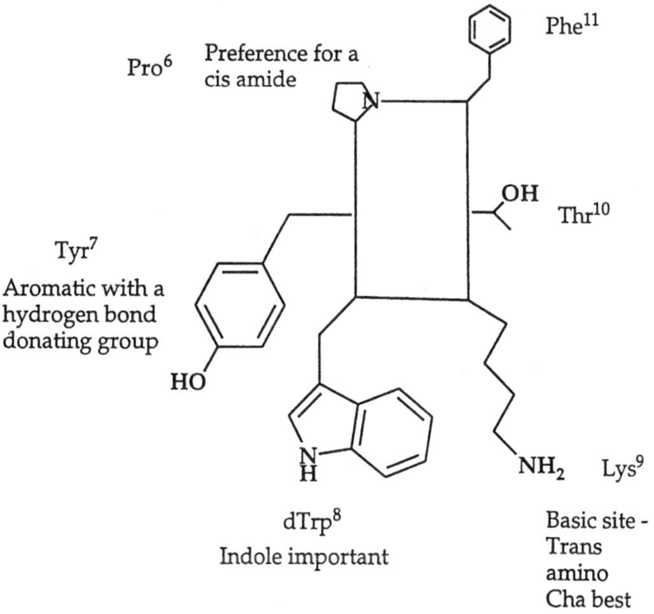

Pro^6   Preference for a cis amide

Phe^{11}

OH   **Thr^{10}**

Tyr^7

Aromatic with a hydrogen bond donating group

HO

dTrp^8

Indole important

NH$_2$ **Lys^9**

Basic site - Trans amino Cha best

3

In **3,** the size of the cyclic peptide's side chains are proportional to their importance from SAR studies of similar peptides. CHARMm dynamics (Brooks et al., 1983) was performed, and the averaged model was minimized to produce the hexapeptide probe used in the SQ searches and the identification of **4** (Yang et al., 1998). Both the NMR data and molecular dynamics indicate that multiple conformations are attainable, but the conformational preferences and SAR of the cyclic peptides clearly implicate the $\beta II'$ turn around $dTrp^8$-Lys^9 and the close proximity of these side chains as essential for biological activity. The conformation used in the initial probe is depicted in Figure 17.5. Although not shown, further refinements have been made to the model to reflect additional SAR and structural information made available since its first development. Comparison of **3** and L-054,522 highlights the degree of chemical and structural homology (Fig. 17.5).

Our three-dimensional search method, SQ (Miller et al., 1995), using the side chains of the Tyr^7-$dTrp^8$-Lys^9 in the modeled hexapeptide as the probe, identified **4** among other compounds in flexibases of our sample collection (Yang et al., 1998). This compound is a potent agonist ($K_i = 122$ nM) at the

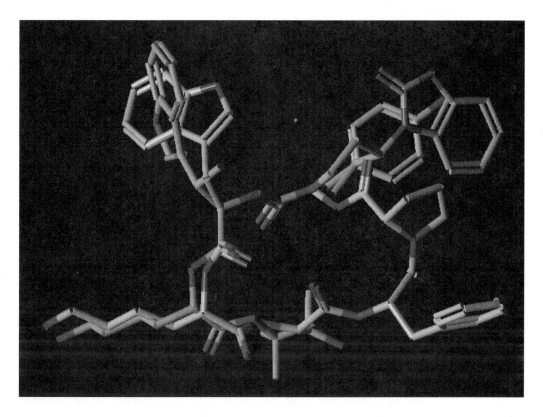

Figure 17.5. Comparison of the NMR-derived structure of **3** (carbon atoms in pink) with a low energy conformation of L-054,522 (carbon atoms in green). In both structures, nitrogens are blue and oxygens are red. (See color plates.)

subtype 2 of the human somatostatin receptor (hSSTR2) and shows high selectivity (>60×) over the other subtypes. Optimization of this class of compounds led to L-054,522, which is a highly potent hSSTR2 agonist (K_i = 0.01 nM) with >3,000-fold selectivity for this subtype. L-054,522 has a potent inhibitory effect on growth hormone release from rat primary pituitary cells and glucagon release from isolated rat pancreatic islets.

4   L-054,522

The stereopair in Figure 17.5 compares L-054,522 with the structure derived for the cyclic peptide **3**. Automated overlay of the d(βMe)Trp^8-Lys^9 of **3** with the equivalent functionality in L-054,522 places the *N*-(4-piperidinyl)benzimidazol-2-one close to Tyr^7 in accord with the SAR of these two positions. The conformation shown in Figure 17.5 for L-054,522 is also consistent with the highly shielded γ-CH$_2$-Lys found in the NMR of this compound and in **3** shielding implicates ring-current effects due to the proximity of the aromatic of dTrp to the γ-CH$_2$-Lys protons. It is interesting to note that the epimeric methyl and desmethyl analogs of L-054,522 are less active and do not show this upfield γ-CH$_2$-Lys shift.

5.C. Neurokinin Antagonists

Neurokinin-1 receptor (NK$_1$) antagonists have been postulated to have therapeutic potential in the treatment of various inflammatory disorders, migraine, emesis, and CNS disorders. CP 96,345 (Fig. 17.1) was the first nonpeptidyl NK$_1$ antagonist identified (Snider et al., 1991), and it was initially suggested that the protonated quinuclidine nitrogen might participate in a salt bridge with an acidic residue in the transmembrane domain of the receptor (Lowe et al., 1992). However, a closer examination of the receptor structure and mutagenesis experiments failed to confirm this expectation. Subsequent modeling and mutagenesis experiments predicted that the protonated nitrogen oriented toward the extracellular domain of the receptor and might be involved in a hydrogen bonding interaction (Fong et al., 1994). This prediction was supported by the demonstration that a cyclohexyl analog (L-741,344 in Fig. 17.1, top center) maintained high affinity for the NK$_1$ receptor with the flexible amide side chain presumably replacing the hydrogen bonding interaction of the protonated nitrogen of CP-96,345 (Mills et al., 1995).

Similar modeling and mutagenesis experiments suggested that the acylated amine moiety of the *N*-acyl-L-tryptophan NK$_1$ antagonist L-732,138 (Fig. 17.6) orients to the extracellular domain, predicting that bulky, polar substitutions might be tolerable at this position (Cascieri et al., 1994). This information was utilized in the design of the methylquinuclidine-substituted NK$_1$ antagonist L-743,310 (Fig. 17.6), a potent, peripherally active compound with poor brain penetration (Tattersall et al., 1996). This compound has been utilized to help

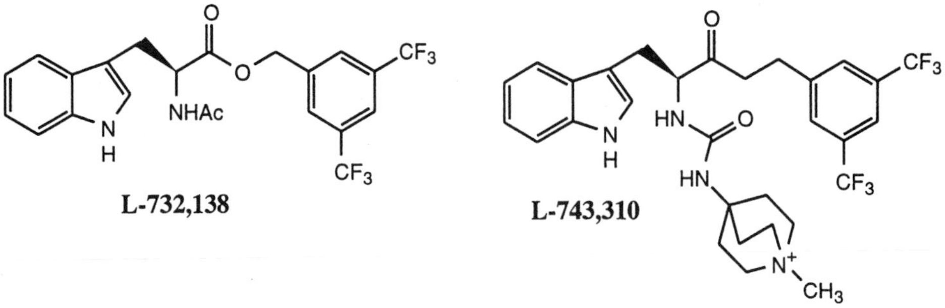

L-732,138   **L-743,310**

Figure 17.6. Tryptophan-based NK$_1$ antagonists.

determine if the pharmacological effects of NK_1 receptor blockade are mediated via peripheral or CNS NK_1 receptors. Therefore, in this program, the combination of mutagenesis and modeling experiments has been a powerful tool to help increase structural diversity and manipulate ligand properties.

6. CONCLUSIONS AND FUTURE DIRECTIONS

The development of methodologies to expedite the discovery of novel therapeutic agents is a dynamic process. We have attempted to focus on several aspects of the current technology for the identification and optimization of lead structures and for characterization of their pharmacological properties. The structural characterization of GPCRs is rapidly advancing and should further enhance capabilities to utilize structure-based approaches in the design and discovery of pharmacological agents. We have not discussed the power of combinatorial chemistry and rapid analog synthesis in this process, but this technology is a great advance in generating structural diversity and in experimentally testing design hypotheses generated by modeling approaches. The prediction of pharmacophore design from primary sequence information of newly discovered GPCRs remains a challenge, but advances to date have made it possible to believe the goal is achievable.

REFERENCES

Ariens E, Beld A, Rodrigues de Miranda J, Simonis A (1979): The Receptors: A Comprehensive Treatise. New York: Plenum.

Baldwin JM (1993): The probable arrangement of the helices in G protein–coupled receptors. EmBO J 12:1693–1703.

Baldwin JM (1994): Structure and function of receptors coupled to G proteins. Curr Opin Cell Biol 6:180–190.

Baldwin JM, Schertler GF, Unger VM (1997): An alpha-carbon template for the transmembrane helices in the rhodopsin family of G-protein–coupled receptors. J Mol Biol 272:144–164.

Bohm G (1996): New approaches in molecular structure prediction. Biophys Chem 59:1–32.

Borman S (1992): New 3D search and *de novo* design techniques aid drug development. Chem Eng News 70:18–26.

Brooks B, Bruccoleri R, Olafson B, States D, Swaminathan S, Karplus M (1983): CHARMm: A program for macromolecular energy, minimization and dynamics calculations. J Comput Chem 4:187–217.

Bush B, Sheridan R (1993): PATTY: A programmable atom typer and language for automatic classification of atoms in molecular databases. J Chem Inf Comput Sci 33:756–762.

Campbell A, Sykes B (1993): The 2-dimensional transferred nuclear overhauser effect—Theory and practice. Annu Rev Biophys Biomol Struct 22:99–122.

Carhart R, Smith D, Venkataraghavan R (1985): Atom pairs as molecular features in structure–activity studies: Definition and applications. J Chem Inf Comput Sci 25:64–73.

Cascieri MA, Fong TM, Strader CD (1995): Molecular characterization of a common binding site for small molecules within the transmembrane domain of G-protein coupled receptors. J Pharmacol Toxicol Methods 33:179–185.

Cascieri MA, MacLeod AM, Underwood DJ, Shiao L-L, Ber E, Sadowski S, Yu H, Merchant KJ, Swain CJ, Strader CD, Fong TM (1994): Characterization of the interaction of N-acyl-L-tryptophan benzyl ester neurokinin antagonists with the human neurokinin-1 receptor. J Biol Chem 269:6587–6591.

Chambers MS, Baker R, Billington DC, Knight AK, Middlemiss DN, Wong EH (1992): Spiropiperidines as high-affinity, selective sigma ligands. J Med Chem 35:2033–2039.

Chicchi GG, Graziano MP, Koch G, Hey P, Sullivan K, Vicario PP, Cascieri MA (1997): Alterations in receptor activation and divalent cation activation of agonist binding by deletion of intracellular domains of the glucagon receptor. J Biol Chem 272:7765–7769.

Christie B, Henry D, Guner O, Moock T (1990): MACCS-3D: A tool for three-dimensional drug design. In Raitt D (ed): Online Information 90; 14th International Online Information Meeting Proceedings. Oxford: Learned Information, pp 137–161.

Collings P (1990): Liquid Crystals. Princeton, NJ: Princeton University Press.

Creuzet F, McDermott A, Gebhard R, van der Hoef K, Spijker-Assink M, Herzfeld J, Lugtenburg J, Levitt M, Griffin R (1991): Determination of membrane protein structure by rotational resonance NR: Bacteriorhodopsin. Science 251:783–786.

Derwent Inc. (1991): Standard Drug File. Alexandria, VA: Derwent Inc.

DesJarlais RL, Sheridan RP, Seibel GL, Dixon JS, Kuntz ID, Venkataraghavan R (1988): Using shape complimentarity as an initial screen in designing ligands for a receptor binding site of known 3D. J Med Chem 31:722–729.

Dezube M, Sugg EE, Birkemo LS, Croom DK, Dougherty RW Jr, Ervin GN, Grizzle MK, James MK, Johnson MF, Mosher JT, et al (1995): Modification of receptor selectivity and functional activity in cholecystokinin peptoid ligands. J Med Chem 38:3384–3390.

Dyson HJ, Rance M, Houghten RA, Wright PE, Lerner RA (1988): Folding of immunogenic peptide fragments of proteins in water solution. II. The nascent helix. J Mol Biol 201:201–217.

Dyson HJ, Wright PE (1995): Antigenic peptides. FASEB J 9:37–42.

Edidin M (1997): Lipid microdomains in cell surface membranes. Curr Opin Struct Biol 7:528–532.

Evans BE, Bock MG, Rittle KE, DiPardo RM, Whitter WL, Veber DF, Anderson PS, Freidinger FM (1986): Design of potent, orally effective, nonpeptidal antagonists of the peptide hormone cholecystokinin. Proc Natl Acad Sci USA 83:4918–4922.

Evans BE, Leighton JL, Rittle KE, Gilbert KF, Lundell GF, Gould NP, Hobbs DW, DiPardo RM, Veber DF, Pettibone DJ (1992): Orally active, nonpeptide oxytocin antagonists. J Med Chem 35:3919–3927.

Evans BE, Rittle KE, Bock MG, DiPardo RM, Freidinger RM, Whitter WL, Lundell GF, Veber DF, Anderson PS, Chang RS, et al (1988): Methods for drug discovery: Development of potent, selective, orally effective cholecystokinin antagonists. J Med Chem 31:2235–2246.

Fong TM, Yu H, Cascieri MA, Underwood D, Swain CJ, Strader CD (1994): Interaction of glutamine 165 in the fourth transmembrane segment of the human neurokinin-1 receptor with quinuclidine antagonists. J Biol Chem 269:14957–14961.

Freedman NJ, Lefkowitz RJ (1996): Desensitization of G protein–coupled receptors. Recent Prog Horm Res 51:319–353.

George ST, Berrios M, Hadcock JR, Wang HY, Malbon CC (1988): Receptor density and cAMP accumulation: Analysis in CHO cells exhibiting stable expression of a cDNA that encodes the beta 2-adrenergic receptor. Biochem Biophys Res Commun 150:665–672.

Groblewski T, Maigret B, Larguier R, Lombard C, Bonnafous JC, Marie J (1997): Mutation of Asn111 in the third transmembrane domain of the AT1A angiotensin II receptor induces its constitutive activation. J Biol Chem 272:1822–1826.

Guner O, Henry D, Pearlman R (1992): Use of flexible queries for searching conformationally flexible molecules in databases of three-dimensional structures. J Chem Inf Comput Sci 32:101–109.

Hamm H, Gilchrist A (1996): Heterotrimeric G proteins. Curr Opin Cell Biol 8:189–196.

Hebert TE, Moffett S, Morello JP, Loisel TP, Bichet DG, Barret C, Bouvier M (1996): A peptide derived from a beta(2)-adrenergic receptor transmembrane domain inhibits both receptor dimerization and activation. J Biol Chem 271:16384–16392.

Henderson R, Baldwin JM, Ceska TA, Zemlin F, Beckmann E, Downing KH (1990): Model for the structure of bacteriorhodopsin based on high-resolution electron cryomicroscopy. J Mol Biol 213:899–929.

Herzyk P, Hubbard R (1995): Automated method for modeling seven-helix transmembrane receptors from experimental data. Biophys J 69:2419–2442.

Heymann B, Muller D, Mitsuoka K, Engel A (1997): Electron and atomic force microscopy of membrane proteins. Curr Opin Struct Biol 7:543–549.

Hibert MF, Trumpp-Kallmeyer S, Bruinvels A, Hoflack J (1991): Three-dimensional models of neurotransmitter G-binding protein–coupled receptors. Mol Pharmacol 40:8–15.

Hirschmann R, Yao W, Cascieri MA, Strader CD, Maechler L, Cichy-Knight MA, Hynes J Jr, van Rijn RD, Sprengeler PA, Smith ABR (1996): Synthesis of potent cyclic hexapeptide NK-1 antagonists. Use of a minilibrary in transforming a peptidal somatostatin receptor ligand into an NK-1 receptor ligand via a polyvalent peptidomimetric. J Med Chem 39:2441–2448.

Hoflack J, Trumpp-Kallmeyer S, Hibert M (1994): Re-evaluation of bacteriorhodopsin as a model for G protein–coupled receptors. Trends Pharmacol Sci 15:7–9.

Jones G, Willet P (1995): Docking small-molecule ligands into active sites. Curr Opin Biotechnol 6:652–656.

Kearsley S, Sallamack S, Fluder E, Andose J, Mosley R, Sheridan R (1996): Chemical similarity using physiochemical property descriptors. J Chem Inf Comp Sci 36:118–127.

Kearsley S, Smith G (1992): An alternative method for the alignment of molecular structures: Maximizing electrostatic and steric overlap. Tetrahedron Comput Methodol 3:615–633.

Kearsley S, Underwood D, Sheridan RP, Miller M (1994): FlexiBases: A way to enhance the use of molecular docking methods. J Comput Aided Mol Design 8:565–582.

Kimura Y, Vassylyev DG, Miyazawa A, Kidera A, Matsushima M, Mitsuoka K, Murata K, Hirai T, Fujiyoshi Y (1997): Surface of bacteriorhodopsin revealed by high-resolution electron crystallography. Nature 389:206–211.

Kivlighn SD, Huckle WR, Zingaro GJ, Rivero RA, Lotti VJ, Chang RS, Schorn TW, Kevin N, Johnson RG Jr, Greenlee WJ, et al (1995): Discovery of L-162,313: A nonpeptide that mimics the biological actions of angiotensin II. Am J Physiol 268:R820–R823.

Kuntz I (1992): Structure-based strategies for drug design and discovery. Science 257:1078–1082.

Lengauer T, Rarey M (1996): Computational methods for biomolecular docking. Curr Opin Struct Biol 6:402–406.

Lowe JA III, Drozda SE, Snider RM, Longo KP, Zorn SH, Morrone J, Jackson ER, McLean S, Bryce DK, Bordner J, Nagahisa A, Kania Y, Suga O, Tsuchiya M (1992): The discovery of (2S, 3S)-*cis*-2-(diphenylmethyl)-*N*-[(2-methoxyphenyl)methyl]-1-azabicyclo[2.2.2]-octan-3-amine as a novel, nonpeptide substance P antagonist. J Med Chem 35:2591–2600.

Martin Y (1992): 3D database searching in drug design. J Med Chem 35:2145–2154.

Martin Y, Willet P (1990): Searching databases of three-dimensional structures. In Lipkowitz, Boyd (eds): Reviews in Computational Chemistry. New York: VCH, pp 213–256.

Meunier J-C, Mollereau C, Toll L, Suadeau C, Moisand C, Alvinerie P, Butour J-L, Guillemot J-C, Ferrara P, Monserrat B, Mazarguil H, Vassart G, Permentier M, Costentin J (1995): Isolation and structure of the endogenous agonist of opioid receptor-like ORL1 receptor. Nature 377:532–535.

Mierke D, Pattaroni C, Delaet N, Toy A, Goodman M, Tancredi T, Motta A, Temussi P, Moroder L, Bovermann G, Wunsch E (1990): Cyclic hexapeptides related to somatostatin. Int J Peptide Prot Res 36:418–432.

Miller M, Culberson J, Kearsley S, Prendergast K (1995): Drug Discovery and Design Using Molecular Superposition. Chicago: American Chemical Society.

Miller M, Kearsley S, Underwood D, Sheridan R (1994a): FLOG: A system to select "quasi-flexible" ligands complementary to a receptor of known three-dimensional structure. J Comput Aided Mol Design 8:153–174.

Miller M, Sheridan R, Kearsley S, Underwood D (1994b): Advances in automated docking applied to human immunodeficiency virus type 1 protease. Methods Enzymol 241:354–370.

Mills SG, MacCoss M, Underwood D, Shah SK, Finke PF, Miller DJ, Budha RJ, Cascieri MA, Sadowski S, Strader CD (1995): 1,2,3-Trisubstituted cyclohexyl substance P antagonists: Significance of the ring nitrogen in piperidine-based NK-1 receptor antagonists. Biorg Med Chem Lett 5:1345–1350.

Mills SG, Wu MT, MacCoss M, Budha RJ, Dorn CP Jr, Cascieri MA, Sadowski S, Strader CD, Greenlee WJ (1993): 1,4-Diacylpiperazine-2-(*S*)-[(*N*-aminoalkyl)carboxamides] as novel, potent substance P receptor antagonists. Biorg Med Chem Lett 3:2707–2712.

Moock T, Henry D, Ozkabak A, Alamgir M (1994): Conformational searching in ISIS/3D databases. J Chem Inf Comput Sci 34:184–189.

Moores S (1991): Sequence dependence of protein isoprenylation. J Biol Chem 266: 14603–14610.

Mosley R, Miller M, Kearsley S, Prendergast K, Underwood D (1995): New lead discovery in drug development. In Witten M (ed): Computational Medicine, Public Health and Biotechnology. Singapore: World Scientific Publishing Co Pty Ltd, pp 101–125.

Mouritsen OG, Jorgensen K (1997): Small-scale lipid-membrane structure: Simulation versus experiment. Curr Opin Struct Biol 7:518–527.

Newman-Tancredi A, Conte C, Chaput C, Verriele L, Millan MJ (1997): Agonist and inverse agonist efficacy at human recombinant serotonin 5-HT1A receptors as a function of receptor: G-protein stoichiometry. Neuropharmacology 36:451–459.

Ni F, Carpenter KA, Ripoll DR, Sanderson SD, Hugli TE (1996): Stabilization of an isolated helical capping box in solution by hydrophobic interactions: Evidence from the

NMR study of bioactive peptides from the C-terminus of human C5a anaphylatoxin. Biopolymers 38:31–41.

Ni F, Scheraga H (1994): Use of the transfered nuclear overhauser effect to determine the conformations of ligands bound to proteins. Acct Chem Res 27:257–264.

Patchett A, Nargund R, Tata J, Chen M-H, Barakat K, Johnson D, Cheng K, Chan W, Butler B, Hickey G, Jacks T, Schleim K, Pong S-S, Chaung L, Chen H, Frazier E, Leung K, Chiu S, Smith R (1995): Design and biological activities of L-163191 (MK-0677): A potent, orally active growth hormone secretagogue. Proc Natl Acad Sci USA 92:7001–7005.

Pebay-Peyroula E, Rummel G, Rosenbusch JP, Landau EM (1997): X-ray structure of bacteriorhodopsin at 2.5 angstroms from microcrystals grown in lipidic cubic phases. Science 277:1676–1681.

Portoghese PS (1989): Bivalent ligands and the message-address concept in the design of selective opioid receptor antagonists. Trends Pharmacol Sci 10:230–235.

Prendergast K, Adams K, Greenlee WJ, Nachbar RB, Patchett AA, Underwood D (1994): Derivation of a 3D pharmacophore model for the angiotensin-II site one receptor. J Comput Aided Mol Design 8:491–512.

Reiss Y, Stradley S, Gierasch L, Brown M, Goldstein J (1991): Sequence requirement for peptide recognition by rat brain p21^{ras} protein farnesyl transferase. Proc Natl Acad Sci USA 88:732–736.

Rini J, Schulze-Gahmen U, Wilson I (1992): Structural evidence for induced fit as a mechanism for antibody–antigen recognition. Science 255:959–965.

Rosenfeld R, Vajda S, DeLisi C (1995): Flexible docking and design. Annu Rev Biophys Biomol Struct 24:677–700.

Rutenber E, Fauman E, Keenan R, Fong S, Furth P, de Montellano P, Meng E, Kuntz I, DeCamp D, Salto R, Rose J, Craik C, Stroud R (1993): Structure of a non-peptide inhibitor complexed with HIV-1 protease. J Biol Chem 268:15343–15346.

Sakurai T, Amemiya A, Ishii M, Matsuzaki I, Chemielli RM, Tanaka H, Williams SC, Richardson JA, Kozlowski GP, Wilson S, Arch JR, Buckingham RE, Haynes AC, Carr SA, Annan RS, McNulty DE, Liu WS, Terrett JA, Elshourbagy NA, Bergsma DJ, Yanagisawa M (1998): Orexins and orexin receptors: A family of hypothalamic neuropeptides and G protein–coupled receptors that regulate feeding behavior. Cell 92:573–585.

Samanen J, Regoli D (1994): Structure–activity relationships of peptide angiotensin II receptor agonists and antagonists. In Ruffolo R Jr (ed): Angiotensin II Receptors, vol 2. Boca Raton, FL: CRC Press, pp 11–97.

Saunders M, Houk K, Wu Y-D, Clark Still W, Lipton M, Chang G, Guida W (1990): Conformations of cycloheptadecane. A comparison of methods for conformational searching. J Am Chem Soc 112:1419–1427.

Saxton MJ, Jacobson K (1997): Single-particle tracking: Applications to membrane dynamics. Annu Rev Biophys Biomol Struct 26:373–399.

Schwyzer R (1986): Molecular mechanism of opioid receptor selection. Biochemistry 25:6335–6342.

Sheridan R, Miller M, Underwood D, Kearsley S (1995): Chemical similarity using geometric atom pair descriptors. J Chem Inf Comp Sci 36:128–136.

Sheridan R, Nachbar R, Bush B (1994): Extending the trend vector: The trend matrix and sample-based partial least squares. J Comput Aided Mol Design 8:323–340.

Sheridan R, Venkataraghavan R (1987): New methods in computer-aided drug design. Acct Chem Res 20:322–329.

Shoichet B, Stroud R, Santi D, Kuntz I, Perry K (1993): Structure based discovery of inhibitors of thymidylate synthase. Science 259:1445–1450.

Siciliano S, Rollins T, DeMartino J, Konteatis Z, Malkowitz L, Van Riper G, Bondy S, Rosen H, Springer M (1994): Two-site binding of C5a by its receptor: An alternative binding paradigm for G protein–coupled receptors. Proc Natl Acad Sci USA 91: 1214–1218.

Simons K, Ikonen E (1997): Functional rafts in cell membranes. Nature 387:569–572.

Snider RM, Constantine JW, Lowe JA III, Longo KP, Lebel WS, Woody HA, Drozda SE, Desai MC, Vinick FJ, Spencer RW, Hess H-J (1991): A potent nonpeptide antagonist of the substance P (NK-1) receptor. Science 251:435–437.

Steinhoff H-J, Mollaaghababa R, Altenbach C, Hideg K, Krebs M, Khorana H, Hubbell W (1994): Time-resolved detection of structural changes during the photocycle of spin-labeled bacteriorhodopsin. Science 266:105–107.

Strader C, Fong T, Tota M, Underwood D, Dixon R (1994): Structure and function of G protein coupled receptors. Annu Rev Biochem 63:101–132.

Tattersall FD, Rycroft W, Francis B, Pearce D, Merchant K, MacLeod AM, Ladduwahetty T, Keown L, Swain C, Baker R, Cascieri M, Ber E, Metzger J, MacIntyre DE, Hill RG, Hargreaves RJ (1996): Tachykinin NK-1 receptor antagonists act centrally to inhibit emesis induced by the chemotherapeutic agent cisplatin in ferrets. Neuropharmacology 35:1121–1129.

Underwood D (1995): Protein structures from domain packing—A game of twenty questions? Biophys J 69:2183–2184.

Underwood D, Prendergast K (1997): Getting it together: Signal transduction in G-protein receptors by association of receptor domains. Chem Biol 4:239–248.

Underwood D, Strader C, Rivero R, Patchett A, Greenlee W, Prendergast K (1994): Structural model of antagonist binding to the angiotensin-II, AT1 subtype, G-protein coupled receptor. Chem Biol 1:211–221.

Unger VM, Hargrave PA, Baldwin JM, Schertler GF (1997): Arrangement of rhodopsin transmembrane alpha-helices. Nature 389:203–206.

Unger VM, Schertler GF (1995): Low resolution structure of bovine rhodopsin determined by electron cryo-microscopy. Biophys J 68:1776–1786.

Van Gunsteran W, Luque F, Timms D, Torda A (1994): Molecular mechanics in biology—From structure to function taking account of solvation. Annu Rev Biophys Biomol Struct 23:847–886.

Veber D (1991): Design and discovery in the development of peptide analogs. In Peptides, Chemistry and Biology: Proceedings of the Twelfth American Peptide Symposium, Cambridge, MA.

Wang S, He C, Hashemi T, Bayne M (1997): Cloning and expression characterization of a novel galanin receptor. J Biol Chem 272:31949–31952.

Ward P, Ewan GB, Jordan CC, Ireland SJ, Hagan RM, Brown JR (1990): Potent and highly selective neurokinin antagonists. J Med Chem 33:1848–1851.

Whaley BS, Yuan N, Birnbaumer L, Clark RB, Barber R (1994): Differential expression of the beta-adrenergic receptor modifies agonist stimulation of adenylyl cyclase: A quantitative evaluation. Mol Pharmacol 45:481–489.

Willett P (1995): Genetic algorithms in molecular recognition and design. Trends Biotechnol 13:516–521.

Williamson M, Madison V (1990): Three-dimensional structure of Porcine C5a(desArg) from H nuclear magnetic resonance data. Biochemistry 29:2895–2905.

Wong TC, Lee CM, Guo W, Chang DK (1993): Conformational study of two substance P hexapeptides by two-dimensional NMR. Int J Peptide Prot Res 41:185–195.

Wright PE, Dyson HJ, Lerner RA (1988): Conformation of peptide fragments of proteins in aqueous solution: Implications for initiation of protein folding. Biochemistry 27:7167–7175.

Wright PE, Dyson HJ, Lerner RA, Riechmann L, Tsang P (1990): Antigen–antibody interactions: An NMR approach. Biochem Pharmacol 40:83–88.

Wu MT, Ikeler TJ, Ashton WT, Chang RSL, Lotti VJ, Greenlee WJ (1993): Synthesis and structure–activity relationships of a novel series of non-peptide AT-2–selective angiotensin II receptor antagonists. Biorg Med Chem Lett 3:2023–2028.

Yang L, Berk S, Rohrer S, Mosley R, Underwood D, Guo L, Arison B, Birzin E, Hayes E, Mitra S, Parmar R, Cheng K, Wu J-T, Butler B, Foor F, Pasternak A, Patchett A (1998): Synthesis and biological activities of potent peptidomimetics selective for somatostatin receptor subtype-2. Proc Natl Acad Sci USA 95:10836–10841.

Young JK, Anklin C, Hicks RP (1994): NMR and molecular modeling investigations of the neuropeptide substance P in the presence of 15 mM sodium dodecyl sulfate micelles. Biopolymers 34:1449–1462.

Zhang M, Wong TC (1993): Solution conformation study of substance P methyl ester and [Nle10]-neurokinin A (4-10) by NMR spectroscopy. Biopolymers 33: 1901–1908.

Zhang X, Boyar W, Toth MJ, Wennogle L, Gonnella NC (1997): Structural definition of the C5a C terminus by two-dimensional nuclear magnetic resonance spectroscopy. Proteins 28:261–267.

INDEX

α carbon atoms, 277, 280
$α_{1b}$-AR, mutagenesis
  activation, mutation and agonist-induced,
    178–179
  conserved amino acids, potential common role of,
    179–180
  conserved E/DRY sequence, mechanistic role of,
    171–174
  helices movement and intracellular loops,
    174–178
  methodology, 169–170
  MD analysis, R and R* search, 170–171
Accessibility method, cysteine-susceptibility, 22–24
Acetylation, 208
Additivity analysis, two-dimensional mutagenesis,
  51–52
Adenosine receptors, 137
Adenylyl cyclase, 47–48, 191, 217–218
Agarose gel electrophoresis, 158
Agglutination, 147
Agonist binding, substituted-cysteine accessibility
  method, 37
α helices, two-dimensional crystals, 273–274
Alkoxides, 127
Allosteric binding, site identification
  degree of cooperativity, assessment of, 54–55
  muscarinic receptors, mutations that affect,
    55–56
ALLSPACE, 261
Aluminum, 130
Amino acid(s)
  sequence, rhodopsin electron crystallography,
    277–280
  synthesis, 342–346
  peptide synthesis, 207–209
  unnatural
    biosynthetic incorporation, 340–349
    characteristics of, generally, 335–337
    defined, 14
Amplitude
  image processing, 254, 258, 260
  merging with phases

from tilted images to calculate three-dimen-
  sional map, 267–273
from untilted images for projection map,
  261–267
Analog binding, 75
Analysis, *see specific types of analyses*
Angiotensin, function of, 375
Angiotensin II antagonists, 388–391
Antihistamines, 43
Arginine, 387
Aromatic(s)
  clusters, 34–35
  privileged structures, 387
Asparagine, 387
Astigmatism, image processing, 258–259
Atomic probes, *see* Metal cations, as atomic probes
ATP, 48
AUTOCORREL, 257
AVRGAMPHS, 263

$β_2$-adrenergic receptors
  fluorescence spectroscopy analysis, conforma-
    tional changes
    constitutively active mutant receptor (CAM),
      analysis of, 329–331
    expression and purification, 318–320
    purified, analysis of, 320–325
    site-selective fluorescent labeling, 325–329
  -Gsα fusion protein, as analysis model
    adenylyl cyclase studies, 198–199
    agonist competition studies, 192–195
    fusion proteins, 200
    general properties of, 191–192
    GTPase studies, 195–196
    GTPγS binding studies, 196–198
    interaction methodology
      adenylyl cyclase assay, 191
      generally, 186–187
      GTPase assay, 190
      GTPγ binding assay, 190
      materials, 189
      receptor binding assay, 190

overview, 186, 188
Sf9 cells and, 188–189
signal transduction, 199
stoichiometry, 188, 199
mutant receptors, generation of, 328–329
Bacteriorhodopsin
characteristics of, generally, 66, 111
rhodopsin compared with, 274–276
bar1 mutations, 147, 149
Benzodiazepine, 385
Beta blockers, 43
Binding assays, thryotropin-releasing hormone
receptor (TRH-R), 63
Binding sites, mapping, 46–47
Biogenic amine binding site, mapping of
conservation pattern analysis, 52–54
deletion mutagenesis, scanning, 48–49
pharmacophore map, construction of, 49–50
specific point mutations, 50–51
two-dimensional mutagenesis, 51–52
Black, James, Sir, 43
BOXIMAGE, 257–258
Bruker AMX-600 spectrometer, 359

Cadmium, 124, 130
Calcium, 129
cAMP, 48
Carboxy termini, 208, 215
Cardiovascular disease, 43
CCP4 program, 254, 261, 266, 271
CCUNBEND, 257
cDNA, 318
cGMP, 87, 207
CHARMM, 67, 75, 169
Chimeric receptors, 3, 5, 14, 16
Chloride, 129
Chromatography
high performance liquid chromatography
(HPLC), 209–210, 342, 346
ion exchange, 239
Chromophore, rhodopsin and, 86, 93–95
Circular dichroism (CD), 356
CLN2, 151
Cloning, 206
Complexation entropy, 125, 133
Conformation, thyrotropin-releasing hormone re-
ceptor (TRH-R), 77–78
Conformational Memories, 79
Congenital night blindness, 98
Conservation pattern, 52–54
Constitutively active receptor mutants, as research
probes

α_{1b}-AR, mutagenesis
activation, mutation and agonist-induced,
178–179
conserved amino acids, potential common role
of, 179–180
conserved E/DRY sequence, mechanistic role
of, 171–174
helices movement and intracellular loops,
174–178
methodology, 169–170
MD analysis, R and R* search, 170–171
receptor sites, identification of, 168
Contact sites, synthetic peptides, 206, 220
Contour plots, three-dimensional maps, 271,
273
Contrast transfer function (CTF)
in electron microscope, 248
image processing, 258–259, 263
Copper, 124, 130
COSY, 359–360, 367
cRNA, 339
Cross-correlation map, two-dimensional crystals,
257
Crosslinking
synthetic peptides, 220–221
second-site suppressors, 161
split receptors
disulfide, 111–112, 116–117
oxidative, 114–117
Cryomicroscopy, low-dose
contrast transfer function, 248
electron microscope, 246
low-dose imaging, 247
optical diffraction, data evaluation, 248
specimens
tilted, spot scan for imaging, 247
vitrification of, 246
two-dimensional crystals
Fourier transform properties, 250–254,
268
preliminary characterization of, 248–250
Crystallography, comparison of electron *vs.* x-ray,
236–238
Cys, defined, 2
Cysteine(s)
endogenous, 23–24
site-directed spin-labeling studies, 298–299
substitution, *see* Substituted-cysteine accessibility
method (SCAM)
Cystoplasmic loops and tails, structure of, 306–307,
309–310
Cytosol, 170

DAP-Q, 222
Datamax software, 324
Deletion mutagenesis, 3, 13–14, 48–49
Density gradients, two-dimensional crystals, 243
Density maps
α helices, arrangement of, 273–274
bacteriorhodopsin, rhodopsin compared with, 274–276
intracellular G protein binding region, 277
retinal binding site, 276
Diffraction
electron, 280, 282
optical, 245
rhodopsin electron crystallography, 263–264
x-ray, 236
Distance measurement, SDSL, 293
Disulfide crosslinks, split receptors, 111–112, 116–117
DNA replication, 154
Dopamine D2 ligands, 32–33
Dopamine receptors, 137–138
Drug discovery
development methods, 376–380
examples
angiotensin II antagonists, 388–391
neurokinin antagonists, 394–395
somatostatin agonists, 391–394
GPCR structure and models, 380–383
ligand and receptor-ligand models
endogenous ligand structure, 383–384
privileged structure, 384–388
D2 receptor construct, epitope-tagged, 25

E. coli, 154–155, 158
Eisenman series, 122–125, 128
Electron crystallography, x-ray crystallography distinguished from, 236–238
Electron microscope, 246
Electron microscopy, two-dimensional crystals, 236, 238, 243–244, 246
Electron paramagnetic resonance (EPR) spectroscopy
defined, 291
qualitative description of, 293–294
instruments, 297
spectrum
basis of, 294
nitroxide, 294
spin label information, as source for
accessibility, 296–297
mobility, 295–296
two spin labels, distance measurement, 297

Electrophysiology, 348–349
Endogenous cysteines, 23–24
Enthalpy, 133
Entropy, 125, 133
Epinephrine, 375
EPR, see Electron paramagnetic resonance (EPR) spectroscopy
EXTEND, 266
Extracellular loops (ECLs), hyrotropin-releasing hormone receptor (TRH-R), 69, 71, 75–77

FAR1 gene, 151
FFTRANS, 257
Field emission microscope, 280–281
FLAG epitope, 25, 317–318
Fourier analysis, 238
Fourier filtering methods, 256
Fourier transform, generally
infrared (FTIR) spectroscopy, applications of, 85, 93, 95, 99, 357, 362, 380
three-dimensional crystals, 253
two-dimensional crystals, 250–254, 260
Free energy, 133
FUS1-lacZ gene
detection of, 149–150
expression, liquid assay for, 150
Fusion proteins, uses of, 200. See also specific types of fusion proteins

Gain-of-function mutations, 160
Gβγ, 207
GDP, 1, 97, 217, 382
Global suppressors, 160
Glutamines, 96–97, 208, 387
Gold, 129–130
Gonadotropin-releasing hormone receptor (GnRH-R), 80
GPA2 gene, 146
GPCR-specific kinases (GRK), 336
G protein-effector interface, 224
G-protein-coupled receptors (GPCR)
activation of, 43
analysis, split receptors as tools for, 109–117
binding sites, mapping, see Thyrotropin-releasing hormone receptor (TRH-R)
constitutive activity, 98
defined, 1
endogenous agonists, 44
function, analysis of, 47–48
generic model, 66–67
peptide ligands and, 79–80

protein interactions, 5–7
ligand binding site in, 3–4
rhodopsin, 85–100, 110–118
structures, generally
  atomic scale probes, metal-ions as, 121–138
  generally, 356–358, 380–383
  static and dynamic, 2–3
  two-state model, 98
  3D, 60–61, 169
Growth hormone (GH), 386
GTP, 187–188, 382
GTPase, 97, 190, 195–196, 213–215
GTPγS binding studies, 190, 196–198,
  215–216

HALFSTAT, 266
Halobacterium, 239
Halobacterium halobium, 99–100
Halorhodopsin, 236
Harvesting cells, 25
H-bond, 4
Helical arrangements, 238, 277
Helical orientation, 7–8
Helical wheel plots, amino acid sequence,
  277–279
Helices, generally
  arrangement, 273–274, 305–306
  three-dimensional packing of, 306
Helix packing, 66
High-affinity ligand binding, 4
High affinity sites, 80
High performance liquid chromatography (HPLC)
  amino acid synthesis, 342, 346
  peptide synthesis, 209–210
3-H-*N*-methylspiperone binding, 25–26
Holotransducin, 89, 91
Hydrolysis, peptide receptors, 213–215
5-hydroxytryptamine (5-HT), 48

Image processing, electron microscopy, 238–239
Infections, fluorescence spectroscopy analysis,
  319–320
Inositol phosphate assay, 63–64
Insertion, 13–14
Intact cells studies, peptides in, 215
Intracellular G protein binding region, 277
Intracellular loops (ICLs)
  activation of G protein, synthetic peptides, 219,
    223
  characteristics of, generally, 69, 71
Inverse ionic radius, 125
Ion exchange chromatography, 239

Ionization site, metal-protein interaction, 124
Irving-Williams series, 124–125, 127

Kinases, *see specific types of kinases*
Kinetics, fluorescent nucleotide binding, 217

LABEL, 268
LATLINE, 269
Leukotrienes, 375
Ligand models, development of
  endogenous structure, 383–384
  privileged structures, 384–387
Ligand-receptor complexes, molecular architecture
    investigation by fluorescence, 350–351
Loop-gap resonator, 297–298
Loss-of-function mutations, 159–160
Low-resolution maps, 279–280
Lysine, 86, 93–95, 275–276

Magnesium, 129
Manganese, 124
Mapping, *see specific types of maps*
Mass spectroscopy, peptide synthesis, 210
Mastoparan, 219
Membrane incorporation, 44–45
Membrane proteins
  alternative study approaches, 358–359
  characteristics of, generally, 236, 283
Mercury, 129
Metabotropic glutamate receptors, 3
Metal-binding
  kinetic analysis of, 134–137
  pharmacological relevance of, 130
  screening for, 128–130
  thermodynamics of, 130–133
Metal cations, as atomic probes
  complexation behavior, influential factors,
    125–127
  electrostatics, Eisenman series, 122–124
  entropy effects, generally, 125, 133
  ionization potential, Irving-Williams series,
    124–125
  metal-binding
    kinetic analysis of, 134–137
    pharmacological relevance of, 130
    screening for, 128–130
    thermodynamics of, 130–133
  metal-protein interactions
    inductive *vs.* deductive approaches, 127–128
    metal binding properties, 128–130
    molecular mechanisms, null pharmacological
      analyses of, 133–134

potential coordination site residues, targeting by superfamily sequence comparisons and sequence subtractions, 137–138

Metal-protein interactions
inductive *vs.* deductive approaches, 127–128
metal binding properties, 128–130
molecular mechanisms, null pharmacological analyses of, 133–134

Methanethiosulfonate (MTS), 22

Methionine sulfoxide, 208

MK-0677, 387

MMBOX, 258, 264

Molecular dynamics simulations, 76

Monte Carlo/stochastic dynamics, 74, 78, 378, 380

MRC image processing software, 254

MTS ethylammonium (MTSEA), 22–23, 26, 28

MTS ethylsulfonate (MTSES), 22, 33–34

MTS ethyltrimethylammonium (MTSET), 22–23, 28, 33–34

MTS reagents, use and reactions of, 26

Multidimensional nuclear magnetic resonance (NOSEY) spectroscopy, 100

Muscarinic receptors, allosteric modulation, 55–56

Mutagenesis, *see specific types of proteins*
cRNA, 339
multiple-residue substitution, 13–14
using PCR and plasmid recombination, 156–159
PCR-based, 11–13
random
in vitro, 152
oligonucleotide-directed, 153–154, 159
site-directed, 7, 24, 144, 154
site-specific, 110
ssDNA-based site-directed, 9–11
suppression, 348–349
techniques, *see* Mutagenesis techniques
TRH-R (thyrotropin-releasing hormone receptor (TRH-R), 61–62
two-dimensional, 51–52

Mutagenesis techniques, overview
data interpretation, 7
future perspectives, 14, 17
mutation design
GPCR, *see* GPCR
receptor activation mechanism, 4–5
receptor-G protein interactions, 5–7
protocols
chimeric receptors, 14, 16
deletion, 13–14
helical orientation, prediction of, 7–8
insertion, 13–14

multiple-residue substitution mutagenesis, 13–14
PCR-based mutagenesis, 11–13
ssDNA-based site-directed mutagensis, uracil-replacement method, 9–11

Mutant isolates, verification of, 158–159

Mutant receptors
generation of, 328–329
library creation of, 152–153

Mutations, *see specific types of proteins*

Neurokinin-1 receptor (NK_1) antagonists, 394–395

Neurokinin-2 (NK_2) receptors, 337, 350

Neuropeptide Y, 375

Nickel, 124

Nitrobenzdioxazol iodoacetamide (IANBD), 320–322, 324–325, 327, 331

NOESY, 359, 361

Nonadditivity principle, 4

NPLT, 266

Nuclear magnetic resonance (NMR)
defined, 236, 356
drug discovery and, 378, 383–384
future research directions, 370
overview, 359–360
peptides, from GPCRs, 360–361
rhodopsin
carboxy-terminal β sheet, 361–363
cytoplasmic loops, 364–367
intradiscal face of, 370
verification of results, overview, 367–370
small peptides, difficulty with, 369
structure refinement, 360

Nucleotide binding, peptide receptors, 213–215

Oligonucleotide-directed mutagenesis, 153–155, 159

Oligonucleotides, 11

Oocytes, microinjection of, 348–349

Optical diffraction, two-dimensional crystals, 245

ORIGTILT, 261, 264, 269

Overview pictures, 243–245

Oxidative crosslinks, split receptors, 114–117

pdCpa, 340–342, 347–348

Peptides, generally
activation of G protein, receptor-mimicking
fluorescent nucleotide binding kinetics, protocol for, 217
implications of, generally, 215–217
overview, 217–219

characterization of
  care and storage of peptides, 210
  identity and purity, determination of,
    209–210
design and synthesis of, 207–209
future research directions, 225
G protein functions and, generally, 206
ligands, 79–80
receptor, 206–207
receptor-G protein coupling, inhibition of
  intact cell studies, 215
  peptide protocol, 211–213
  receptor-stimulated nucleotide binding and hy-
    drolysis, 213–215
receptor-mimicking, 215–219
as research tool, overview, 205–215
storage of, 210
structural studies, 220–222
synthetic
  data interpretation and caveats, 223–224
  receptor-mimicking, activation of G protein,
    215–217
  rhodopsin-derived, 96
  therapeutic utility, potential, 224–225
Peptidomimetic compounds, 225
Pharmacological specificity, substituted-cysteine accessibility method, 37–38
Pharmacophore map, construction of, 49–50
Phenoxides, 127
Phenylalanine, 387
Pheromone response pathway, in yeast
  assays
    growth arrest, 151–152
    mating efficiency, 147–149
    pheromone-responsive reporter genes,
      149–150
  overview, 142–144
Phosphor Imager System, 91
Phosphorylation, 357
Photoisomerization, 95
Piperazine, 386
Plaque purification, 318
Plasmid recombination, 156–157
PLC-β, 207
Polymerase chain reaction (PCR)
  mutagenesis, 11–13, 153, 156–159
  virus stocks, 329
p1 map, 216
Potassium, 129
Privileged structure, defined, 384
Projection formation, 147
Projection map, rhodopsin electron crystallography

amplitude, scaling using diffraction data,
  263–264
calculation from list of amplitudes and phases,
  266
CTF correction, refining, 263
merging data from different images, 261–263
projection structures, interpretation of, 267
resolution limit, evaluation of, 264
Protein-protein interactions, 223
Pseudo-noble-gas configuration (PNGC), 124–125,
  127
Purification, β_2-adrenergic receptors, fluorescence
  spectroscopy analysis, 318–320, 329. *See
  also specific types of proteins*

QUADSEARCH, 257
Quantitative mating assay, yeast cells, 147–148

R, MD analysis, 170–171
R*, 170–171, 357
Radioligand binding, 45–47
Random mutagenesis
  in vitro, 154
  oligonucleotide-directed, 153–155
Receptor activation
  conserved amino acids, 179–180
  E/DRY sequence, mechanistic role of, 171–173
  helices movement, 174–178
  intracellular loops, 174–178
  mutation and agonist-induced, comparison be-
    tween, 178–179
  substituted-cysteine acessibility method (SCAM)
    agonist binding, transduction into, 37
    overview, 35–37
  via mutagenesis, 4–5
Receptor binding assay, 190
Receptor-effector coupling, 224–225
Receptor-G protein interactions, 5–7
Receptor kinases, 224
Receptor mutagenesis
  membrane incorporation, determination of,
    44–45
  radioligand binding, data analysis and, 45–47
  receptor function, analysis of, 47–48
Reciprocal space, 251
Restricted analogs, TRH-R, 78
Restriction mapping, 24
Retinal binding site, 276
Retinitis pigmentosa, 98
RGS proteins, 146
Rhodopsin
  absorption maximum, 86

biochemistry of system, 303–304
bovine, generally, 87
characteristics of, generally, 31, 61, 235
chromophore-protein interactions, 93–95
cytoplasmic loops of, 358, 364–365
intracellular loops, 223
methodology
  crude ROS membranes, preparation of, 88–89
  overview, 87–88
  retinal holotransducin and transducin subunits, purification of, 89–91
  ROS binding assay, 92–93
  solubilized, 89
  transducin activation assays, 91–92
  UW-ROS membranes, preparation of, 89
mutant opsins, constitutive activity of, 98–99
nuclear magnetic resonance (NMR), *see* Nuclear magnetic resonance (NMR), rhodopsin
photoactivation
  molecular mechanism of, 95–97
  generally, 86
recombinant, overexpression of, 86
site-directed spin-labeling (SDSL) studies
  EPR spectroscopy, principles of, 293–298
  future directions, 310–311
  outcome of, 302–310
  overview, 292–293
  typical steps in, 298–302
split receptors
  current research, generally, 117
  disulfide crosslinks, 111–112
  future research, 117
  methodology, 112–117
structural models of, 99–100
structure of, 303
transducin, coupling to, 97–98
two-dimensional crystals, electron-crystallographic analysis
  data collection, with low-dose electron cryomicroscopy, 245–254
  preparation of, 239–245
  x-ray crystallography compared with, 236–238
ROS membranes
  binding assay, 92–93
  preparation of
    crude, 88–89
    urea-washed (UW-ROS), 89

S. cerevisiae, 144
Saccharomyces kluyvei, 144
SCALIMAMP3D, 264
Scanning

image processing, 254–256
nitroxide, SDSL, 292–293
spot, 247
Schild analysis, 134
Screening libraries, yeast cells
  ligands that interact with GPCRs, 145–146
  proteins that mediate/regulate cellular signaling pathways, 146
  random G protein and effector mutations, 145
  random receptor mutations, 144–145
SDS-PAGE, 109, 113, 115
Second-site suppressors, 159–160
Sequence analysis, 3–4
Serotonin, 375
Sf9 cells, 188–189, 191–193, 318, 329
SHAKE algorithm, 76
Signal transduction, 199, 225
Silent mutations, 10
Silver, 129
Simulated annealings, 76, 78
Site-directed spin-labeling (SDSL) studies, rhodopsin
  cysteine mutations
    consequence on protein structure, assessment of, 300–301
    EPR studies, 301–302
    expression, harvesting, purification, and reconstitution of, 299–300
    spin-labeling efficiency, 301
  efficiency, determination of, 300
  electron paramagnetic resonance (EPR) spectroscopy
    defined, 291
    instruments, 297
    qualitative description of, 293–294
    spectrum, 294
    spin label information, as source for, 295–297
  future directions, 310–311
  instrumentation
    EPR instruments, 297
    loop-gap resonator, 297–298
  outcome of studies, overview, 302–310
  protein conformational changes, measurement by, 293
  protein structures
    determination using, 292–293
    model, construction of, 302
  typical steps in, 298–302
Small molecule receptor binding sites, mapping
  allosteric modulation, site identification, 54–56
  biogenic amine binding, 48–54
  receptor mutagenesis, 44–48

Sodium, 128–129

Somatostatin agonists, 391–394

SPEX Fluoromax spectrofluorometer, 324

Spin labeling, site-directed, *see* Site-directed spin-labeling (SDSL) studies

Spin labels
  accessibility, 296–297
  defined, 291
  mobility, 295–296
  two spin labels, distance measurement, 297

Spiropiperidine, 387

Split receptors
  defined, 109
  design and synthesis of, 112–113
  disulfide crosslinking, of, 112
  strategy for, 113
  types of, overview, 111

Split rhodopsins, expression and reconstitution of, 113–114

Spot scanning, 247

Standard Drug File (SDF), 388

STE2 gene, 157

Stoichiometry, 188, 199

Structure-activity relationships (SAR) studies, drug discovery, 391–392

Substance P, 375

Substituted-cysteine accessibility method (SCAM)
  assumptions, interpretation of, 27–28
  endogenous cysteines, 23–24
  future perspectives
    agonist binding, transduction into receptor activation, 37
    pharmacological specificity, structural basis of, 37–38
  methodology
    accessibility method, 22–24
    epitope-tagged D2 receptor construct, construction of, 25
    harvesting cells, 25
    ³-H-*N*-methylspiperone binding, 25–26
    MTS reagents, use and reactions of, 26
    protection, 26–27
    site-directed mutagenesis, 24
    transfection, transient and stable, 25
  overview, 22–24
  reaction detection, 23
  results interpretation
    altered binding mechanisms, 28–29
    binding site, aromatic cluster and, 34–35
    classic mutagenesis, distinguished from SCAM, 32

cysteine substitution, 30
  dopamine D2 ligands, differentiated by sensitivity to modification of engineered cysteines, 33
  electrostatic potential, 33–34
  MTSEA reactions, MTSET compared with, 34
  receptor activation, conformational changes associated with, 35–37
  SCAM, assumption of, 27–28
  secondary structure, 30–32

Sup-tRNA, 337–338

Surface representation, three-dimensional maps, 271, 273

SYBYL program, 360

TANGL, 267–268

Taq DNA, polymerase, 156

TAXA, 268

Thermodynamics
  drug discovery, 380
  metal binding, 130, 133

Three-dimensional map, rhodopsin electron crystallography
  calculation of, 269, 271
  contour plots, 271–272
  data anisotropy, evaluation of, 269
  lattice lines, fitting, 269–270
  merging data from tilted images, 267–269
  resolution, evaluation of, 269
  surface representations, 271–272

Thyrotropin-releasing hormone receptor (TRH-R)
  binding pocket, determination of, 72–73
  biologically active conformation of TRH, 77–79
  defined, 61
  extracellular loops, modeling, 75–77
  intramolecular interactions, determination of, 69–71
  mutant receptors, construction and characterization of
    competition binding assays, 63–64
    mutagenesis, 61–62
    transient transfection, 62–63
  transmembrane binding pocket, modeling, 73–75
  transmembrane bundle, modeling
    GPCR model, generic, 66–67
    overview, 64–65
    realistic helices, construction of, 65
    templates, formation of, 66
    TRH-R model, construction of, 67–69

TM domains, 65, 67
TopoSim, 389
Transducin, ROS membranes
   activation assays, 91–92
   coupling to, 97–98
   holotransducin, 89, 91
Transfection
   substituted-cysteine accessiblity method, 25
   Thryotropin-releasing hormone receptor (TRH-R), 62–63
tRNA, engineering and construction of
   chemically misacylated, preparation of, 346–348
   cRNA, mutagenesis and preparation of, 339
   dinucleotide-amino acid conjugates, synthesis of, 346
   dinucleotide pdCpA, preparation and synthesis of, 340–342
   materials and methodology, 338–339
   protected amino acids, synthesis of, 342–346
Tryptophan, 216, 387
T7 RNA polymerase, 338
Two-dimensional crystals, rhodopsin electron crystallography
   amino acid sequence, assignment of
      α carbon atoms, prediction of position of, 280
      helical wheel plots, 277–278
      low-resolution maps, 279–280
      structural constraints, 278–279
   density maps
      α helices, arrangement of, 273–274
      bacteriorhodopsin, rhodopsin compared with, 274–276
      intracellular G protein binding region, 277
      retinal binding site, 276
   future directions, 280–283
   image processing
      amplitudes and phases, extraction from images, 254–258
      astigmatism, 258–149
      contrast transfer function (CTF), 258–259
      plane group, determination of, 260–261
   low-dose electron cryomicroscopy, data collection
      contrast transfer function (CTF), 248
      electron microscope, 246
      Fourier transform properties, 250–254
      low-dose imaging, 247
      optical diffraction, data evaluation, 248, 250
      preliminary characterization of, 248–250
      specimens, 246–247

optical diffraction, 250
preparation methods
   crystal induction, by selective extraction of membranes, 241–243
   membrane proteins, 239
   purification of rhodopsin, 239–240
   reconstitution experiments, 240
   screening for two-dimensional crystals, 243–245
projection map, merging amplitudes and phases, 261–267
symmetry, in crystals, 248, 250, 260–261
three-dimensional maps, merging amplitudes and phases, 267–273
TWOFILE, 257
Tyrosine, 387

Ultraviolet (UV)-visible difference spectroscopy, 85
Unbending, low-dose electron cryomicroscopy, 256–258
Unnatural amino acids, biosynthetic incorporation
   applications, 349
   ligand-receptor complexes, investigation of molecular architecture by fluorescence, 350–351
   misacylated suppressor tRNA, engineering and construction of
      chemically misacylated suppressor tRNAs, preparation of, 346–348
      cRNA, mutagenesis and preparation of, 339
      dinucleotide-amino acid conjugates, synthesis of, 346
      dinucleotide pdCpA, preparation and synthesis of, 340–342
      materials and methodology, 338–339
      protected amino acids, synthesis of, 342–346
   overview, 335–337
   suppression mutagenesis, microinjection of oocytes and electrophysiology, 348–349
URA3 gene, 152
Uracil replacement, ssDNA-based site-directed mutagenesis, 9–11

Virus
   amplification, fluorescence spectroscopy analysis, 318
   polymerase chain reaction (PCR), 329
Visual disease, 98

Western blots, 111, 113
Wild-type receptors, 177

X-ray crystallography
   drug discovery and, 378
   electron crystallography distinguished from,
      236–238
Xenopus laevis, 337, 348
XIMDISP, 268

Yeast
   GPCR structure
      biochemical analyses, 159
      cellular signaling pathways, proteins that medi-
        ate and/or regulate, 146
      ligands, 145–146

      mutant receptor libraries, creation of, 152–159
      random G protein and effector mutations, 145
      random receptor mutations, screening libraries,
        144–145
      second-site suppressor approach to, 159–161
   pheromone response pathway
      assays of, 146–152
      pheromone-responsive reporter genes, 149
      overview, 142–144
   transformations, 158
   two-hybrid system, 146

Zinc, 130, 134, 137